P9-AQS-332

Geometry of Manifolds

PERSPECTIVES IN MATHEMATICS, Vol. 8

J. Coates and S. Helgason, editors

Geometry of Manifolds

Edited by

K. Shiohama

Department of Mathematics
Faculty of Science
Kyushu University
Japan

ACADEMIC PRESS, INC.
Harcourt Brace Jovanovich, Publishers
Boston San Diego New York
Berkeley London Sydney
Tokyo Toronto

ACADEMIC PRESS, INC.
1250 Sixth Avenue, San Diego, CA 92101

United Kingdom Edition published by
ACADEMIC PRESS INC. (LONDON) LTD.
24-28 Oval Road, London NW1 7DX

Library of Congress Cataloging-in-Publication Data

Geometry of manifolds.

(Perspectives in mathematics ; vol. 8)
Includes bibliographies.
1. Geometry, Differential. 2. Manifolds (Mathematics)
I. Shiohama, K. (Katsuhiro), Date— . II. Series.
QA649.G47 1989 516.3′6 89-15157
ISBN 0-12-640170-5 (alk. paper)

Printed in the United States of America
89 90 91 92 9 8 7 6 5 4 3 2 1

Contents

Contributors

The numbers in parentheses refer to the pages on which the authors' contributions begin.

Hiroshi Akiyama (3), *Department of Applied Mathematics, Faculty of Engineering, Shizuoka University, Hamamatsu 432, Japan*

Kazuo Akutagawa (59), *Department of Mathematics, Nippon Bunri University, Oita 870-03, Japan*

Hideo Doi (447), *Department of Mathematics, Faculty of Science, Hiroshima University, Hiroshima 730, Japan*

Hideya Hashimoto (71), *Department of Mathematical Science, Graduate School of Science and Technology, Niigata University, Niigata 950-21, Japan*

Ryosuke Ichida (247), *Department of Mathematics, Yokohama City University, 22-2 Seto Kanazawa-ku, Yokohama 236, Japan*

Akira Ikeda (383), *Department of Mathematics, Faculty of School Education, Hiroshima University, Shinonome Hiroshima 734, Japan*

Jin-ichi Itoh (275), *Kyushu Institute of Technology, Iizuka 820, Japan*

Mitsuhiro Itoh (453), *Institute of Mathematics, University of Tsukuba 305, Japan*

Yasuhiko Kamiyama (477), *Department of Mathematics, Faculty of Science, University of Tokyo, Hongo, Tokyo 113, Japan*

Jyoichi Kaneko (165), *Department of Mathematics, College of General Education, Kyushu University, Ropponmatsu, Fukuoka 810, Japan*

Toyoko Kashiwada (95), *Shinkawa 5-6-6-406, Mitaka-shi, Tokyo, Japan*

Masao Maeda (281), *Department of Mathematics, Faculty of Education, Yokohama National University, 156 Tokiwadai, Hodogaya-ku, Yokohama 240, Japan*

Hiromichi Matsunaga (483), *Department of Mathematics, Faculty of Science, Shimane University, Matsue, Japan*

Reiko Miyaoka (181), *Department of Mathematics, Tokyo Institute of Technology, Meguroku, Tokyo 152, Japan*

Tadashi Nagano (111), *Department of Mathematics, Sophia University, Tokyo 102, Japan*

Takeyuki Nagasawa (499), *Mathematical Institute, Tohoku University, Sendai 980, Japan*

Hisashi Naito (419), *Department of Mathematics, Nagoya University, Nagoya 464, Japan*

Takayuki Okai (447), *Department of Mathematics, Faculty of Science, Hiroshima University, Hiroshima 730, Japan*

Takashi Okayasu (427), *Department of Mathematics, Faculty of Science, Hirosaki University, Hirosaki 036, Japan*

Yukio Otsu (295), *Department of Mathematics, Faculty of Science, Kyushu University, Fukuoka 812, Japan*

Tetsuya Ozawa (181), *Department of Mathematics, Nagoya University, Chikusaku, Nagoya 464, Japan*

Takashi Sakai (303), *Department of Mathematics, Faculty of Science, Okayama University, Okayama 700, Japan*

Hajime Sato (191), *Department of Mathematics, Nagoya University, Nagoya 464-01, Japan*

Takuji Sato (129), *Faculty of Technology, Kanazawa University, Kanazawa 920, Japan*

Katsuhiro Shiohama (317, 345), *Department of Mathematics, Faculty of Science, Kyushu University, Fukuoka 812, Japan*

Takashi Shioya (351), *Department of Mathematics, Faculty of Science, Kyushu University, Fukuoka 812, Japan*

Makiko Sumi (111), *Department of Mathematics, Sophia University, Tokyo 102, Japan*

M. Tanaka (317), *Department of Mathematics, Faculty of Science, Tokai University, Hiratsuka, 259-12 Japan*

Yoshiharu Taniguchi (141), *Department of Mathematics, Osaka University, Toyonaka Osaka 560, Japan*

Masaaki Umehara (151), *Institute of Mathematics, University of Tsukuba, Tsukuba, Ibaraki 305, Japan*

Hajime Urakawa (435), *Department of Mathematics, College of General Education, Tohoku University, Kawauchi, Sendai 980, Japan*

Yoshihide Watanabe (29), *Department of Applied Mathematics, Faculty of Engineering, Hiroshima University, Higashi-Hiroshima 724, Japan*

Kotaro Yamada (151), *Department of Mathematics, Faculty of Science and Technology, Keio University, Yokohama 223, Japan*

Keizo Yamaguchi (191), *Department of Mathematics, Hokkaido University, Sapporo 060, Japan*

Tadashi Yamaguchi (365), *Department of Mathematics, College of General Education, Kyushu University, Ropponmatsu, Fukuoka 810, Japan*

Takao Yamaguchi (345), *Department of Mathematics, College of General Education, Kyushu University, Fukuoka 810, Japan*

Tomoaki Yatsui (239), *Department of Mathematics, Hokkaido University, Sapporo 060, Japan*

Akira Yoshioka (39), *Department of Mathematics, Faculty of Science and Technology, Science University of Tokyo, Noda, Chiba 287, Japan*

Preface

Two hundred and fifty-five people participated in the Thirty-Fifth Symposium on Differential Geometry, sponsored by the Japanese Ministry of Education through the 'Grant-in-Aid for Scientific Research,' 25-30 July 1988, at Shinshu University, Matsumoto, Japan. The initial impetus for this symposium came from a burst of recent activity in various topics in differential geometry which was mainly brought up by young geometers including graduate students. There were eight survey lectures followed by sixty-five one-hour talks in nine parallel sessions.

This volume is a collection of original papers contributed by the speakers and participants of the Matsumoto symposium. All papers were refereed by the Editorial Board, all the members of which are organizers of six parallel sessions. The titles of these sessions are Dynamical Systems and Geometry (Chapter I) Geometry of Submanifolds and Tensor Geometry (Chapter II), Lie Sphere Geometry (Chapter III), Riemannian Geometry (Chapter IV), The Geometry of Laplace Operator (Chapter V) and Yang-Mills Connections (Chapter VI).

The Dynamical Systems and Geometry Session was organized by T. Iwai. A modified Maslov index is used to give eigenvalues for the Bochner-Laplacian. Nonstandard analysis is applied to give the heat kernels on compact Riemannian manifolds. Hamiltonian structure and integrability are studied for evolution equations.

The Geometry of Submanifolds and Tensor Geometry Session was organized by N. Innami, K. Ogiue and R. Takagi. Nagano and Sumi made a survey on the structure of symmetric spaces and their totally geodesic submanifolds. The Solvability for the Dirichlet problem at infinity for harmonic mappings between Hadamard manifolds is proved and harmonic mappings of 2-tori into 2-spheres are studied. Some conditions for a Kaehler submanifold of a complex space form to be homogeneous are obtained and the structure of 6-dimensional submanifolds of the octanion space is discussed.

The Lie Sphere Geometry Session was organized by H. Sato and K. Yamaguchi. The paper of H. Sato and K. Yamaguchi settles the foundation of Lie contact geometry. Lie contact geometry is a generalization of Lie sphere geometry to general contact manifolds. It is also related to the twistor geometry of Penrose. Counter-examples to the Cecil-Ryan conjecture are established. The authors construct Dupin hypersurfaces which are not Lie equivalent to any isoparametric hypersurfaces. A relation is given between Dupin hypersurfaces and the wave front sets of simple progressing solutions of wave equations.

The Riemannian Geometry Session was organized by T. Sakai. Concerning curvature and topology, the following results are presented: A condition for complete open manifolds of nonnegative Ricci curvature to be diffeomorphic to $\mathbf{R}^n$; topological decomposition of certain compact manifolds of nonnegative Ricci curvature; a differentiable sphere theorem for positively curved manifolds with large diameter; a new estimate of the norm of stable Jacobi fields. Concerning the theory of surfaces, an isodiametric inequality for S^2 and an isoperimetric inequality for infinitely connected complete open surfaces are presented. The ideal boundaries for complete open surfaces with infinite total curvature and the size of the set of poles on open convex surfaces are discussed. A Morse theoretic argument is developed to obtain topological restrictions for p-convex domains.

The Geometry of Laplace Operator session was organized by H. Urakawa. New type isospectral examples about p-forms are presented. Two lens spaces are constructed in such a way that they are isospectral for all k-forms ($0 \leqq k \leqq p$) but non-isospectral for $(p + 1)$-forms. By using a variant of the Bochner technique, holomorphicity of pluriharmonic maps of certain Kaehler manifolds into Riemann surfaces is shown. Instability theorems for harmonic maps and minimal submanifolds are presented in terms of the Dirichlet boundary problem of Laplacian. The inequalities about the bottom of the spectrum of a complete non-compact Riemannian manifold are discussed.

The Yang-Mills Connection Session was organized by M. Itoh. An asymptotical stability of the Yang-Mills gradient flow near a product connection is presented. The compactification of the moduli space of 1-instanton on HP^n is obtained. Theorems on the second Betti number of the instanton moduli spaces on S^4 is obtained. Group-equivariant Yang-Mills theory is given in terms of instanton examples and infinitesimal deformations of the instanton moduli spaces. The simple connectivity of ALE hyperkaehler 4-spaces and the geometry of instanton moduli spaces on them are presented.

The Editorial Board consists of:

Iwai, T. Department of Applied Mathematics and Physics, Faculty of Engineering, Kyoto University, Kyoto, 606-Japan.
Ogiue, K. Department of Mathematics, Faculty of Science, Tokyo Metropolitan University, Tokyo, 158-Japan.
Sakai, T. Department of Mathematics, Faculty of Science, Okayama University, Okayama, 700-Japan.
Sato, H. Department of Mathematics, College of General Education, Nagoya University, Nagoya, 464-Japan.
Takagi, R. Department of Mathematics, Faculty of Science, Chiba University, Chiba, 260-Japan.
Urakawa, H. Department of Mathematics, College of General Education, Tohoku University, Sendai, 980-Japan.
Yamaguchi, K. Department of Mathematics, Faculty of Science, Hokkaido University, Sapporo, 060-Japan.

We would like to express our thanks to all contributors for their cooperation. We also like to express our thanks to Professors Saito, Yokota, Abe and other mathematicians in Shinshu University for their kind preparation for the symposium and for providing facilities.

Katsuhiro Shiohama

Chapter I
Dynamical Systems and Geometry

APPLICATIONS OF NONSTANDARD ANALYSIS TO STOCHASTIC FLOWS AND HEAT KERNELS ON MANIFOLDS

Hiroshi Akiyama

Nonstandard analysis enables us to use infinitesimals and often brings us simpler treatments of stochastic processes (cf. [3], [11], [12]). In this paper we apply nonstandard analysis to stochastic flows (related to heat equations) and heat kernels on manifolds. In §1, we give some preliminaries from nonstandard analysis by adopting as our basic framework the nonstandard set theory UNST presented by T. Kawai [7]. Section 2 is devoted to nonstandard construction of certain stochastic processes, including a Brownian motion, on a compact Riemannian manifold. In §3, Itô's forward and backward formulas are established for (local) cross sections of fiber bundles by using nonstandard analysis. The forward formula in §3 is a nonstandard version of [2] (cf. [10]). The formulas thus obtained are applied to forward and backward heat equations

---

This research was partially supported by Grant-in-Aid for Scientific Research (No. 63740114), Ministry of Education, Science and Culture. Part of this work was carried out at Kyoto University.

ISBN 0-12-640170-5

for cross sections of a vector bundle and to stochastic flows of projective transformations. Finally in §4, motivated by S. Watanabe's work [14] giving probabilistic expressions of heat kernels by using an analysis of Wiener functionals (the Malliavin calculus; see also [15], [6], [1]), we give a nonstandard probabilistic representation of the heat kernel for a heat equation for differential forms with values in a vector bundle.

1. PRELIMINARIES FROM NONSTANDARD ANALYSIS

We begin with some preliminaries from the nonstandard set theory UNST presented by T. Kawai [7].

The language of UNST is given by adding three constant symbols $\mathfrak{U}$, $\mathcal{I}$ and $*$ to the language of ZFC (the Zermelo-Fraenkel set theory with the axiom of choice). Let letters a, b, ..., A, B, ... denote variables of UNST. If $a\in\mathfrak{U}$ [resp. $a\in\mathcal{I}$], a is called *usual* [resp. *internal*]. For a formula φ in ZFC (that is, a formula in UNST without $\mathfrak{U}$, $\mathcal{I}$ and $*$), we denoted by $^{\mathfrak{U}}\varphi$ [resp. $^{\mathcal{I}}\varphi$] the formula in UNST obtained from φ by restricting the scope of all variables in φ to $\mathfrak{U}$ [resp. $\mathcal{I}$].

The axioms of UNST are the following (1)–(9):

(1) If φ is an axiom of ZFC, then $^{\mathfrak{U}}\varphi$ is an axiom of UNST.

(2) Each axiom of ZFC different from the axiom of regularity is an axiom of UNST.

(3) (Axiom of regularity in a restricted form)

$\forall A\ [A\neq 0 \wedge A\cap\mathcal{I}=0 \longrightarrow \exists x\in A\ [x\cap A=0]]$. ($0=\phi$: the empty set.)

(4) $* : \mathfrak{U} \longrightarrow \mathcal{I}$ (map). When $a \in \mathfrak{U}$, we write ${}^*a$ for $*(a)$.

(5) (Transitivity of $\mathcal{I}$) $\forall A\ \forall B\ [A \in B \wedge B \in \mathcal{I} \longrightarrow A \in \mathcal{I}]$.

(6) (Transitivity of $\mathfrak{U}$) $\forall A\ \forall B\ [A \in B \wedge B \in \mathfrak{U} \longrightarrow A \in \mathfrak{U}]$.

(7) $\forall A\ \forall B\ [A \subset B \wedge B \in \mathfrak{U} \longrightarrow A \in \mathfrak{U}]$.

(8) (Transfer principle) Let $\varphi(x_1, \ldots, x_n)$ be an n-ary formula in ZFC (the free variables of φ are among $x_1, \ldots, x_n$). Then

$$\forall x_1, \ldots, x_n \in \mathfrak{U}\ [{}^{\mathfrak{U}}\varphi(x_1, \ldots, x_n) \longleftrightarrow {}^{\mathcal{I}}\varphi({}^*x_1, \ldots, {}^*x_n)].$$

(9) (Saturation principle) Define $(D : \mathfrak{U}\text{-}size) \equiv \exists F\ [F : \mathfrak{U} \longrightarrow D$ (onto map)]. Let $\varphi(a, b, x_1, \ldots, x_n)$ be an $(n+2)$-ary formula in ZFC (the free variables of φ are among $a, b, x_1, \ldots, x_n$). Then

$$\forall D\colon \mathfrak{U}\text{-size}\ \ \forall x_1, \ldots, x_n \in \mathcal{I}$$
$$\left[\forall d \in \mathcal{I} \begin{bmatrix} d \text{ is finite} \wedge d \subset D \\ \longrightarrow \exists b \in \mathcal{I}\ \forall a \in d\ {}^{\mathcal{I}}\varphi(a, b, x_1, \ldots, x_n) \end{bmatrix} \longrightarrow \exists B \in \mathcal{I}\ \forall A \in D \cap \mathcal{I}\ {}^{\mathcal{I}}\varphi(A, B, x_1, \ldots, x_n)\right].$$

It is shown in [7] that UNST is a conservative extension of ZFC. Moreover, in UNST it holds that (Extension principle)

$$\forall A \in \mathcal{I}\ \forall B \in \mathcal{I}\ \forall a\ \forall f$$
$$\begin{bmatrix} a \subset A \wedge a : \mathfrak{U}\text{-size} \wedge f : a \longrightarrow B \text{ (map)} \\ \longrightarrow \exists F \in \mathcal{I}\ [F : A \longrightarrow B \text{ (map)} \wedge \forall x \in a\ [F(x) = f(x)]] \end{bmatrix}.$$

(Here $F \in \mathcal{I}$ means that the graph of F is an internal subset of $A \times B$.) Also, $\exists! \mathcal{S}\ \forall y\ [y \in \mathcal{S} \longleftrightarrow \exists x \in \mathfrak{U}\ [y = {}^*x]]$, where $\exists!$ means that "there exists a unique". If $x \in \mathcal{S}$, x is called *standard*. Since the map $* : \mathfrak{U} \longrightarrow \mathcal{I}$ is injective by the transfer principle, for $a \in \mathfrak{U}$, we often identify a

with ${}^*a$ (and use a instead of ${}^*a$) when we do not need to take account of the set-structure of a. If $f : A \longrightarrow B$ is a map with $A, B, f \in \mathfrak{U}$, then on the transfer principle we have ${}^*f : {}^*A \longrightarrow {}^*B$ and ${}^*(f(x)) = {}^*f({}^*x)$ for all $x \in A$; we often write $f : {}^*A \longrightarrow {}^*B$ rather than ${}^*f : {}^*A \longrightarrow {}^*B$.

Let $N \in \mathfrak{U}$ be the non-negative integers. Let $\mathbf{Z} \in \mathfrak{U}$ and $\mathbf{R} \in \mathfrak{U}$ be the integers and the real numbers, respectively. For $n \in N$, it holds that $n = {}^*n$ in UNST, where $1 = \{0\}$, $2 = 1 \cup \{1\}$, ..., $n+1 = n \cup \{n\} = \{0,1,2,\ldots,n\}$. For $r \in R$, we identify r with ${}^*r$. Then $N \subsetneqq {}^*N$, $\mathbf{Z} \subsetneqq {}^*\mathbf{Z}$, $R \subsetneqq {}^*R$. If $x \in {}^*R$ [resp. $x \in {}^*\mathbf{Z}$], x is called a *hyperreal* number [resp. a *hyperinteger*]. A hyperreal number $x \in {}^*\mathbf{R}$ is called *finite* if $|x| < n$ for some $n \in N$, *infinite* if it is not finite. A hyperreal number x is called *infinitesimal* if $|x| < 1/n$ for all $n \in N - \{0\}$. Each $K \in {}^*N - N$ is infinite; $1/K$ is non-zero and infinitesimal. An internal set $A \in \mathcal{I}$ is called *hyperfinite* (or **-finite*) if it holds that $\exists n \in {}^*N\ \exists f \in \mathcal{I}\ [f : n \longrightarrow A \text{ (bijection)}]$.

We need basic notions from nonstandard topology. Let $X \in \mathfrak{U}$ be a topological space. Regard it as a subset of ${}^*X$. The *monad* of a point $a \in X$ is defined by

$$\mathrm{Mon}(a) := \bigcap \{{}^*A;\ A (\subset X) \text{ is an open neighborhood of } a\}.$$

When $x \in \mathrm{Mon}(a)$, we write $x \approx a$. (If $x, y \in {}^*R^n$ with $n \in N - \{0\}$ and $|x-y|$ is infinitesimal where $|\cdot|$ denotes Euclidean norm in ${}^*R$, we write $x \approx y$.) A point $x \in {}^*X$ is called *near-standard* if $x \in \mathrm{Mon}(a)$ for some $a \in X$. If X is compact, then every point of ${}^*X$ is near-standard. If X is a Hausdorff space and $x \in {}^*X$ is near-standard, then there exists a unique $a \in X$ with $x \approx a$; such $a \in X$ is called the *standard part* of x and is denoted by ${}^{\circ}x$. (If $r \in {}^{*}R-\{0\}$ is positive [resp. negative] and infinite, we write ${}^{\circ}r = +\infty$ [resp. ${}^{\circ}r = -\infty$]). If $f \in \mathfrak{U}$ is a map from a topological space $X(\in \mathfrak{U})$ to another $Y(\in \mathfrak{U})$, then f is continuous at $a \in X$ if and only if $x \approx a$ implies $f(x) \approx f(a)$ (cf. [7], [12], [13]).

Now we prepare a uniform Loeb probability space. Let $K_0 \in {}^{*}N-N$ and put $K = K_0!$ $(:= K_0{}^{*}!)$ (factorial in ${}^{*}N$). Set $T = \{t_i = i/K \ ; \ i \in {}^{*}N, \ i \le K\}$. Let $k \in N-\{0\}$ and set $\Omega = \{-1, 1\}^{T \times \{1,2,\ldots,k\}} (\in \mathcal{I})$, which is hyperfinite. Denote by $\mathcal{A}$ the internal algebra of all internal subsets of Ω and define $\nu : \mathcal{A} \longrightarrow {}^{*}[0,1]$ $(= \{x \in {}^{*}R \ ; \ 0 \le x \le 1\})$ by $\nu(A) = |A|/|\Omega|$ for all $A \in \mathcal{A}$, where $|\cdot|$ denotes internal cardinarity. Define ${}^{\circ}\nu : \mathcal{A} \longrightarrow [0,1] (= \{x \in R; \ 0 \le x \le 1\})$ by ${}^{\circ}\nu(A) = {}^{\circ}(\nu(A))$, $A \in \mathcal{A}$. Then ${}^{\circ}\nu$ is a finitely additive measure on $\mathcal{A}$. The saturation principle implies that if $A_i \in \mathcal{A}$ $(i \in N)$ and $A = \bigcup_{i \in N} A_i \in \mathcal{A}$ then there exists a number $m \in N$ such that $A = \bigcup_{i=0}^{m} A_i$. Thus ${}^{\circ}\nu$ is countably additive on $\mathcal{A}$ and is extended uniquely to a probability measure $\tilde{\nu}$ on the σ-algebra $\sigma(\mathcal{A})$ generated by $\mathcal{A}$. The completion of the probability space $(\Omega, \sigma(\mathcal{A}), \tilde{\nu})$ is called the (uniform) *Loeb probability space* of $(\Omega, \mathcal{A}, \nu)$ and is denoted by $(\Omega, L(\mathcal{A}), \nu_L)$ (see [11], [12]).

Let $\{e_1, \ldots, e_k\}$ be the canonical basis of R^k and set $\Delta t = 1/K$. Define a hyperfinite random walk $w : {}^{*}[0,1] \times \Omega \longrightarrow {}^{*}R^k$ by

$$w(t,\omega) = \sum_{\alpha=1}^{k} w^{\alpha}(t,\omega) e_{\alpha}, \qquad (t,\omega) \in {}^{*}[0,1] \times \Omega,$$

$$w^{\alpha}(t,\omega) = \left(\sum_{\{s\in T;\, s\leq t\}} \omega^{\alpha}(s) + (Kt-[Kt])\omega^{\alpha}\left(\frac{[Kt]+1}{K}\right)\right)\sqrt{\Delta t} ,$$

where $\omega^{\alpha}(\cdot) = \omega(\cdot,\alpha)$, and $[Kt]$ stands for the greatest hyperinteger less than or equal to Kt. For later use put $\Delta w^{\alpha}_{t_i}(\omega) = w^{\alpha}(t_{i+1},\omega) - w^{\alpha}(t_i,\omega)$, so that $|\Delta w^{\alpha}_{t_i}(\omega)| = \sqrt{\Delta t}$.

Proposition 1.1 ([3]). Set

$$b(t,\omega) = \sum_{\alpha=1}^{k} b^{\alpha}(t,\omega)e_{\alpha} = {}^{\circ}(w(t,\omega)), \qquad (t,\omega)\in[0,1]\times\Omega.$$

Then $b(t,\omega)$ is continuous in t and finite for almost all ω, and (a continuous and finite version of) $(b(t,\omega))_{t\in[0,1]}$ is a Brownian motion, called *Anderson's Brownian motion*. [See [3], [12] for applications to stochastic integrals.]

2. HEAT EQUATIONS AND NONSTANDARD CONSTRUCTION OF A BROWNIAN MOTION ON A COMPACT RIEMANNIAN MANIFOLD

Let (M, g) be an n-dimensional, compact, connected, C^{∞} Riemannian manifold ($n\in N-\{0\}$), and let $\pi : O(M) \longrightarrow M$ be the orthonormal frame bundle over M. (They are in $\mathfrak{U}$.) We consider the connection in $O(M)$ induced from g in a natural manner. For $\xi\in R^n$, let $B(\xi)$ be the basic vector field on $O(M)$ corresponding to ξ. (In [9], $B(\xi)$ is called the standard horizontal vector field, but we keep the term "standard" for use in the sense of §1.) Let Z_0

be a C^∞ vector field on M, and let A_0 be the horizontal lift of Z_0 to $O(M)$. As stated in §1, we write A_0 instead of ${}^*A_0$, and for $\xi \in {}^*R^n$, we write $B(\xi)$ instead of ${}^*B(\xi)$. Let $(\Omega, L(\mathscr{A}), \nu_L)$ be as in §1 with $k = n$.

For each $\omega \in \Omega$, consider the following ordinary differential equation on ${}^*(O(M))$:

$$\frac{dr_t}{dt} = \left(B\left(\frac{dw(t,\omega)}{dt}\right) + A_0\right)_{r_t}, \qquad t \in {}^*[0,1], \tag{2.1}$$

where $dw(t,\omega)/dt = \sum_{\alpha=1}^{n} (dw^\alpha(t,\omega)/dt)e_\alpha$. Since $w(\cdot,\omega)$ is "hyper"-piecewise smooth, the equation (2.1) has a correct meaning. The solution of (2.1) with the condition $r_0 = r \in {}^*(O(M))$ is denoted by $r_t(r,\omega)$ or $r_t(r)$ (with ω missing).

Let $\mathbb{E}[\cdot]$ denote expectation and Δ_g be the Laplace-Beltrami operator.

Lemma 2.1. Let $x_t(r,\omega) = \pi(r_t(r,\omega))$, and let $h : M \longrightarrow R$ be a C^∞ function. Then

$$u(t,x) = \mathbb{E}[h({}^\circ x_t(r,\omega))], \quad t \in (0,1), \; x \in M, \; r \in \pi^{-1}(x), \tag{2.2}$$

is well-defined and satisfies the heat equation

$$\frac{\partial u}{\partial t} = \left(\frac{1}{2}\Delta_g + Z_0\right)u, \qquad \lim_{t \downarrow 0} u(t,\cdot) = h. \tag{2.3}$$

Proof. For $t \in (0,1)$, take $t_j \in T$ such that $t_j \le t < t_{j+1}$. Then for $r \in O(M)$ and a C^∞ function $f : O(M) \longrightarrow R$, regarding f as $f : {}^*(O(M)) \longrightarrow {}^*R$ (cf. §1), we have

$$f(r_t(r,\omega)) - f(r) \approx \sum_{\alpha=1}^{n} \sum_{i=0}^{j-1} (B(e_\alpha))_{r_{t_i}(r,\omega)}[f]\cdot\Delta w^{\alpha}_{t_i}(\omega)$$

$$+ \sum_{i=0}^{j-1} \left(\frac{1}{2} \sum_{\alpha=1}^{n} (B(e_\alpha))^2 + A_0 \right)_{r_{t_i}(r,\omega)} [f]\cdot\Delta t \;. \qquad (2.4)$$

Here we have used the fact that terms including $B(e_\alpha)B(e_\beta)[f]$ with $\alpha\neq\beta$ are infinitesimal; cf. the proof of [3, Theorem 37]. We can deduce from (2.4) that

$$\mathbb{E}[f({}^{\circ}r_t(r))]$$

$$= f(r) + \int_0^t \left(\frac{1}{2}\sum_{\alpha=1}^{n}(B(e_\alpha))^2 + A_0\right)_r \mathbb{E}[f({}^{\circ}r_s(\cdot))]\; ds. \qquad (2.5)$$

As is easily seen, $\mathbb{E}[h({}^{\circ}x_t(r))]$ does not depend on the choice of $r\in\pi^{-1}(x)$ $(x\in M)$. Thus we can define $u(t,x)$ as in (2.2). Put $f = h\circ\pi$ in (2.5). Then we see that u satisfies (2.3).

<u>**Corollary 2.2**</u>. In the case $Z_0 = 0$, the stochastic process $({}^{\circ}x_t(r))_{t\in[0,1]}$, $r\in O(M)$, is a Brownian motion on (M, g) such that ${}^{\circ}x_0(r) = \pi(r)$.

3. ITÔ'S FORMULA FOR (LOCAL) CROSS SECTIONS OF FIBER BUNDLES. NONSTANDARD APPROACH

Let $\pi_E : E \longrightarrow M$ be a C^∞ fiber bundle over an n-dimensional C^∞ manifold M with $n\in\mathbb{N}-\{0\}$, let σ be a

C^∞ (local) cross section of E defined on an open set V of M, and let Y_λ ($\lambda = 0, 1, \ldots, k$; $k\in N-\{0\}$) be projectable C^∞ vector fields on E, so that there exist C^∞ vector fields X_λ, $\lambda = 0, 1, \ldots, k$, on M such that each Y_λ is π_E-related to X_λ. (They are all in $\mathfrak{U}$.) For convenience we set $w_t^0 \equiv t$, $\Delta w_{t_i}^0 \equiv \Delta t$. Let $(\Omega, L(\mathscr{A}), \nu_L)$ be as in §1. We consider for each $\omega\in\Omega$ the following ordinary differential equation on ${}^*E$:

$$\frac{d\eta_t}{dt} = \sum_{\lambda=0}^{k} \frac{dw^\lambda(t,\omega)}{dt} (Y_\lambda)_{\eta_t} . \tag{3.1}$$

We assume that for each $q\in{}^*E$ the solution $\eta_{s,t}(q,\omega)$, $0\le s\le t\le 1$, of (3.1) satisfying the initial condition $\eta_s = q$ exists. We often write $\eta_{s,t}(q,\omega)$ simply as $\eta_{s,t}(q)$ by omitting ω. Then we have $\eta_{s,t} = \eta_t\circ\eta_s^{-1}$ where $\eta_t \equiv \eta_{0,t}$.

Lemma 3.1. For $x\in M$, $\pi_E\circ\eta_{s,t}(q)$ does not depend on the choice of $q\in\pi_E^{-1}(x)$ and thus $\theta_{s,t}(x) := \pi_E\circ\eta_{s,t}(q)$, $(q\in\pi_E^{-1}(x)$, $x\in M)$, is well-defined. Moreover, $\theta_{s,t}(x)$ is the solution of the following equation (3.2) such that $\theta_{s,s}(x) = x$;

$$\frac{d\theta_t}{dt} = \sum_{\lambda=0}^{k} \frac{dw^\lambda(t,\omega)}{dt} (X_\lambda)_{\theta_t} . \tag{3.2}$$

Proof. Immediate.

Now define $\sigma_{s,t} := \eta_{s,t}^{-1}\circ\sigma\circ\theta_{s,t}$ and put

$$D_\lambda \sigma(x) := \sigma_*(X_\lambda)_x - (Y_\lambda)_{\sigma(x)} \in T_{\sigma(x)}E, \quad x \in V,$$

where $\sigma_* : T_xM(=T_xV) \longrightarrow T_{\sigma(x)}E$, $x \in V$, is the differential of σ. Then we obtain the following.

Theorem 3.2 (Forward and backward Itô's formulas for $\sigma_{s,t}$). Assume that ${}^O\eta_{s,t}$ defines a stochastic flow of diffeomorphisms of E (in the sense of [10]; cf. [5]). Let $x \in V$. Let $f : E \longrightarrow \mathbf{R}$ be a C^∞ function. Then,

(1) when $0 \le s \le t < \inf\{u \in [0,1] \ ; \ {}^O\theta_{s,u}(x) \notin V\}$ ($:= 1$ if $\{\cdots\} = \phi$) and $s, t \in \mathbf{R}$, it holds that

$$f({}^O\sigma_{s,t}(x)) - f(\sigma(x))$$

$$= \sum_{\alpha=1}^{k} \int_s^t ({}^O\eta_{s,u}^{-1})_* D_\alpha \sigma({}^O\theta_{s,u}(x))[f] \cdot db^\alpha(u) \quad \text{(Itô integral)}$$

$$+ \int_s^t \Big(({}^O\eta_{s,u}^{-1})_* D_0 \sigma({}^O\theta_{s,u}(x))[f]$$

$$+ \frac{1}{2} \sum_{\alpha=1}^{k} \{ (X_\alpha)_{{}^O\theta_{s,u}(x)} [D_\alpha \sigma [f \circ {}^O\eta_{s,u}^{-1}]]$$

$$- D_\alpha \sigma({}^O\theta_{s,u}(x))[Y_\alpha [f \circ {}^O\eta_{s,u}^{-1}]]\} \Big) \, du ,$$

(2) when $\sup\{u \in [0,1] \ ; \ {}^O\theta_{u,t}(x) \notin V\}$ ($:= 0$ if $\{\cdots\} = \phi$) $< s \le t \le 1$ and $s, t \in \mathbf{R}$, it holds that

$$f({}^O\sigma_{s,t}(x)) - f(\sigma(x))$$

$$= \sum_{\alpha=1}^{k} \int_s^t D_\alpha{}^{\circ}\sigma_{u,t}(x)[f]\cdot \hat{d}b^{\alpha}(u) \quad \text{(backward Itô integral)}$$

$$+ \int_s^t \Big(D_0{}^{\circ}\sigma_{u,t}(x)[f] + \frac{1}{2}\sum_{\alpha=1}^{k}\{(X_\alpha)_x[D_\alpha{}^{\circ}\sigma_{u,t}[f]]$$

$$- D_\alpha{}^{\circ}\sigma_{u,t}(x)[Y_\alpha[f]]\}\Big)\, du \; .$$

Proof. We prove only (2) here, since the proof of (1) is similar (see also [2]). Let $u_i \in T$, $u_i < u < u_{i+1} = u_i + \Delta t < t$ (in ${}^*R$) and $\Delta u = \Delta t$. We first note that η_u^{-1} and θ_u^{-1} respectively satisfy

$$\frac{d\eta_u^{-1}}{du} = -\sum_{\lambda=0}^{k}(\Delta w_{u_i}^{\lambda}/\Delta u)((\eta_u^{-1})_* Y_\lambda)_{\eta_u^{-1}} \; ,$$

$$\frac{d\theta_u^{-1}}{du} = -\sum_{\lambda=0}^{k}(\Delta w_{u_i}^{\lambda}/\Delta u)((\theta_u^{-1})_* X_\lambda)_{\theta_u^{-1}} \; .$$

Since $\sigma_{u,t} = \eta_{u,t}^{-1}\circ\sigma\circ\theta_{u,t} = \eta_u\circ\eta_t^{-1}\circ\sigma\circ\theta_t\circ\theta_u^{-1}$, we have

$$\frac{\partial f(\sigma_{u,t}(x))}{\partial u} = -\sum_{\lambda=0}^{k}(\Delta w_{u_i}^{\lambda}/\Delta u)\Big((\sigma_{u,t})_*(X_\lambda)_x - (Y_\lambda)_{\sigma_{u,t}(x)}\Big)[f]$$

$$= -\sum_{\lambda=0}^{k}(\Delta w_{u_i}^{\lambda}/\Delta u)D_\lambda\sigma_{u,t}(x)[f],$$

$$\frac{\partial^2 f(\sigma_{u,t})}{\partial u^2}$$

$$= -\sum_{\lambda=0}^{k}(\Delta w_{u_i}^{\lambda}/\Delta u)\{\frac{\partial}{\partial u}[X_\lambda[f\circ\sigma_{u,t}]] - \frac{\partial}{\partial u}[(Y_\lambda)_{\sigma_{u,t}}[f]]\}$$

$$= -\sum_{\lambda=0}^{k}(\Delta w_{u_i}^{\lambda}/\Delta u)\{X_\lambda\left[\frac{\partial}{\partial u}[f\circ\sigma_{u,t}]\right]$$

$$+\sum_{\mu=0}^{k}(\Delta w_{u_i}^{\mu}/\Delta u)D_\mu\sigma_{u,t}[Y_\lambda[f]]\}$$

$$= \sum_{\lambda,\mu=0}^{k}(\Delta w_{u_i}^{\lambda}/\Delta u)(\Delta w_{u_i}^{\mu}/\Delta u)\{X_\lambda[D_\mu\sigma_{u,t}[f]]-D_\mu\sigma_{u,t}[Y_\lambda[f]]\}.$$

Therefore

$$f(\sigma_{u_i,t}) - f(\sigma_{u_{i+1},t})$$

$$= \sum_{\alpha=1}^{k} D_\alpha\sigma_{u_{i+1},t}[f]\ \Delta w_{u_i}^{\alpha} + \Big(D_0\sigma_{u_{i+1},t}[f]$$

$$+ \frac{1}{2}\sum_{\alpha=1}^{k}\{X_\alpha[D_\alpha\sigma_{u_{i+1},t}[f]] - D_\alpha\sigma_{u_{i+1},t}[Y_\alpha[f]]\}\Big)\Delta u$$

$$+ \frac{1}{2}\sum_{\substack{\alpha,\beta=1\\ \alpha\neq\beta}}^{k}\{X_\alpha[D_\beta\sigma_{u_{i+1},t}[f]] - D_\beta\sigma_{u_{i+1},t}[Y_\alpha[f]]\}\Delta w_{u_i}^{\alpha}\,\Delta w_{u_i}^{\beta}$$

$$+ \text{(the remainder)} = I_{1,i} + I_{2,i} + I_{3,i} + I_{4,i}\,.$$

Consider the sum $\sum_{\{i;u_i>s,\ u_{i+1}<t\}}(f(\sigma_{u_i,t})-f(\sigma_{u_{i+1},t}))$ and take up the standard part. Note that the contributions to the sum by $I_{3,i}$ and $I_{4,i}$ are still infinitesimal (cf. the proof of [3, Theorem 37]). This completes the proof.

We now consider the following case (#):

(#) E is a C^∞ real vector bundle associated with a principal fiber bundle $P(M, G, \pi_P)$ and each Y_λ is induced from a C^∞ vector field on P invariant under the right translation by any element of G.

For $\lambda = 0, 1, \ldots, k$, let $\{\eta_\varepsilon^{(\lambda)}\}$ [resp. $\{\theta_\varepsilon^{(\lambda)}\}$] be the flow on E [resp. M] generated by Y_λ [resp. X_λ]. For $\sigma \in C^\infty(E) := \{C^\infty$ (global) cross sections of $E\}$, $x \in M$, and $\lambda = 0, 1, \ldots, k$, we define $L_\lambda \sigma \in C^\infty(E)$ by (cf. [2])

$$L_\lambda \sigma(x) := \lim_{\varepsilon \to 0} \frac{(\eta_\varepsilon^{(\lambda)})^{-1} \circ \sigma \circ \theta_\varepsilon^{(\lambda)}(x) - \sigma(x)}{\varepsilon} \quad (\in E_x = \pi_E^{-1}(x)).$$

<u>**Corollary 3.3**</u>. Under the situation (#), let σ be a C^∞ (global) cross section of E and assume that ${}^o\eta_{s,t}$ defines a stochastic flow of diffeomorphisms of E. Then for $s, t \in [0,1]$ with $s \le t$ it holds that

$$\begin{aligned} {}^o\sigma_{s,t} - \sigma &= \sum_{\alpha=1}^{k} \int_s^t {}^o\eta_{s,u}^{-1}(L_\alpha \sigma({}^o\theta_{s,u}(\cdot))) \cdot db^\alpha(u) \\ &\quad + \int_s^t {}^o\eta_{s,u}^{-1}\Big(\Big(\{\tfrac{1}{2} \sum_{\alpha=1}^{k} (L_\alpha)^2 + L_0\}\sigma\Big)({}^o\theta_{s,u}(\cdot))\Big) \cdot du \\ &= \sum_{\alpha=1}^{k} \int_s^t L_\alpha {}^o\sigma_{u,t} \cdot \hat{d}b^\alpha(u) + \int_s^t \{\tfrac{1}{2} \sum_{\alpha=1}^{k} (L_\alpha)^2 + L_0\} {}^o\sigma_{u,t} \cdot du. \end{aligned}$$

Proof. Let $\pi_{E^*} : E^* \longrightarrow M$ be the dual vector bundle of E. Let κ be an arbitrary C^∞ cross section of E^*. Define a C^∞ function $f : E \longrightarrow \mathbf{R}$ by $f(q) = {}_{E^*}(\kappa(\pi_E(q)), q)_E$, $q \in E$, where ${}_{E^*}(\ ,\ )_E$ stands for the canonical pairing between $E_x^* = \pi_{E^*}^{-1}(x)$ and E_x for arbitrary $x \in M$. We prove only the backward formula here (for the forward formula, see also [2]). Note that

$$D_\lambda{}^O\sigma_{u,t}(x)[f] = \frac{d}{d\varepsilon}\,{}_{E^*}(\kappa(x),(\eta_\varepsilon^{(\lambda)})^{-1}\circ{}^O\sigma_{u,t}\circ\theta_\varepsilon^{(\lambda)}(x))_E\Big|_{\varepsilon=0}$$

$$= {}_{E^*}(\kappa(x),\ L_\lambda{}^O\sigma_{u,t}(x))_E = f\circ L_\lambda{}^O\sigma_{u,t}(x). \tag{3.3}$$

For $\lambda,\ \nu = 0,\ 1,\ \ldots,\ k$, since $\eta_\varepsilon^{(\lambda)} : E_x \longrightarrow E_{\theta_\varepsilon^{(\lambda)}(x)}$ is linear (by the assumption for Y_λ) for each $x\in M$, we have

$$D_\nu{}^O\sigma_{u,t}(x)[Y_\lambda[f]]$$

$$= \frac{\partial}{\partial\varepsilon_1}\frac{\partial}{\partial\varepsilon_2}\ f(\eta_{\varepsilon_2}^{(\lambda)}\circ(\eta_{\varepsilon_1}^{(\nu)})^{-1}\circ{}^O\sigma_{u,t}\circ\theta_{\varepsilon_1}^{(\nu)}(x))\ \Big|_{\varepsilon_1=\varepsilon_2=0}$$

$$= \frac{d}{d\varepsilon_2}\,{}_{E^*}(\kappa(\pi_E(\eta_{\varepsilon_2}^{(\lambda)}\circ L_\nu{}^O\sigma_{u,t}(x))),\eta_{\varepsilon_2}^{(\lambda)}\circ L_\nu{}^O\sigma_{u,t}(x))_E\Big|_{\varepsilon_2=0}$$

$$\left[\begin{array}{l} \text{by}\quad \pi_E(\eta_{\varepsilon_2}^{(\lambda)}\circ(\eta_{\varepsilon_1}^{(\nu)})^{-1}\circ{}^O\sigma_{u,t}\circ\theta_{\varepsilon_1}^{(\nu)}(x)) = \theta_{\varepsilon_2}^{(\lambda)}(x) \\ \quad = \pi_E(\eta_{\varepsilon_2}^{(\lambda)}\circ L_\nu{}^O\sigma_{u,t}(x)) \end{array}\right]$$

$$= \frac{d}{d\varepsilon_2}\ f(\eta_{\varepsilon_2}^{(\lambda)}(L_\nu{}^O\sigma_{u,t}(x)))\Big|_{\varepsilon_2=0} = (Y_\lambda)_{L_\nu{}^O\sigma_{u,t}(x)}[f].$$

Use (3.3) and the following to get the backward formula.

$$(X_\lambda)_x[D_\nu{}^O\sigma_{u,t}[f]] - D_\nu{}^O\sigma_{u,t}(x)[Y_\lambda[f]]$$

$$= (L_\nu{}^O\sigma_{u,t})_*(X_\lambda)_x[f] - (Y_\lambda)_{L_\nu{}^O\sigma_{u,t}(x)}[f]$$

$$= D_\lambda(L_\nu{}^O\sigma_{u,t})(x)[f] = {}_{E^*}(\kappa(x),\ L_\lambda(L_\nu{}^O\sigma_{u,t})(x))_E\ .$$

Corollary 3.4. Under the same conditions as in Corollary 3.3, $v_{s,t}(x) = \mathbb{E}[{}^{O}\sigma_{s,t}(x)]$ satisfies

$$\frac{\partial v_{s,t}}{\partial t} = \left\{ \frac{1}{2} \sum_{\alpha=1}^{k} (L_\alpha)^2 + L_0 \right\} v_{s,t} , \qquad v_{s,s} = \sigma ;$$

$$\frac{\partial v_{s,t}}{\partial s} = - \left\{ \frac{1}{2} \sum_{\alpha=1}^{k} (L_\alpha)^2 + L_0 \right\} v_{s,t} , \qquad v_{t,t} = \sigma .$$

The following is an application of Theorem 3.2.

Theorem 3.5. Suppose that a projective structure $\mathcal{P}$ (in the sense of [8]) on M is given and that ${}^{O}\theta_t$ defines a stochastic flow of diffeomorphisms of M. Then with respect to $\mathcal{P}$, $({}^{O}\theta_t)_{t\in[0,1]}$ is a stochastic flow of projective transformations if and only if each X_λ $(\lambda = 0, 1, \ldots, k)$ is an infinitesimal projective transformation.

Proof. $\mathcal{P}$ is identified with a cross section σ of a certain fiber bundle associated with the bundle of second order frames over M ([8]). Then ${}^{O}\theta_t$ preserves $\mathcal{P}$ if and only if ${}^{O}\theta_t^{\#}\sigma = \sigma$, where ${}^{O}\theta_t^{\#}\sigma$ is the "stochastic deformation" of σ by ${}^{O}\theta_t^{-1}$ (cf. [2]). By Theorem 3.2, we see that this condition is satisfied if and only if each X_λ is an infinitesimal projective transformation.

4. NONSTANDARD REPRESENTATIONS OF HEAT KERNELS

Let M be a C^∞ compact manifold, $\pi_E : E \longrightarrow M$ a C^∞

real vector bundle, $\pi_{E^*} : E^* \longrightarrow M$ the dual vector bundle of E, and $|\Lambda|(M)$ the C^∞ real line bundle of densities over M (cf. [4]). Let η be an E^*-valued random variable defined on a complete probability space, and put $F = \pi_{E^*} \circ \eta$, which is an M-valued random variable. In the following, we write the evaluation between E^* [resp. ${}^*E^*$] and E [resp. ${}^*E$] as the multiplication sign " $\square$ "; for example, if $\sigma \in C^\infty(E) = \{C^\infty$ cross sections of $E\}$, then $\eta \square (\sigma \circ F) (= {}_{E^*}(\eta, \sigma \circ F)_E)$ is a real-valued random variable. We assume $\mathbb{E}[|\eta \square (\sigma \circ F)|] < \infty$ for all $\sigma \in C^\infty(E)$. For any C^∞ real vector bundle $E' \longrightarrow M$, we let $E \boxtimes E' \longrightarrow M \times M$ [resp. $E \otimes E' \longrightarrow M$] denote the vector bundle whose fiber over $(x,y) \in M \times M$ [resp. $x \in M$] is $E_x \otimes E'_y$ [resp. $E_x \otimes E'_x$].

We denote both $\int_M$ and ${}^*\int_{{}^*M}$ by $\int$. Observe that there exists a $\hat{\delta}^E \in {}^*(C^\infty(E \boxtimes (E^* \otimes |\Lambda|(M))))$, (so that $\hat{\delta}^E(x,y) \in {}^*E_x \otimes {}^*(E^* \otimes |\Lambda|(M))_y$ for $x, y \in {}^*M$), such that

$$\int \hat{\delta}^E(x,\cdot) \square \sigma \approx \sigma(x) \ (\in E_x \subset {}^*E_x), \qquad \sigma \in C^\infty(E), \quad x \in M.$$

Note that $\hat{\delta}^E(x,\cdot) \square \sigma \in {}^*E_x \otimes {}^*(C^\infty(|\Lambda|(M)))$, $\int \hat{\delta}^E(x,\cdot) \square \sigma \in {}^*E_x$.

If we take up a *positive* density $\rho \in C^\infty(|\Lambda|(M))$, then there exists a $\hat{\delta}^{E,\rho} \in {}^*(C^\infty(E \boxtimes E^*))$ such that

$$\hat{\delta}^E(x,\cdot) = \hat{\delta}^{E,\rho}(x,\cdot) \otimes \rho , \qquad x \in {}^*M.$$

Moreover, there exists a non-negative $\hat{\delta}_\rho \in {}^*(C^\infty(M \times M))$ such that

$$h(x) \approx \int \hat{\delta}_\rho(x,\cdot)h\rho = \int \hat{\delta}_\rho(\cdot,x)h\rho$$

for every C^∞ function $h : M \longrightarrow \mathbf{R}$ and $x \in M$, where $C^\infty(M\times M)$ is the space of $\mathbf{R}$-valued C^∞ functions on $M\times M$.

Lemma 4.1. Let η, F, $\hat{\delta}^E$, ρ, $\hat{\delta}^{E,\rho}$ and $\hat{\delta}_\rho$ be as above. Let $\mathbb{E}[\ |F=\cdot]$ denote conditional expectation under $F = \cdot$. Then for $\sigma \in C^\infty(E)$,

$$(1)\quad \mathbb{E}[\eta\square(\sigma\circ F)] \approx \int \mathbb{E}[\eta\square\hat{\delta}^E(F,\cdot)]\square\sigma = \int\{\mathbb{E}[\eta\square\hat{\delta}^{E,\rho}(F,\cdot)]\square\sigma\}\rho,$$

$$(2)\quad \mathbb{E}[\eta\square(\sigma\circ F)] \approx \int \mathbb{E}[\hat{\delta}_\rho(F,\cdot)]\mathbb{E}[\eta\square(\sigma\circ F)|F=\cdot]\rho.$$

Proof. Part (1) is proved by noting that

$$\mathbb{E}[\eta\square(\sigma\circ F)] \approx \mathbb{E}[\eta\square(\int \hat{\delta}^E(F,\cdot)\square\sigma)] = \mathbb{E}[\int(\eta\square\hat{\delta}^E(F,\cdot))\square\sigma].$$

Part (2) is shown as follows:

$$\mathbb{E}[\eta\square(\sigma\circ F)] \approx \mathbb{E}[\int \hat{\delta}_\rho(F,\cdot)\mathbb{E}[\eta\square(\sigma\circ F)|F=\cdot]\rho]$$

$$= \int \mathbb{E}[\hat{\delta}_\rho(F,\cdot)]\mathbb{E}[\eta\square(\sigma\circ F)|F=\cdot]\rho\ .$$

Let (M, g) be as in §2, and let Q be a C^∞ Riemannian vector bundle of rank $\ell\in\mathbb{N}-\{0\}$ over M. Set $E = \bigoplus_{m=0}^{n} \{(\Lambda^m T^*M)\otimes Q\}$ (Whitney sum), $(T^*M = (TM)^*)$. Let ∇^E and $\nabla^{T^*M\otimes E}$ denote linear connections in E and $T^*M\otimes E$, respectively, induced naturally from the Levi-Civita connection and the given connection in Q, and let "Trace" denote the trace operator with respect to the fiber metric in $T^*M\otimes T^*M\otimes E$ induced naturally from g and the given fiber metric in Q. We consider the following heat equation for Q-valued differential forms:

$$\frac{\partial\sigma}{\partial t} = \frac{1}{2}\Delta_Q\sigma \left(:= \frac{1}{2}\,\mathrm{Trace}\,(\nabla^{T^*M\otimes E}\nabla^E\sigma)\right), \quad t \in (0,1),$$

$$\lim_{t\downarrow 0}\sigma(t,\cdot) = \sigma_0 \in C^\infty(E). \tag{4.1}$$

We denote by $dv_g\in C^\infty(|\Lambda|(M))$ the Riemannian volume density; it is a positive density. Let $O(M)$ [resp. $O(Q)$] be the bundle of orthonormal frames for TM [resp. Q] over M. Denote by P the fiber product of $O(M)$ and $O(Q)$. Let π_1 and π_2 denote the projections from P onto $O(M)$ and $O(Q)$ respectively, and let $\pi_3 : O(M) \longrightarrow M$ be the projection. Then P is a principal fiber bundle over M with structure group $O(n)\times O(\ell)$ and projection $\tilde{\pi} = \pi_3\circ\pi_1$. As usual, we regard each $p\in P$ as a diffeomorphism of $\bigoplus_{m=0}^{n}\{(\Lambda^m R^n)\otimes R^\ell\}$ onto E_x or onto E_x^*, where $x = \tilde{\pi}(p)$.

Let γ_1 [resp. γ_2] be the connection form on $O(M)$ [resp. $O(Q)$]. Define a connection form γ and an R^n-valued 1-form θ on P, respectively, by

$$\gamma = \pi_1^*\gamma_1 \oplus \pi_2^*\gamma_2 : TP \longrightarrow \mathfrak{o}(n)\oplus\mathfrak{o}(\ell) \text{ (Lie algebra)},$$

$$\theta(Y) = p_1^{-1}(\tilde{\pi}_*Y) \in R^n, \qquad Y \in T_pP, \quad p \in P, \quad p_1 = \pi_1(p),$$

where $\pi_i^*\gamma_i = \gamma_i\circ(\pi_i)_*$, $i=1,2$. Then for each $\xi \in R^n$ there exists uniquely a C^∞ vector field $\tilde{B}(\xi)$ on P such that $\gamma(\tilde{B}(\xi)) = 0$, $\theta(\tilde{B}(\xi)) = \xi$. For $\xi \in {}^*\mathbf{R}^n$, we write $\tilde{B}(\xi)$ instead of ${}^*\tilde{B}(\xi)$.

Let $(\Omega, L(\mathcal{A}), \nu_L)$ be as in §1 with $k = n$. Consider, for each ω, the following ordinary differential equation on ${}^*P$:

$$\frac{dp_t}{dt} = \tilde{B}\left(\frac{dw(t,\omega)}{dt}\right)_{p_t}, \qquad t \in {}^*[0,1]. \tag{4.2}$$

Let $p_t(p, \omega)$ be the solution of (4.2) starting from $p \in {}^*P$ at time $t = 0$. Set, for $p \in P$ and $t \in [0,1]$,

$$X_t(p, \omega) := {}^{\circ}(\tilde{\pi}(p_t(p, \omega))) = \tilde{\pi}({}^{\circ}p_t(p, \omega)).$$

(Since both of P and M are compact, $p_t(t, \omega)$ and $\tilde{\pi}(p_t(p, \omega))$ are both near-standard.) Henceforth we omit ω. For $t \in (0,1)$, define $\tau_t = {}^{\circ}p_t(p)\circ p^{-1} : E_x \longrightarrow E_{X_t(p)}$ and $\tau_t' = {}^{\circ}p_t \circ p^{-1} : E_x^* \longrightarrow E_{X_t(p)}^*$, where $x = \tilde{\pi}(p)$, $p \in P$. Then the stochastic process $(X_t(p))_{t \in [0,1]}$ is a Brownian motion (starting from $\tilde{\pi}(p)$ at time $t=0$) on M, and τ_t [resp. τ_t'] defines the stochastic parallel displacement of fibers of E [resp. E^*] along the Brownian curve $(X_{\cdot}(p))$.

__Lemma 4.2__. The solution of (4.1) is given by

$$\sigma(t,x) = \mathbb{E}[\tau_t^{-1}\sigma_0(X_t(p))], \quad p \in \tilde{\pi}^{-1}(x),\ x \in M,\ t \in (0,1). \tag{4.3}$$

Proof. Let $f_{\sigma_0} : P \longrightarrow \bigoplus_{m=0}^{n} \{(\Lambda^m \mathbf{R}^n) \otimes \mathbf{R}^{\ell}\}$ be the $O(n)\times O(\ell)$-contravariant $\bigoplus_{m=0}^{n} \{(\Lambda^m \mathbf{R}^n) \otimes \mathbf{R}^{\ell}\}$-valued C^∞ function on P associated with σ_0 ; that is, $f_{\sigma_0}(p) = p^{-1}(\sigma_0(\tilde{\pi}(p)))$, $p \in P$ ([9]). Let $\{e_1, \ldots, e_n\}$ be the canonical basis of R^n. As in §2 we have for $t \in (0,1)$

$$\mathbb{E}[f_{\sigma_0}({}^O p_t(p))] - f_{\sigma_0}(p)$$

$$= \int_0^t \frac{1}{2} \sum_{\alpha=1}^{n} (\tilde{B}(e_\alpha))^2 \mathbb{E}[f_{\sigma_0}({}^O p_s(\cdot))](p)\, ds ,$$

from which it follows that

$$\mathbb{E}[\tau_t^{-1}\sigma_0(X_t(p))] - \sigma_0(\tilde{\pi}(p))$$

$$= \int_0^t \frac{1}{2} \Delta_Q \, \mathbb{E}[\tau_s^{-1}\sigma_0(X_s(\cdot))](\tilde{\pi}(p))\, ds .$$

From this, we can deduce that (4.3) is the (unique) solution of (4.1). Here we note that $\mathbb{E}[\tau_t^{-1}\sigma_0(X_t(p))]$ does not depend on the choice of $p \in \tilde{\pi}^{-1}(x)$ and that (4.3) is well-defined.

We note that the map $C^\infty(E) \ni \sigma_0 \longmapsto \mathbb{E}[\tau_t^{-1}\sigma_0(X_t(p))] \in E_x$ is in $\mathfrak{U}$ (see §1).

Theorem 4.3. The heat kernel $e_0(t,x,y)$ with respect to dv_g for the equation (2.3) with $Z_0 = 0$ is given by

$$e_0(t,x,y) = {}^O\mathbb{E}[\hat{\delta}_{dv_g}(X_t(p),y)],\ t \in (0,1),\ x,y \in M,\ p \in \tilde{\pi}^{-1}(x).$$

Proof. Let $t \in (0,1)$, $x,y \in M$, $p \in \tilde{\pi}^{-1}(x)$. By Lemma 2.1

and Corollary 2.2, the solution of (2.3) with $Z_0 = 0$ is

$$u(t,x) = \mathbb{E}[h(X_t(p))] = {}^{\circ}\mathbb{E}[\int \hat{\delta}_{dv_g}(X_t(p),\cdot)h\, dv_g]$$

$$= {}^{\circ}\int \mathbb{E}[\hat{\delta}_{dv_g}(X_t(p),\cdot)]h\, dv_g .$$

This shows that if $\mathbb{E}[\hat{\delta}_{dv_g}(X_t(p),y)]$ is near-standard then its standard part is the heat kernel with respect to dv_g for the equation (2.3) with $Z_0 = 0$. Since the heat kernel $e_0(t,x,y)$ is known to exist, it holds that

$$\mathbb{E}[\hat{\delta}_{dv_g}(X_t(p),y)] = \int \hat{\delta}_{dv_g}(\cdot,y)e_0(t,x,\cdot)dv_g \approx e_0(t,x,y).$$

Therefore $\mathbb{E}[\hat{\delta}_{dv_g}(X_t(p),y)]$ is near-standard and it holds that ${}^{\circ}\mathbb{E}[\hat{\delta}_{dv_g}(X_t(p),y)] = e_0(t,x,y)$.

Now we can give a nonstandard representation of the heat kernel with respect to dv_g for the equation (4.1).

Theorem 4.4. The heat kernel with respect to dv_g for the heat equation (4.1) is given by

$$e(t,x,y) = {}^{\circ}\mathbb{E}[\hat{\delta}_{dv_g}(X_t(p),y)]\cdot\mathbb{E}[\tau_t^{-1}|X_t(p)=y]$$

$$(= e_0(t,x,y)\mathbb{E}[\tau_t^{-1}|X_t(p)=y]),\quad t\in(0,1),\ x,y\in M,\ p\in\tilde{\pi}^{-1}(x), \tag{4.4}$$

where the map $\mathbb{E}[\tau_t^{-1}|X_t(p)=y] \in \mathrm{Hom}(E_y,E_x)$ is defined in such a way that

$$\mathbb{E}[\tau_t^{-1}|X_t(p)=y]\sigma_0(y) = \mathbb{E}[\tau_t^{-1}\sigma_0(y)|X_t(p)=y]$$

$$= \mathbb{E}[\tau_t^{-1}\sigma_0(X_t(p))|X_t(p)=y] \in E_x , \quad \sigma_0 \in C^\infty(E),$$

and thus $e(t,x,y) \in \mathrm{Hom}(E_y,E_x)$ for each $t \in (0,1)$.

Proof. Using the fiber-wise norms defined by the induced metrics in E and E^*, we have

$$\mathbb{E}[|\tau_t'\xi\Box\sigma_0(X_t(p))|] \le \mathbb{E}[|\tau_t'\xi|_{E^*_{X_t(p)}} \cdot |\sigma_0(X_t(p))|_{E_{X_t(p)}}]$$

$$\le \max_{y\in M}|\sigma_0(y)|_{E_y} \cdot \mathbb{E}[|\tau_t'\xi|_{E^*_{X_t(p)}}] = \max_{y\in M}|\sigma_0(y)|_{E_y} \cdot |\xi|_{E^*_x} < \infty$$

for $\xi\in E^*_x$, $p\in\tilde{\pi}^{-1}(x)$, $x\in M$, $t\in(0,1)$, $\sigma_0\in C^\infty(E)$. Fixing t, x and p, we regard $X_t(p)$, $\tau_t'\xi$ and dv_g, respectively, as F, η and ρ. Apply (2) of Lemma 4.1. Then we have

$$\xi\Box\sigma(t,x) = \mathbb{E}[(\tau_t'\xi)\Box\sigma_0(X_t(p))] \qquad \text{[by (4.3)]}$$

$$\approx \int \mathbb{E}[\hat{\delta}_{dv_g}(X_t(p),\cdot)]\mathbb{E}[(\tau_t'\xi)\Box\sigma_0(X_t(p))|X_t(p)=\cdot]\, dv_g$$

$$= \xi\Box\left(\int \mathbb{E}[\hat{\delta}_{dv_g}(X_t(p),\cdot)]\mathbb{E}[\tau_t^{-1}|X_t(p)=\cdot]\sigma_0\, dv_g\right).$$

Thus we have the following, which yields (4.4);

$$\sigma(t,x) = \int {}^{O}\mathbb{E}[\hat{\delta}_{dv_g}(X_t(p),y)]\mathbb{E}[\tau_t^{-1}|X_t(p)=y]\sigma_0(y)dv_g(y).$$

We can also give another expression for $e(t,x,y)$:

Theorem 4.5. For each $t \in (0,1)$, x, $y \in M$, and $p \in \tilde{\pi}^{-1}(x)$, consider the map

$$^{O}\mathbb{E}[(\tau_t')\square\hat{\delta}^{E,dv}(X_t(p),y)] \in \mathrm{Hom}(E_x^*, E_y^*), \qquad (dv = dv_g),$$

$$E_x^* \ni \xi \longmapsto {}^{O}\mathbb{E}[(\tau_t'\xi)\square\hat{\delta}^{E,dv}(X_t(p),y)] \in E_y^* .$$

Let $\iota : \mathrm{Hom}(E_x^*, E_y^*) \cong \mathrm{Hom}(E_y, E_x)$ be the canonical isomorphism. Then the heat kernel $e(t,x,y)$ is rewritten as

$$e(t,x,y) = \iota({}^{O}\mathbb{E}[(\tau_t')\square\hat{\delta}^{E,dv}(X_t(p),y)]).$$

Proof. Let $\xi \in E_x^*$. Then by (1) of Lemma 4.1 and Lemma 4.2, we have

$$\begin{aligned}
\xi\square\sigma(t,x) &= \mathbb{E}[(\tau_t'\xi)\square\sigma_0(X_t(p))] \\
&= {}^{O}\!\!\int \{\mathbb{E}[(\tau_t'\xi)\square\hat{\delta}^{E,dv}(X_t(p),\cdot)]\square\sigma_0\}dv_g \\
&= \int \{{}^{O}\mathbb{E}[(\tau_t')\square\hat{\delta}^{E,dv}(X_t(p),\cdot)](\xi)\square\sigma_0\}dv_g \\
&= \xi\square\left(\int \iota({}^{O}\mathbb{E}[(\tau_t')\square\hat{\delta}^{E,dv}(X_t(p),\cdot)])\sigma_0 \, dv_g\right).
\end{aligned}$$

Hence we have

$$\sigma(t,x) = \int \iota({}^{O}\mathbb{E}[(\tau_t')\square\hat{\delta}^{E,dv}(X_t(p),y)])\sigma_0(y)dv_g(y) .$$

This implies that $\iota({}^{O}\mathbb{E}[(\tau_t')\square\hat{\delta}^{E,dv}(X_t(p),y)]) = e(t,x,y)$.

References

[1] H. Akiyama: Geometric aspects of Malliavin's calculus on vector bundles, J. Math. Kyoto Univ., **26** (1986), 673-696.

[2] H. Akiyama: On Itô's formula for certain fields of geometric objects, J. Math. Soc. Japan, **39** (1987), 79-91.

[3] R.M. Anderson: A non-standard representation for Brownian motion and Itô integration, Israel J. Math., **25** (1976), 15-46.

[4] V. Guillemin and S. Sternberg: Geometric Asymptotics, Math. Surveys No. 14, Amer. Math. Soc., Providence, R.I., 1977.

[5] N. Ikeda and S. Watanabe: Stochastic Differential Equations and Diffusion Processes, Kodansha/ North-Holland, Tokyo/Amsterdam, 1981.

[6] N. Ikeda and S. Watanabe: An introduction to Malliavin's calculus, Stochastic Analysis (K. Itô, ed.), Proc. Taniguchi Internat. Symp. SA., Katata and Kyoto, 1982, North-Holland/ Kinokuniya, Amsterdam/ Tokyo, 1984, pp. 1-52.

[7] T. Kawai: Nonstandard analysis by axiomatic method, Southeast Asian Conference on Logic (C.-T. Chong and M.J. Wicks, eds.), Proc. Logic Conf., Singapore, 1981, Elsevier (North-Holland), 1983, pp. 55-76.

[8] S. Kobayashi and T. Nagano: On projective connections, J. Math. Mech., **13** (1964), 215-235.

[9] S. Kobayashi and K. Nomizu: Foundations of Differential Geometry, I, Interscience, New York, 1963.

[10] H. Kunita: Stochastic differential equations and stochastic flows of diffeomorphisms, École d'Été de Probab. de Saint-Flour XII-1982 (P.L. Hennequin, ed.), Lecture Notes in Math., Vol. 1097, Springer, Berlin, 1984, pp. 143-303.

[11] P.A. Loeb: Conversion from nonstandard to standard measure spaces and applications in probability theory, Trans. Amer. Math. Soc., **211** (1975), 113-122.

[12] P.A. Loeb: An introduction to nonstandard analysis and hyperfinite probability theory, Probabilistic Analysis and Related Topics, Vol. 2 (A.T. Bharucha-Reid, ed.), Academic Press, New York, 1979, pp. 105-142.

[13] M. Saito: Ultraproducts and Non-standard Analysis (in Japanese), Tokyo Tosyo, Tokyo, 1976 and 1987.

[14] S. Watanabe: Lectures on Stochastic Differential Equations and Malliavin Calculus, Springer, Berlin, 1984.

[15] S. Watanabe: Analysis of Wiener functionals (Malliavin calculus) and its applications to heat kernels, Ann. Probab., **15** (1987), 1-39.

Department of Applied Mathematics
Faculty of Engineering
Shizuoka University
Hamamatsu 432, Japan

Hamiltonian Structure and Formal Complete Integrability of Third-order Evolution Equations of Not Normal Type

YOSHIHIDE WATANABE

Abstract. Certain third-order evolution equations of not normal type are presented which admit the second Hamiltonian structure due to Fuchssteiner-Fokas, Gel'fand-Dorfman et al., so that those equations are proved to be formally completely integrable.

Introduction.

One of the most marvelous features of the soliton equations such as the KdV equation is the existence of infinitely many conserved quantities. Moreover these conserved quantities are in involution with respect to a certain Poisson bracket, that is, they commute with one another with respect to the Poisson bracket. This shows that the soliton equations are infinite-dimensional analogues of completely integrable Hamiltonian systems in classical mechanics.

In classical Hamiltonian mechanics, if a system with n-degrees of freedom admits n independent first integrals in involution, then it is called completely integrable. Famous Liouville's theorem asserts that a completely integrable Hamiltonian equation can be transformed into a simple canonical form in the so-called action-angle coordinates by a suitable canonical transformation so that it can be solved by quadrature. Suppose that an evolution equation admits infinitely many conserved quantities in involution with respect to a certain Poisson bracket, it is called *formally completely integrable* on the analogy of classical Hamiltonian mechanics. However, this formal complete integrability does not imply the actual solvability of the equation. It is only one aspect of actually integrable evolution equations such as the soliton equations but it is, in author's view, of great importance and it has given rise to a variety of new ideas which have been playing a central role in the geometrical study of nonlinear evolution equations.

Our interest lies in the search for evolution equations which are formally completely integrable. Such equations admit infinitely many pairwise commuting infinitesimal Hamiltonian flows; this infinitesimal symmetry is interpreted as the Lie-Bäcklund symmetry. We have already classified third-order polynomial evolution equations of certain class admitting nontrivial Lie-Bäcklund symmetries, obtaining evolution equations which have not been familiar in *soliton theory*([**FW1,2**]). The purpose of this paper is to prove some of the above-mentioned equations to be formally completely integrable indeed. The method we mainly use is the theory of the second Hamiltonian structure (or the bi-Hamiltonian structure) developed by Fuchssteiner-Fokas([**FF**]), Gel'fand-Dorfman([**GD1,2**]), Kupershmidt-Wilson([**KW**]) et al..

The contents of the paper are as follows:

Section 1 which is divided into two subsection is devoted to preliminaries. In subsection 1-A, we give the basic notations and definitions related to Lie-Bäcklund symmetry and conserved quantity of evolution equations. In subsection 1-B, according to Gel'fand-Dorfman ([**GD1**]), we give the definition of the Hamiltonian

ISBN 0-12-640170-5

operator and the Hamiltonian pair by introducing the Schouten bracket and then provide some examples of Hamiltonian pairs. Further, we quote a theorem due to Fuchssteiner-Fokas ([**FF**]), which is used in the next section.

Section 2 is the body of the paper. If one can find the second Hamiltonian operator for the evolution equation written in a Hamiltonian form, the formal complete integrability of the equation follows rather automatically from the theorem of Fuchssteiner-Fokas. However, the calculation is so long and tedious that it is performed on the computer algebra system REDUCE. We give a list of evolution equations which are proved to be formally completely integrable. In addition, we present their Hamiltonian operators and their second Hamiltonian operators.

1. Preliminaries.

A. Lie-Bäcklund symmetries and conserved densities. Let $R[u]$ denote the differential ring of polynomials of $u_0, u_1, \dots,$ with the derivation D satisfying $Du_j = u_{j+1} (j = 0, 1, 2, \dots)$. We denote by $R(u)$ the quotient field of $R[u]$; it naturally becomes a differential field with the derivation D. We write u for u_0 and abbreviate the derivation $\partial/\partial u_i$ as $\partial_i (i = 0, 1, 2, , \dots)$. We assign to $f \in R(u)$ a formal vector field

$$\partial_f = \sum_{j \geq 0} (D^j f), \tag{1.1}$$

which is called the evolution derivation associated with f. Evolution derivations satisfy the following commutation relation:

$$[\partial_f, \partial_g] = \partial_f \partial_g - \partial_g \partial_f = \partial_{[f,g]}, \tag{1.2}$$

where the bracket $[\,,\,]$ in $R(u)$ is defined by

$$[f, g] = \partial_f g - \partial_g f. \tag{1.3}$$

Let us consider an evolution equation for $u(x, t)$

$$u_t = H(u, u_1, u_2, \dots, u_m) \quad (H \in R(u)), \tag{1.4}$$

where $u_t = (\partial/\partial t)u(x, t)$ and $u_j = (\partial/\partial x)^j u(x, t)$. The (Lie-Bäcklund) symmetry of Eq.(1.4) is a solution of the differential equation

$$(\partial_H - D(H))f = 0, \tag{1.5}$$

where $D(H)$ is the differential operator defined by

$$D(H) = \sum_{j \geq 0} (\partial_j H) D^j. \tag{1.6}$$

In terms of the bracket introduced by (1.3) Eq.(1.4) can be written as $[H, f] = 0$, and the set of all symmetries of Eq.(1.4) forms a Lie algebra L_H with this bracket.

It is obvious that $u_1, H \in L_H$; they are called trivial symmetries. An integro-differential operator $\mathcal{D}$ on $R(u)$ is said to be a recursion operator for symmetries of Eq.(1.4) if it satisfies $\mathcal{D}L_H \subset L_H$. If $\mathcal{D}$ satisfies

$$[\mathcal{D}, \partial_H - D(H)] = 0, \tag{1.7}$$

it becomes a recursion operator for symmetries of Eq.(1.4); this result is due to Olver([**O**]). Applying the recursion operator recursively to trivial symmetries, we can get an infinite series of symmetries. These symmetries commute with one another if the recursion operator satisfies

$$\mathcal{D} \cdot [\mathcal{D}, \partial_f - D(f)] = [\mathcal{D}, \partial_{\mathcal{D}f} - D(\mathcal{D}f)] \tag{1.8}$$

for an arbitrary $f \in R(u)$; such a recursion operator is called hereditary([**F**]).

Next we shall define conserved densities. An element $f \in R(u)$ is said to be a conserved density of Eq.(1.4) if there exists $g \in R(u)$ such that the following relation holds:

$$\partial_H f = Dg. \tag{1.9}$$

Under the rapidly decreasing or the periodic boundary condition of Eq.(1.4), the integral with respect to x of a conserved density is constant along the flow of the equation. If $f = Dh$ for some $h \in R(u)$, then f automatically becomes a conserved density of an arbitrary evolution equation since ∂_H commutes with D. Such a conserved densities are sometimes called trivial. To describe this triviality, we introduce an equivalence relation $\sim$ as follows: For $f, g \in R(u)$, we say that f and g are equivalent and write $f \sim g$ if there exists $h \in R(u)$ such that $f - g = Dh$. Then the condition (1.9) is equivalent to

$$\partial_H f \sim 0. \tag{1.9'}$$

We denote by $\tilde{R}(u)$ the set of all equivalence classes of $R(u)$ and write $\tilde{f}$ or $\int f dx$ for the equivalence class containing f. The equivalence class $\tilde{f}$ of a conserved density f is said to be a conserved quantity.

B. The Schouten bracket and Hamiltonian structure. We start with some basic notations and definitions, and then give the definition of the Hamiltonian operator.

We introduce the paring $<,>$ in $R(u)$ with values in $\tilde{R}(u)$ by

$$< f, g > = \int fg dx \quad (f, g \in R(u)). \tag{1.10}$$

Let us consider an integro-differential operator $\mathcal{P}$ of finite order of the form

$$\mathcal{P} = \sum_{j=-\infty}^{n} p_j D^j \quad (p_j \in R(u) \text{ and } n : \text{integer}).$$

We define its formal adjoint $\mathcal{P}^*$ to be

$$\mathcal{P}^* = \sum_j (-1)^j D^j \cdot p_j = \sum_j \sum_{k \geq 0} (-1)^j (D^k p_j) \binom{j}{k} D^{j-k}.$$

Then the following formula holds

$$< \mathcal{P}f, g > = < f, \mathcal{P}^* g > \tag{1.11}$$

for arbitrary $f, g \in R(u)$. We note that $\mathcal{P}f$ or $\mathcal{P}^*g$ is not always well-defined when $\mathcal{P}$ is an integral operator. In this case we must restrict operands f and g in $R(u)$ to those for which $\mathcal{P}f$ and $\mathcal{P}^*g$ are well-defined. For example, we consider only the operands of the form $f^{(n)} = D^n f$ $(f \in R(u))$ for the operator $\mathcal{P} = pD^{-n}$ $(p \in R(u))$. This is always the case with the integral operators.

We define the Schouten bracket $[\![\mathcal{P}, \mathcal{Q}]\!]$ of operators $\mathcal{P}$ and $\mathcal{Q}$ as a trilinear form on $R(u)$ with values in $\tilde{R}(u)$ as follows:

$$\begin{aligned} [\![\mathcal{P}, \mathcal{Q}]\!](f, g, h) = &< \mathcal{P}'_{\mathcal{Q}f} g, h > + < \mathcal{Q}'_{\mathcal{P}f} g, h > \\ &+ \text{cyclic sum for } (f, g, h), \quad f, g, h \in R(u), \end{aligned} \tag{1.12}$$

where $\mathcal{P}'_f$ $(f \in R(u))$ denotes the formal Fréchet differential of $\mathcal{P} = \sum p_j D^j$ on f defined by

$$\mathcal{P}'_f = \sum (\partial_f p_j) D^j.$$

Let $\mathcal{H}$ denote a skew symmetric integro-differential operator, that is, $\mathcal{H}$ satisfies $\mathcal{H}^* = -\mathcal{H}$. The operator $\mathcal{H}$ is said to be a Hamiltonian operator if it satisfies

$$[\![\mathcal{H}, \mathcal{H}]\!](f, g, h) = 0 \tag{1.13}$$

for all $f, g, h \in R(u)$. With a Hamiltonian operator $\mathcal{H}$ we associate the Poisson bracket $\{\ ,\ \}_{\mathcal{H}}$ in $\tilde{R}(u)$ as follows:

$$\{\tilde{f}, \tilde{g}\}_{\mathcal{H}} = < \frac{\delta f}{\delta u}, \mathcal{H} \frac{\delta g}{\delta u} >, \tag{1.14}$$

where $\delta / \delta u$ denotes the Euler operator defined by

$$\frac{\delta}{\delta u} = \sum_j (-1)^j D^j \cdot \partial_j. \tag{1.15}$$

We note that the right-hand side of (1.14) is independent of the choice of representatives f and g of $\tilde{f}$ and $\tilde{g}$, since $\delta / \delta u \cdot D = 0$. It is obvious that the Poisson bracket defined by (1.14) is skew symmetric since $\mathcal{H}$ is skew symmetric. Further, it follows from (2.4) that the Poisson bracket satisfies the Jacobi identity, so that the vector space $\tilde{R}(u)$ becomes a Lie algebra.

Let $\mathcal{H}$ and $\mathcal{K}$ be Hamiltonian operators. The pair $(\mathcal{H}, \mathcal{K})$ is called a Hamiltonian pair if the Schouten bracket of them vanishes, that is, $[\![\mathcal{H}, \mathcal{K}]\!](f, g, h) = 0$ for any

$f, g, h \in R(u)$. If $(\mathcal{H}, \mathcal{K})$ is a Hamiltonian pair, then $a\mathcal{H} + b\mathcal{K}$ becomes a Hamiltonian operator as well for arbitrary $a, b \in \mathbb{R}$. It is well known that a skew symmetric operator with constant coefficients always becomes a Hamiltonian operator. Hamiltonian operators whose coefficients are linear in u_i have been characterized by Gel'fand-Dorfmann ([**GD3**]). However, as far as we know, there has been no systematic studies on Hamiltonian operators whose coefficients are not linear in u_i. We shall deal with such Hamiltonian operators in the next section. Here we give typical examples:

$$\mathcal{H}(p) = p(u)D + D \cdot p(u) = 2p(u)D + p'(u)u_1, \tag{1}$$

$$\mathcal{K}(q, r) = q(u)u_1 D^{-1} \cdot (r(u)u_1), \tag{2}$$

where p, q, r are elements of $R(u)$ depending only on $u = u_0$. and $'$ denotes the differentiation with respect to u. $(\mathcal{H}(p), \mathcal{H}(q))$ and $(\mathcal{H}(p), \mathcal{K}(q, r))$ are proved to be Hamiltonian pairs. In particular,

$$\mathcal{H} = u^m D + D \cdot u^m = 2u^m D + mu^{m-1}u_1 \tag{1'}$$

and

$$\mathcal{K} = u_1 D^{-1} \cdot u_1 \tag{2'}$$

form a Hamiltonian pair so that their linear combination with constant coefficients becomes a Hamiltonian operator.

An evolution equation $u_t = H$ $(H \in R(u))$ is said to be written in a Hamiltonian form if H can be written as

$$H = \mathcal{H}\frac{\delta h}{\delta u} \tag{1.16}$$

with a suitable Hamiltonian operator $\mathcal{H}$ and a conserved density h of the equation. In the sequel, we assume that the evolution equation under consideration is written in the Hamiltonian form as (1.16). Then $\tilde{f} \in \tilde{R}(u)$ becomes a conserved density if $\{\tilde{h}, \tilde{f}\}_{\mathcal{H}} = 0$ and the set of all conserved quantities forms a Lie algebra $(C_H, \{\ ,\ \}_{\mathcal{H}})$. Moreover, $\mathcal{H}\frac{\delta}{\delta u}$ induces a Lie algebra homomorphism from the Lie algebra $(C_H, \{\ ,\ \}_{\mathcal{H}})$ of conserved quantities to the Lie algebra $(L_H, [\ ,\ ])$ of symmetries.

An evolution equation written in a Hamiltonian form is said to be formally completely integrable if it admits an infinite series $\{\tilde{f}_j\,; j \in \mathbb{N}\}$ of conserved quantities in involution:

$$\{\tilde{f}_i, \tilde{f}_j\}_{\mathcal{H}} = 0, \quad i, j = 1, 2, \ldots.$$

Formally completely integrable equations also admit an infinite series of commuting symmetries $\{\mathcal{H}\frac{\delta}{\delta u} f_j\,; j \in \mathbb{N}\}$. On the other hand, the following theorem due to Fuchssteiner-Fokas([**FF**]) (also see Gel'fand-Dorfmann [**GD2**]) gives a sufficient condition for an evolution equation to be formally completely integrable.

Theorem. *Consider an evoluton equation written in the Hamiltonian form with a Hamiltonian operator $\mathcal{H}$ and a conserved density h:*

$$u_t = H = \mathcal{H}\frac{\delta h}{\delta u} \quad (H, h \in R(u)).$$

Suppose that the equation has a recursion operator $\mathcal{D}$ for symmetries of the form $\mathcal{D} = \mathcal{K} \cdot \mathcal{H}^{-1}$ where $\mathcal{K}$ is a Hamiltonian operator such that $(\mathcal{H}, \mathcal{K})$ is a Hamiltonian pair. Then the following statements hold:

(1) *$\mathcal{D}$ is a hereditary operator so that the infinite series of symmetries $\mathcal{D}^i H$ $(i = 1, 2, \ldots)$ commute with one another.*

(2) *Set $S_n = \mathcal{H}^{-1}\mathcal{D}^n H$. Then, there exists $h_n \in R(u)(n = 0, 1, 2, \ldots, h_0 = h)$ such that $S_n = \frac{\delta}{\delta u} h_n$ and $\tilde{h}_n$ become conserved quantities in involution with respect to the Poisson bracket associated with the Hamiltonian operator $\mathcal{H}$. Consequently the equation becomes formally completely integrable.*

Remark. The Hamiltonian operator $\mathcal{K}$ in the above theorem is called the second Hamiltonian operator of the equation.

2. Formal complete integrability of third-order equations.

On the basis of the theorem referred to in the previous section we shall prove, in this section, that some of the third-order equation obtained in [**FW1,2**] are formally completely integrable. The proof is performed rather automatically; its procedure is divided into the following three steps:

(1) To find a recursion operator;
(2) To find a Hamiltonian operator $\mathcal{H}$ and a conserved density h with which the equation is written in a Hamiltonian form;
(3) To find the second Hamiltonian operator $\mathcal{K}$.

The first step is already performed in [**FW1**]. The second step is usually carried out by inspection. However, if we consider the case when the equation under consideration takes the potential form of another equation which is written in a Hamiltonian form, then the Hamiltonian form of the former equation is automatically derived from that of the latter by the procedure described below. Finally, the third step is performed as follows: Set $\mathcal{K} = \mathcal{D} \cdot \mathcal{H}$ and prove that $\mathcal{K}$ is a Hamiltonian operator and that $(\mathcal{H}, \mathcal{K})$ is a Hamiltonian pair. Usually, calculation of the Schouten bracket is very cumbersome, so we have used the computer program which we had made for the calculation of the bracket on the computer algebra REDUCE.

Before giving the list of formally completely integrable equations, we make brief remarks on evolution equations in the potential form and on their Hamiltonian structure. Let us consider an evolution equation of the form

$$u_t = DG, \quad G(u, u_1, u_2, \ldots, u_n) \in R(u). \tag{2.1}$$

Introducing a new dependent variable v by setting $u = v_1$, we obtain the evolution equation with respect to v as follows:

$$v_t = \bar{G} = G(v_1, v_2, \ldots, v_{n+1}), \tag{2.2}$$

where $v_t = (\partial/\partial t)v(x,t)$, $v_i = (\partial/\partial x)^i v(x,t)$ and $\bar{G}$ denote the function obtained from G by the replacement: $u_j = v_{j+1} (j = 0, 1, \dots, n)$. Equation (2.2) is said to be in the potential form of Eq.(2.1). For example, $v_t = v_1^3 v_3$ is in the potential form of $u_t = u^3 u_3 + 3u^2 u_1 u_2 = D(u^3 u_2)$. Suppose that Eq.(2.1) is written in a Hamiltonian form:

$$u_t = H = \mathcal{H}\frac{\delta h}{\delta u}.$$

Then $\bar{G}$ can be written as

$$\bar{G} = D^{-1}\bar{H} = D^{-1}\bar{\mathcal{H}}\overline{\frac{\delta h}{\delta u}}$$

where $\bar{\mathcal{H}}$ is obtained from $\mathcal{H}$ by the replacement $u_j = v_{j+1}$ in its coefficients. By the well-known formula

$$\frac{\delta \bar{h}}{\delta v} = -D\overline{\frac{\delta h}{\delta u}}$$

we have the expression

$$\bar{G} = -D^{-1} \cdot \bar{\mathcal{H}} \cdot D^{-1}\frac{\delta \bar{h}}{\delta v}.$$

It follows from the general result of Kupershmidt-Wilson([**KW**]) that the operator $\hat{\mathcal{H}} = D^{-1} \cdot \bar{\mathcal{H}} \cdot D^{-1}$ becomes a Hamiltonian operator so that Eq.(2.2) can be written in a Hamiltonian form.

Now, we shall state the main results of this paper.

Theorem 2.1. *The following five evolution equations are formally completely integrable:*

(I-1) $$u_t = H_I^{(1)} = u_1^3 u_3 + \alpha u_1^4,$$

(I-2) $$u_t = H_I^{(2)} = u_1^3 u_3 + \alpha u_1^3,$$

(II-2) $$u_t = H_{II}^{(2)} = u^3 u_3 + 3u^2 u_1 u_2 + 4\alpha u^3 u_1,$$

(II-3) $$u_t = H_{II}^{(3)} = u^3 u_3 + 3u^2 u_1 u_2 + 3\alpha u^2 u_1,$$

(IV-2) $$u_t = H_{IV}^{(2)} = u^3 u_3 + \frac{3}{2}u^2 u_1 u_2 + \alpha u^2 u_1.$$

Remark 2.1. We use the same numbers for the equations as those in [**FW1,2**], because this numbering shows that some of the numbers are missing. Missing equations are Eqs.(II-1),(III) and (IV-1). In [**FW1**], we found hereditary recursion operators except for Eq.(IV-1). However we were not able to find conserved densities or Hamiltonian form for those three equations.

In order to prove this theorem, we have only to write the equations in the Hamiltonian form and to find out the second Hamiltonian operator. Below, we shall give the Hamiltonian form and the second Hamiltonian operator for the five equations referred to in the theorem.

(1). Equation (I-1) can be written in the Hamiltonian form

$$u_t = H_I^{(1)} = -u_1 D^{-1} \cdot u_1 \frac{\delta}{\delta u}\left(\frac{u_1^2 u_3 + 2\alpha u_1^3}{4}\right) = \mathcal{H}\frac{\delta}{\delta u} h_I^{(1)},$$

where $\mathcal{H}_I = -u_1 D^{-1} \cdot u_1$ is a Hamiltonian operator. The second Hamiltonian operator for the equation is

$$\begin{aligned}\mathcal{K}_I^{(1)} &= -\frac{1}{2}(u_1^4 D + D \cdot u_1^4) - 2(H_I^{(1)} D^{-1} + D^{-1} \cdot H_I^{(1)}) \\ &= -u_1^4 - 2u_1^3 u_2 - 2(u_1^3 u_3 D^{-1} + D^{-1} \cdot u_1^3 u_3) - 2\alpha(u_1^4 D^{-1} + D^{-1} \cdot u_1^4);\end{aligned}$$

$(\mathcal{H}_I, \mathcal{K}_I^{(1)})$ forms a Hamiltonian pair and $\mathcal{D}_I^{(1)} = \mathcal{K}_I^{(1)} \cdot \mathcal{H}_I^{-1}$ is the hereditary recursion operator for Eq.(I-1).

(2). Equation (I-2) can be written in the Hamiltonian form

$$u_t = H_I^{(2)} = -u_1 D^{-1} \cdot u_1 \frac{\delta}{\delta u}\left(\frac{u_1^2 u_3}{4} + \alpha u_1^2\right) = \mathcal{H}\frac{\delta}{\delta u} h_I^{(2)},$$

The second Hamiltonian operator for the equation is

$$\begin{aligned}\mathcal{K}_I^{(2)} &= -\frac{1}{2}(u_1^4 D + D \cdot u_1^4) - 2(H_I^{(2)} D^{-1} + D^{-1} \cdot H_I^{(2)}) \\ &= -u_1^4 D - 2u_1^3 u_2 - 2(u_1^3 u_3 D^{-1} + D^{-1} \cdot u_1^3 u_3) - 2\alpha(u_1^3 D^{-1} + D^{-1} \cdot u_1^3);\end{aligned}$$

$(\mathcal{H}_I, \mathcal{K}_I^{(2)})$ forms a Hamiltonian pair and $\mathcal{D}_I^{(2)} = \mathcal{K}_I^{(2)} \cdot \mathcal{H}_I^{-1}$ is the hereditary recursion operator for Eq.(I-2).

(3). Equation (II-2) can be written in the Hamiltonian form

$$u_t = H_{II}^{(2)} = (u_2 D + u u_1 - u_1 D^{-1} \cdot u_1)\frac{\delta}{\delta u}\left(\frac{u^2 u_2 + 2\alpha u^3}{4}\right) = \mathcal{H}_{II}\frac{\delta}{\delta u} h_{II}^{(2)},$$

where $\mathcal{H}_{II} = u^2 D + u u_1 - u_1 D^{-1} \cdot u_1 = (u^2 D + D \cdot u^2)/2 - u_1 D^{-1} \cdot u_1$ is a Hamiltonian operator. The second Hamiltonian operator for the equation is

$$\begin{aligned}\mathcal{K}_{II}^{(2)} &= \frac{1}{2}\{(u^4 D^3 + D^3 \cdot u^4) - 12(u^2 u_1^2 D + D \cdot u^2 u_1^2) + 4\alpha(u^4 D + D \cdot u^4)\} \\ &= u^4 D^3 + 6u^3 u_1 D^2 + 6(u^3 u_2 + u^2 u_1^2) D + 2(u^3 u_3 + 3u_2 u_1 u_2) \\ &\quad + 4\alpha(u^4 D + 2u^3 u_1);\end{aligned}$$

$(\mathcal{H}_{II}, \mathcal{K}_{II}^{(2)})$ forms a Hamiltonian pair and $\mathcal{D}_{II}^{(2)} = \mathcal{K}_{II}^{(2)} \cdot \mathcal{H}_{II}^{-1}$ is the hereditary recursion operator for Eq.(II-2).

(4). Equation (II-3) can be written in the Hamiltonian form

$$u_t = H_{II}^{(3)} = (u_2 D + uu_1 - u_1 D^{-1} \cdot u_1)\frac{\delta}{\delta u}(\frac{u^2 u_2}{4} + \alpha u^2) = \mathcal{H}_{II}\frac{\delta}{\delta u}h_{II}^{(3)},$$

The second Hamiltonian operator for the equation is

$$\begin{aligned}\mathcal{K}_{II}^{(3)} &= \frac{1}{2}\{(u^4 D^3 + D^3 \cdot u^4) - 12(u^2 u_1^2 D + D \cdot u^2 u_1^2) + 4\alpha(u^3 D + D \cdot u^3)\} \\ &= u^4 D^3 + 6u^3 u_1 D^2 + 6(u^3 u_2 + u^2 u_1^2)D + 2(u^3 u_3 + 3u_2 u_1 u_2) \\ &\quad + 4\alpha(u^3 D + \frac{3}{2}u^2 u_1);\end{aligned}$$

$(\mathcal{H}_{II}, \mathcal{K}_{II}^{(3)})$ forms a Hamiltonian pair and $\mathcal{D}_{II}^{(3)} = \mathcal{K}_{II}^{(3)} \cdot \mathcal{H}_{II}^{-1}$ is the hereditary recursion operator for Eq.(II-3).

(5). Equation (IV-2) can be written in the Hamiltonian form

$$u_t = H_{IV}^{(2)} = (u_3 D + \frac{3}{2}u^2 u_1)\frac{\delta}{\delta u}(\frac{uu_2}{2} + \frac{2}{3}\alpha u) = \mathcal{H}_{IV}\frac{\delta}{\delta u}h_{IV}^{(2)},$$

where $\mathcal{H}_{IV} = u^3 D + \frac{3}{2}u^2 u_1 = \frac{1}{2}(u^3 D + D \cdot u^3)$ is a Hamiltonian operator. The second Hamiltonian operator for the equation is

$$\begin{aligned}\mathcal{K}_{IV}^{(2)} &= \frac{1}{2}\{(u^5 D^3 + D^3 \cdot u^5) - \frac{71}{4}(u^3 u_1^2 D + D \cdot u^3 u_1^2) \\ &\quad + u_1 D^{-1} \cdot H_{IV}^{(2)} + H_{IV}^{(2)} D^{-1} \cdot u_1 + \frac{4}{3}\alpha(u^4 D + D \cdot u^4)\} \\ &= u^5 D^3 + \frac{15}{2}u^4 u_1 D^2 + (\frac{15}{2}u^4 u_2 + \frac{49}{4}u^2 u_1^2)D \\ &\quad + \frac{5}{2}u^4 u_3 + \frac{49}{4}u^3 u_1 u_2 + \frac{27}{8}u^2 u_1^3 \\ &\quad + \frac{1}{2}\{u_1 D^{-1} \cdot (u^3 u_3 + \frac{3}{2}u^2 u_1 u_2) + (u^3 u_3 + \frac{3}{2}u^2 u_1 u_2)D^{-1} \cdot u_1\} \\ &\quad + \alpha\{\frac{4}{3}(u^4 D + 2u^3 u_1) + u_1 D^{-1} \cdot u^2 u_1 + u^2 u_1 D^{-1} \cdot u_1\};\end{aligned}$$

$(\mathcal{H}_{IV}, \mathcal{K}_{IV}^{(2)})$ forms a Hamiltonian pair and $\mathcal{D}_{IV}^{(2)} = \mathcal{K}_{IV}^{(2)} \cdot \mathcal{H}_{IV}^{-1}$ is the hereditary recursion operator for Eq.(IV-2).

Remark 2.2. Equations (I-1) and (I-2) are in the potential form of Eqs.(II-2) and (II-3), respectively, so that the former equations can be put in the Hamiltonian form using the latter equations through the procedure described above. For example, the Hamiltonian operators $\mathcal{H}_I$ and $\mathcal{K}_I^{(1)}$ can be obtained from $\mathcal{H}_{II}$ and $\mathcal{K}_{II}^{(2)}$ as follows:

$$\mathcal{H}_I = -D^{-1} \cdot \bar{\mathcal{H}}_{II} \cdot D^{-1}, \quad \mathcal{K}_I^{(1)} = -D^{-1} \cdot \bar{\mathcal{K}}_{II}^{(2)} \cdot D^{-1},$$

and for the conserved densities the following relation holds:

$$h_I^{(1)} = \bar{h}_{II}^{(2)}.$$

Here the over bar $^-$ denotes the replacement of u_j by u_{j+1}.

Acknowledgement. I would like to express my gratitude to professor A. Fujimoto for his encouragement in carring out this work. I am also greateful to associate professor T.Iwai, the editor of this session, for carefully reading the manuscript and offering valuable comments.

REFERENCES

[**F**] B. Fuchssteiner, *Application of hereditary symmetries to nonlinear evolution equations*, Nonlinear Anal., Theory, Method and Appl. **3(6)** (1979), 849–862.

[**FF**] B. Fuchssteiner and A.S. Fokas, *Symplectic structures, their Bǎcklund transformations and hereditary symmetries*, Physica **4D** (1981), 47–66.

[**FW1**] A. Fujimoto and Y. Watanabe, *Polynomial evolution equations of not normal type admitting nontrivial symmetries*, Preprint.

[**FW2**] A. Fujimoto and Y. Watanabe, *Classification of the third-order polynomial evolution equations of not normal type admitting nontrivial symmetries*, Preprint.

[**GD1**] I.M. Gel'fand and I.Ya Dorfman, *Hamiltonian operators and algebraic structures related to them*, Funct. Anal. Appl. **13(4)** (1979), 248–262.

[**GD2**] I.M. Gel'fand and I.Ya Dorfman, *The Schouten bracket and Hamiltonian operators*, Funct. Anal. Appl. **14(3)** (1980), 223–226.

[**GD3**] I.M. Gel'fand and I.Ya Dorfman, *Hamiltonian operators and infinite-dimensional Lie algebra*, Funct. Anal. Appl. **15(3)** (1981), 173–187.

[**KW**] B.A. Kupershmidt and G. Wilson, *Modifying Lax equations and the second Hamiltonian structure*, Invent. Math. **19(5)** (1981), 403–436.

[**O**] P.J. Olver, *Evolution equations possessing infinitely many symmetries*, J. Math. Phys. **18(6)** (1977), 1212–1215.

Yoshihide WATANABE
Department of Applied Math.
Faculty of Engineering
Hiroshima Univirsity
Higashi-Hiroshima, 724
Japan

The quasi-classical calculation of eigenvalues for the Bochner-Laplacian on a line bundle

by Akira Yoshioka

1. Introduction.

Concerning the operators acting on functions on a manifold, the quasi-classical calculation of eigenvalues is formulated by means of differential geometrical materials, that is, Lagrangian submanifolds, action integrals and Maslov indices. For details and results on that calculation, see Maslov [10], Colin de Verdiere [3], Leray [9], Weinstein [11], Ii [5], Yoshioka [13].

In this paper, we extend the method of quasi-classical calculations to be applicable to the Bochner-Laplacian, an operator acting on sections of a complex line bundle $E \to M$. To do this, we must set up Hamiltonian mechanics on T^*M with a modified symplectic structure σ, referred to as the charged symplectic structure (cf. [12]), instead of the canonical one. Here σ is given by adding the curvature form of a linear connection in the bundle E to the canonical symplectic form. The main difficulty arising with the σ is that the canonical 1-form θ of T^*M cannot define an element of the 1st cohomology class of a Lagrangian submanifold of (T^*M, σ) in general, so that the action integral using θ is not available. Thus the Maslov quantization condition fails to work in the original form. An alternative quantization condition in (T^*M, σ) will be defined through introducing a principal U(1) bundle B over T^*M and a connection form on B with the curvature σ. By means of this quantization condition, the

ISBN 0-12-640170-5

quasi-classical calculation is reformulated in parallel with the one in [11].

Let $P \to M$ be the principal $U(1)$ bundle associated with E. It is here to be noted that (T^*M, σ) is viewed, through the $U(1)$ action, as the reduced phase space of $(T^*P, d\theta_P)$, where $d\theta_P$ is the canonical symplectic structure on T^*P. Then a relation will be brought about between the Maslov quantization condition in $(T^*P, d\theta_P)$ and the proposed quantization condition in (T^*M, σ). Thus the method of quasi-classical calculation is extended to (T^*M, σ). An application is made to the Bochner-Laplacian associated with the harmonic connection on the line bundle over CP^n. The eigenvalue problem of that operator is exactly solved in [8]. We show that for the Bochner-Laplacian the quasi-classical eigenvalues coincide with the exact eigenvalues in [8] within an additive constant.

2. The Bochner-Laplacian and quantization condition.

In this section, we define the quantization condition to formulate the method of quasi-classical calculation for the eigenvalue problem of the Bochner-Laplacian.

We begin with the Bochner-Laplacian (for details, see [8]). Let $\pi : (E, \langle\ ,\ \rangle) \to (M, g_M)$ be a C^∞ Hermitian line bundle over a C^∞ Riemannian manifold with Hermitian structure $\langle\ ,\ \rangle$. We denote the space of all C^∞ sections of E by $C^\infty(E)$. Let $\tilde{d}$ be a metric connection in $(E, \langle\ ,\ \rangle)$. The Bochner-Laplacian D is the operator acting on $C^\infty(E)$ in the following form; with respect to a local unitary frame of E and local coordinates of M, D is expressed as

$$D = -\sum_{j,k=1}^{n} g^{jk}(\nabla_j + ia_j)(\nabla_k + ia_k),$$

where $n = \dim M$, $\omega = i\sum a_j dx_j$ is a local connection form of $\tilde{d}$, ∇ is the Levi-Civita connection defined by g, and ∇_j stands for $\nabla_{\partial/\partial x^j}$. We denote the curvature form of $\tilde{d}$ by $i\Omega$.

The classical mechanical system associated with D is given as follows. Let θ_M be the canonical 1-form of T^*M and $\pi_M : T^*M \to M$ be the canonical projection. We consider the symplectic manifold (T^*M, σ) whose symplectic form is given by $\sigma = d\theta_M + \pi_M^*\Omega$. Let $H(x; \xi) = |\xi|^2$ be the energy Hamiltonian of g, where $x \in M$, $\xi \in T_x^*M$. The associated Hamiltonian system is the triple (T^*M, σ, H).

Consider a Lagrangian submanifold L of (T^*M, σ). The Maslov quantization condition is then written as

$$\text{(MQ)} \qquad \frac{1}{2\pi}\int_c \theta_M - \frac{1}{4}\langle \mu_L, [c]\rangle \in \mathbf{Z}, \qquad \text{for any } [c] \in H_1(L; \mathbf{Z}),$$

where μ_L is the Maslov class of L. From the definition of σ with $\Omega \neq 0$, $d\theta_M$ does not vanish on L in general, so that the action integral $\int_c \theta_M$ is not well defined by the homology class of $[c] \in H_1(L; \mathbf{Z})$. Thus, (MQ) is not effective for L. We define an alternative quantization condition as follows. Let $\nu_P : P \to M$ be the principal U(1) bundle associated with $(E, \langle\ ,\ \rangle)$. The metric connection $\tilde{d}$ yields the connection β on P such that $d\beta = \nu_P^*\Omega$. We have the pull-back bundle $\nu_B : B \to T^*M$ by $B = \pi_M^{-1}P$. π_M is lifted naturally to $\hat{p} : B \to P$ satisfying $\nu_P \hat{p} = \pi_M \nu_B$. Set a 1-form $\alpha = \nu_B^*\theta_M + \hat{p}^*\beta$. Then α defines a connection in B whose curvature form projects to σ, i.e., $d\alpha = \nu_B^*\sigma$. For a Lagrangian submanifold $L \subset (T^*M, \sigma)$, $d\alpha$ vanishes on $\nu_B^{-1}(L)$ since $\sigma | L = 0$. Hence α defines an element of $H^1(\nu_B^{-1}(L); \mathbf{R})$.

Now, we define: L satisfies the quantization condition if and only if

(Q) $\frac{1}{2\pi}\int_c \alpha - \frac{1}{4}\langle \nu_B^* \mu_L, [c]\rangle \in \mathbf{Z}$, for $\forall [c] \in H_1(\nu_B^{-1}(L); \mathbf{Z})$.

If $E = M\times\mathbf{C}$ and $\tilde{d}$ is flat, $D = \Delta_g$. Moreover, $B = T^*M\times U(1)$, $\alpha = \nu_B^*\theta_M + d\lambda/i\lambda$, $(\lambda \in U(1))$ and $\nu_B^{-1}(L) = L\times U(1)$. (Q) is trivial for the generator defined by $pt\times U(1)$. In this case (Q) is equivalent to (MQ). Hence, the quantization condition (Q) can be considered as a generalization of the Maslov one. Replacing (MQ) by (Q), we formulate the method of quasi-classical calculation in parallel with usual one (cf. [11]). A compact Lagrangian submanifold is called a quasi-classical state (a QCS) if it satisfies (Q). A QCS L is called a quasi-classical eigenstate (a QCE) of H when $H\,|\,L = E$(const.). E is called a quasi-classical eigenvalue of H.

3. U(1) reduction and quantization conditions.

It is well-known that (T^*M, σ) is thought of as the U(1) reduction of $(T^*P, d\theta_P)$ (cf. [1], [7]). In this section, we investigate in the above framework how the bundle B with the connection form α are linked with the cotangent bundle T^*P with θ_P, and study a relation between the Maslov class of a Lagrangian submanifold in (T^*M, σ) and that for $(T^*P, d\theta_P)$. Then it turns out that (Q) is obtained, in a sense, by U(1) reducing the Maslov quantization condition for $(T^*P, d\theta_P)$.

First, we recall the U(1) reduction of $(T^*P, d\theta_P)$ briefly. The action of structure group U(1) on P is lifted to be an exact symplectic action on $(T^*P, d\theta_P)$, which is a free action, of course. We denote the moment map of this action by $\psi : T^*P \to \mathbf{R}$. Consider a submanifold $F = \psi^{-1}(0)$. Since ψ is U(1) invariant, U(1) acts freely on F. We then have a principal U(1) bundle $\nu_F : F \to F/U(1) \sim T^*M$. We note that $\nu_F^*\theta_M = \iota_F^*\theta_P$, where $\iota_F : F \to T^*P$ is the inclusion map. For every $c \in \mathbf{R}$, we equip

T^*M with a symplectic form as $\sigma_c = d\theta_M + \pi_M^* c\Omega$. The symplectic manifold (T^*M, σ_c) is then obtained as the reduced phase space of $(T^*P, d\theta_P)$ as follows: Using the connection form β on P, we define a map $\tau_c : T^*P \to T^*P$ by $\tau_c(x; \xi) = (x; \xi - c\beta(x))$, where $x \in P$, $\xi \in T_x^*P$ and we consider $\beta(x) \in T_x^*P$. τ_c commutes with the U(1) action. It further holds that

$$\tau_c^* \theta_P = \theta_P - \pi_P^* c\beta, \tag{3.1}$$

where $\pi_P : T^*P \to P$ is the canonical projection. Consider a submanifold $\psi^{-1}(c)$ with the inclusion map $\iota_c : \psi^{-1}(c) \to T^*P$. Then τ_c induces a map of $\psi^{-1}(c)$ to $F = \psi^{-1}(0)$. Set $\nu_c = \nu_F \tau_c$.

Proposition 3.1. (U(1) reduction)

(i) $\nu_c : \psi^{-1}(c) \to T^*M$ is a principal U(1) bundle.

(ii) $\nu_c^* \sigma_c = \iota_c^* d\theta_P$.

Now, we relate B, α to T^*P, θ_P. We set $c = 1$ in Proposition 3.1. Define a map $\kappa : F \to B$ by $\kappa(x; \xi) = (\nu_F(x; \xi), \pi_P \iota_F(x; \xi))$. Then κ is a bundle isomorphism such that $\nu_B \kappa = \nu_F$. Hence,

$$\kappa^* \alpha = \iota_F^*(\theta_P + \pi_P^* \beta). \tag{3.2}$$

Define a map of $\psi^{-1}(1)$ to B as $\Phi = \kappa \tau_1$. Then, (3.1) and (3.2) give the following.

Proposition 3.2.

(i) $\Phi : \psi^{-1}(1) \to B$ is a bundle isomorphism such that $\nu_1 = \nu_B \Phi$.

(ii) $\Phi^* \alpha = \iota_1^* \theta_P$.

Moreover, the symplectic structures $\sigma_m = d\theta_M + \pi_M^* m\Omega$, $m \in Z$, are of special interest. Let ρ_m be the U(1) representation on **C** such that $\rho_m(\lambda)w = \lambda^m w$, $\lambda \in U(1)$, $w \in \mathbf{C}$. We then have a complex line bundle E_m over M, where E_m is the associated line bundle with P by the representation ρ_m. E_m is naturally equipped with the Hermitian structure $\langle\, ,\, \rangle_m$. The connection β in P induces a metric connection $\tilde{d}_m$ in $(E_m, \langle\, ,\, \rangle_m)$ with the curvature $im\Omega$. We denote in turn by $\nu_{Pm} : P_m \to M$ the principal U(1) bundle associated with $(E_m, \langle\, ,\, \rangle_m)$. $\tilde{d}_m$ yields the connection β_m in P_m whose curvature form is $m\Omega$. As in Section 2, we define the principal U(1) bundle $\nu_{Bm} : B_m \to T^*M$ and the connection α_m in B_m such that $d\alpha_m = \nu_{Bm}^* \sigma_m$. Note that $B_1 = B$, $\alpha_1 = \alpha$, etc., given in Section 2. We denote by R_m the U(1) action on the bundle P_m. The fundamental vector field γ_m is then given by $\gamma_m(x) = (d/dt)|_{t=0}\, R_m(e^{it})x$, $x \in P_m$. R_m is lifted to be an exact symplectic action $\hat{R}_m$ on T^*P_m. The moment map of the action $\hat{R}_m$ is denoted by ψ_m . If we write transition functions of the bundle $P(=P_1\ )$ as $\kappa_{\alpha\beta}(x) \in U(1)$, $x \in U_\alpha \cap U_\beta \subset M$, those of P_m are given by $(\kappa_{\alpha\beta}(x))^m \in U(1)$. Hence, gluing maps $(x, w) \to (x, w^m)$, $(x, w) \in U_\alpha \times U(1)$, we get a bundle homomorphism $\varphi_m : P_1 \to P_m$ such that $\varphi_m R_1(\lambda) = R_m(\lambda)^m \varphi_m$, $\lambda \in U(1)$. Differentiated, this equation yields

$$\varphi_{m*}\gamma_1 = m\gamma_m. \tag{3.3}$$

Suppose $m \neq 0$. Since the tangent map φ_{m*x} is a linear isomorphism of T_xP_1 to T_yP_m $(y = \varphi_m(x))$, the dual map of φ_{m*x}^{-1} yields a linear isomorphism of $T_x^*P_1$ to $T_y^*P_m$. Hence, φ_m gives rise to a map $\hat{\varphi}_m : T^*P_1 \to T^*P_m$ with the property $\pi_{Pm}\hat{\varphi}_m = \varphi_m\pi_P$. By means of $\hat{\varphi}_m$, we can relate B_m, α_m to T^*P, θ_P as follows.

Lemma 3.3. (i) $\hat{\varphi}_m^*\theta_{Pm} = \theta_P$.

(ii) $\hat{\varphi}_m^*\psi_m = (1/m)\psi_1$.

Proof. (i) is obvious by the definition of $\hat{\varphi}_m$. Recall $\psi_m(x; \xi) = \langle \xi, \gamma_m(x) \rangle$ (the pairing of $\xi \in T_x^*P_m$ with $\gamma_m(x) \in T_xP_m$). Then (3.3) shows (ii).

$\hat{\varphi}_m$ induces a map $\hat{\varphi}_m : \psi^{-1}(m) \to \psi_m^{-1}(1)$. Applying Proposition 3.1 to $(T^*P_m, d\theta_{Pm})$ with $c = 1$, we have a principal U(1) bundle $\nu_1^{(m)} : \psi_m^{-1}(1) \to T^*M$ such that $\nu_1^{(m)*}\sigma_m = \iota_1^{(m)*}d\theta_{Pm}$, where $\iota_1^{(m)} : \psi_m^{-1}(1) \to T^*P_m$ is the inclusion map. Applying Proposition 3.2 to $\psi_m^{-1}(1)$ and B_m with α_m, we have a bundle isomorphism $\Phi^{(m)} : \psi_m^{-1}(1) \to B_m$ satisfying $\nu_1^{(m)} = \nu_{Bm}\Phi^{(m)}$ and $\Phi^{(m)*}\alpha_m = \iota_1^{(m)*}\theta_{Pm}$. Note that $\nu_1^{(m)}\hat{\varphi}_m = \nu_m$, which is given in Proposition 3.1 with $c = m$. Set a map of $\psi^{-1}(m)$ to B_m as $\Phi_m = \Phi^{(m)}\hat{\varphi}_m$. Then, we get the following.

Proposition 3.4. (i) $\Phi_m : \psi^{-1}(m) \to B_m$ is a bundle homomorphism such that $\nu_m = \nu_{Bm}\Phi_m$ and $\Phi_m\hat{R}_1(\lambda) = R_m(\lambda)^m\Phi_m$, $\lambda \in U(1)$.

(ii) $\Phi_m^*\alpha_m = \iota_m^*\theta_P$, where $\iota_m : \psi^{-1}(m) \to T^*P$ is the inclusion map.

Remark. In case of $m = 0$, $B_m = T^*M \times U(1)$ and $\alpha_m = \theta_M + d\lambda/i\lambda$, $\lambda \in U(1)$, so that we may take $\Phi_m(x; \xi) = (\nu_F(x; \xi), 1)$.

In what follows, we investigate relations of the respective Maslov classes for the symplectic manifolds $(T^*P, d\theta_P)$ and (T^*M, σ_m). For the definition of the Maslov class and index, see [4]. We write a point of T^*P as $t = (x; \xi)$ where $x \in P$, $\xi \in T_x^*P$ and that of T^*M as $\tilde{t} = (\tilde{x}; \tilde{\xi})$ where $\tilde{x} \in M$, $\tilde{\xi} \in T_{\tilde{x}}^*M$. Hence a tangent vector at t is expressed as $\zeta = (\zeta_1; \zeta_2)$, where $\zeta_1 \in T_xP$ and $\zeta_2 \in T_\xi T_x^*P$. Using the connection form β, we define a vector field on T^*P by $\hat{\beta}(t) = (0; \beta(x)) \in T_tT^*P$, where we consider $\beta(x) \in T_\xi T_x^*P$ by means of the linear space structure of T_x^*P. We denote the Hamiltonian vector field of the moment map ψ by X_ψ. From Proposition 3.1, the tangent map $\nu_{c*t} : T_t\psi^{-1}(c) \to T_{\tilde{t}}T^*M$ is onto, where $\tilde{t} = \nu_c(t)$. The kernel is spanned by $X_\psi(x)$. Define the subspace of $T_t\psi^{-1}(c)$ to be $N_t = \{\zeta \in T_t\psi^{-1}(c) \mid (\pi_P^*\beta(t))(\zeta) = 0\}$. Then, $T_t\psi^{-1}(c) = \mathbf{R}X_\psi(t) + N_t$ (a direct sum). The restriction of ν_{c*t} to N_t induces a linear isomorphism to $T_{\tilde{t}}T^*M$, which is denoted by $\hat{\nu}$. Let $\tilde{\mathcal{L}}_f(\tilde{t})$ be the Lagrangian subspace of $\tilde{V}_{\tilde{t}} = (T_{\tilde{t}}T^*M, \sigma_c(\tilde{t}))$, tangent to the fibre $\pi_M^{-1}(\tilde{x})$, and $\tilde{\mathcal{L}}_b(\tilde{t})$ be a Lagrangian subspace of $\tilde{V}_{\tilde{t}}$ transversal to $\tilde{\mathcal{L}}_f(\tilde{t})$. Let $\mathcal{L}_f(t)$ be the Lagrangian subspace of $V_t = (T_tT^*P, d\theta_P(t))$ tangent to the fibre $\pi_P^{-1}(x)$. Then, $\mathcal{L}_f(t) = \mathbf{R}\hat{\beta}(t) + \hat{\nu}^{-1}\tilde{\mathcal{L}}_f(\tilde{t})$ (a direct sum). Set a subspace of V_t as $\mathcal{L}_b(t) = \mathbf{R}X_\psi(t) + \hat{\nu}^{-1}\tilde{\mathcal{L}}_b(\tilde{t})$ (a direct sum). Then, $\mathcal{L}_b(t)$ is the Lagrangian subspace transversal of $\mathcal{L}_f(t)$. It may be assumed we have Riemannian metrices g_M on M and g_P on P, where g_P is U(1) invariant and $\nu_P : (P, g_P) \to (M, g_M)$ is a Riemannian submersion. In accordance with these Riemannian structures and the above-mentioned direct sum decompositions together with the transversality, identifications $T_{\tilde{t}}T^*M = \mathbf{C}^n$ and $T_tT^*P = \mathbf{C}^{n+1}$ can be made in such a manner that $\hat{\nu}^{-1}$ induces naturally a map $\hat{\nu}^{-1} : \mathbf{C}^n \to \mathbf{C}^{n+1}$ to be expressed as $z \to (0, z)$, $z \in \mathbf{C}^n$, where $n = \dim M$.

Suppose L is a Lagrangian submanifold of (T^*M, σ_c). Then, $\hat{L} = \nu_c^{-1}(L) \subset \psi^{-1}(c)$ is a Lagrangian submanifold of $(T^*P, d\theta_P)$. We then have a principal U(1) bundle $\nu_c : \hat{L} \to L$. For each $t \in \hat{L}$, $T_t\hat{L} = \mathbf{R}X_\psi(t) + \hat{\nu}^{-1}T_{\tilde{t}}L$

(a direct sum) where $\tilde{t} = \nu_c(t)$. Thus, if W is a unitary matrix such that $T_{\tilde{t}}L = W\mathscr{L}_b(\tilde{t})$, a unitary matrix $\hat{W}$ with $T_t\hat{L} = \hat{W}\mathscr{L}_b(t)$ is given by $\hat{W} = \begin{bmatrix} 1 & 0 \\ 0 & W \end{bmatrix}$. Let μ_L and $\hat{\mu}_L$ be the Maslov classes of L and $\hat{L}$, respectively. Then, we have the following.

Proposition 3.5. $\hat{\mu}_L = \nu_c^*\mu_L$.

Now, we are in a position to make clear the relation between the Maslov quantization condition in $(T^*P, d\theta_P)$ and the quantization condition (Q) in (T^*M, σ_m), $m \in \mathbf{Z}$. To this end, we employ the principal U(1) bundle $\nu_{Bm} : B_m \to T^*M$ together with the connection α_m such that $d\alpha_m = \nu_{Bm}^*\alpha_m$. Let L be a Lagrangian submanifold in (T^*M, σ_m). We set $\tilde{L} = \nu_{Bm}^{-1}(L)$. Then, $\nu_{Bm} : \tilde{L} \to L$ is a principal U(1) bundle. Note that $\tilde{L}$ is trivial because $\alpha_m | \tilde{L}$ is a connection with the zero curvature. If u is the generator of $H_1(\tilde{L}: \mathbf{Z})$ that is given by the U(1) action, the quantization condition (Q) is trivial for u. Hence, L satisfies the quantization condition (Q) if and only if (Q) holds for any class of the form s_*a, $a \in H_1(L; \mathbf{Z})$, where s is a section of the bundle $\tilde{L} \to L$. Now we set $\hat{L} = \nu_m^{-1}(L)$. Then, $\hat{L}$ is a Lagrangian submanifold of $(T^*P, d\theta_P)$, and further one has $\Phi_m : \hat{L} \to \tilde{L}$. Accordingly, Propositions 3.4 and 3.5 with c = m imply that

$$\frac{1}{2\pi}\int_{\hat{a}}\theta_P - \frac{1}{4}\langle\hat{\mu}_L, \hat{a}\rangle = \frac{1}{2\pi}\int_{\tilde{a}}\alpha_m - \frac{1}{4}\langle\nu_{Bm}^*\mu_L, \tilde{a}\rangle,$$

for any $\hat{a} \in H_1(\hat{L}: \mathbf{Z})$, where $\tilde{a} = s_*\nu_{m*}\hat{a}$. Thus, we get the following.

Theorem 3.6. L satisfies the quantization condition (Q) if and only if $\nu_m^{-1}(L)$ satisfies the Maslov quantization condition in $(T^*P, d\theta_P)$.

4. Example.

In this section, we work out quasi-classical eigenvalues of the Bochner-Laplacian associated with the harmonic connection on the line bundle over CP^n.

The harmonic connection is given as follows (cf. [8]). We provide $C^{n+1} = \{ z = (z_0,\dots,z_n) \}$ with the Hermitian inner product $\langle z, z' \rangle = \sum_{j=0}^n z_j \overline{z'_j}$ and the real inner product $\langle z, z' \rangle_{\mathbf{R}} = \mathrm{Re}\langle z, z' \rangle$. Consider the $2n + 1$ dimensional sphere with radius 2,

$$S^{2n+1}_{[2]} = \{ z \in C^{n+1} \mid \langle z, z \rangle_{\mathbf{R}} = 4 \},$$

which is endowed with the canonical Riemannian metric g_S induced from $\langle \, , \, \rangle_{\mathbf{R}}$. $U(1) = \{ \lambda \in \mathbf{C} \mid | \lambda | = 1 \}$ acts on $S^{2n+1}_{[2]}$ as $R(\lambda) z = \lambda z$, $\lambda \in U(1)$, $z \in S^{2n+1}_{[2]}$. Making the quatient space, we get the principal fibre bundle (Hopf fibre bundle),

$$\nu_P : S^{2n+1}_{[2]} \to CP^n.$$

We fix the Riemannian metric g on CP^n so that ν_P may be a Riemannian submersion. Define the connection on $S^{2n+1}_{[2]}$ by

$$\beta = g_S(\gamma, *)/ | \gamma |^2,$$

where γ is the fundamental vector field on $S^{2n+1}_{[2]}$ of the action R. Its curvature form is denoted by Ω. For every $m \in \mathbf{Z}$, we consider the U(1) representation ρ_m on $\mathbf{C}$ such that $\rho_m(\lambda) w = \lambda^m w$, $\lambda \in U(1)$, $w \in \mathbf{C}$. We then have the Hermitian line bundle $(E_m, \langle \, , \, \rangle_m)$ associated with $S^{2n+1}_{[2]}$ by ρ_m. The metric connection $\tilde{d}_m$ induced by β is called the harmonic connection in $(E_m, \langle\, ,\, \rangle_m)$. We denote by D_m the Bochner-Laplacian associated with $\tilde{d}_m$. The eigenvalues and their multiplicities are known already.

Proposition 4.1. ([8]). The eigenvalues of D_m are

$$\lambda_k^{(m)} = (k + |m|/2)(k + |m|/2 + n) - m^2/4, \qquad k = 0, 1, 2, \ldots$$

and the multiplicity of $\lambda_k^{(m)}$ is

$$\binom{k + |m| + n}{k + |m|}\binom{k + n}{k} - \binom{k + |m| + n - 1}{k + |m| - 1}\binom{k + n - 1}{k - 1}$$

Consider the cotangent bundle $\pi : T^*\mathbf{C}P^n \to \mathbf{C}P^n$. We denote the energy Hamiltonian of g by H. Let $T^*\mathbf{C}P^n$ be endowed with the symplectic structure $\sigma_m = d\theta + \pi^* m\Omega$, where θ is the canonical 1-form of $T^*\mathbf{C}P^n$. The classical mechanical system against D_m is the triple $(T^*\mathbf{C}P^n, \sigma_m, H)$. Tracing back the projection $\nu_P : S^{2n+1}_{[2]} \to \mathbf{C}P^n$, we consider the symplectic manifold $(T^*S^{2n+1}_{[2]}, d\theta_S)$, where θ_S is the canonical 1-form on $T^*S^{2n+1}_{[2]}$. Let $\psi : T^*S^{2n+1}_{[2]} \to \mathbf{R}$ be the moment map associated with the action lifted from R. Then, $(T^*\mathbf{C}P^n, \sigma_m)$ is the reduced phase space, namely, one has a principal U(1) bundle $\nu_m : \psi^{-1}(m) \to T^*\mathbf{C}P^n$ together with $\nu_m^*\sigma_m = \iota_m^* d\theta_S$, where $\iota_m : \psi^{-1}(m) \to T^*S^{2n+1}_{[2]}$ is the inclusion map. We now identify $T^*S^{2n+1}_{[2]}$ with $TS^{2n+1}_{[2]}$ by means of g_S. Hence, any point of $T^*S^{2n+1}_{[2]}$ is expressed as a pair $(z; \xi)$ with $z \in \mathbf{C}^{n+1}$, $|z| = \langle z, z\rangle_{\mathbf{R}}^{1/2} = 2$, and $\xi \in \mathbf{C}^{n+1}$, $\langle z, \xi\rangle_{\mathbf{R}} = 0$. Setting $z^{(j)} = (z_0, z_1, \ldots, z_j, 0, \ldots, 0)$, $\xi^{(j)} = (\xi_0, \xi_1, \ldots, \xi_j, 0, \ldots, 0)$, $j = 0, 1, 2, \ldots n$, we introduce U(1) invariant functions on $T^*S^{2n+1}_{[2]}$:

$$H_j(z; \xi) = (|z^{(j)}|^2 |\xi^{(j)}|^2 - \langle z^{(j)}, \xi^{(j)}\rangle_{\mathbf{R}}^2)/4,$$

$$K_j(z;\zeta) = (\bar{z}_j\zeta_j - z_j\bar{\zeta}_j)/2i, \qquad j = 0, 1, \dots, n.$$

Note that H_n is the energy Hamiltonian of g_S and $H_0 = \frac{1}{4}K_0^2$. Further, the moment map ψ is expressed as $\sum_{j=0}^{n} K_j = \psi$.

Proposition 4.2. (cf. [6]).

(i) H_j, K_j (j=0, 1,..., n) are in involution in $(T^*S^{2n+1}_{[2]}, d\theta_S)$.

(ii) For every $(z;\zeta) \in T^*S^{2n+1}_{[2]}$, it holds

$$2\sqrt{H_j(z;\zeta)} \geqq 2\sqrt{H_{j-1}(z;\zeta)} + |K_j(z;\zeta)|, \qquad (j = 1,\dots, n). \tag{4.1}$$

Moreover, dH_0, dH_j, dK_j (j= 1,..., n) are linearly independent at $(z;\zeta) \in T^*S^{2n+1}_{[2]}$ if and only if $(z;\zeta)$ satisfies the strict inequalities in (4.1).

The U(1) invariant functions H_j and K_j (j = 1,..., n) are reduced to the functions $\widetilde{H}_j$ and $\widetilde{K}_j$, resectively, on $(T^*\mathbf{C}P^n, \sigma_m)$, which are also in involution. We remark that

$$\widetilde{H}_n - m^2/4 = H. \tag{4.2}$$

For 2n real numbers $(h_1,\dots, h_n, k_1,\dots, k_n) = (h, k) \in \mathbf{R}^{2n}$, we set

$$L(h_1,\dots, h_n, k_1,\dots, k_n)$$

$$= \{\, t \in T^*\mathbf{C}P^n \mid \widetilde{H}_j(t) = h_j,\ \widetilde{K}_j(t) = k_j,\quad j = 1, 2,\dots, n \,\}.$$

Owing to the Liouville-Arnold theorem (cf. [2]), L(h, k)'s are Lagrangian submanifolds diffeomorphic to 2n-torus when $d\widetilde{H}_j(t)$, $d\widetilde{K}_j(t)$ $(j = 1, 2,\dots, n)$ are linearly independent. Thus, it becomes feasible to work out the quantization condition (Q) for $L(h, k)$ to calculate quasi-classical eigenvalues of H. To this end, we set

$$\hat{L}(h_1,\dots, h_n, k_1,\dots, k_n, k_0)$$

$$= \{ (z; \xi) \in T^*S^{2n+1}_{[2]} \mid H_j(z; \xi) = h_j, \quad K_j(z; \xi) = k_j,$$

$$K_0(z; \xi) = k_0, \; j = 1, 2,\dots, n \},$$

where $(h_1,\dots, h_n, k_1,\dots, k_n, k_0) = (h, k, k_0) \in \mathbf{R}^{2n+1}$. In view of Proposition 4.2, $\hat{L}(h, k, k_0)$ is a Lagrangian submanifold if and only if

$$2\sqrt{h_j} > 2\sqrt{h_{j-1}} + |k_j|, \qquad j = 1, 2,\dots, n. \tag{4.3}$$

For $j = 0$, one has $2\sqrt{h_0} = |k_0|$ because of $H_0 = \frac{1}{4} K_0^2$. With these conditions, the equation $\psi = \sum_{j=0}^n K_j$ implies that $\hat{L}(h, k, k_0) = \nu_m^{-1}(L(h, k))$, where

$$m = \sum_{j=0}^n k_j. \tag{4.4}$$

Hence Theorem 3.6 allows us to work out the Maslov quantization condition for $\hat{L}(h, k, k_0)$ with (4.3) and (4.4), instead of the quantization condition (Q) for $L(h, k)$.

We proceed with determining the generators of $H_1(\hat{L}(h, k, k_0): \mathbf{Z})$ in explicit form, and thereby calculate action integrals and the Maslov indices. It is easy to check that there exists a point $\hat{p} = (\hat{z}; \hat{\xi}) \in \hat{L}(h, k, k_0)$ such that $\langle \hat{z}^{(j)}, \hat{\xi}^{(j)} \rangle_{\mathbf{R}} = 0$, $j = 1, 2, \ldots, n$. Consider the Hamiltonian vector fields of ψ, K_j, H_j $(j = 1, 2, \ldots, n)$, which are mutually commuting and span $T\hat{L}(h, k, k_0)$ altogether. We denote the phase flows of ψ, K_j, H_j by $g_0(t)$, $g_j(t)$, $f_j(t)$, $(j = 1, 2, \ldots, n)$, respectively. The respective integral curves through $\hat{p}$ are as follows:

$$g_0(t)(\hat{p}) = (e^{it}\hat{z};\ e^{it}\hat{\xi}), \tag{4.5}$$

$$g_j(t)(\hat{p}) = (z(t);\ \xi(t)), \quad (j = 1, 2, \ldots, n), \quad \text{where}$$

$$z_j(t) = e^{it}\hat{z}_j, \qquad \xi_j(t) = e^{it}\hat{\xi}_j,$$

$$z_l(t) = \hat{z}_l, \qquad \xi_l(t) = \hat{\xi}_l \qquad (l = 0, 1, \ldots, n,\ l \neq j). \tag{4.6}$$

$$f_j(t)(\hat{p}) = (z(t);\ \xi(t)), \quad (j = 1, 2, \ldots, n), \quad \text{where}$$

$$z_l(t) = \hat{z}_l \cos(\sqrt{h_j}\ t) + \hat{\xi}_l (| \hat{z}^{(j)} |^2 / 2\sqrt{h_j}) \sin(\sqrt{h_j}\ t),$$

$$\xi_l(t) = \hat{\xi}_l \cos(\sqrt{h_j}\ t) - \hat{z}_l (| \hat{\xi}^{(j)} |^2 / 2\sqrt{h_j}) \sin(\sqrt{h_j}\ t),$$

$$\text{for } l = 0, 1, \ldots, j, \text{ and}$$

$$z_l(t) = \hat{z}_l, \qquad \xi_l(t) = \hat{\xi}_l \qquad \text{for } l \geqq j + 1. \tag{4.7}$$

We define $\varphi : \mathbf{R}^{2n+1} \to \hat{L}(h, k, k_0)$ by

$$\varphi(s_0, s_1, \ldots, s_n, t_1, \ldots, t_n)$$

$$= g_0(s_0)g_1(s_1)\dots g_n(s_n)f_1(t_1/\sqrt{h_1})\dots f_n(t_n/\sqrt{h_n})(\hat{p}). \tag{4.8}$$

Then φ satisfies $\varphi(s + s', t + t') = \varphi(s, t)\varphi(s', t')$, $(s, t) = (s_0, s_1,\dots, s_n, t_1,\dots, t_n) \in \mathbf{R}^{2n+1}$. Because of periodicity, $\Gamma = \varphi^{-1}(\hat{p})$ proves to be a discrete subgroup of $\mathbf{R}^{2n+1}$. The expressions (4.5)-(4.8) imply then that Γ has the generators $a_1,\dots, a_n$, $b_0, b_1,\dots, b_n$ defined by

$$a_j : t_j = s_{j+1} = t_{j+1} = \pi, \quad \text{others} = 0, \qquad (j=1, 2,\dots, n-1),$$

$$a_n : t_n = s_0 = \pi, \quad \text{others} = 0,$$

$$b_l : s_l = 2\pi, \quad \text{others} = 0, \qquad (l = 0, 1,\dots, n).$$

Thus the φ induces the diffeomorphism of the quatient group $\mathbf{R}^{2n+1}/\Gamma = T^{2n+1}$ to $\hat{L}(h, k, k_0)$, $\varphi : T^{2n+1} \to \hat{L}(h, k, k_0)$. The cycles in T^{2n+1} defined by

$$\hat{a}_j(t) = t\, a_j, \qquad \hat{b}_l(t) = t\, b_l,$$

where $t \in [0, 1]$ and $j = 1, 2,\dots, n$, $l = 0, 1,\dots, n$, provide generators $[\varphi \hat{a}_j]$, $[\varphi \hat{b}_l]$ of $H_1(\hat{L}(h, k, k_0): \mathbf{Z})$. On these cycles, the canonical 1-form θ_S is integrated to give

$$\int_{a(j)} \theta_S = (2\sqrt{h_j} + k_{j+1} + 2\sqrt{h_{j+1}})\pi, \qquad (j = 1, 2,\dots, n - 1),$$

$$\int_{a(n)} \theta_S = (2\sqrt{h_n} + m)\pi,$$

$$\int_{b(0)} \theta_S = 2\pi m,$$

$$\int_{b(l)} \theta_S = 2\pi k_l, \qquad (l = 1, 2,\dots, n),$$

where $a(j)$ and $b(l)$ ($j = 1, 2,\dots, n$, and $l = 0, 1,\dots, n$), stand for $\varphi\hat{a}_j$ and for $\varphi\hat{b}_l$, respectively. We are now at the last stage to calculate the Maslov indices. Since $g_j(t)$ ($j = 0, 1,\dots, n$) are the lifts of diffeomorphisms in $S^{2n+1}_{[2]}$ to the cotangent bundle, the Maslov indices along these flows are equal to zero. Further, (4.7) shows that $f_j(t)(\hat{p})$ is a geodesic flow in T^*S^{2j+1} ($j = 1, 2,\dots, n$). As is well known, the Maslov index of a geodesic flow is equal to the Morse index of the geodesic curve in the base manifold (cf. [11]). Therefore, all the Maslov indices are found out to be $\langle\mu, [\varphi\hat{a}_j]\rangle = 4j + 2$, ($j = 1, 2,\dots, n - 1$), $\langle\mu, [\varphi\hat{a}_n]\rangle = 2n$, and $\langle\mu,[\varphi\hat{b}_l]\rangle = 0$, ($l = 0, 1,\dots, n$), where μ is the Maslov class of $\hat{L}(h, k, k_0)$. Thus, the Maslov quantization condition for $\hat{L}(h, k, k_0)$ turns out to be

$$\sqrt{h_j} + k_{j+1}/2 + \sqrt{h_{j+1}} - (j + 1/2) \in \mathbf{Z}, \quad (j = 1, 2,\dots, n-1),$$

$$\sqrt{h_n} + m/2 - n/2 \in \mathbf{Z},$$

$$k_l \in \mathbf{Z}, \quad (l = 1, 2,\dots, n). \tag{4.9}$$

On account of (4.4), the condition (4.9) is equivalent to

$$2\sqrt{h_j} = 2\gamma_j + |m| + j - \textstyle\sum_{q=j+1}^{n} |l_q|, \quad (j = 1, 2,\dots, n - 1),$$

$$2\sqrt{h_n} = 2\gamma_n + |m| + n,$$

$$k_l = p_l, \quad (l = 0, 1,\dots, n), \tag{4.10}$$

where γ_j $(j = 1, 2,\ldots, n)$, p_l $(l = 0, 1,\ldots, n)$ are integers with $\sum_{l=0}^{n} p_l = m$. Substituting (4.10) into (4.3), we get

$$\gamma_n \geqq \gamma_{n-1} \geqq \cdots \geqq \gamma_1 \geqq \frac{1}{2}(| p_0 | + | p_1 | + \cdots + | p_n | - m). \tag{4.11}$$

Hence, we have

Lemma 4.3. The Lagrangian submanifold $\hat{L}(h, k, k_0)$ satisfies the Maslov quantization condition if and only if its parameters satisfy (4.10) and (4.11).

In conclusion, the quasi-classical eigenvalues are calculated through (4.2) and (4.10). Note here that γ_n is allowed to take values $\gamma_n = 0, 1, 2,\ldots$. Thus, we get the following.

Theorem 4.4. The quasi-classical eigenvalues of H for D_m are

$$\tilde{\lambda}_k^{(m)} = (k + | m | /2)(k + | m | /2 + n) - m^2/4 + n^2/4,$$

$$k = 0, 1, 2,\ldots.$$

References

[1] R. Abraham and J. E. Marsden, Foundations of Mechanics, Benjamin, London, Amsterdam, 1978.

[2] V. I. Arnold, Mathematical methods of classical mechanics, Springer-Verlag, New York, Heidelberg, Berlin, 1978.

[3] Y. Colin de Verdiere, Quasi-modes sur les varietes riemanniennes compactes, Inv. Math, 43(1977), 15-52.

[4] V. Guillemin and S. Sternberg, Geometric Asymptotics, Math. Surveys, 14, A.M.S., 1977.

[5] K. Ii, On the multiplicities of the spectrum for quasi-classical mechanics on spheres, Tôhoku Math. J. 30(1978), 517-524.

[6] K. Ii and S. Watanabe, Complete integrability of the geodesic flows on symmetric spaces, Advanced Studies in Pure Math. 3(1984), Geometry of Geodesics and Related Topics, 105-124.

[7] M. Kummer, On the construction of the reduced phase space of a Hamiltonian system with symmetry. Indiana Univ. Math. J., 30 (1981), 281-291.

[8] R. Kuwabara, Spectrum of the Schrodinger operator on a line bundle over complex projective spaces, Tôhoku Math. J. 40(1988), 199-211.

[9] J. Leray, Lagrangian analysis and quantum mechanics, the MIT Press, Cambridge, Massachusetts, London, England, 1981.

[10] V. P. Maslov, Theorie des perturbations et methodes asymptotics, Dunod, Guthier-Villars, Paris, 1972.

[11] A. Weinstein, Quasi-classical mechanics on spheres, Symposia Math., 24(1974), 25-32.

[12] N. Woodhous, Geometric quantization, Claredon Press, Oxford, 1980.

[13] A. Yoshioka, Maslov's quantization conditions for the bound states of the hydrogen atom, Tokyo J. Math. 9(1986), 415-437.

Akira YOSHIOKA
Department of Mathematics
Faculty of Science and Technology
Science University of Tokyo
Noda, Chiba 287
Japan

Chapter II

Geometry of Submanifolds and Tensor Geometry

THE DIRICHLET PROBLEM AT INFINITY FOR HARMONIC MAPPINGS BETWEEN HADAMARD MANIFOLDS

Kazuo Akutagawa

ABSTRACT

In this paper, we consider a Dirichlet problem at infinity for harmonic maps from a complete simply connected Riemannian manifold of negative sectional curvature into an Hadamard manifold, and prove the existence and uniqueness for solutions of the Dirichlet problem. This is a generalization of the Anderson and Sullivan existence and uniqueness theorem of the asymptotic Dirichlet problem for the Laplace-Beltrami operator Δ.

1. Introduction

Let X and Y be Hadamard manifolds, i.e., complete simply connected Riemannian manifolds of nonpositive sectional curvature. There is a well-known compactification $\bar{X} =$

ISBN 0-12-640170-5

$X \cup X(\infty)$ of X (cf. [3]) giving a homeomorphism of $(X, X(\infty))$ with the Euclidean pair (B^n, S^{n-1}), where $\dim X = n$. One can state:

Asymptotic Dirichlet problem for Δ. Given a C^0 map $\phi : X(\infty) \longrightarrow Y$, find $u \in C^\infty(X,Y) \cap C^0(\bar{X},Y)$ satisfying

$$\begin{cases} \Delta u = 0 & \text{in } X \\ u = \phi & \text{on } X(\infty), \end{cases} \qquad (*)$$

where $\Delta u = \text{Trace}\, \nabla du$, a cross-section of $u^{-1}TY$, and ∇ denotes the covariant derivative in the bundle $T^*X \otimes u^{-1}TY$.

The object of this paper is to prove the following.

THEOREM A. Let X be a complete simply connected Riemannian manifold of negative sectional curvature K_X, $-b^2 \leq K_X \leq -a^2 < 0$ $(0 < a < b)$. Let Y be an Hadamard manifold. Then the asymptotic Dirichlet problem for Δ is uniquely solvable for any $\phi \in C^0(X(\infty), Y)$.

In [1] and [12], Anderson and Sullivan proved the following theorem for the so-called asymptotic Dirichlet problem for Δ (see also Anderson-Schoen [2]).

THEOREM B (Anderson, Sullivan). Let X be a complete simply connected Riemannian manifold of negative sectional curvature K_X, $-b^2 \leq K_X \leq -a^2 < 0$. Let ψ be a C^0 function on $X(\infty)$. Then there exists a unique harmonic function $f \in C^\infty(X) \cap C^0(\bar{X})$ such that $f|_{X(\infty)} = \psi$.

Theorem A could be thought of a generalization of Theorem B, which is of essential use in our proof of Theorem A.

In particular, if Y is the n-dimensional simply connected hyperbolic space form $\mathbb{H}^n$ of constant curvature -1, the asymptotic Dirichlet problem for $\mathbb{A}$ is related to the theory of entire spacelike hypersurfaces of constant mean curvature in the (n+1)-dimensional Minkowski space $\mathbb{L}^{n+1}$ (see section 4).

2. Preliminaries

Let X be an Hadamard manifold. The Cartan-Hadamard theorem may be used to define a geometric compactification of X by means of geodesic rays (cf. [3]). The sphere at infinity $X(\infty)$ of X is defined to be the set of asymptotic classes of geodesic rays in X; two rays γ_1 and γ_2 are asymptotic if $\mathrm{dist}_X(\gamma_1(t),\gamma_2(t))$ is bounded function in t, $t \geq 0$. There is a natural topology on $\bar{X} = X \cup X(\infty)$, so-called the cone topology, given as follows: For each pole $o \in X$ and $v \in T_oX$, let $C_o(v,\delta)$ be the cone about v of angle δ, i.e.,

$$C_o(v,\delta) = \{ x \in X ; \measuredangle_o(v,T_{\overline{ox}}) < \delta \},$$

where $T_{\overline{ox}}$ is the tangent vector to the geodesic ray through o and x, and $\measuredangle_o$ denotes angle in T_oX. Let $T_o(v,\delta,R)$ be the truncated cone of radius R, i.e., $T_o(v,\delta,R) = C_o(v,\delta) \setminus$

$\bar{B}_o(R)$, where $\bar{B}_o(R)$ is the closure of $\{ x \in X ; \text{dist}_X(x,o) < R \}$. Then the domains $T_o(v,\delta,R)$ together with $B_x(r)$, $x \in X$ and $r > 0$ from a local basis for the cone topology. It turns out that this topology is independent of the choice of $o \in X$. In general, there is no natural (independent of o) smooth structure on $X(\infty)$.

3. Proof of Theorem A

Choose fixed poles $o \in X$, $\tilde{o} \in Y$ and identify $X(\infty)$ with the collection of geodesic rays emanating from o.

(Uniqueness) Let u_1 and u_2 be solutions of (*). By the maximum principle of Jäger-Kaul [6] for solutions of the Dirichlet problem with image contained in a convex ball, we have

$$\sup_{x \in B_o(R)} \text{dist}_Y(u_1(x),u_2(x)) \leq \sup_{x \in \partial B_o(R)} \text{dist}_Y(u_1(x),u_2(x)) \quad (1)$$

for all $R > 0$. From $u_1|_{X(\infty)} = \phi = u_2|_{X(\infty)}$, letting $R \longrightarrow \infty$ in (1), we obtain $u_1 = u_2$ on $\bar{X}$.

Let $\zeta : [0,1] \longrightarrow [0,\infty]$ be a fixed homeomorphism. The map $E_\zeta : B(1) \subset T_oX \longrightarrow X$ given by $E_\zeta(v) = \exp_o(\zeta(\|v\|)\cdot v)$ is a homeomorphism from the unit ball $B(1)$ in T_oX onto X. Then E_ζ extends to a homeomorphism from the unit sphere $U_oX = \partial B(1)$ onto $X(\infty)$. By the identification $X(\infty) \simeq U_oX$, we can introduce a C^∞ structure (depending on o) on $X(\infty)$.

The proof of existence for the solutions of (*) is divided into three steps. In step one, under the condition

$\phi \in C^{\infty}(X(\infty),Y)$ we construct the solution u of (*). In step two, using Theorem B we show $u|_{X(\infty)} = \phi$. In step three, under the general situation $\phi \in C^0(X(\infty),Y)$, using C^{∞} approximation of ϕ, we solve the asymptotic Dirichlet problem for Δ.

(Existence) Step I. We first assume that $\phi \in C^{\infty}(X(\infty),Y)$. Extend ϕ radially along rays from o to a map on $\bar{X} \setminus \{o\}$, with boundary values ϕ on $X(\infty)$. From the existence and uniqueness theorems of the Dirichlet problem for Δ [4], [5] and [6], for each $R > 0$ there exists a unique map $u_R \in C^{\infty}(\bar{B}_o(R),Y)$ satisfying

$$\begin{cases} \Delta u_R = 0 & \text{in } B_o(R) \\ u_R = \phi & \text{on } \partial B_o(R). \end{cases} \tag{2}$$

Choose a constant $R_0 > 0$ satisfying $\text{Im}\,\phi \subset B_{\tilde{o}}(R_0) = \{ y \in Y ; \text{dist}_Y(\tilde{o},y) < R_0 \}$. Since $\partial B_{\tilde{o}}(R_0)$ is convex, we note that

$$\text{Im}\, u_R \subset B_{\tilde{o}}(R_0) \qquad \text{for any } R > 0. \tag{3}$$

It then follows from a-priori estimates for the gradient of harmonic maps in [7], [8] that for each α $(0 < \alpha < 1)$ and $R > L > 0$

$$\|u_R\|_{C^{2,\alpha}(B_o(L/3))} \leqq C \cdot \sup_{\substack{x \in B_o(L/3) \\ z \in B_x(L/3)}} \text{dist}_Y(u_R(x),u_R(z))/L,$$

where $\|\cdot\|_{C^{2,\alpha}(B_o(L/3))}$ denotes $C^{2,\alpha}$-Hölder norm in $B_o(L/3)$ and $C = C(\alpha,L,R_0,b, \sup_{y \in B_{\tilde{o}}(R_0)} |K_Y|, \dim X, \dim Y)$. Hence from (3)

$$\|u_R\|_{C^{2,\alpha}(B_o(L/3))} \leq 2C\cdot R_0/L \qquad \text{for} \quad R > L > 0. \tag{4}$$

According to (3), (4) and Ascoli-Arzelà's theorem, there exist a subsequence $\{u_{R_i}\}_{i\in\mathbb{N}}$ of $\{u_R\}_{R>0}$ and a map $u \in C^{2,\alpha/2}(X,Y)$ such that u_{R_i} converges uniformly to u on every compact domain in X when $i \longrightarrow \infty$. From (2), u is harmonic in X and then $u \in C^{\infty}(X,Y)$.

Step II. We shall show that $u|_{X(\infty)} = \phi$. Choose a point $x_\infty = [\{\exp_o tv\}_{t\geq 0}] \in X(\infty)$ and a small constant $\varepsilon > 0$, where $v \in U_oX$ and $[\{\exp_o tv\}_{t\geq 0}]$ denotes the asymptotic class of the ray $\{\exp_o tv\}_{t\geq 0}$. Put $\rho(\cdot) = \mathrm{dist}_Y^2(\cdot,\phi(x_\infty))$. Since u_R is harmonic in $B_o(R)$ and ρ is convex in Y, the composition $(\rho o u_R)(\cdot) = \mathrm{dist}_Y^2(u_R(\cdot),\phi(x_\infty))$ is subharmonic in $B_o(R)$. By Theorem B, there exists a unique function $f \in C^\infty(X) \cap C^0(\bar{X})$ such that

$$\begin{cases} \Delta f = 0 & \text{in} \quad X \\ f = \rho o \phi & \text{on} \quad X(\infty). \end{cases} \tag{5}$$

From (5), we then obtain that there exists a constant $R_1 > 0$ such that

$$|(\rho o \phi)(x) - f(x)| < \varepsilon^2/2 \qquad \text{for all} \quad x \in X \setminus \bar{B}_o(R_1). \tag{6}$$

From $f(x_\infty) = 0$, we also obtain that there exist a constant $R_2 > 0$ and a small constant $\delta > 0$ such that

$$|f(x)| < \varepsilon^2/2 \qquad \text{for all} \quad x \in T_o(v,\delta,R_2). \tag{7}$$

Since for each $R > 0$ $\rho o u_R$ is subharmonic in $B_o(R)$ and f is harmonic in X, $\rho o u_R - f$ is subharmonic in $B_o(R)$ and hence by the maximum principle

$$\sup_{x\in B_o(R)} |(\rho o u_R)(x) - f(x)| \leq \sup_{x\in\partial B_o(R)} |(\rho o u_R)(x) - f(x)|. \quad (8)$$

It then follows from (6)-(8) that for each $R > R_3 = \max\{R_1, R_2\}$

$$\begin{aligned} \mathrm{dist}_Y^2(u_R(x),\phi(x_\infty)) &= (\rho o u_R)(x) \\ &\leq |(\rho o u_R)(x) - f(x)| + |f(x)| \\ &\leq \sup_{z\in\partial B_o(R)} |(\rho o \phi)(z) - f(z)| + |f(x)| \\ &\leq \varepsilon^2/2 + \varepsilon^2/2 = \varepsilon^2 \end{aligned}$$

for all $x \in T_o(v,\delta,R_3) \cap B_o(R)$, and hence

$$\mathrm{dist}_Y(u_R(x),\phi(x_\infty)) \leq \varepsilon \quad (9)$$

for all $x \in T_o(v,\delta,R_3) \cap B_o(R)$. Applying (9) to the subsequence $\{u_{R_i}\}_{i\in\mathbb{N}}$ in Step I and letting $R_i \longrightarrow \infty$, u satisfies

$$\mathrm{dist}_Y(u(x),\phi(x_\infty)) \leq \varepsilon \qquad \text{for all } x \in T_o(v,\delta,R_3).$$

This implies that $\lim_{x\to x_\infty} u(x) = \phi(x_\infty)$.

Step III. Finally, for a C^0 map $\phi \in C^0(X(\infty),Y)$, we now show the existence of the solution of (*). Take a sequence $\{\phi_i\}_{i\in\mathbb{N}}$ of C^∞ maps on $X(\infty) \simeq U_oX$ which converges uniformly to ϕ. From arguments similar to Step I and Step II, for every $i \in \mathbb{N}$ there exists a unique harmonic map $u_i \in C^\infty(X,Y) \cap C^0(\bar{X},Y)$ such that $u_i|_{X(\infty)} = \phi_i$ and that for each $R > 0$

$$\|u_i\|_{C^{2,\alpha}(B_o(R/3))} \leq 2C \cdot R_4/R, \tag{10}$$

where $\bigcup_{j \in \mathbb{N}} \operatorname{Im} \phi_j \subset B_{\tilde{o}}(R_4)$. According to (10) and Ascoli-Arzelà's theorem, there exist a subsequence $\{u_{\nu_i}\}_{i \in \mathbb{N}}$ of $\{u_i\}_{i \in \mathbb{N}}$ and a harmonic map $u \in C^\infty(X,Y)$ such that u_{ν_i} converges uniformly to u on every compact domain in X when $\nu_i \longrightarrow \infty$. We then obtain that for each $R > 0$

$$\begin{aligned}\sup_{x \in B_o(R)} \operatorname{dist}_Y(u(x),u_{\nu_i}(x)) &\leq \sup_{\substack{x \in B_o(R) \\ j > i}} \operatorname{dist}_Y(u_{\nu_j}(x),u_{\nu_i}(x)) \\ &\leq \sup_{\substack{x \in \partial B_o(R) \\ j > i}} \operatorname{dist}_Y(u_{\nu_j}(x),u_{\nu_i}(x)).\end{aligned} \tag{11}$$

Letting $R \longrightarrow \infty$ in (11), we see that u satisfies

$$\sup_{x \in X} \operatorname{dist}_Y(u(x),u_{\nu_i}(x)) \leq \sup_{\substack{x \in X(\infty) \\ j > i}} \operatorname{dist}_Y(\phi_{\nu_j}(x),\phi_{\nu_i}(x)). \tag{12}$$

From (12), we have that for each $x_\infty \in X(\infty)$

$$\begin{aligned}\operatorname{dist}_Y(u(x),\phi(x_\infty)) &\leq \operatorname{dist}_Y(u(x),u_{\nu_i}(x)) \\ &+ \operatorname{dist}_Y(u_{\nu_i}(x),\phi_{\nu_i}(x_\infty)) + \operatorname{dist}_Y(\phi_{\nu_i}(x_\infty),\phi(x_\infty)) \\ &\leq 2 \cdot \sup_{\substack{z \in X(\infty) \\ j > i}} \operatorname{dist}_Y(\phi_{\nu_j}(z),\phi_{\nu_i}(z)) + \operatorname{dist}_Y(u_{\nu_i}(x),\phi_{\nu_i}(x_\infty))\end{aligned} \tag{13}$$

for all $x \in X$. Since ϕ_{ν_i} converges uniformly to ϕ on

$X(\infty) \simeq U_o X$ and $u_{\nu_i}|_{X(\infty)} = \phi_{\nu_i}$, the inequality (13) implies that $\lim_{x \to x_\infty} u(x) = \phi(x_\infty)$.

This completes the proof of Theorem A.

COROLLARY. Let X be a complete simply connected manifold of negative sectional curvature K_X, $-b^2 \leq K_X \leq -a^2 < 0$. Let ϕ be a C^0 map from $X(\infty)$ into $\mathbb{H}^n$. Then there exists a unique harmonic map $u \in C^\infty(X,\mathbb{H}^n) \cap C^0(\bar{X},\mathbb{H}^n)$ such that $u|_{X(\infty)} = \phi$.

4. Concluding remarks

The asymptotic Dirichlet problem for Δ was motivated by the following remarks.

1. PROPOSITION (cf. Milnor [9] for n = 2). Let X be a spacelike hypersurface in $\mathbb{L}^{n+1}$. Let $G : X \longrightarrow \mathbb{H}^n = \{ (x^1,\cdots,x^n,x^{n+1}) \in \mathbb{L}^{n+1} ; (x^1)^2 + \cdots + (x^n)^2 - (x^{n+1})^2 = -1, x^{n+1} < 0 \}$ denote the Gauss map of X and let $H = \text{Trace} \langle \nabla_\bullet G, \bullet \rangle : X \longrightarrow \mathbb{R}$ denote the mean curvature of X. Then

$$\Delta G = \text{grad } H.$$

In particular, the mean curvature H of X is constant if and only if the Gauss map G of X is harmonic.

In Proposition, since for each point $x \in X$ the tangent

spaces $T_x X$ and $T_{G(x)}\mathbb{H}^n$ are parallel in $\mathbb{L}^{n+1}$, we regard the vector $(\Delta G)_x$ as a vector in $T_x X$. Proposition is a Lorentzian version of the result established in [10], [11].

2. An entire spacelike hypersurface X of constant mean curvature H (> 0) in $\mathbb{L}^{n+1}$ can be represented as the graph of an entire solution $u \in C^2(\mathbb{R}^n)$ of the following

$$(1-\|Du\|^2)\Sigma_{i=1}^n D_i D_i u + \Sigma_{i,j=1}^n D_i u D_j u D_i D_j u = H(1-\|Du\|^2)^{3/2}$$
$$\|Du(x)\| < 1 \quad \text{for} \quad x \in \mathbb{R}^n.$$

In [13], Treibergs proved the following.

" The set of equivalence classes of entire spacelike hypersurfaces of constant mean curvature in $\mathbb{L}^{n+1}$ under the relation that their solutions are equivalent if they have the same projective boundaries at infinity coincides with convex homogeneous functions whose gradient has norm one whenever defined. "

He also pointed out that the second fundamental form of X is positive semi-definite and hence, by the Gauss formula for spacelike hypersurfaces in $\mathbb{L}^{n+1}$, X has nonpositive sectional curvature. We then note

" The Gauss map of an entire spacelike hypersurface of constant mean curvature in $\mathbb{L}^{n+1}$ is a harmonic map from an n-dimensional Hadamard manifold into $\mathbb{H}^n$. "

References

[1] M. T. Anderson, The Dirichlet problem at infinity for manifolds of negative curvature, J. Diff. Geom. **18** (1983), 701-721.

[2] M. T. Anderson and R. Schoen, Positive harmonic functions on complete manifolds of negative curvature, Ann. of Math. **121** (1985), 429-461.

[3] P. Eberlein and B. O'Neill, Visibility manifolds, Pacific J. Math. **46** (1973), 45-109.

[4] R. Hamilton, Harmonic maps of manifolds with boundary, Lecture Notes in Math. Vol. **471**, Springer, New York, 1975.

[5] S. Hildebrandt, H. Kaul and K.-O. Widman, Harmonic mappings into Riemannian manifolds with non-positive sectional curvature, Math. Scand. **37** (1975), 257-263.

[6] W. Jäger and H. Kaul, Uniqueness and stability of harmonic maps and their Jacobi fields, Manuscripta Math. **28** (1979), 269-291.

[7] J. Jost, Harmonic mappings between Riemannian manifolds, Proceedings Centre Math. Anal., ANU-Press, Canberra, 1984.

[8] J. Jost and H. Karcher, Geometrische Methoden zur Gewinnung von a-priori-Schranken für harmonische Abbildungen, Manuscripta Math. **40** (1982), 27-77.

[9] T. K. Milnor, Harmonic maps and classical surface theory in Minkowski 3-space, Trans. Amer. Math. Soc. **280** (1983), 161-185.

[10] E. A. Ruh, Asymptotic behaviour of non-parametric

minimal hypersurfaces, J. Diff. Geom. **4** (1970), 509-513.

[11] E. A. Ruh and J. Vilms, The tension field of the Gauss map, Trans. Amer. Math. Soc. **149** (1970), 569-573.

[12] D. Sullivan, The Dirichlet problem at infinity for a negatively curved manifold, J. Diff. Geom. **18** (1983), 723-732.

[13] A. E. Treibergs, Entire spacelike hypersurfaces of constant mean curvature in Minkowski space, Invent. Math. **66** (1982), 39-56.

Department of Mathematics
Nippon Bunri University
Oita 870-03
Japan

Oriented 6-dimensional submanifolds in the octonians; II

By Hideya Hashimoto

§1. Introduction.

We say that an almost Hermitian manifold $N = (N, J, \langle\ ,\ \rangle)$ is *Kähler*, *nearly Kähler* and *quasi-Kähler* if $\nabla J = 0$, $(\nabla_X J)(X) = 0$ and $(\nabla_X J)(Y) + (\nabla_{JX} J)(JY) = 0$ for any vector fields X and Y on N, respectively. "Kähler" implies "nearly Kähler" and "nearly Kähler" implies "quasi-Kähler" ([4]). The 6-dimensional unit sphere is a nearly Kähler manifold which is not Kähler([3]). Riemannian 3-symmetric spaces are homogeneous quasi-Kähler manifold ([8]). Several authors have studied submanifolds with these properties ([2],[3],[5],[6],[7],[8],[10],[11]). R. Bryant ([1]) has studied oriented 6-dimensional submanifolds in the octonians from $(\mathbb{O}, Spin(7))$ geometry and obtained some interesting and important results. In this paper, we prove the following Theorems A and B as an application of his formulas.

Theorem A. Let (M^6, Ψ) be an oriented

ISBN 0-12-640170-5

6-dimensional submanifold in the octonians $\mathbb{O}$ where Ψ is the immersion. M^6 is a non-Kähler nearly Kähler manifold if and only if M^6 is locally isometric to a 6-dimensional round sphere and Ψ is totally umbilic.

Theorem B. Let (M^6, Ψ) be an oriented 6-dimensional, non-nearly Kähler, quasi-Kähler submanifold in the octonians $\mathbb{O}$ where Ψ is the immersion. If its normal connection is flat, then M^6 is locally isometric to $S^2 \times \mathbb{R}^4$ and the immersion Ψ is locally of the form $\Psi = \psi \times id : S^2 \times \mathbb{R}^4 \longrightarrow \mathbb{R}^3 \times \mathbb{R}^4 (\simeq \mathbb{R}^7) \subset \mathbb{O}$ where $\mathbb{R}^3$ is a 3-dimensional Euclidean subspace closed under the vector cross product in $\mathbb{O}$ and (S^2, ψ) is a 2-dimensional round sphere in $\mathbb{R}^3$ with totally umbilical immersion ψ.

A. Gray proved in [6] that M^6 is isometric to a 6-dimensional sphere if M^6 is an oriented compact nearly Kähler Einstein submanifolds in the octonians with positive sectional curvature. Theorem A states that some assumptions are superfloues. Since any oriented hypersurface of the 7-dimensional sphere cannot be a Kähler manifold, we have the following.

Corollary.([10.Corollary 4.1]) Let (M^6, Ψ) be an oriented hypersurface of the 7-dimensional sphere in the octonians. Then, M^6 is a nearly Kähler manifold if and only if it is totally umbilic.

In this paper, we adopt the same notational convention as in [1] and all the manifolds are assumed to be connected and of class C^∞ unless otherwise stated. Throughout this article, we denote by (M^6, Ψ) an oriented 6-dimensional submanifold in the octonians. The author would like to express his hearty thanks to Professor K. Sekigawa and Professor N. Innami for their encouragement and many valuable suggestions.

§2. Preliminaries.

2.1. We denote by $M_{p\times q}(\mathbb{C})$ the set of pxq complex matrices and $[a] \in M_{3\times 3}(\mathbb{C})$ is given by

$$[a]:= \begin{bmatrix} 0 & a_3 & -a_2 \\ -a_3 & 0 & a_1 \\ a_2 & -a_1 & 0 \end{bmatrix}$$

where $a = {}^t(a_1,a_2,a_3) \in M_{3\times 1}(\mathbb{C})$. Then, we have

$$[a]b + [b]a = 0, \tag{2.1}$$

$$[Aa] = \mathrm{tr}A[a] - {}^tA[a] - [a]A \tag{2.2}$$

where $a,b \in M_{3\times 1}(\mathbb{C})$ and $A \in M_{3\times 3}(\mathbb{C})$. We denote by $\mathbb{O}$ the octonians and $\langle\ ,\ \rangle$ the canonical inner product of $\mathbb{O}$ (for details see [1] or [9]). For any $x \in \mathbb{O}$, we denote by $\bar{x}$ the conjugate of x. We remark that the octonians may be regarded as the direct sum $\mathbb{H}\oplus\mathbb{H}$ where $\mathbb{H}$ is the quaternions.

Let $S^6 = \{ u \in \mathrm{Im}\mathbb{O} \mid \langle u,u\rangle = 1 \}$. We may use $u \in S^6$

to define a map $J_u : \mathbb{O} \longrightarrow \mathbb{O}$ given by $J_u(x) = xu$ for any $x \in \mathbb{O}$. Then, each J_u is an orthogonal almost complex structure on $\mathbb{O}$. It is known that the Lie group Spin(7) is isomorphic to the subgroup of SO(8) generated by the set $\{J_u \in SO(8) \mid u \in S^6\}$ ([9]).

2.2. The geometry of $(\mathbb{O}, \mathrm{Spin}(7))$.

In this section, we shall recall the structure equations of $(\mathbb{O},\mathrm{Spin}(7))$ established by R. Bryant([1]). We extend the action of Spin(7) complex linealy on $\mathbb{C}\otimes_{\mathbb{R}}\mathbb{O}$. We denote a basis (called standard basis) of $\mathbb{C}\otimes_{\mathbb{R}}\mathbb{O}$ by $N := (1-\sqrt{-1}\varepsilon)/2$, $E_1 := iN$, $E_2 := jN$, $E_3 := kN$, $\overline{N} := (1+\sqrt{-1}\varepsilon)/2$, $\overline{E}_1 := i\overline{N}$, $\overline{E}_2 := j\overline{N}$, $\overline{E}_3 := k\overline{N}$ where $\varepsilon = (0,1) \in \mathbb{H}\oplus\mathbb{H}$ $(\simeq \mathbb{O})$ and $\{1,i,j,k\}$ is the canonical basis of the quaternions $\mathbb{H}$. A basis $(n,f,\overline{n},\overline{f})$ of $\mathbb{C}\otimes_{\mathbb{R}}\mathbb{O}$ is said to be *admissible* if there exists $g \in \mathrm{Spin}(7) \subset M_{8\times 8}(\mathbb{C})$ so that $(n,f,\overline{n},\overline{f}) = (N,E,\overline{N},\overline{E})g$. We identify Spin(7) with the admissible basis. Then, we have

Fact 2.1. ([1, Proposition 1.2]) There exist left invariant 1-forms on Spin(7); ρ with values in $\mathbb{R}$; θ, $\mathfrak{h}$ with values in $M_{3\times 1}(\mathbb{C})$ and κ with values in 3×3 skew Hermitian matrices satisfying $\mathrm{tr}\kappa + \sqrt{-1}\rho = 0$,

$$d(n,f,\overline{n},\overline{f}) = (n,f,\overline{n},\overline{f}) \begin{pmatrix} \sqrt{-1}\rho & -{}^t\overline{\mathfrak{h}} & 0 & -{}^t\theta \\ \mathfrak{h} & \kappa & \theta & [\overline{\theta}] \\ 0 & -{}^t\overline{\theta} & -\sqrt{-1}\rho & -{}^t\overline{\mathfrak{h}} \\ \overline{\theta} & [\theta] & \overline{\mathfrak{h}} & \overline{\kappa} \end{pmatrix}$$

$$= (n,f,\bar{n},\bar{f})\phi$$

where ϕ satisfies $d\phi = -\phi\wedge\phi$, or equivalently,

$$d(\sqrt{-1}\rho) = {}^t\bar{\mathfrak{h}}\wedge\mathfrak{h} + {}^t\theta\wedge\bar{\theta}, \tag{2.3}$$

$$d\mathfrak{h} = -\mathfrak{h}\wedge(\sqrt{-1}\rho) - \kappa\wedge\mathfrak{h} - [\bar{\theta}]\wedge\bar{\theta}, \tag{2.4}$$

$$d\theta = -\kappa\wedge\theta + \theta\wedge(\sqrt{-1}\rho) - [\bar{\theta}]\wedge\bar{\mathfrak{h}}, \tag{2.5}$$

$$d\kappa = \mathfrak{h}\wedge{}^t\bar{\mathfrak{h}} - \kappa\wedge\kappa + \theta\wedge{}^t\bar{\theta} - [\bar{\theta}]\wedge\theta. \tag{2.6}$$

Let $\mathcal{F} := \mathbb{O}\times Spin(7)$ and $x:\mathcal{F} \longrightarrow \mathbb{O}$ denote the projection onto the first factor. We regard $\mathcal{F}$ as the space of pairs $(y;(n,f,\bar{n},\bar{f}))$ consisting of a base point $y \in \mathbb{O}$ and admissible basis $(n,f,\bar{n},\bar{f})$ at that point. Then, we have

<u>Fact 2.2.</u> ([1, Proposition 1.3]) There exists the dual basis $(\nu,\omega,\bar{\nu},\bar{\omega})$ of $(n,f,\bar{n},\bar{f})$ on $\mathcal{F}$ so that

$$dx = (n,f,\bar{n},\bar{f})\begin{pmatrix}\nu\\ \omega\\ \bar{\nu}\\ \bar{\omega}\end{pmatrix} = (n,f,n,f)\psi$$

where ψ satisfies $d\psi = -\phi\wedge\psi$, or equivalently,

$$d\nu = -\sqrt{-1}\rho\wedge\nu + {}^t\bar{\mathfrak{h}}\wedge\omega + {}^t\theta\wedge\bar{\omega}, \tag{2.7}$$

$$d\omega = -\mathfrak{h}\wedge\nu - \kappa\wedge\omega - \theta\wedge\bar{\nu} - [\bar{\theta}]\wedge\bar{\omega}. \tag{2.8}$$

2.3. Oriented 6-dimensional submanifolds in $\mathbb{O}$.

In this section, we shall recall the structure equations for the 6-dimensional submanifolds in $\mathbb{O}$ established by Bryant ([1]). We set

$$\mathcal{F}_{\Psi}(M^6) := \left\{ (y,(n,f,\bar{n},\bar{f})) \in \mathcal{F} \;\middle|\; \begin{array}{l} y = \Psi(p) \quad \text{and} \quad -2\sqrt{-1}n\wedge\bar{n} = T_p^{\perp}M^6 \\ \text{for any} \quad p \in M^6 \end{array} \right\}.$$

Then, by [1,Proposition 2.2], the frame $(f,\bar{f})$ is the unitary basis of $\mathbb{C}\otimes_{\mathbb{R}}\mathbb{O}$ with respect to the almost complex structure $J_{2\sqrt{-1}n\times\bar{n}}$ where $\times$ is the *vector cross product* which is defined by $x\times y = (\bar{y}x - \bar{x}y)/2$ for any x, $y \in \mathbb{O}$. Hence, we see that $p:\mathcal{F}_{\Psi}(M^6) \longrightarrow M^6$ be a right U(3) bundle over M^6. Then, we have

Fact 2.3. ([1, Theorem 3.1]) M^6 inherits a U(3) structure (also called almost Hermitian structure), $\nu = 0$ and we have the structure equations;

$$dx = f\omega + \bar{f}\bar{\omega}, \tag{2.9}$$

$$df = -n\,{}^t\bar{\mathfrak{h}} + f\kappa - \bar{n}\,{}^t\bar{\theta} + \bar{f}[\theta] \quad \text{(Gauss formula)}, \tag{2.10}$$

$$dn = n(\sqrt{-1}\rho) + f\mathfrak{h} + \bar{f}\bar{\theta} \quad \text{(Weingarten formula)}, \tag{2.11}$$

$$d\omega = -\kappa\wedge\omega - [\bar{\theta}]\wedge\bar{\omega}, \tag{2.12}$$

(and the equations gotten from these by conjugation).

By (2.9) and (2.10), the second fundamental form II is given by

$$\mathrm{II} = -2\mathrm{Re}\{({}^t\bar{\mathfrak{h}}\circ\omega + {}^t\theta\circ\bar{\omega})n\}, \tag{2.13}$$

where $\circ$ is the symmetric tensor product. Differentiating the equation $\nu = 0$ on $\mathcal{F}_\Psi(M^6)$ and by (2.7), there exists 3×3 matrices of functions A, B, C on $\mathcal{F}_\Psi(M^6)$ (with complex values) satisfying

$$A = {}^tA, \quad C = {}^tC,$$
$$\begin{pmatrix} \mathfrak{h} \\ \theta \end{pmatrix} = \begin{pmatrix} \bar{B} & \bar{A} \\ {}^tB & \bar{C} \end{pmatrix}\begin{pmatrix} \omega \\ \bar{\omega} \end{pmatrix}. \tag{2.14}$$

By (2.13), (2.14), the mean curvature vector H is given by

$$H = -(2/3)\{(\mathrm{tr}B)n + (\mathrm{tr}\bar{B})\bar{n}\}. \tag{2.15}$$

We shall express the induced Riemannian connection ∇ with respect to the adapted frames. Any vector field Y on M^6 is represented by

$$Y = f\alpha + \bar{f}\bar{\alpha} \tag{2.16}$$

where α is an $M_{3\times 1}(\mathbb{C})$-valued function on M^6. By (2.10) and (2.16), we have

$$\nabla_X Y = fd\alpha(X)+(f\kappa(X)+\bar{f}[\theta(X)])\alpha+\bar{f}d\bar{\alpha}(X)+(\bar{f}\bar{\kappa}(X)+f[\bar{\theta}(X)])\bar{\alpha}. \tag{2.17}$$

We extend the Riemannian connection ∇ complex linealy and continue to denote by the same letters. For the sake of later uses, we recall the following:

Fact 2.4. ([1,Theorems 3.3 and 3.13]) The induced U(3)-structure of (M^6, Ψ) is almost Kähler (that is, $d\Omega = 0$) if and only if $trB = 0$ and $C = 0$. It is also Kähler.

Fact 2.5. ([1, The proof of Theorem 3.3]) The induced U(3)-structure of (M^6, Ψ) is quasi-Kähler if and only if $C = 0$.

The following Lemma 2.1 plays an essential role in the proof of Theorem A.

Lemma 2.1. The induced U(3)-structure of (M^6, Ψ) is nearly Kähler if and only if $B = (1/3)(trB)I_3$ and $C = 0$.

Proof: Since a nearly Kähler manifold is a quasi-kähler manifold, from Fact 2.5, we have $C = 0$. Any vector field X on M^6 represented by

$$X = f\alpha + \bar{f}\bar{\alpha} \tag{2.18}$$

where α is a $M_{3\times1}(\mathbb{C})$ valued function on M^6. By (2.10), (2.17) and (2.18), we get

$$0 = (\nabla_X J)(X) = \nabla_X(JX) - J(\nabla_X X)$$

$$= \nabla_X(J(f\alpha + \bar{f}\bar{\alpha})) - J(\nabla_X(f\alpha + \bar{f}\bar{\alpha}))$$

$$= \nabla_X(\sqrt{-1}(f\alpha - \bar{f}\bar{\alpha})) - J(\nabla_X(f\alpha) + \nabla_X(\bar{f}\bar{\alpha}))$$

$$= \sqrt{-1}\{ fd\alpha(X) + (f\kappa(X)+\bar{f}[\theta(X)])\alpha$$

$$- \bar{f}d\bar{\alpha}(X) - (\bar{f}\bar{\kappa}(X)+f[\bar{\theta}(X)])\bar{\alpha}\}$$

$$- J\{ fd\alpha(X) + (f\kappa(X)+\bar{f}[\theta(X)])\alpha$$

$$+ \bar{f}d\bar{\alpha}(X) + (\bar{f}\bar{\kappa}(X)+f[\bar{\theta}(X)])\bar{\alpha}\}$$

$$= 2\sqrt{-1}\{\bar{f}[\theta(X)]\alpha - f[\bar{\theta}(X)]\bar{\alpha}\}$$

for any $X\in\mathfrak{X}(M^6)$. Since $\{f_i, \bar{f}_i\}$ is linealy independent, we get

$$0 = [\theta(X)]\alpha \tag{2.19}$$

for any $X\in\mathfrak{X}(M^6)$. From Fact 2.5, (2.14) and (2.18), we get

$$\theta(X) = {}^tB\omega(X) = {}^tB\alpha. \tag{2.20}$$

By (2.19) and (2.20), we get

$$[{}^t B\alpha]\alpha = 0$$

for any $\alpha \in M_{3\times 1}(\mathbb{C})$. From this, we have finally

$${}^t B = \lambda I_3$$

where $\lambda = (\mathrm{tr}B)/3$. □

§3. Proof of Theorem A.

From Lemma 2.1 and (2.14), we get

$$\theta = \lambda\omega, \quad \mathfrak{h} = \overline{\lambda}\omega + \overline{A}\overline{\omega} \tag{3.1}$$

where $\lambda = (\mathrm{tr}B)/3$. Differentiating $(3.1)_1$ and taking account of (2.4) and (2.12), we get

$$-\kappa\wedge\theta + \theta\wedge(\sqrt{-1}\rho) - [\overline{\theta}]\wedge\overline{\mathfrak{h}} = d\lambda\wedge\omega - \lambda\{\kappa\wedge\omega + [\overline{\theta}]\wedge\overline{\omega}\}. \tag{3.2}$$

By (3.1) and (3.2), we get

$$-\kappa\wedge(\lambda\omega) + \lambda\omega\wedge(\sqrt{-1}\rho) - [\overline{\lambda}\overline{\omega}]\wedge(\lambda\overline{\omega} + A\omega)$$

$$= d\lambda\wedge\omega - \lambda\{\kappa\wedge\omega + [\overline{\lambda}\overline{\omega}]\wedge\overline{\omega}\}.$$

From this, we get

$$(d\lambda + \sqrt{-1}\lambda\rho)\wedge\omega = -[\bar{\lambda}\bar{\omega}]\wedge A\omega,$$

or equivalently

$$(d\lambda + \sqrt{-1}\lambda\rho)\wedge\begin{pmatrix}\omega^1\\ \omega^2\\ \omega^3\end{pmatrix} = -\bar{\lambda}\begin{pmatrix}0 & \bar{\omega}^3 & -\bar{\omega}^2\\ -\bar{\omega}^3 & 0 & \bar{\omega}^1\\ \bar{\omega}^2 & -\bar{\omega}^1 & 0\end{pmatrix}\wedge\begin{pmatrix}a^1_j\omega^j\\ a^2_j\omega^j\\ a^3_j\omega^j\end{pmatrix} \qquad (3.3)$$

where $A = (a^i_j)$. Let U be a connected component of the subset $\{ p\in M^6 \mid |trB| \neq 0 \}$ of M^6. Since M^6 is a non-Kähler nearly Kähler manifold, U is an non-empty open subset in M^6. In fact, Fact 2.4 states that M^6 is a Kähler manifold if $trB = 0$ on M^6. This is a contradiction. By (3.3), we get

$$(d\lambda + \sqrt{-1}\lambda\rho)\wedge\omega^1 = -\bar{\lambda}(\bar{\omega}^3\wedge a^2_j\omega^j - \bar{\omega}^2\wedge a^3_j\omega^j). \qquad (3.4)$$

From this, we get

$$(\bar{\omega}^3\wedge a^2_j\omega^j - \bar{\omega}^2\wedge a^3_j\omega^j)\wedge\omega^1 = 0$$

on U. Since $\{\omega^i, \bar{\omega}^i\}$ is linealy independent, we get

$$a^2_2 = a^3_2 = a^2_3 = a^3_3 = 0$$

on U. By the same calculation and (3.3), we get

$$A = 0 \qquad (3.5)$$

on U. By Lemma 2.1, (2.13)~(2.15) and (3.5), we have

$$\mathrm{II} = -2\mathrm{Re}(2\lambda({}^t\bar{\omega}\circ\omega)n) = -{}^t\bar{\omega}\circ\omega\{2((\mathrm{tr}B)n + (\mathrm{tr}\bar{B})\bar{n})/3\} = g\otimes H$$

on U. Hence, each point of U is an umbilical point. Therefore U is an open and closed, non-empty subset in M^6. Thus, we have $U = M^6$, since M^6 is connected. This completes the proof of Theorem A.

§4. Proof of Theorem B.

From Fact 2.5 and (2.14), we get

$$\theta = {}^tB\omega, \quad \mathfrak{h} = \bar{B}\omega + \bar{A}\bar{\omega}. \tag{4.1}$$

Differentiating $(4.1)_1$ and taking account of (2.5) and (4.1), we get

$$\kappa\wedge{}^tB\omega - {}^tB\omega\wedge(\sqrt{-1}\rho) + [{}^t\bar{B}\bar{\omega}]\wedge(B\bar{\omega} + A\omega) \tag{4.2}$$

$$= {}^tB\kappa\wedge\omega + {}^tB[{}^t\bar{B}\bar{\omega}]\wedge\bar{\omega} - d({}^tB)\wedge\omega.$$

We compare in (0.2) parts of both sides of (4.2), we get

$$[{}^t\bar{B}\bar{\omega}]\wedge B\bar{\omega} = {}^tB[{}^t\bar{B}\bar{\omega}]\wedge\bar{\omega}. \tag{4.3}$$

On one hand, by (2.2), we get

$$[{}^t\bar{B}\bar{\omega}] = (\mathrm{tr}\bar{B})[\bar{\omega}] - \bar{B}[\bar{\omega}] - [\bar{\omega}]{}^t\bar{B}. \tag{4.4}$$

By (2.2) and (4.4), we get

$${}^tB[{}^t\bar{B}\bar{\omega}]\wedge\bar{\omega} = {}^tB\{(\mathrm{tr}\bar{B})I_3 - \bar{B}\}[\bar{\omega}]\wedge\bar{\omega} - {}^tB[\bar{\omega}]\wedge{}^t\bar{B}\bar{\omega}$$

$$= {}^tB\{(\mathrm{tr}\bar{B})I_3 - \bar{B}\}[\bar{\omega}]\wedge\bar{\omega} - {}^tB[{}^t\bar{B}\bar{\omega}]\wedge\bar{\omega}.$$

From this, we get

$${}^tB[{}^t\bar{B}\bar{\omega}]\wedge\bar{\omega} = (1/2){}^tB\{(\mathrm{tr}\bar{B})I_3 - \bar{B}\}[\bar{\omega}]\wedge\bar{\omega}. \tag{4.5}$$

By (2.2) and (4.4), we get

$$[{}^t\bar{B}\bar{\omega}]\wedge B\bar{\omega} = \{(\mathrm{tr}\bar{B})I_3 - \bar{B}\}[\bar{\omega}]\wedge B\bar{\omega} - [\bar{\omega}]{}^t\bar{B}\wedge B\bar{\omega} \tag{4.6}$$

$$= \{(\mathrm{tr}\bar{B})I_3 - \bar{B}\}[B\bar{\omega}]\wedge\bar{\omega} - [{}^t\bar{B}B\bar{\omega}]\wedge\bar{\omega}$$

$$= (1/2)\{((\mathrm{tr}\bar{B})I_3 - \bar{B})((\mathrm{tr}B)I_3 - {}^tB)$$

$$-((\mathrm{tr}\,{}^t\bar{B}B)I_3 - {}^tB\bar{B})\}[\bar{\omega}]\wedge\bar{\omega}.$$

By (4.3), (4.5) and (4.6), we get

$$2(\mathrm{tr}\bar{B}){}^tB + (\mathrm{tr}B)\bar{B} - (2{}^tB\bar{B} + \bar{B}{}^tB) \tag{4.7}$$

$$=\{(\mathrm{tr}B)(\mathrm{tr}\bar{B}) - (\mathrm{tr}\,{}^tB\bar{B})\}I_3.$$

Comparing (4.7) with its conjugation, we get

$$(\mathrm{tr}\bar{B})B = (\mathrm{tr}B)\,{}^t\bar{B}. \tag{4.8}$$

Hence, $(\mathrm{tr}\bar{B})B$ is a hermitian matrix. Then, there exists a unitary matrix U so that

$$ {}^t\bar{U}((\mathrm{tr}\bar{B})B)U = \Lambda \tag{4.9}$$

at each point of M^6, where $\Lambda = \begin{pmatrix} \alpha_1 & 0 & 0 \\ 0 & \alpha_2 & 0 \\ 0 & 0 & \alpha_3 \end{pmatrix}$ and $\alpha_i \in \mathbb{R}$.

By (4.7) and (4.9), we get

$$3(\mathrm{tr}B)(\mathrm{tr}\bar{B})\Lambda - 3\Lambda^2 \tag{4.10}$$

$$= (\mathrm{tr}B)(\mathrm{tr}\bar{B})\,\{(\mathrm{tr}B)(\mathrm{tr}\bar{B}) - (\mathrm{tr}\,{}^t\bar{B}B)\}I_3.$$

We set $k := (\mathrm{tr}B)(\mathrm{tr}\bar{B})$, $\ell := \mathrm{tr}({}^t\bar{B}B)$. Then, by (4.10), each eigenvalue α_i $(1 \leq i \leq 3)$ satisfies the following equation;

$$3x^2 - 3kx + k(k-\ell) = 0. \tag{4.11}$$

By (4.11), $(\mathrm{tr}\bar{B})B$ has at most two distinct eigenvalues. The following two cases are possible; (1) $\alpha_1 = \alpha_2 = \alpha_3$ or (2) one of α_1, α_2, α_3 is different from the others. Since M^6 is a non-nearly Kähler manifold, (1) is impossible. In the case (2), we assume that $\alpha_1 = \alpha \neq \alpha_2 = \alpha_3 = \beta$. Let U_0 be a connected component of the set $\{ p \in M^6 \mid \alpha(p) \neq \beta(p) \}$. Then, we see that U_0 is a non-empty open subset in M^6 because the solutions of (4.11) are continuous. By (4.9), we get

$$
{}^t\bar{U}((\mathrm{tr}\bar{B})B)U = \begin{pmatrix} \alpha & 0 & 0 \\ 0 & \beta & 0 \\ 0 & 0 & \beta \end{pmatrix} \tag{4.12}
$$

on U_0. By (4.11) and (4.12), we get

$$
\alpha + 2\beta = k, \quad \alpha + \beta = k
$$

on U_0. From this, we have

$$
\beta = 0 \tag{4.13}
$$

on U_0. By (4.12) and (4.13), we get

$$
{}^t\bar{U}BU = \begin{pmatrix} \mathrm{tr}B & 0 & 0 \\ 0 & 0 & 0 \\ 0 & 0 & 0 \end{pmatrix} \tag{4.14}
$$

on U_0. Since $\alpha \neq \beta$, we see that $\mathrm{tr}B \neq 0$ on U_0. Hence, the multiplicities of the eigenvalues $\mathrm{tr}B$ and 0 of B are constant on U_0. So, we can choose a differntiable unitary frame field on U_0 which diagonize the matrix B. Thus, if $(n, f, \bar{n}, \bar{f})$ is such a frame field on U_0, we have

$$
B = \begin{pmatrix} \mathrm{tr}B & 0 & 0 \\ 0 & 0 & 0 \\ 0 & 0 & 0 \end{pmatrix} \tag{4.15}
$$

on U_0. We now return to the equation $\theta = {}^t B\omega = B\omega$ armed with this new information. We get,

$$d\theta = dB \wedge \omega + B d\omega \tag{4.16}$$

on U_0. By (2.5),(2.12),(2.14) and (4.16), we get

$$\{\kappa B - B\kappa + dB + (\sqrt{-1}\rho)B + [{}^t\bar{B}\bar{\omega}]A\} \wedge \omega = 0. \tag{4.17}$$

We set $\kappa = (\kappa^i_j)$, $A = (a^i_j) \in M_{3\times 3}(\mathbb{C})$.

<u>Lemma 4.1.</u> The following equalities hold on U_0.

$$\{d(\mathrm{tr}B) + \sqrt{-1}(\mathrm{tr}B)\rho + (\mathrm{tr}B)^2(\bar{a}^3_1\omega^2 - \bar{a}^2_1\omega^3)/(\mathrm{tr}\bar{B})\} \wedge \omega^1 = 0, \tag{4.18}$$

$$\begin{aligned} &(\mathrm{tr}B)\kappa^2_1 + a^3_1(\mathrm{tr}\bar{B})\bar{\omega}^1 = 0, \\ &(\mathrm{tr}B)\kappa^3_1 - a^2_1(\mathrm{tr}\bar{B})\bar{\omega}^1 = 0, \end{aligned} \tag{4.19}$$

$$A = \begin{pmatrix} a^1_1 & a^1_2 & a^1_3 \\ a^2_1 & 0 & 0 \\ a^3_1 & 0 & 0 \end{pmatrix}, \qquad B = \begin{pmatrix} \mathrm{tr}B & 0 & 0 \\ 0 & 0 & 0 \\ 0 & 0 & 0 \end{pmatrix}. \tag{4.20}$$

Proof: By (4.15) and (4.17), we get

$$\begin{pmatrix} d(trB)+(trB)\sqrt{-1}\rho & -(trB)\kappa_2^1 & -(trB)\kappa_3^1 \\ (trB)\kappa_1^2+a_1^3(tr\bar{B})\bar{\omega}^1 & a_2^3(tr\bar{B})\bar{\omega}^1 & a_3^3(tr\bar{B})\bar{\omega}^1 \\ (trB)\kappa_1^3-a_1^2(tr\bar{B})\bar{\omega}^1 & -a_2^2(tr\bar{B})\bar{\omega}^1 & -a_3^2(tr\bar{B})\bar{\omega}^1 \end{pmatrix} \wedge \begin{pmatrix} \omega^1 \\ \omega^2 \\ \omega^3 \end{pmatrix} = 0 \tag{4.21}$$

on U_0. Since $\{\omega^i, \bar{\omega}^i\}$ is linealy independent and $trB \neq 0$ on U_0, by (4.21), we have

$$\{d(trB) + \sqrt{-1}(trB)\rho\}\wedge\omega^1-(trB)(k_2^1\wedge\omega^2 + \kappa_3^1\wedge\omega^3) = 0, \tag{4.22}$$

$$\{(trB)\kappa_1^2 + a_1^3(tr\bar{B})\bar{\omega}^1\}\wedge\omega^1 = 0,$$

$$\{(trB)\kappa_1^3 - a_1^2(tr\bar{B})\bar{\omega}^1\}\wedge\omega^1 = 0, \tag{4.23}$$

$$a_2^2 = a_3^2 = a_2^3 = a_3^3 = 0, \tag{4.24}$$

on U_0. By (4.23), we see that there exist functions μ, η on U_0 satisfying

$$(trB)\kappa_1^2 + a_1^3(tr\bar{B})\bar{\omega}^1 = \mu\omega^1,$$

$$(trB)\kappa_1^3 - a_1^2(tr\bar{B})\bar{\omega}^1 = \eta\omega^1. \tag{4.25}$$

Since κ is a skew hermitian matrix, by (4.25), we get

$$\kappa_2^1 = -\bar{\kappa}_1^2 = -(\bar{\mu}\bar{\omega}^1 - \bar{a}_1^3(trB)\omega^1)/(tr\bar{B}),$$

$$\kappa_3^1 = -\bar{\kappa}_1^3 = -(\bar{\eta}\bar{\omega}^1 + \bar{a}_1^2(trB)\omega^1)/(tr\bar{B}), \tag{4.26}$$

on U_0. By (4.22) and (4.26), we get

$$\{d(\mathrm{tr}B) + \sqrt{-1}(\mathrm{tr}B)\rho - (\mathrm{tr}B)^2(\bar{a}^3_1\omega^2 - \bar{a}^2_1\omega^3)/(\mathrm{tr}\bar{B})\}\wedge\omega^1$$

$$+ (\mathrm{tr}B)\{\bar{\mu}\bar{\omega}^1\wedge\omega^2 + \bar{\eta}\bar{\omega}^1\wedge\omega^3\} = 0. \tag{4.27}$$

Hence, we have $\mu = \eta = 0$ on U_0. We get the desired equalities. □

Lemma 4.2. The normal connection of M^6 is flat, then we have $A = 0$ on U_0.

Proof: Since the codimension of M^6 in $\mathbb{O}$ is two, the flatness of normal connection is equivalent to $d(\sqrt{-1}\rho) = 0$. By (2.3), we get

$$0 = {}^t\bar{\mathfrak{h}}\wedge\mathfrak{h} + {}^t\theta\wedge\bar{\theta}, \tag{4.28}$$

on M^6. On the other hand, by (2.14) and (4.20), we get

$$\mathfrak{h} = \begin{pmatrix} (\mathrm{tr}\bar{B})\omega^1 + \Sigma\bar{a}^1_i\bar{\omega}^i \\ \bar{a}^1_2\bar{\omega}^1 \\ \bar{a}^1_3\bar{\omega}^1 \end{pmatrix}, \quad \theta = \begin{pmatrix} \mathrm{tr}B\omega^1 \\ 0 \\ 0 \end{pmatrix}, \tag{4.29}$$

on U_0. By (4.28) and (4.29), we get

$$0 = {}^t\bar{\mathfrak{h}}\wedge\mathfrak{h} + {}^t\theta\wedge\bar{\theta}$$

$$\equiv \left(\sum_{i=1}^{3} |a_i^1|^2\right)\omega^1\wedge\bar{\omega}^1 \qquad \text{mod } \{\omega^1\wedge\bar{\omega}^2,\ \omega^1\wedge\bar{\omega}^3,\ \bar{\omega}^i\wedge\bar{\omega}^j\}.$$

From this, we have $A = 0$ on U_0. □

From Lemma 4.1 and Lemma 4.2, we have

$$A = 0, \qquad B = \begin{pmatrix} \mathrm{tr}B & 0 & 0 \\ 0 & 0 & 0 \\ 0 & 0 & 0 \end{pmatrix} \tag{4.30}$$

on U_0. By (2.13), (2.14) and (4.30), the second fundamental form $\mathbb{I}$ is given by

$$\mathbb{I} = -2\mathrm{Re}\{2(\omega^1\circ\bar{\omega}^1)(\mathrm{tr}B)n\}. \tag{4.31}$$

Let D and $D^{\perp}$ be the holomorphic distributions on U_0 corresponding to the eigenvalues $\mathrm{tr}B$ and 0 of B, respectively. Let $\{f_1,\bar{f}_1\}$ (resp, $\{f_2,f_3,\bar{f}_2,\bar{f}_3\}$) be the local unitary basis of D (resp, $D^{\perp}$). From Lemma 4.2, (2.10),(2.11), (4.19) and (4.30), we get

$$df_1 = -((\mathrm{tr}B)n + (\mathrm{tr}\bar{B})\bar{n})\bar{\omega}^1 + f_1\kappa_1^1,$$

$$\begin{aligned} df_2 &= f_2\kappa_2^2 + f_3\kappa_2^3 - \bar{f}_3(\mathrm{tr}B)\omega^1, \\ df_3 &= f_2\kappa_3^2 + f_3\kappa_3^3 + \bar{f}_2(\mathrm{tr}B)\omega^1. \end{aligned} \tag{4.32}$$

$$dn = n(\sqrt{-1}\rho) + \mathrm{tr}\bar{B}(f_1\omega^1 + \bar{f}_1\bar{\omega}^1)$$

By (4.32), we get

$$\nabla_X f_1 \in D, \quad \nabla_X f_i \in D^{\perp} \ (i = 2,3) \tag{4.33}$$

for any $X \in T_p M^6$. From (4.32) and (4.33), we see that D and $D^{\perp}$ are parallel distributions on U_0 and each leaf of D (resp, $D^{\perp}$) is a totally umbilical (resp. totally geodesic) submanifold of U_0 in $\mathbb{O}$.

<u>Lemma 4.3.</u> The mean curvature vector field of M^6 is parallel in the normal bundle on U_0.

Proof: We denote by $\nabla^{\perp}$ the normal connection of M^6 in $\mathbb{O}$. By (4.31), we get

$$\mathrm{II} = -\omega^1 \circ \bar{\omega}^1 (3H).$$

From this, we get

$$\mathrm{II}(f_1, \bar{f}_1) = -(3/2)H, \quad \mathrm{II}(f_1, f_i) = \mathrm{II}(f_i, \bar{f}_j) = 0,$$

$$(i,j = 2,3). \tag{4.34}$$

By $(4.32)_1$ and $(4.34)_1$, we get

$$(\nabla_{f_1} \mathrm{II})(f_1, \bar{f}_1) = -(3/2)\nabla^{\perp}_{f_1}(H) - \{\kappa^1_1(f_1) + \bar{\kappa}^1_1(f_1)\}\mathrm{II}(f_1, \bar{f}_1)$$

$$= -(3/2)\nabla^{\perp}_{f_1}(H). \tag{4.35}$$

On one hand, by Codazzi equation and (4.31), we get

$$(\nabla_{f_1}\mathrm{II})(f_1,\bar{f}_1) = (\nabla_{\bar{f}_1}\mathrm{II})(f_1,f_1) \tag{4.36}$$

$$= \nabla^{\perp}_{\bar{f}_1}(\mathrm{II}(f_1,f_1)) - 2\mathrm{II}(\nabla_{\bar{f}_1}f_1,\ f_1) = -2\kappa^1_1(\bar{f}_1)\mathrm{II}(f_1,f_1) = 0.$$

By (4.35) and (4.36), we have

$$\nabla^{\perp}_{f_1}(H) = 0 \tag{4.37}$$

on U_0. By (4.32), we get

$$(\nabla_{f_i}\mathrm{II})(f_1,\bar{f}_1) \tag{4.38}$$

$$= \nabla^{\perp}_{f_i}(\mathrm{II}(f_1,\bar{f}_1)) - \mathrm{II}(\nabla_{f_i}f_1,\ \bar{f}_1) - \mathrm{II}(f_1,\nabla_{f_i}\bar{f}_1)$$

$$= -(3/2)\nabla^{\perp}_{f_i}(H) - \{\kappa^1_1(f_i) + \bar{\kappa}^1_1(f_i)\}\mathrm{II}(f_1,\bar{f}_1)$$

$$= -(3/2)\nabla^{\perp}_{f_i}(H) \quad (i = 2,3).$$

On one hand, by Codazzi equation, (4.31) and (4.33), we get

$$(\nabla_{f_i}\mathrm{II})(f_1,\bar{f}_1) = (\nabla_{f_1}\mathrm{II})(f_i,\bar{f}_1) \tag{4.39}$$

$$= \nabla^{\perp}_{f_1}(\mathrm{II}(f_i,\bar{f}_1)) - \mathrm{II}(\nabla_{f_1}f_i,\ \bar{f}_1) - \mathrm{II}(f_i,\nabla_{f_1}\bar{f}_1) = 0,$$

$(i = 2,3)$. By (4.38) and (4.39), we have

$$\nabla^{\perp}_{f_i}(H) = 0 \qquad (i = 2,3). \tag{4.40}$$

By (4.37) and (4.40), we get the conclusion. □

Hence, it follows that U_0 is locally of the form $S^2\times\mathbb{R}^4$. From Lemma 4.3, U_0 is a non-empty open and closed subset in M^6. Since M^6 is connected, $U_0 = M^6$. Again, by (4.32) and (4.33), we see that the immersion Ψ is locally of the form $\Psi = \psi\times id: S^2\times\mathbb{R}^4 \longrightarrow \mathbb{R}^3\times\mathbb{R}^4(\simeq\mathbb{R}^7)\subset\mathbb{O}$. This completes the proof of Theorem B.

Finally, we shall write down the quasi-Kähler structure of $S^2\times\mathbb{R}^4$ in $\mathrm{Im}\mathbb{O}$ (the pruely imaginary part of $\mathbb{O}$). We regard $\mathrm{Im}\mathbb{O}$ as the direct sum of $\mathrm{Im}\mathbb{H}\oplus\mathbb{H}$. Let S^2 be a 2-dimensional unit sphere in $\mathrm{Im}\mathbb{H}$ and x a position vector of S^2 in $\mathrm{Im}\mathbb{H}$. Then, the almost complex structure J at $(x,y)\in S^2\times\mathbb{R}^4 = S^2\times\mathbb{H}\subset\mathrm{Im}\mathbb{O}$ is given by

$$JX = X\times x$$

for any $X\in T_{(x,y)}(S^2\times\mathbb{R}^4)$. If $A\in T_xS^2\subset\mathrm{Im}\mathbb{H}$ and $V\in T_y\mathbb{R}^4$, then we have

$$(\nabla_A J)(X) = X\times A, \qquad (\nabla_V J)(X) = 0 \tag{4.41}$$

for any $X \in T_{(x,y)}(S^2 \times \mathbb{R}^4)$. By (4.41), we see that

$$(\nabla_A J)(T_x S^2) \subset T_x S^2, \quad (\nabla_A J)(T_y \mathbb{R}^4) \subset T_y \mathbb{R}^4 \quad \text{and} \quad \nabla_V J = 0,$$

for any $A \in T_x S^2$ and $V \in T_y \mathbb{R}^4$.

Refereces

[1] R.L.Bryant, Submanifolds and special structures on the octonians,J.Diff.Geom.,17(1982),185-232.

[2] E.Calabi, Construction and properties of some 6-dimensional manifolds,Trans.Amer.Math.Soc., 87(1958), 407-438.

[3] T. Fukami and S. Ishihara, Almost Hermitian structure on S^6,Tohoku Math.J., 7(1955), 151-156.

[4] A.Gray, Minimal varieties and almost hermitian submanifolds. Mich.Math.J.,12(1965),273-287.

[5] A.Gray, Vector cross products on manifolds, Trans.Amer. Math.Soc.,141(1969),465-504.

[6] A.Gray, Six dimensional almost complex manifolds defined by means of three fold vector cross products. Tohoku Math.J., 21(1969), 614-620.

[7] A.Gray, Riemannian manifolds with geodesic symmetries of order 3,J.Diff.Geom.,7(1972),343-369.

[8] A.Gray, Curvature identities for Hermitian and almost Heremitian manifolds, Tohoku Math.J., 28(1976), 601-612.

[9] R.Harvey and H.B.Lawson.Jr,Calibrated geometries,Acta

Math, 148(1982),47-157.

[10] H.Hashimoto, Some 6-dimensional oriented submanifold in the octonians,Math.Rep.Toyama Univ,11(1988),1-19.

[11] K.Yano and T.Sumitomo, Differential Geometry on Hypersurfaces in a Cayley space. Proc.of the Royal Society of Edinburgh.,66(1964),216-231.

Department of Mathematical Science
Graduate school of Science and Technology
Niigata University
Niigata 950-21
Japan

Notes on Tricerri-Vanhecke's decomposition of curvature tensors

Toyoko Kashiwada

Introduction.

In [9], Tricerri and Vanhecke established a $\mathfrak{U}(n)$-irreducible decomposition of curvature tensors over a Hermitian vector space. Our purpose of this paper is to study some properties on an almost Hermitian manifold by means of their decomposition.

1. Preliminaries

Let (V,g) be an m-dimensional real vector space with inner product g. A (1,3)-tensor R is called a <u>curvature tensor</u> over V if it satisfies the properties:

(i) $R(x,y)z = - R(y,x)z$,

(ii) $g(R(x,y)z,w) = - g(R(x,y)w,z)$ $(= R(x,y,z,w))$,

(iii) $R(x,y)z + R(y,z)x + R(z,x)y = 0$,

for any $x,y,z,w \in V$.

We denote by $\mathcal{R}(V)$ the vector space of all curvature tensors over V. It has the natural inner product induced from g. The Ricci tensor $\rho(R)$

ISBN 0-12-640170-5

and the scalar curvature $\tau(R)$ of $R \in \mathcal{R}(V)$ are defined as

$$\rho(R)(x,y) = \sum_{i=1}^{m} g(R(e_i,x)y,e_i), \qquad \tau(R) = \sum_{i=1}^{m} \rho(R)(e_i,e_i)$$

where $\{e_i\}_{i=1,2,..,m}$ is an orthonormal basis of (V,g).

As is well known ([5],[7]), $\mathcal{R}(V)$ is orthogonally decomposed as

$$\mathcal{R}(V) = \mathcal{L}\{\pi_1\} \oplus \mathcal{R}_W \oplus \mathcal{R}_E(V),$$

where

$$\mathcal{L}\{\pi_1\} = \{a\pi_1 \mid a \in R\} \quad (\pi_1(x,y)z = g(y,z)x - g(x,z)y),$$

$$\mathcal{R}_W(V) = \{R \in \mathcal{R}(V) \mid \rho(R) = 0\},$$

$$\mathcal{R}_E(V) = \{\varphi(S) \mid S \text{ is a traceless symmetric } (0,2)\text{-tensor}\}$$

$$(\varphi(S)(x,y,z,w) = g(x,w)S(y,z) - g(x,z)S(y,w) + g(y,z)S(x,w) - g(y,w)S(x,z)),$$

$$\mathcal{R}_W(V) \oplus \mathcal{R}_E(V) = \{R \in \mathcal{R}(V) \mid \tau(R) = 0\}.$$

$\mathcal{R}_E(V)$-component of $R \in \mathcal{R}(V)$ is represented as $\frac{1}{m-2}\varphi(\rho(R) - \frac{\tau(R)}{m}g)$, and R is Einstein iff its $\mathcal{R}_E(V)$-component vanishes.

Next, let (V,J,g) be a 2n-dimensional real vector space with a complex structure J and a hermitian product g. Let the subspaces $\mathcal{R}_i(V)$ (i = 1, 2,3), $\mathcal{R}_1(V) \subset \mathcal{R}_2(V) \subset \mathcal{R}_3(V) \subset \mathcal{R}(V)$, be

$$\mathcal{R}_1(V) = \{R \in \mathcal{R}(V) \mid R(x,y,z,w) = R(x,y,Jz,Jw)\},$$

$$\mathcal{R}_2(V) = \{R \in \mathcal{R}(V) \mid R(x,y,z,w) = R(Jx,Jy,z,w) + R(Jx,y,Jz,w) + R(Jx,y,z,Jw)\},$$

$$\mathcal{R}_3(V) = \{R \in \mathcal{R}(V) \mid R(x,y,z,w) = R(Jx,Jy,Jz,Jw)\},$$

and let $\mathcal{R}_i^{\perp}(V)$ $(i = 1,2,3)$ be the orthogonal complements of $\mathcal{R}_i(V)$ in $\mathcal{R}_{i+1}(V)$ $(\mathcal{R}_4(V) = \mathcal{R}(V))$. Namely, $\mathcal{R}_{i+1}(V) = \mathcal{R}_i(V) \oplus \mathcal{R}_i^{\perp}(V)$, $\mathcal{R}(V) = \mathcal{R}_1(V) \oplus \mathcal{R}_1^{\perp}(V) \oplus \mathcal{R}_2^{\perp}(V) \oplus \mathcal{R}_3^{\perp}(V)$. L_i $(i = 1,2,3)$ are the operators defined on $\mathcal{R}_{i+1}(V)$ as, for $x,y,z,w \in V$,

$$L_1(R)(x,y,z,w) = \frac{1}{2}\{R(Jx,Jy,z,w) + R(y,Jz,Jx,w) + R(Jz,x,Jy,w)\},$$

$$L_2(R)(x,y,z,w) = \frac{1}{2}\{R(x,y,z,w) + R(Jx,Jy,z,w) + R(Jx,y,Jz,w) + R(Jx,y,z,Jw)\},$$

$$L_3(R)(x,y,z,w) = R(Jx,Jy,Jz,Jw).$$

$\mathcal{R}_i(V)$ and $\mathcal{R}_i^{\perp}(V)$ are characterized in [9] as

$$\mathcal{R}_i(V) = \{R \in \mathcal{R}_{i+1}(V) \mid L_i(R) = R\},$$
$$\mathcal{R}_i^{\perp}(V) = \{R \in \mathcal{R}_{i+1}(V) \mid L_i(R) = -R\}. \tag{1.1}$$

For $R \in \mathcal{R}(V)$, $\rho^*(R)$ and $\tau^*(R)$ are defined as

$$\rho^*(R)(x,y) = \sum_{i=1}^{2n} g(R(e_i,x)Jy,Je_i) \quad \left(= -\frac{1}{2}\sum_{i=1}^{2n} g(R(e_i,Je_i)x,Jy)\right),$$
$$\tau^*(R) = \sum_{i=1}^{2n} \rho^*(R)(e_i,e_i).$$

Then, it holds that

$$\rho^*(R)(x,y) = \rho^*(R)(Jy,Jx), \tag{1.2}$$

and $\rho^*(R)$ is symmetric if and only if $\rho^*(R)$ is hybrid.

In [9], the following $\mathcal{U}(n)$-irreducible decomposition of (V,J,g) has been obtained:

$$\mathcal{W}_1 = \left\{ \begin{array}{c} R \in \mathcal{R}_1(V) \text{ with constant holomorphic} \\ \text{sectional curvature} \end{array} \right\} = \mathcal{L}\{\pi_1 + \pi_2\},$$

$$\mathcal{W}_2 = (\mathcal{W}_1 \oplus \mathcal{W}_3)^{\perp} \text{ in } \mathcal{R}_1(V) = (\varphi + \psi)(\mathcal{V}_1),$$

$$\mathcal{W}_3 = \{R \in \mathcal{R}_1(V) \mid \rho(R) = 0\},$$

$$\mathcal{W}_4 = \mathcal{L}\{3\pi_1 - \pi_2\},$$

$$\mathcal{W}_5 = (\mathcal{W}_4 \oplus \mathcal{W}_6)^{\perp} \text{ in } \mathcal{R}_1^{\perp}(V) = (3\varphi - \psi)(\mathcal{V}_1) \quad (= \{0\} \text{ if } n = 2),$$

$$\mathcal{W}_6 = \{R \in \mathcal{R}_1^{\perp}(V) \mid \rho(R) = 0\} \quad (= \{0\} \text{ if } n = 2,3),$$

$$\mathcal{W}_7 = \mathcal{R}_2^{\perp}(V),$$

$$\mathcal{W}_8 = \{R \in \mathcal{W}_{10}^{\perp} \text{ in } \mathcal{R}_3^{\perp}(V) \mid \rho^*(R) = 0\} = \varphi(\mathcal{V}_2),$$

$$\mathcal{W}_9 = \{R \in \mathcal{W}_{10}^{\perp} \text{ in } \mathcal{R}_3^{\perp}(V) \mid \rho(R) = 0\} = \psi(\mathcal{V}_3),$$

$$\mathcal{W}_{10} = \{R \in \mathcal{R}_3^{\perp}(V) \mid \rho(R) = \rho^*(R) = 0\} \quad (= \{0\} \text{ if } n = 2),$$

where

$$\pi_2(x,y)z = -\Omega(y,z)Jx + \Omega(x,z)Jy + 2\Omega(x,y)Jz \qquad (\Omega(x,y) = g(x,Jy))$$

$$\psi(S)(x,y,z,w) = \Omega(x,w)S(y,Jz) - \Omega(x,z)S(y,Jw) + \Omega(y,z)S(x,Jw) - \Omega(y,w)S(x,Jz) - 2\Omega(x,y)S(z,Jw) - 2\Omega(z,w)S(x,Jy)$$

for a (0,2)-tensor S, and

$$\mathcal{V}_1 = \{\text{Traceless hybrid symmetric (0,2)-tensors}\},$$

$$\mathcal{V}_2 = \{\text{Pure symmetric (0,2)-tensors}\},$$

$$\mathcal{V}_3 = \{\text{Pure skew-symmetric (0,2)-tensors}\}.$$

We denote by $p_i(R)$ the $\mathcal{W}_i$-component of $R \in \mathcal{R}(V)$: $R = \sum_{i=1}^{10} p_i(R)$. Especially,

$$p_2(R) = \frac{1}{16(n+2)} (\varphi + \psi)(\rho(R) + J\rho(R) + 3(\rho^*(R) + J\rho^*(R)) - \frac{\tau(R)+3\tau^*(R)}{n} g),$$

$$p_5(R) = \frac{1}{16(n-2)} (3\varphi - \psi)(\rho(R) + J\rho(R) - (\rho^*(R) + J\rho^*(R)) - \frac{\tau(R)-\tau^*(R)}{n} g) \quad (n \neq 2),$$

$$p_8(R) = \frac{1}{4(n-1)} \varphi(\rho(R) - J\rho(R)),$$

$$p_9(R) = \frac{1}{4(n+1)} \psi(\rho^*(R) - J\rho^*(R)),$$

where we defined JS by JS(x,y) = S(Jx,Jy) for (0,2)-tensor S.

$\mathcal{R}_E(V)$-component over a hermitian vector space is expressed as

$$\mathcal{R}_E(V) = \varphi(\mathcal{V}_1) \oplus \varphi(\mathcal{V}_2) \subset \mathcal{W}_2 \oplus \mathcal{W}_5 \oplus \mathcal{W}_8$$

and, since

$$\rho(p_2(R) + p_5(R)) = \frac{1}{2} (\rho(R) + J\rho(R) - \frac{\tau(R)}{n} g),$$

$$\rho(p_8(R)) = \frac{1}{2} (\rho(R) - J\rho(R)),$$

R is Einstein if and only if $\rho(p_2(R) + p_5(R)) = 0$ and $\rho(p_8(R)) = 0$ (i.e., $p_8(R) = 0$).

2. The holomorphic sectional curvature

For $R \in \mathcal{R}(V)$, its holomorphic sectional curvature with respect to $x \in V$ is defined by $H(x) = \dfrac{R(x,Jx,Jx,x)}{|x|^4}$.

Proposition 2.1. For $R \in \mathcal{R}_1^{\perp}(V) \oplus \mathcal{R}_2^{\perp}(V) \oplus \mathcal{R}_3^{\perp}(V)$, the holomorphic sectional curvature vanishes.

Proof. R is denoted as $R = R_1^{\perp} + R_2^{\perp} + R_3^{\perp}$ $(R_i^{\perp} \in \mathcal{R}_i^{\perp}(V))$. By the definition of L_i, we have $L_i(R_i^{\perp})(x,Jx,Jx,x) = R_i^{\perp}(x,Jx,Jx,x)$ for $i = 1,2,3$. On the other hand, by (1.1), $L_i(R_i^{\perp}) = - R_i^{\perp}$. Then the holomorphic sectional curvature of R_i, and hence, of R, vanishes. Q.E.D.

Corollary 2.2. $R \in \mathcal{R}(V)$ has constant holomorphic sectional curvature (i.e., $H(x) = c$ for any $x \in V$) if and only if $p_2(R) = p_3(R) = 0$.

3. The Bochner curvature tensor

The Bochner curvature tensor B(R) of $R \in \mathcal{R}(V)$ is, by the definition in [9], the $\mathcal{W}_3 \oplus \mathcal{W}_6 \oplus \mathcal{W}_7 \oplus \mathcal{W}_{10}$ -component:

$$B(R) = p_3(R) + p_6(R) + p_7(R) + p_{10}(R).$$

Remark. In [11], the Bochner component of $\mathcal{R}_3(V)$ has been defined by

$$\mathcal{B}'(V) = \{R \in \mathcal{R}_3(V) \mid \rho(R) + 3\rho^*(R) = 0, \quad \tau(R) = \tau^*(R) = 0\}.$$

The relation between B(R) and $\mathcal{B}'(V)$-component B'(R) of $R \in \mathcal{R}_3(V)$ is

$$B'(R) = B(R) + p_5(R).$$

As is well known, a Riemannian manifold is conformally flat iff the Riemannian curvature tensor R satisfies $R(x,y,z,w) = 0$ for any orthogonal

4-vectors $\{x,y,z,w\}$. Corresponding to this fact, we prove in this section the following theorem.

Theorem 3.1. For $R \in \mathcal{R}(V)$ ($\dim V \geqq 8$), $B(R)$ vanishes if and only if

(*) $R(x,y,z,w) = 0$ for any anti-holomorphic 4-vectors $\{x,y,z,w\}$.[1)]

In gereral, it can be seen that

Lemma 3.2. For $R \in \mathcal{R}(V)$, the following (1) — (3) are equivalent:

(1) $R = 0$.

(2) The sectional curvature of R is zero.

(3) (i) The holomorphic sectional curvature is zero,

(ii) The sectional curvature of any anti-holomorphic section is zero,

and (iii) $R(x,Jx,x,y) = 0$ for any anti-holomorphic vectors $\{x,y\}$.

Therefore, we have

Lemma 3.3. (a) For $R \in \mathcal{R}_1(V)$, $R = 0$ iff (3)(i) in Lemma 3.2 holds.

(b) For $R \in \mathcal{R}_1^{\perp}(V) \oplus \mathcal{R}_2^{\perp}(V)$, $R = 0$ iff (3)(ii) in Lemma 3.2 holds.

(c) For $R \in \mathcal{R}_3^{\perp}(V)$, $R = 0$ iff (3)(ii) and (iii) in Lemma 3.2 hold.

Proof. (a) comes from the fact that a sectional curvature of $R \in \mathcal{R}_1(V)$ can be represented by holomorphic sectional curvatures([1]).

(b): From $L_i(R_i^{\perp})(x,Jx,x,y) = R_i^{\perp}(x,Jx,x,y)$ for $R_i^{\perp} \in \mathcal{R}_i^{\perp}(V)$ ($i=1,2$) and (1.1), for $R \in \mathcal{R}_1^{\perp}(V) \oplus \mathcal{R}_2^{\perp}(V)$, $R(x,Jx,x,y) = 0$ holds for any x,y. Hence,

1) Anti-holomorphic 4-vectors $\{x,y,z,w\}$ means that x,y,z,w,Jx,Jy,Jz,Jw are orthogonal to each other.

on account of Proposition 2.1, (3)(i) and (iii) in Lemma 3.2 hold for any $R \in \mathcal{R}_1^{\perp}(V) \oplus \mathcal{R}_2^{\perp}(V)$, necessarily.

(c) follows from Proposition 2.1 and Lemma 3.2. Q.E.D.

Lemma 3.4. If $\dim V \geqq 8$, and $R \in \mathcal{R}(V)$ satisfies (*), then, for different $|i|, |j|, |k|$,[1)]

$$R(e_i,e_k,e_k,e_i) - R(e_i,Je_k,Je_k,e_i) = R(e_j,e_k,e_k,e_j) - R(e_j,Je_k,Je_k,e_j) \tag{3.1}$$

holds for any orthonormal J-basis $\{e_1,\ldots,e_n,e_{1*}(=Je_1),\ldots,e_{n*}(=Je_n)\}$.

Especially, in the case $R \in \mathcal{R}_3(V)$, it holds

$$R(e_i,e_j,e_j,e_i) = R(e_i,Je_j,Je_j,e_i). \tag{3.2}$$

Proof. The assumption (*) means

$$R(e_i,e_j,e_k,e_h) = 0 \tag{3.3}$$

for different $|i|, |j|, |k|, |h|$ of every orthonormal J-basis $\{e_1,\ldots,e_n, e_{i*},\ldots,e_{n*}\}$. Taking a new J-basis $\{e'_1,\ldots,e'_n,e'_{1*},\ldots,e'_{n*}\}$ for fixed i,j, defined by

$$\begin{cases} e'_i = (e_i + e_j)/\sqrt{2} \\ e'_j = (-e_i + e_j)/\sqrt{2} \\ e'_k = e_k \qquad (|i| \neq |j|,\ |k| \neq |i|,\ |j|), \end{cases} \tag{3.4}$$

from the equation $R(e'_k,e'_i,e'_j,e'_h) = 0$, we have

1) $|a| = a$, $|a^*| = a$, $a = 1,2,..,n$.

$$R(e_k,e_i,e_i,e_h) = R(e_k,e_j,e_j,e_h) \tag{3.5}$$

for different $|i|,|j|,|k|,|h|$. Hence, $R(e_i,e_a,e_a,e_j) = R(e_i,e_h,e_h,e_j) = R(e_i,e_{a*},e_{a*},e_j)$ $(h \neq a,a*)$ holds, namely,

$$R(e_i,e_k,e_k,e_j) = R(e_i,Je_k,Je_k,e_j) \tag{3.6}$$

for different $|i|,|j|,|k|$. From the equation (3.6) with respect to the exchanged basis by (3.4), we obtain (3.1).

Next, replacing e_j by Je_j in (3.1), we have

$$R(e_i,e_k,e_k,e_i) - R(e_i,Je_k,Je_k,e_i)$$

$$= R(Je_j,e_k,e_k,Je_j) - R(Je_j,Je_k,Je_k,Je_j).$$

Then if $R \in \mathcal{R}_3(V)$, $R(e_i,e_k,e_k,e_i) - R(e_i,Je_k,Je_k,e_i) = 0$. Q.E.D.

<u>Lemma 3.5.</u> Let the components $\mathcal{R}_1(V)$, $\mathcal{R}_i^{\perp}(V)$ $(i = 1,2,3)$ of $R \in \mathcal{R}(V)$ be $R_i^{\perp}$ $(i = 0,1,2,3)$, respectively. Then R satisfies the property (*) iff $R_i^{\perp}$ satisfies (*) for any i $(i = 0,1,2,3)$.

Proof. We assume R satisfies (*). From the definition of L_i $(i = 1,2,3)$, we know $L_i(R)$ also satisfies (*). So, because of

$$R_0^{\perp} = \frac{1}{8}(I + L_1)(I + L_2)(I + L_3)(R)$$

$$R_1^{\perp} = \frac{1}{8}(I - L_1)(I + L_2)(I + L_3)(R)$$

$$R_2^{\perp} = \frac{1}{4}(I - L_2)(I + L_3)(R)$$

$$R_3^{\perp} = \frac{1}{2}(I - L_3)(R),$$

$R_i^{\perp}$ satisfies (*). The converse is trivial. Q.E.D.

Since π_1, π_2, $\varphi(S)$ and $\psi(S)$ vanish for any anti-holomorphic 4-vectors $\{x,y,z,w\}$, a curvature tensor belonging to $\mathcal{W}_1 \oplus \mathcal{W}_2 \oplus \mathcal{W}_4 \oplus \mathcal{W}_5 \oplus \mathcal{W}_8 \oplus \mathcal{W}_9$ -component vanishes for such 4-vectors. Hence, by virtue of Lemma 3.5, Theorem 3.1 is equivalent to the fact that a curvature tensor R belonging to $\mathcal{W}_3, \mathcal{W}_6, \mathcal{W}_7$ or $\mathcal{W}_{10}$ satisfies (*) iff $R = 0$.

Proof of Theorem 3.1. By virtue of Lemma 3.3, it is enough to prove

(A) If $R \in \mathcal{W}_3$ satisfies (*), then the holomorphic sectional curvature vanishes.

(B) If $R \in \mathcal{W}_6$ or $R \in \mathcal{W}_7$ satisfies (*), then the sectional curvature of any anti-holomorphic section vanishes.

(C) If $R \in \mathcal{W}_{10}$ satisfies (*), then the sectional curvature of any anti-holomorphic section vanishes and $R(x,Jx,x,y) = 0$ for any anti-holomorphic 2-vectors $\{x,y\}$.

Proof of (A): Taking the basis (3.4), we have from (3.2),

$$R(e_i,Je_i,Je_i,e_i) + R(e_j,Je_j,Je_j,e_j) = 8R(e_i,e_j,e_j,e_i) \tag{3.7}$$

for $|i| \neq |j|$. Hence, taking summation in (3.7) for $j = 1,..,n,1^*,..,n^*$ except $|j| = |i|$, we have, for each i,

$$-(n+2)R(e_i,Je_i,Je_i,e_i) = \sum_{a=1}^{n} R(e_a,Je_a,Je_a,e_a) \quad (=\mu)$$

where we have used $\rho(R) = 0$ because of $R \in \mathcal{W}_3$. Furthermore taking the summation for i, we obtain $\mu = 0$, namely, $R(e_a,Je_a,Je_a,e_a) = 0$.

Proof of (B): Assume $R \in \mathcal{W}_6$. Taking summation in (3.5) for j except $|j| = |i|,|k|,|h|$, on account of $\rho(R) = 0$ and (3.6), we get

$$2(n-2)R(e_k,e_i,e_i,e_h) = -\{R(e_k,Je_k,Je_k,e_h) + R(e_k,Je_h,Je_h,e_h)\}.$$

So, because of $R(x,Jx,x,y) = 0$ for anti-holomorphic vecrors $\{x,y\}$ as is mentioned in the proof of Lemma 3.3 (b), $R(e_i,e_k,e_k,e_j) = 0$ holds for different $|i|,|j|,|k|$. From this equation with respect to the basis exchanged by (3.4), we have

$$R(e_i,e_k,e_k,e_i) = R(e_j,e_k,e_k,e_j).$$

Hence, from the summation for j, making use of Proposition 2.1 and (3.2), we can get $R(e_i,e_k,e_k,e_i) = 0$ for $|i| \neq |k|$.

Next, if we assume $R \in \mathfrak{V}_7$, it follows for any $x,y,z,w \in V$,

$$R(Jx,y,z,w) = R(x,Jy,z,w) \tag{3.8}$$

In fact, $L_2(R) = - R$ implies

$$3R(Jx,y,z,w) = R(x,Jy,z,w) + R(x,y,Jz,w) + R(x,y,z,Jw),$$

and exchanging x and y, adding them, we get the above equation.

Then, by virtue of (3.8) and the Bianchi's identity, we have $R(x,Jy,Jy,x) = - R(x,y,y,x)$. Because of (3.2), this means $R(x,y,y,x) = 0$ for anti-holomorphic $\{x,y\}$.

Proof of (C): From $L_3(R) = - R$ and (3.1), it follows

$$\begin{aligned} &R(e_i,e_j,e_j,e_i) - R(e_i,Je_j,Je_j,e_i) \\ &\quad = R(e_k,e_j,e_j,e_k) - R(e_k,Je_j,Je_j,e_k) \\ &\quad = R(e_k,e_j,e_j,e_k) + R(Je_k,e_j,e_j,Je_k) \\ &\quad = 2R(e_k,e_j,e_j,e_k) + R(Je_k,e_i,e_i,Je_k) - R(e_k,e_i,e_i,e_k) \\ &\quad = 2R(e_k,e_j,e_j,e_k) - R(e_k,Je_i,Je_i,e_k) - R(e_k,e_i,e_i,e_k). \end{aligned}$$

Using the same properties again in the second term, we get lastly

$$R(e_i,Je_j,Je_j,e_i) = R(e_i,e_k,e_k,e_i) - R(e_j,e_k,e_k,e_j).$$

Hence, taking the summation for k except $|k|=|i|, |j|$, and making use of Proposition 2.1 and $\rho(R) = 0$, we obtain $R(e_i,Je_j,Je_j,e_i) = 0$, namely, the sectional curvature for any anti-holomorphic section vanishes.

Next, we prove the second part of (C). For anti-holomorphic vectors $\{x,y\}$, from the vanishing sectional curvature of the anti-holomorphic section $\{Jx + y, x + Jy\}$, it follows that

$$R(Jx,x,x,y) + R(x,y,y,Jy) = 0. \tag{3.9}$$

For an orthonormal J-basis $\{e_i\}$, putting in (3.9) as $x = (\pm e_i + e_j)/\sqrt{2}$, $y = e_k$ for different $|i|,|j|,|k|$, and taking the sum of the two equations, we have

$$R(Je_j,e_i,e_i,e_k) + 2R(Je_i,e_j,e_i,e_k) + R(e_j,e_i,Je_i,e_k) + R(e_j,e_k,e_k,Je_k) = 0.$$

Hence, by virtue of $\rho(R) = \rho^*(R) = 0$, it follows that

$$2R(Je_j,e_j,e_j,e_k) - (n-1)R(e_j,e_k,e_k,Je_k) = 0,$$

and by (3.9), this implies $R(Je_j,e_j,e_j,e_k) = 0$ for $|j| \neq |k|$. Q.E.D.

4. The case of 4-dimension

Let * be the Hodge operator of $\wedge^2$ with respect to the orientation

$\{e_1, Je_1, e_2, Je_2\}$. If we consider $R \in \mathcal{R}(V)$ as the curvature operator $\Lambda^2 \to \Lambda^2$, it is known ([7],[9]) that $\mathcal{R}_W(V)$ is split as $\mathcal{R}_W(V) = \mathcal{R}_W^+(V) \oplus \mathcal{R}_W^-(V)$,

$$\mathcal{R}_W^+(V) = \{R \in \mathcal{R}_W(V) \mid R* = R\} = \mathcal{L}\{\pi_1 - \pi_2\} \oplus \mathcal{W}_7 \oplus \mathcal{W}_9$$

$$\mathcal{R}_W^-(V) = \{R \in \mathcal{R}_W(V) \mid R* = -R\} = \mathcal{W}_3 \,.$$

$\mathcal{R}_W(V)$-component $C(R)$ of $R \in \mathcal{R}(V)$ is strictly represented as

$$C(R) = \frac{\tau(R) - 3\tau^*(R)}{48}(\pi_1 - \pi_2) + p_3(R) + p_7(R) + p_9(R).$$

A 4-dimensional almost Hermitian manifold (M^4, J, g) is called <u>self-dual</u> (resp., <u>anti-self-dual</u>) if its Weyl tensor $C(R)$ belongs to $\mathcal{R}_W^+(T_p)$ (resp., $\mathcal{R}_W^-(T_p)$) at each point $p \in M$. Then (M^4, J, g) is self-dual iff $p_3(R) = 0$, and anti-self-dual iff $\tau(R) - 3\tau^*(R) = 0$, $p_7(R) = p_9(R) = 0$ for the Riemannian curvature tensor R. From this, on account of Corollary 2.2, we can say

<u>Theorem 4.1.</u>(Cf.[3]) A 4-dimensional almost Hermitian manifold is of (point wise) constant holomorphic sectional curvature if and only if it is self-dual and $p_2(R) = 0$.

Remark. For $n = 2$, $\rho(R) + J\rho(R) - \frac{\tau(R)}{2} g = \rho^*(R) + J\rho^*(R) - \frac{\tau^*(R)}{2} g$ holds. Then, the following (i) — (iii) are equivalent:

(i) $p_2(R) = 0$.

(ii) $\rho(R) + J\rho(R) = \frac{\tau(R)}{2} g$.

(iii) $\rho^*(R) + J\rho^*(R) = \frac{\tau^*(R)}{2} g$.

Next, we assume (M^4, J, g) to be integrable. In this case, $p_7(R) = 0$ holds ([9]), the kaehler form satisfies $d\Omega = \alpha \wedge \Omega$ $(\alpha = \delta\Omega \circ J)$ and

$$\Lambda(\Omega) d\alpha = 0. \tag{4.1}$$

If α is closed, (M^4,J,g) is a locally conformal Kaehler manifold.

Now, in a 4-dimensional Hermitian manifold, the following holds ([6]):

$$
\begin{aligned}
&2\{R(u,x,Jy,z) + R(u,x,y,Jz)\}\\
&= g(x,z)\{(\nabla_u\alpha)Jy + \tfrac{1}{2}\alpha(Jy)\alpha(u) - \tfrac{1}{2}\Omega(u,y)|\alpha|^2\}\\
&\quad - g(u,z)\{(\nabla_x\alpha)Jy + \tfrac{1}{2}\alpha(Jy)\alpha(x) - \tfrac{1}{2}\Omega(x,y)|\alpha|^2\}\\
&\quad - g(x,y)\{(\nabla_u\alpha)Jz + \tfrac{1}{2}\alpha(Jz)\alpha(u) - \tfrac{1}{2}\Omega(u,z)|\alpha|^2\}\\
&\quad + g(u,y)\{(\nabla_x\alpha)Jz + \tfrac{1}{2}\alpha(Jz)\alpha(x) - \tfrac{1}{2}\Omega(x,z)|\alpha|^2\}\\
&\quad + \Omega(x,z)\{(\nabla_u\alpha)(y) + \tfrac{1}{2}\alpha(u)\alpha(y)\} - \Omega(u,z)\{(\nabla_x\alpha)(y) + \tfrac{1}{2}\alpha(x)\alpha(y)\}\\
&\quad - \Omega(x,y)\{(\nabla_u\alpha)(z) + \tfrac{1}{2}\alpha(u)\alpha(z)\} + \Omega(u,y)\{(\nabla_x\alpha)(z) + \tfrac{1}{2}\alpha(x)\alpha(z)\}.
\end{aligned}
$$

Let $\{y,z\}$ be anti-holomorphic orthonormal vectors and $\{e_1,e_{1*},e_2,e_{2*}\}$ be the orthonormal basis $\{y,Jy,z,Jz\}$. Putting $u = e_i$ and $x = Je_i$ (i= 1,1*, 2,2*) in (4.2), and taking the sum of the four equations, we obtain

$$2\{\rho^*(z,y) - \rho^*(y,z)\} = d\alpha(Jy,Jz) - d\alpha(y,z). \tag{4.3}$$

On the other hand, if we notice that

$$*(e_i\wedge e_j) = -Je_i\wedge Je_j \quad (|i| \neq |j|), \qquad *(e_1\wedge e_{1*}) = e_2\wedge e_{2*},$$

we can say

Lemma 4.2. If a skew-symmetric 2-tensor T of a 4-dimensional Hermitian vector space (V^4,J,g) satisfies

(i) $T(Jx,Jy) = T(x,y)$ $(x,y\in V)$ and (ii) $\Lambda(\Omega)T = 0$,

then $*T = -T$ holds. The converse is also true.

Now, it has been shown in [9] that in a locally conformal Kaehler manifold, $p_9(R)$ $(p_6(R), p_7(R), p_{10}(R))$ of the Riemannian curvature tensor vanishes, but the converse is not true in general. As a restricted case, we can say

<u>Theorem 4.3.</u>(Cf.[3]) A 4-dimensional compact Hermitian manifold with vanishing $p_9(R)$ is a locally conformal Kaehler manifold.

Proof. From the explicit representation of $\mathcal{W}_9$-component of R, we know $p_9(R) = 0$ iff $\rho^*(R)(x,y) - \rho^*(Jx,Jy) = 0$, i.e., by (1.2), $\rho^*(R)$ is symmetric. Therefore, by virtue of (4.1),(4.3) and Lemma 4.2, we know $p_9(R) = 0$ iff $-d\alpha = *d\alpha$ $(=\delta * \alpha)$. Then, by the compactness, $d\alpha = 0$.

Q.E.D.

Remark. From Lemma 4.2, we can see that except for generalized Hopf manifold (i.e., $\nabla\alpha = 0$), there exists no 4-dimensional compact Hermitian manifold which satisfies the properties: (i) $\nabla\alpha$ is skew-symmetric and (ii) $(\nabla\alpha)(Jx,Jy) = (\nabla\alpha)(x,y)$. (Cf. [10] τ_4-case.)

Bibliography

[1] R.I.Bishop and S.I.Goldberg, Some implications of the generalized Gauss-Bonnet theorem, Trans. Amer. Math. Soc., 112(1964), 508-535.

[2] T.Kashiwada, Some characterizations of vanishing Bochner curvature tensor, Hokkaido Math. J., 3(1974), 290-296.

[3] T.Koda, Self-dual and anti-self-dual hermitian surfaces, Kodai Math. J., 10(1987), 335-342.

[4] H.Mori, On the decomposition of generalized K-curvature tensor fields, Tôhoku Math. J., 25(1973), 225-235.

[5] K.Nomizu, On the decomposition of generalized curvature tensor fields, Differential geometry in honor of K.Yano, 1972, 335-345.

[6] K.Sekigawa, On some 4-dimensional compact almost Hermitian manifolds, J. Ramanujan Math. Soc., 2(2)(1987), 101-116.

[7] I.M.Singer and J.A.Thorpe, The curvature of 4-dimensional Einstein spaces, Global analysis, Papers in honor of K.Kodaira, Princeton University Press, Princeton, 1969, 355-365.

[8] S.Tanno, 4-dimensional conformally flat Kähler manifolds, Tôhoku Math. J., 24(1972), 501-504.

[9] F.Tricerri and L.Vanhecke, Curvature tensors on almost Hermitian manifolds, Trans. Amer. Math. Soc., 267(1981), 356-398.

[10] F.Tricerri and I.Vaisman, On some 2-dimensional Hermitian manifold, Math. Z., 192(1986), 205-216.

[11] L.Vanhecke, The Bochner curvature tensor on almost hermitian manifolds, Geometriae Dedicata, 6(1977), 389-397.

[12] ———— , The Bochner curvature tensor on almost hermitian manifolds, Rent. Sem. Mat. Univer. Politecn. Torino, 34(1975-1976), 21-38.

Shinkawa 5-6-6-406

Mitaka-shi, Tokyo, Japan

The Structure of the Symmetric Space with Applications

By Tadashi Nagano and Makiko Sumi

Introduction. We wish to discuss new aspects of the structure theory of the compact symmetric spaces, as opposed to the classification theory. It is expected that the theory will make it easier to solve some types of problems on symmetric spaces, as illustrated by applications we will explain. Among them is the problem of finding all the totally geodesic spheres in the compact symmetric spaces (§ 3).

The theory of symmetric spaces has a long history since E.Cartan, and the structure has been studied by many people. This article is partly a survey; we will include our own results [CN-1,2] and [N-1,2]. Briefly, our idea is to use the connected components of the fixed point set $F(s_o, M)$ together with their "orthogonal complements" as the building blocks of the structure of the symmetric space; here s_o is the point-symmetry of M at a point o (Definition. 1.1).

Proofs will be omitted for most statements; see [H] and [KNo] for a general reference.

1. The framework of the theory.

We first recall basic facts and definitions on symmetric spaces and we will explain results in [CN-1,2] and [N-1,2] without proofs.

ISBN 0-12-640170-5

1.1 **Definition.** A symmetric space M is a smooth manifold with the point-symmetry s_x for every point x of M such that (1) the smooth map s_x: $M \to M$ is involutive ($s_x{}^2 = 1_M$), (2) the point $\{x\}$ is open in the fixed point set $F(s_x, M) = \{y \in M; s_x(y) = y\}$ of s_x and (3) there is an affine connection ∇ which is left invariant by all the point symmetries s_x, $x \in M$.

1.2 Remarks. (0) Examples of symmetric spaces include the euclidean space, the sphere and the other space form, the projective spaces, the Grassmannian manifolds and the compact Lie groups (1.6 ①), which justify through studies of symmetric spaces. (i) The invariant connection ∇ is unique, since s_x acts as -1 times the identity on the tangent space T_xM for every point x of M. (ii) The condition (3) in the above definition is naturally void on each zero dimensional component of M. (iii) If every point symmetry preserves a pseudo-Riemannian structure, then ∇ is its Levi-Civita connection by (i)
and M is called a pseudo-Riemannian symmetric space; similarly it is called Riemannian when the point symmetries preserve a Riemannian metric. A compact symmetric space is necessarily Riemannian.

1.3 **Definition.** A smooth map f: $M \to N$ of a symmetric space into another is called a homomorphism if f commutes with the point symmetries; i.e. if one has $f \circ s_x = s_{f(x)} \circ f$ for every point x of M.

1.4 Remarks. (i) Obviously the symmetric spaces and the homomorphisms make a category, on which we will work; thus a subspace means a symmetric space contained in another whose inclusion map is a homomorphism. The fixed point set $F(t, M)$ of an automorphism t of M is a subspace, for example. (ii) In case M is connected, f: $M \to N$ is a homomorphism if and only if f is totally geodesic. (iii) If M is connected, then every point symmetry s_x is an automorphism; hence the automorphism group $Aut(M)$ is transitive on M.
Its identity component $G := Aut(M)_{(1)}$ is also transitive in this case.

1.5 **Proposition.** An automorphism of a subspace M of M extends to that of M if it lies in $Aut(M\ )_{(1)}$.

to that of M if it lies in Aut(M $)_{(1)}$.

1.6 **Proposition.** Let f: M → N be a homomorphism. Then the image f(M) of a connected subspace M of M is a subspace of N, and the inverse image f^{-1}(N) of a subspace N of N is a subspace of M. In particular every geodesic has no self-intersection (i.e. it is homeomorphic with either a line or a circle).

1.7 Problems. A fundamental problem is to determine all the symmetric spaces and all the homomorpisms between them. All the connected Riemannian symmetric spaces are known, and the other connected symmetric spaces are known at least on the level of the Lie algebra [BM]. Many works have been done for homomorphisms, and we intend to explain a few results of ours. A basic and most important special case is the study of the homomorphisms of the space O(1) = {1, − 1} of just two points into a compact connected symmetric space M ([C], [CN-2], [N-1] and [N-3]). The rest of this section will be devoted to some results along this line.

1.7a In a simplest case the domain of the definition may be a finite set Σ whose point symmetries are all the idenitity map of Σ. One could ask the greatest possible cardinality of Σ from which there is a monomorphism of Σ into a given connected symmetric space M [CN-3]. We call the cardinality the 2-number of M and denote it by $\#_2M$. An obvious meaning of $\#_2M$ is that it is an obstruction to an existence of a monomorphism of a connected space M into another space N, since one has $\#_2M \leqq \#_2N$ in case it exists. The 2-number has something to do with the space; in fact, if M is hermitian, then $\#_2M$ equals the Euler number of M, for example, and hence every hermitian symmetric subspace of a hermitian symmetric space has a smaller Euler number. If M is a symmetric R-space, then $\#_2M$ equals dim H(M, Z_2), the dimension of the homology group of M over the field Z_2, as Takeuchi has recently proven
[T-2]. 〔Thus it equals the infimum of the absolute total curvature of M for the immersions into the Cartesian spaces in the sense of Chern-Lashof. 〕 But, if M is a compact Lie group (1.16 ①), then $\#_2M$ equals the r_2(M)-th power of 2, where r_2(M) is the 2-rank of M (in the sense ofBorel-Serre).

Determination of all the homomorphisms: $O(1) = \{1, -1\} \rightarrow M$ amounts to finding the fixed point set $F(s_o, M) = \{x \in M;\ s_o(x) = x\}$ of s_o acting on M, or rather its connected components. Every homomorphisms: $O(1) \rightarrow M$ extends to a homomorphism: $U(1)$=circle $\rightarrow M$ if M is connected.

1.8 **Definitions.** Every connected component of the fixed point set $F(s_o, M)$ is called a polar of the point o in M, whether or not M is connected. We denote by $M^+(p)$ the polar which contains the point p; see [C],[CN-2], and [N-1].

1.9 Remarks. The concept of the polar is important in the study of compact symmetric spaces. To quote a few examples, ① a certain type of symmetric spaces (including R-spaces) admit stratifications by disc bundles over the polars, since those spaces admit Morse functions whose critical points are exacty the points on the polars, which make the nondegenerate critical submanifolds [N-2]. [The Morse function is actually a spherical function, and therefore it gives rise to a G-equivariant embedding of M into a Cartesian space and would give the infimum of the total absolute curvature mentioned in (1.7a) which is due to Kobayasi-Takeuchi.] This suggests that the polars have a close relationship with the whole space M. Indeed ② the Euler number, say, of M equals the sum of those of the polars if s_o is homotopic to the identity (in general, the sum equals the Lefschetz number of course). Also easy to see is the fact that ③ M is orientable if and only if all the polars are even dimensional. Much less easy to see is the fact ④ that the topological index (signature) of M is the sum of those of the polars too if the polars are appropriately oriented;this is a non-trivial consequence of the Atiyah-Singer index theorem. There is a strange fact: $\mathrm{ind}\ (G_r(\mathbb{C}^m)) = \chi\ (G_r(\mathbb{R}^m))$, which may be proven by means of the Hodge index theorem. Since this last theorem has a quaternion-Kähler analog [NT], one has a similar strange fact too.

The polars have been completely determined together with the closely related meridians to be defined below; see [CN-1] and [N-1].

Theoretical importance of these concepts lies in the theorem:

any pair (M^+, M^-) of a polar and a meridian for it in a compact symmetric space M determines M (globally); thus it can replace the root system R(M) and has more information. The table of all the congruence classes of the pairs (M^+, M^-) are more convenient for many practical purposes.

1.10 Definition. By <u>the meridian</u> through the point p of $M^+(p)$, denoted by $M^-(p)$, we mean the connected component, $F(s_p \circ s_o, M)_{(p)}$, of $F(s_p \circ s_o, M)$ through p. The tangent space $T_pM^-(p)$ to $M^-(p)$ at p is the orthogonal complement to $T_pM^+(p)$ in T_pM. The <u>G-congruence</u> type of $M^-(p)$ is independent of the choice of the point from $M^+(p)$; that is, the meridians through the points of $M^+(p)$ are carried into each other by the action of G, because each polar $M^+(p)$ is an orbit of the isotropy subgroup K of G at o. We call two subspaces of M are <u>congruent</u> if some member of Aut(M) carries one onto the other.

1.11 **Proposition.** Every meridian $M^-(p)$ has an equal rank to that of M; i.e. a maximal torus of $M^-(p)$ (which is its subspace) is a maximal torus in M.

The root system $R(M^-(p))$ of $M^-(p)$ is read off from R(M) with a certain simple algorism operating on the Dynkin diagram (See Proposition 2.4).

1.12 **Propostition.** The meridian $M^-(p)$ meets every polar $M^+(q)$ including {o} if M is connected.

An immediate corollary: given two involutive members of a compact Lie group which are lying in the same connected component of the group, one commutes with some conjugate of the other.

1.13 Remark. The meridian $M^-(p)$ represents the position of M in the local isomorphism class of M in the following sense. Let π be a covering morphism (viz. a covering map which is a homomorphism): M $\rightarrow$ M of a connected symmetric space M. Then π extends to an exact sequence of homomorphisms: $\{o\} \rightarrow \Gamma \rightarrow M \rightarrow M \rightarrow \{\pi(o)\}$, $o \in M$. This gives rise to another exact sequence: $\{1\} \rightarrow \pi_1(M) \rightarrow \pi_1(M) \rightarrow \Gamma \rightarrow \{1\}$ which contains the fundamental group $\pi_1(M)$. The inverse image Γ of π (o) is not only a subspace but an abelian group having the unit at o. Now restricting π to the meridian $M^-(p)$ gives the exact sequence: $\{o\} \to \Gamma \to M^-(p) \to M^-(\pi(p)) \to \{\pi(o)\}$; one sees $M^-(\pi(p)) = \pi(M^-(p))$ but the point is that we have the same Γ in this sequence.

1.14 **Definition.** A polar $M^+(p)$ of o is called a pole if it is a singleton {p}. By the centrosome C(o,p) for the pair of the point o and its pole p we mean the subset consisting of the midpoints of the geodesic arcs from o to p. C(o,p) may be disconnected.

1.15 **Proposition.** (i) A centrosome C(o,p) is a subspace. (ii) If p is a pole of a point o of M, then there is a unique double covering morphism $\pi : M \to M$ which carries o and p into a single point π (o); and vice versa. (iii) Given the double covering morphism $\pi : M \to M$ with $\pi(o) = \pi(p)$, the polars of M are the π -images of those of M together with the connected components of π (C(o,p)) . (iv) Under any covering morphism $\pi : M \to M$ with $\pi(o) = \pi(p)$, the polars of M project onto those of M .

1.16 Examples. ① The orthogonal group O(m) is a compact symmetric space M as well as any compact Lie group by the point symmetry $s_x(y) = xy^{-1}x$. The fixed point set $F(s_1, O(m))$ is the totality of the involutive members x, $x^2 = 1$. If the trace Tr(x) equals m $-$ 2r, then the polar $M^+(x)$ is isomorphic with $G_r(\mathbb{R}^m)$:= the real Grassmann manifold of the r dimensional linear subspaces of $\mathbb{R}^m$, $0 \leqq r \leqq m$. The corresponding meridian $M^-(x)$ is isomorphic with $SO(r) \times SO(m-r)$. Notice that the isotropy subgroup at the origin 1 of (the identity component of) Aut(M), the automorphism group of the space M = O(m), acts transitively on each polar, which is a general fact; in particular, the congruence class of the meridian $M^-(p)$ for the polar $M^+(p)$ is independent of the point p within $M^+(p)$. ② If M is the unitary group U(m), then clearly the polars are the complex Grassmannian manifolds $G_r(\mathbb{C}^m)$, $0 \leqq r \leqq m$, with the corresponding meridians $\cong U(r) \times U(m-r)$.Notice however that the polars of SU(m) are only those spaces $G_r(\mathbb{C}^m)$ for even r; and similarly for the subgroup SO(m)

of O(m). K. Uhlenbeck, J. Wood, and others showed that a harmonic map: $S^2 \to U(n)$ of the 2-sphere into the unitary group, for instance, is factored into the product of harmonic maps of S^2 into polars of U(n). ③ It goes without saying that the polars of the symplectic group Sp(m) are the quaternion Grassmannians $G_r(H^m)$ with the meridians $Sp(r) \times Sp(m-r)$. But the projection π : Sp(m) → ad(Sp(m)) onto the adjoint group $Sp(m)/\{\pm 1\}$ carries the polars $G_r(H^m)$ and $G_{m-r}(H^m)$ onto a single polar in ad(Sp(m)) = $Sp(m)/\{\pm 1\}$. Hence there must be other polars in $Sp(m)/\{\pm 1\}$, for otherwise the sum of the Euler numbers of the polars would be less than the Lefschetz number 2^m of s_1. The missing polar comes down from the centrosome $C(1, -1) \cong CI(m) := Sp(m)/U(m)$. The meridian $Sp(r) \times Sp(m-r)$ goes down to $Sp(r)\cdot Sp(m-r)$ or
$(Sp(r) \times Sp(m-r))/\{\pm 1\}$. ④ If $M = G_r(\mathbb{C}^m)$, then the polars are $G_a(\mathbb{C}^r) \times G_b(\mathbb{C}^{m-r})$, a+b=r, with the meridians $G_a(\mathbb{C}^{m-2b}) \times G_b(\mathbb{C}^{2b})$. One can find the Euler number of $G_r(\mathbb{C}^m)$ from this quickly. ⑤ If M is a projective space, then a hypersurface is the only polar of a point o and the meridian is "a line" that is a sphere of dimension 1, 2, 4 or 8 according as M is a real, a complex, a quaternion or a Cayley projective space. In general, a homomorphism f: L → M carries an arbitrary pair (L^+, L^-) of a polar and the meridian for it into a counterpart (M^+, M^-), one can determine all the subspaces of the projective spaces immediately.
⑥ If M^+ is a pole of o in M, one has a one-to-one correspondence:
b{o and its poles} ↔ $b(M^+)$, $b \in G$, that is a generalization of the duality in projective geometry. Also, given an involution t of M, denote by M^t a connected component of the fixed point set F(t, M) and one has the set
$\{ b(M^t) \mid b \in G \}$, which is another symmetric space. These lines of geometry is yet to be pursued.
⑦ The space of the oriented planes in $\mathbb{R}^m$, denoted by $G^\circ_2(\mathbb{R}^{m+2})$, is a complex quadratic hypersurface. The maximal subspaces are the polar
$G^\circ_2(\mathbb{R}^m)$, $S^p \times S^q$, p+q=m, and, if m=2n is even, the complex projective space of dimension n [CN-1]. Contrary to this example, the maximal subspaces are not necessarily connected components of the fixed point sets of involutions, in general.
⑧ The stratification mentioned earlier (1.9 ①) is dual to the one made

out of the cut loci (which was found by Sakai and Takeuchi [T-3]) in some sense.

⑨ The symmetric R-spaces which may be defined as the fixed point sets of sort of complex conjugations of hermitian symmetric spaces form a distinguished subcategory of that of all the symmetric spaces. Their classification gives that of the graded simple Lie algebras $\mathfrak{l} = \mathfrak{l}^{(-1)} + \mathfrak{l}^{(0)} + \mathfrak{l}^{(1)}$ (See [KN]). Similarly the classification of the graded simple Lie algebras $\mathfrak{l} = \mathfrak{l}^{(-2)} + \mathfrak{l}^{(-1)} + \mathfrak{l}^{(0)} + \mathfrak{l}^{(1)} + \mathfrak{l}^{(2)}$ can be done by using quaternion Kähler symmetric spaces and fixed point sets of involutions, as J.H. Cheng did.

2. The curvature and the root system.

In this section M denotes a compact connected symmetric space. The curvature K appears in the Jacobi equation, as is well known:

2.1 $$\nabla_{\dot{c}} \nabla_{\dot{c}} v + K(v, \dot{c})\dot{c} = 0,$$

where c is a geodesic and v is a vector field along c. By the fortunate formula $\nabla K=0$, this is a linear differential equation with constant coefficients. Since the linear map: $v \mapsto K(v, \dot{c})\dot{c}$ is symmetric, the space of the solutions v is decomposed into the direct sum of its eigenspaces. Each eigenvalue $\alpha(\dot{c})^2$ is nonnegative, since the sectional curvature SK is so. On each eigenspace one has

2.2 $$(d/dt)^2 v + \alpha(\dot{c})^2 v = 0.$$

To understand these observations on a single geodesic in the context of the whole space, one sees that c is contained in a maximal torus A in M, and one recalls [H] that A is congruent with every other maximal torus and its dimension r(M) is called the <u>rank</u> of M. We write $\mathfrak{a}$ for the tangent space T_oA, which is a vector subspace of $\mathfrak{m} = T_oM$. Given a vector H in $\mathfrak{a}$, one has a unique gedesic c(t) = Exp(tH), for which one has the Jacobi equation 2.2. The map: $H \mapsto \pm\alpha(H)$ is linear. The set, R(M), of these linear forms $\alpha \neq 0$ is a <u>root</u> system in the usual sense [B].

The members of the Lie algebra $\mathfrak{g}$ of G, thought of as vector fields on M and restricted to a gedesic, are solutions of 2.1. And $\mathfrak{g}$ is the direct sum of the subspaces $\mathfrak{g}(\alpha)$ whose members satisfy 2.2. Moreover the point symmetry s_o, $o \in A$, preserves $\mathfrak{g}(\alpha)$ and hence $\mathfrak{g}(\alpha)$ is the direct sum of the eigenspaces $\mathfrak{m}(\alpha)$ and $\mathfrak{k}(\alpha)$ for the eigenvalues -1 and 1 of ad(s_o). Thus we have the decomposition

2.3 $\quad \mathfrak{g} = \mathfrak{k} + \mathfrak{m}, \quad \mathfrak{k} = \mathfrak{k}(0) + \Sigma\, \mathfrak{k}(\alpha)$ and $\mathfrak{m} = \mathfrak{a} + \Sigma\, \mathfrak{m}(\alpha)$,

where the summation ranges over the roots. The first one, $\mathfrak{g} = \mathfrak{k} + \mathfrak{m}$, is called <u>the symmetry decomposion of $\mathfrak{g}$ at the point o</u>. The dimension of $\mathfrak{m}(\alpha)$ is <u>the multiplicity of the root</u> α, denoted by $m(\alpha)$; one notes $\dim \mathfrak{m}(\alpha) = \dim \mathfrak{k}(\alpha)$.

One has obtained the root system R(M), $\mathfrak{a}$ and the multipicity m from the curvature K. The converse is true in two ways ; i.e. first every root system is that of some symmetric space (in contast with the Lie algebras) and second R(M), $\mathfrak{a}$ and m determine K and hence the local structure of M (if M is semisimple as defined below). Thus R(M) with the multipicity carries the same information as K. M is called <u>semisimple</u> $\Leftrightarrow$ R(M) spans $\mathfrak{a}$ (identified with its dual space) $\Leftrightarrow$ $\mathfrak{g}$ is semisimple $\Leftrightarrow$ $[H, \mathfrak{m}] = 0$, $H \in \mathfrak{a}$, implies $H=0$ $\Leftrightarrow$ the de Rham decomposition of M in terms of the holonomy group has no flat part. (The definition also applies to the noncompact case.) If M is a compact connected semisimple symmetric space, then there is a space N such that every connected locally isomorphic space to M is a covering space of N. We call N <u>the bottom space</u> of M and denote N by M^{bottom}, abbreviated to M^{bott} or M^{*}. M^{*} is called the adjoint space in [H], since M^{*} is the adjoint group for a group M. A semisimple space M is called <u>simple</u> $\Leftrightarrow$ M^{*} is not the product of two spaces of positive dimensions $\Leftrightarrow$ R(M) is irreducible.

2,4 **Proposition.** Let M be a simple space. Then the root system $R(M^-)$ of a meridian $M^- \neq M$ is obtained from the Dynkin diagram of R(M) as follows. Express the highest root as a linear combination $\Sigma\, p^i \alpha_i$ of the fundamental roots α_i. There are two cases. Either 1° M^- is a local product of a circle and a semisimple space whose Dynkin diagram is obtained from that of M by deleting the vertex corresponding to $p^i = 1$ or 2° M^- is semisimple and its Dynkin diagram is obtained from the extended Dynkin diagram of M by deleting the vertex corresponding to $p^i = 2$. The multiplicity of each roots is preserved in both cases. The converse is true for the bottom space M^{*}.

Corollary. A meridian is a maximal connected subspace.

The homomorphism $Q: M = G/K \to G: x \mapsto s_x \circ s_o$ defined for a point o may be used to show that every compact connected symmetric space M is a subspace of a covering group of G, since $Q: M \to Q(M)$ is a covering morphism. M^{*} is a subspace of G^{*}. S^n is a subspace of Spin(n+1), the universal covering group of SO(n+1), and not a subspace of SO(n+1), n>1, except when n = 4 or 8. The classification of compact connected symmetric spaces is the determination of the polars of compact Lie groups as a consequence.

The linear subspace $\mathfrak{m}(\alpha) + R\alpha$ of $\mathfrak{m}$ has an interesting geometric properties. The Grassmannian $G_2(\mathfrak{m}(\alpha) + R\alpha)$ of the 2-planes consists of critical points of the sectional curvature $SK:G_2(\mathfrak{m}) \to R$ of M. There is a subspace $\$(\alpha)$ of constant curvature to which $\mathfrak{m}(\alpha) + R\alpha$ is the tangent space at o if 2α is not a root. The sectional curvature of $\$(\alpha)$ depends on the length of α in a simple way. We call $\$(\alpha)$ <u>the root-ψ-sphere for the root</u> α; more generally, we mean by a <u>ψ-sphere</u> a sphere or a real projective space. If 2α is a root, the linear subspace $\mathfrak{m}(2\alpha) + \mathfrak{m}(\alpha) + R\alpha$ is the tangent space to a projective space at o; more precisely it is a complex, quaternion or Cayley projective space respectively if the multiplicity $m(2\alpha)$ is 1, 3 or 7 respectively. The configuration of the the root-ψ-spheres is in an exact geometric correspondence with the root system; the sectional curvature of $\$(\alpha)$ is determined by the length of α and the angle between two roots α and β equals the angle between $\$(\alpha)$ and $\$(\beta)$. Also it is closely related to the differential geometry of M; for instance, the sectional curvature of $\$(\alpha)$ for a longest root α is the maximum of the sectional curvature of M [H].

2.5 Remark on history. Chow [CWL] proved that, if one defines a (discrete) distance d(x,y) for the linear subspaces x and y of $\mathbb{C}^n$ of fixed dimension r by $d(x,y) = \dim x/(x \cap y)$, then the isometries with respect to d are holomorphic or antiholomorphic transformations of $G_r(\mathbb{C}^n)$. This theorem may be forumulated in terms of root- ψ -spheres as follows. Let $\$(\alpha)$ be the root- ψ -sphere for a longest root. Let Σ denote the set of all the ψ -spheres $b\$(\alpha)$ for the members b of G. If $\$(\beta)$ is the root- ψ -sphere for another longest root β, then $\$(\beta)$ is a member of Σ. Let D(x,y) denote the minimum number of the root- ψ -spheres in Σ with which the two points in $G_r(\mathbb{C}^n)$ can be joined with each other in such a way that the first one contains x, the adjacent ones meet each other and the last one contains y. Then one can prove d=D. Since the definition of D makes sense for every compact connected semisimple space, one may well ask if Chow's is true for any such space. S.Peterson generalized it to certain spaces and later M.Takeuchi did it to the symmetric R-spaces [T-1] by using a theorem of N. Tanaka. Incidentally, by taking the sets $\{ x \in M \mid D(o,x) < p \}$ for $0 \leqq p \leqq \mathrm{Max}(D)$, one obtains a stratification of M which coincides with the one mentioned earlier.

The root system can be defined in a more general setting, namely the root system based on the maximal abelian subalgebra $\mathfrak{a}^t$ in the tangent space T_oM^t to $M^t = F(t, M)$ for an involution t, t(0) = o. We do not go into the details about it, but we just mention that a few more Satake diagrams are needed to describe the situation [OS].

3. The involutions and the affine symmetric spaces.

One can derive various conclusions from the studies of the structure of the compact symmetric spaces in the previous sections. We determined the fixed point set for every involution [N-1,3] for one thing. In short, if t is an involution of M which fixes a point o, then the connected component M^t of the fixed point set F(t, M) through o meets polars $M^+(p)$ of o in M at polars $(M^t)^+$ of o in M^t obviously, and those polars

$(M^t)^+$ in $M^+(p)$ make the fixed point set $F(t, M^+(p))$. Thus one finds all the involutions t and their fixed point set by induction on the dimension of M. As to the other components of F(t, M), one may use the following proposition which generalizes (1.12): "the orthogonal space" $F(t \circ s_o)_{(o)}$ to M^t at o meets every component of F(t, M) (at the polars of o in it).

Once t and F(t,M) are all determined, one can quickly classify all the affine symmetric spaces which are connected and have semisimple automorphism groups; the results are global in contrast with [BM].

Let T denote the functor of assigning the tangent bundle T M to a manifold M and the differential T f to a smooth mapping f.

3.1 **Proposition.** If M is an affine symmetric space, then so is T M functorially.

Proof. Immediate; in fact we define the point symmetry at X in $T_x M$ simply by $T s_x$ if X = 0 and otherwise s_X is so defined that T G becomes the connected automorphism group where G is that of M. The invariant affine connection is the tangential to the one for M ([KS] p.150).

3.2 Remark. T M is not semisimple, since T G has the normal subgroup $T_1 G$ which is abelian.

There is another way of making T M a symmetric space. We begin with the bottom space $M^* = G^*/K$. By extending the involution of G^* to its complexification G^{*C}, we obtain $M^{*C} = G^{*C}/K^C$ which may be identified with $T M^*$ and has a natural structure of affine symmetric space. The projection of M onto M^* pulls back the symmetric space structure on T M. Let us denote this symmetric space T M by M^C. For every subspace B of M, T B is a subspace of T M. The normal bundle N B = N (B, M) is another subspace of T M. For every affine symmetric space M^R of semisimple type, there is an involution t of "the compact form" M of M^R such that M^R is a covering space of N B where B is a connected component of the fixed point set F(t, M). This way, the affine symmetric spaces of semisimple type are globally classified.

There is still another way of making T M a symmetric space if M is a symmetric R-space. Thus M admits a Lie transformation group L which contains G as a lower dimensional subgroup. L contains a subgroup $L^{(0)}$ (Cf, 1.16 ⑨) such that $L/L^{(0)}$ may be identified with T M (or rather the cotangent bundle) [N-2] and $L/L^{(0)}$ has a natural structure of symmetric space; this space $L/L^{(0)}$ is isomorphic with $M^{\mathbb{C}}$ in the special case of a hermitian space M. The interest is in that N B as above is now thought of as an open subset of M and the connected automorphism group of N B is a subgroup of L, a fact which generalizes the Borel and Harish-Chandra embedding of a hermitian symmetric space of noncompact type as a bounded domain. The work is still in progess in cooperation with Jiro Sekiguchi.

4. The spheres in the symmetric spaces.

The problem here is to determine all the monomorphisms of the spheres into the connected symmetric spaces M; that is, to determine all the spheres S^m which are totally geodesic submanifolds of M and have constant sectional curvature with respect to the induced metric, up to congruence. We would like to report the results [NS] by way of illustrating our theory.

We may assume that M is compact and that S^m is a maximal sphere (with respect to inclusion) which is a subspace. The case m=0 was already discussed; the problem was then the determination of the polars. The case m=1 was done by E.Cartan himself. Dynkin [D] made a significant contribution to the case where m=3 and M is a group manifold. We will later mention what Wolf did after Y.C. Wong.

Our method, another application of our basic method, is simple and geometric, as we feel; in particular Wolf's index can be explained in a geometric way.

Let S^m be a sphere which is a subspace of M. Let o be a point on S^m. The pole p of o in S^m lies on a polar $M^+(p)$ in M. It is quite easy to see that S^m is contained in the meridian $M^-(p)$ (and p is a pole of o in $M^-(p)$ too). Thus the study is reduced to the case of the space of lower dimension than M unless p is a pole in M. We will therefore concentrate on this exceptional case. We denote by γ the corresponding covering transformation (= the free involutive automorphism which carries o into p) in order to avoid the possible confusion caused by the existence of more than one poles of o.

4.1 **Theorem.** (i) Assume the pole p of a point o in a sphere S^m is a pole γ (o) in M. Then S^m meets the centrosome C(o,p) in the equator S^{m-1} and the G-congruence class of S^m corresponds to that of S^{m-1}; (ii) the pole of a point in S^{m-1} is a pole in C(o,p). Conversely, if S^{m-1} is a sphere in C(o,p) such that (1) S^{m-1} contains the pole γ (x) of every point x on it and (2) its diameter equals the distance $d_M(o,p)$ and if $d_M(o,p)$ equals twice the distance between o and the component $C(o,p)_{(x)}$ through a point x on S^{m-1}, then there exists a sphere S^m which contains o, p, and S^{m-1}.

A few comments may be due. The values of the diameter (equivalently, the sectional curvature) of the spheres in M are finite in number. The theorem says two S^m's in M (satisfying the condition) are G-congruent if and only if their intersections with the centrosome C(o,p) are G-congruent with each other. The converse part is proven by means of the classical conjugate point arguments for geodesics. The theorem tells how to classify all the spheres of the samallest size, which constitute the absólute majority of the spheres fortunately. Wolf [W] determined the spheres in the Grassmann manifolds $G_r(R^n)$, $G_r(C^n)$ and $G_r(R^n)$ satisfying 2r = n which satisfy the pole condition in the theorem, all of which turned to be of the smallest size.

We will briefly explain how the theorem works. Take $G_r(\mathbb{C}^{2r})$, for example. There is a unique pole γ (x) of a point x in it. The centrosome is isomorphic with U(r). Also $G_r(\mathbb{C}^{2r})$ is a component of the centrosome C(1, − 1) in U(2r); and indeed the only one component that contains the pole γ (x) of a point x in it. Thus we have the sequence of monomorphisms:

$U(k) \to G_k(\mathbb{C}^{2k}) \to U(2k) \to \cdots \to U(r) \to G_r(\mathbb{C}^{2r}) \to U(2r) \to \cdots,$

where k is odd. Suppose S^m is a sphere in U(2r) which conains 1 and − 1. Then S^m meets $G_r(\mathbb{C}^{2r})$ in S^{m-1}, which meets in turn U(r) in S^{m-2} and so on, until the sphere meets a unitary group U(m) in a circle S^1, never in S^2 (See Lemma 4.2 below). Conversely, choose S^1 of the smallest size in a unitary group to the left of the space in question within the above diagram, go right following the arrows and increasing the dimension of the sphere by one each time one passes the arrow till the sphere enters the target space, and you will have a desired sphere in the space U(r) or $G_r(\mathbb{C}^{2r})$. This way, one obtains the maximal spheres of the smallest size and containing a point o along with its pole γ (o). Their congruence classes are distinguished by the number m of U(m) which the sphere meets in S^1. Furthermore these classes depend on which pair of components of the centrosome this S^1 meets. This is Wolf's index. Thus their congruence classes are classified by m and Wolf's index.

There is a fantastic sequence of inclusions of centrosomes: O $\to G_r(\mathbb{R}^{2r}) \to$ UI $\to$ CI $\to$ Sp $\to G_r(\mathbb{H}^{2r}) \to$ U II $\to$ O III $\to$ O, where the orthogonal group O(r), the number r suppressed, is the centrosome of the real Grassmann $G_r(\mathbb{R}^{2r})$, next $G_r(\mathbb{R}^{2r})$ is the only relevant component of the centrosome of UI(2r) := U(2r)/O(2r), UI(r) that of CI(r) := Sp(r)/U(r), CI(r) that of the symplectic group Sp(r), Sp(r) that of the quaternion Grassmann $G_r(\mathbb{H}^{2r})$, $G_r(\mathbb{H}^{2r})$ that of U II (2r) := U(4r)/Sp(2r), U II (r) that of O III (2r) := O(4r)/U(2r), and O III (r) that of SO(2r). Using this cyclic sequence of 8 spaces, one can determine the congruence classes of the maximal spheres in the theorem in the same way. [One might remember the beautiful Bott periodicity of the stable homotopy groups of these spaces, based on the same geometric ground.] Notice that most classical spaces appear in the two sequences; to be more precise, there one finds all the local isomorphism classes of the classical spaces which have poles.

Consider SU(n) to get some idea of what the other spheres are. In a preliminary step, one can prove the next general fact for groups.

4.2 **Lemma.** Let M be a compact Lie group. Then (i) every maximal sphere in M has at least 3 dimensions if its dimension is greater than one; and (ii) every 3 dimensional sphere in M is congruent with a subgroup SU(2) of M.

Let S^m be a sphere in SU(n) which passes through 1. If n is odd, S^m is contained in some $SU(2k) \times SU(n-2k)$, $1 \leqq 2k \leqq n$, the semisimple component of the meridian for some polar of 1 in SU(n); it must be obvious that S^m is indeed contained in SU(2k). If n is even, there are spheres which pass through 1 and the pole -1. Here we consider the spheres among them which are maximal and not of the smallest size. Such a sphere is necessarily a three-dimensional, and hence it is congruent with a subgroup SU(2). An example is the "principal" subgroup SU(2) (See [D] or Chap. 8 of [B]). The principal subgroup SU(2) is not contained in any meridian other than SU(n) itself, as one sees easily by means of representation theory. Generally, such a sphere SU(2) is contained in a meridian and an induction argument produces all these spheres; thus such an SU(2) corresponds to an arbitrary decomposition of the number n into the sum $\Sigma_i n_i + p$ of even numbers $n_i > 2$ and a nonnegative even number p in such a way that the sphere SU(2) is contained in the product $\Pi_i SU(n_i) \times SU(p) \subseteq SU(n)$ and "the diagonal" in the product of the principal subgroups of $SU(n_i)$ and a sphere in SU(p) which is of the smallest size; the converse is true. We would like to explain what Theorem 4.1 implies as to the maximal spheres of the smallest size in SU(n) which contain $\{\pm 1\}$. The centrosome is the disjoint union of the complex Grassmannians $G_p(\mathbb{C}^n)$, $p + n/2$ even, which is quite an elementary fact. The Grassmannians $G_p(\mathbb{C}^n)$ has a pole if and only if $2p = n$. Since the spheres in question have dimensions at least three by Lemma 4.2, they meet the centrosome only in $G_p(\mathbb{C}^n)$ for $p = n/2$, and the cyclic reccurrence starts as mentioned earlier, which has a happy ending. This way one obtains a complete classification of the congruence classes of the spheres in SU(n). We summarize as a proposition what we have explained about SU(n).

4.3 **Proposition.** Let S^m be a maximal sphere in SU(n) which contains 1. Then S^m falls into one of the following three cases. Case 1° S^m is contained in some $SU(2k) \subseteq SU(2k) \times SU(n-2k) \subseteq SU(n)$, $1 \leqq 2k < n$, and S^m contains the pole of 1 in SU(2k). 2° n is even and S^m is a sphere of the smallest size which passes through the pole -1. Such an S^m is classified with the method discussed earlier. (One can get the result quickly by raising by one the dimensions of the spheres in $G_{n/2}(\mathbb{C}^n)$ on Wolf's table [W] if one wishes.) 3° The sphere is congruent with SU(2) in the product $\Pi_i SU(n_i) \times SU(p) \subseteq SU(n)$ such that the projection of SU(2) into $SU(n_i)$ is principal and the one into SU(p) is a sphere of the smallest size and contains -1 in SU(p), $n = \Sigma_i n_i + p$, n_i even > 2 and p even.

The task is not too difficult for all the other spaces including the locally isomorphic groups to SU(n), except that we have to rely on Dynkin's result [D] for the case of the group E_7.

Bibliography

[B] N. Bourbaki, Groupes et Algèbres de Lie, Chap. 4 through 8.

[BM] M. Berger, Les espces symétriques non compacts. Ann. Sci. Ecole Norm. Sup. 74(1957), 85-177.

[C] B.Y. Chen, A New Approach to Compact Symmetric Spaces and Applications. Katholicke Universiteit Leuven (1987).

[CN-1] B.Y. Chen and T. Nagano, Totally geodesic submanifolds of symmetric spaces, I. Duke Math. J. 44(1977), 745-755.

[CN-2] B.Y. Chen and T. Nagano, Totally geodesic submanifolds of symmetric spaces, II. Duke Math. J. 45(1978), 405-425.

[CN-3] B.Y. Chen and T. Nagano, A Riemannian invariant and its applications to a problem of Borel and Serre. Trans. Amer.Math. Soc.308(1988), 273-297.

[CWL] W.L. Chow, On the geometry of homogeneous spaces. Ann. of Math. 50(1949), 32-67.

[D] E.B. Dynkin, Semisimple subalgebras of semisimple Lie algebras. Mat. Sb. 30(1952), 349-462.

[G] Gelfand-MacPherson, Geometry in Grassmannians and a generalization of the dilogarithm. Adv. in Math. 44(1982), 279-312.

[H] S. Helgason, Differential Geometry, Lie Groups, and Symmetric Spaces. Academic Press. 1978.

[K] S. Kaneyuki, On orbit structure of compactification of para-hermitian symmetric space. Janan. J. Math. 13(1987), 333-370.

[KN] S. Kobayashi - T. Nagano, On filtered Lie algebras and geometric structures, I. J. Math. Mech. 13(1964), 875-908.

[KNo] S.Kobayashi- K. Nomizu, Foundations of Differential Geometry, II. Interscience Publishers. 1969.

[KS] S.Kobayashi, Theory of connections. Annali di Mat. 43(1957), 119-194.

[N-1] T. Nagano, The involutions of compact symmetric spaces. Tokyo J. of Math. 11(1988), 57-79.

[N-2] ------, Transformation groups on compact symmetric spaces. Trans. Amer. Math. Soc. 118(1965), 428-453.

[N-3] T. Nagano, The involutions of compact symmetric spaces, II. (In preparation.)

[NS] T. Nagano - M.Sumi, The spheres in symmetric spaces as subspaces. (In preparation.)

[NT] T. Nagano - M. Takeuchi, Signature of quaternionic Kaehlerian manifolds. Proc. Japan Acad., 59(1983). 384-386; Cohomology of quaternionic Kaehlerian manifolds. J. Fac. Sci., Univ. Tokyo. 34(1987), 57-63.

[OS] T. Oshima - J. Sekiguchi, The restricted root system of a semisimple symmetric pair. Adv. Studies in Pur Math. 4(1984), 433-497.

[T-1] M. Takeuchi, Basic transformations of symmetric R-spaces. Osaka J. Math.

[T-2] -------, The 2-numbers of symmetric R-spaces. To appear.

[T-3] -------, On conjugate loci and cut loci of compact symmetric spaces, I, II. Tsukuba J. Math., 2(1978), 35-68; 3(1979), 1-29.

[W] J. Wolf, Geodesic spheres in Grassmann manifolds. Illinois J. 7(1963), 447-462.

Tadashi NAGANO	Makiko SUMI
Dept. of Mathematics	Dept. of Mathematics
Sophia University	Sophia University
Tokyo, 102 Japan	Tokyo, 102 Japan

ON SOME COMPACT ALMOST KÄHLER MANIFOLDS WITH CONSTANT HOLOMORPHIC SECTIONAL CURVATURE

By Takuji SATO

§1. Introduction

An almost Hermitian manifold $M=(M,J,\langle\ ,\ \rangle)$ is called an almost Kähler manifold if the corresponding Kähler form of M is closed (equivalently, $\langle(\nabla_X J)Y,Z\rangle+\langle(\nabla_Y J)Z,X\rangle+\langle(\nabla_Z J)X,Y\rangle=0$, for all $X,Y,Z \in \chi(M)$, where $\chi(M)$ denotes the Lie algebra of all differentiable vector fields on M). By the definition, a Kähler manifold is necessarily an almost Kähler manifold. If the almost complex structure J of an almost Kähler manifold M is integrable, then M is a Kähler manifold([12]). A strictly almost Kähler manifold is an almost Kähler manifold whose almost complex structure is not integrable. Several examples of strictly almost Kähler manifolds are known([1], [2], [9], [11]). As for the integrability of an almost complex structure of an almost Kähler manifold, some interesting studies has been done under some curvature conditions([3], [5], [7], [8]). In particular, Sekigawa has proved the following

Theorem A([8]). <u>Let M be a compact Einstein almost Kähler manifold whose scalar curvature is non-negative. Then</u>

ISBN 0-12-640170-5

M is a Kähler manifold.

On the other hand, the present author studied in [6] an almost Kähler manifold with constant holomorphic sectional curvature under the condition

$$G(JX,Y,JZ,W)=G(X,Y,Z,W), \quad \text{for } X,Y,Z,W \in \chi(M). \qquad (*)$$

(The definition of G will be given in §2.)

The purpose of this paper is to prove the following

Theorem B. Let M be a compact almost Kähler manifold of constant holomorphic sectional curvature c. If M satisfies the condition (*) and the constant c is non-negative, then M is a Kähler manifold.

In §2, we shall prepare some elementary equalities and fundamental facts which will be used in the proof. The last section will be devoted to the proof of Theorem B.

The author wishes to express his hearty thanks to Prof. K. Sekigawa for his valuable suggestions.

§2. Preliminaries

Let $M = (M,J,\langle\ ,\ \rangle)$ be an $n(=2m)$-dimensional almost Hermitian manifold with the almost Hermitian structure $(J,\langle,\rangle)$ and Ω the Kähler form of M defined by $\Omega(X,Y)=\langle X,JY\rangle$, for $X,Y \in \chi(M)$. We assume that M is oriented by the volume form $\sigma=((-1)^m/m!)\,\Omega^m$. We denote by ∇, R, r and s the Riemannian connection, the curvature tensor, the Ricci tensor and the scalar curvature of M, respectively. We assume that the curvature tensor R is defined by

$$R(X,Y)Z = \nabla_{[X,Y]}Z - [\nabla_X,\nabla_Y]Z,$$

and

$$R(X,Y,Z,W)=\langle R(X,Y)Z,W\rangle,$$

for $X,Y,Z,W \in \chi(M)$. The Ricci *-tensor r^* of type (0,2), resp. Q^* of type (1,1), is defined by

$$r^*(x,y)=\langle Q^*x,y\rangle=\text{trace of } (z \longmapsto R(Jz,x)Jy),$$

$$= \sum_i R(x,e_i,Jy,Je_i),$$

where $\{ e_i;i=1,\ldots,n \}$ is an arbitrary orthonormal basis of T_pM([10]). The *-scalar curvature s^* is defined by the trace of Q^*. The tensor field r^* satisfies

$$r^*(JX,JY)=r^*(Y,X), \tag{2.1}$$

for $X,Y \in \chi(M)$.

The holomorphic sectional curvature is defined by

$$H(x)=\frac{R(x,Jx,x,Jx)}{\|x\|^4},$$

for $x \in T_pM(p \in M)$ with $x\neq 0$.

We introduce a tensor field G of type (0,4) defined by

$$G(X,Y,Z,W)=R(X,Y,Z,W)-R(X,Y,JZ,JW). \tag{2.2}$$

The tensor field G plays an important role in the study of almost Hermitian manifolds(cf.[4], [6], [10], etc.). For $X,Y,Z,W \in \chi(M)$, G satisfies the following relations:

$$G(X,Y,Z,W)=-G(Y,X,Z,W)=-G(X,Y,W,Z), \tag{2.3}$$

$$G(X,Y,JZ,JW)=-G(X,Y,Z,W), \tag{2.4}$$

$$G(X,Y,JZ,W)=G(X,Y,Z,JW), \tag{2.5}$$

$$G(X,Y,Z,JZ)=0, \tag{2.6}$$

$$G(X,Y,Z,W)+G(JX,JY,JZ,JW)=G(Z,W,X,Y)+G(JZ,JW,JX,JY), \tag{2.7}$$

$$G(X,JY,Z,JW)+G(JX,Y,JZ,W)=G(Z,JW,X,JY)+G(JZ,W,JX,Y). \tag{2.8}$$

For an orthonormal frame $\{E_i\}$ of M,

$$\sum_i G(X,E_i,Y,E_i)=r(X,Y)-r^*(X,Y). \tag{2.9}$$

Let $\{ e_i;i=1,...,n \}$ be an orthonormal basis of T_pM at any point p of M. In the sequel, we shall adopt the following notational convention:

$$R_{ijkl}=R(e_i,e_j,e_k,e_l),$$

$$R_{\bar{i}jkl}=R(Je_i,e_j,e_k,e_l),$$

$$\cdots$$

$$R_{\bar{i}\bar{j}\bar{k}\bar{l}}=R(Je_i,Je_j,Je_k,Je_l),$$

$$G_{ijkl}=G(e_i,e_j,e_k,e_l),$$

$$r_{ij}=r(e_i,e_j), \quad r^*_{ij}=r^*(e_i,e_j),$$

$$J_{ij}=\langle Je_i,e_j\rangle, \quad \nabla_i J_{jk}=\langle(\nabla_{e_i} J)e_j,e_k\rangle,$$

and so on, where the Latin indices run over the range 1,...,n. We get easily

$$\nabla_i J_{\bar{j}\bar{k}}=-\nabla_i J_{jk}. \tag{2.10}$$

In the rest of this paper, we assume that M is an n-dimensional almost Kähler manifold. Then it is well known that M is a quasi-Kähler manifold, i.e.,

$$\nabla_i J_{jk} + \nabla_{\bar{i}} J_{\bar{j}k} = 0. \tag{2.11}$$

Gray has obtained the following

Lemma 2.1([4]). <u>Let M be an almost Kähler manifold. Then</u>

$$G_{ijkl} + G_{\bar{i}\bar{j}\bar{k}\bar{l}} + G_{\bar{i}j\bar{k}l} + G_{i\bar{j}k\bar{l}}$$

$$= -2 \sum_a (\nabla_i J_{ja} - \nabla_j J_{ia})(\nabla_k J_{la} - \nabla_l J_{ka}). \tag{2.12}$$

If M satisfies the condition

$$G_{\bar{i}\bar{j}\bar{k}l} = G_{ijkl}, \tag{*}$$

then it follows that([6])

$$G_{ijkl} = G_{klij}, \tag{2.13}$$

$$G_{\bar{i}\bar{j}\bar{k}\bar{l}} = G_{ijkl}. \tag{2.14}$$

Hence, under the condition (*), (2.12) reduces to

$$G_{ijkl} = -\frac{1}{2} \sum_a (\nabla_i J_{ja} - \nabla_j J_{ia})(\nabla_k J_{la} - \nabla_l J_{ka})$$

$$= -\frac{1}{2} \sum_a (\nabla_a J_{ij}) \nabla_a J_{kl}. \tag{2.15}$$

From (2.9) and (2.15), we have

$$r_{ij} - r^*_{ij} = -\frac{1}{2} \sum_{a,b} (\nabla_a J_{bi}) \nabla_a J_{bj}, \tag{2.16}$$

$$s - s^* = -\frac{1}{2} \|\nabla J\|^2. \tag{2.17}$$

In the previous paper [6], the author has proved the following

Proposition 2.2. <u>Let M be an almost Kähler manifold of constant holomorphic sectional curvature c. If M satisfies the condition (*), then we have</u>

$$R_{ijkl}=\frac{c}{4}(\delta_{ik}\delta_{jl}-\delta_{il}\delta_{jk}+J_{ik}J_{jl}-J_{il}J_{jk}+2J_{ij}J_{kl})$$

$$+\frac{1}{4}(2G_{ijkl}+G_{ikjl}-G_{iljk}), \tag{2.18}$$

$$r_{ij}+3r^*_{ij}=(n+2)c\delta_{ij}, \tag{2.19}$$

$$s+3s^*=n(n+2)c, \tag{2.20}$$

where G_{ijkl} is given by (2.15).

On one hand, Sekigawa has established the following integral formula on a compact almost Kähler manifold which plays an essential role in the proof of Theorem A (and also Theorem B) in §1.

Proposition 2.3([8]). Let M be an n-dimensional compact almost Kähler manifold. Then we have

$$\int_M (f_1-\frac{1}{2}f_2+f_3-2f_4)\sigma=0, \tag{2.21}$$

where f_1, f_2, f_3 and f_4 are functions on M defined by

$$f_1(p)=\sum R_{abij}(R_{\bar{a}\bar{b}ij}-R_{\bar{a}\bar{b}\bar{i}\bar{j}}),$$

$$f_2(p)=\sum R_{a\bar{a}ij}(R_{b\bar{b}ij}-R_{b\bar{b}\bar{i}\bar{j}}),$$

$$f_3(p)=-\sum R_{a\bar{a}ij}(\nabla_{\bar{b}}J_{ik})\nabla_b J_{jk},$$

$$f_4(p)=-\sum R_{abij}(\nabla_{\bar{b}}J_{ik})\nabla_{\bar{a}}J_{jk}, \tag{2.22}$$

at any point p of M.

§3. Proof of Theorem B

Let M be an n-dimensional compact almost Kähler manifold with constant holomorphic sectional curvature c.

Furthermore, we assume that M satisfies the condition (*). We shall evaluate the values of the functions f_1, f_2, f_3 and f_4 in Proposition 2.3.

Lemma 3.1. $f_1=-\frac{1}{2}\|G\|^2$.

<u>Proof.</u> By (2.4), (2.13), (2.14) and (2.22), we have

$$f_1=\sum R_{abij}(R_{\bar{a}\bar{b}ij}-R_{\bar{a}\bar{b}\bar{i}\bar{j}})$$

$$=\sum R_{abij}G_{\bar{a}\bar{b}ij}$$

$$=\frac{1}{2}\sum (R_{abij}G_{\bar{a}\bar{b}ij}+R_{ab\bar{i}\bar{j}}G_{\bar{a}\bar{b}\bar{i}\bar{j}})$$

$$=-\frac{1}{2}\sum (R_{abij}G_{abij}-R_{ab\bar{i}\bar{j}}G_{abij})$$

$$=-\frac{1}{2}\sum G_{abij}G_{abij}$$

$$=-\frac{1}{2}\|G\|^2.$$

Lemma 3.2. $f_2=0$.

<u>Proof.</u> By (2.6),(2.13) and (2.22), we have

$$f_2=\sum R_{a\bar{a}ij}(R_{b\bar{b}ij}-R_{b\bar{b}\bar{i}\bar{j}})$$

$$=\sum R_{a\bar{a}ij}G_{b\bar{b}ij}$$

$$=\sum R_{a\bar{a}ij}G_{ijb\bar{b}}$$

$$=0.$$

Lemma 3.3. $f_3=-\frac{(n+2)}{2}c\|\nabla J\|^2-\|r-r^*\|^2$.

<u>Proof.</u> By Bianchi identity, we have

$$\sum R_{a\bar{a}i\bar{j}}=2r^*_{ij}. \tag{3.1}$$

From (2.10),(2.11),(2.16),(2.17),(2.19),(2.22) and (3.1), we have

$$f_3=-\sum R_{a\bar{a}ij}(\nabla_{\bar{b}}J_{ik})\nabla_b J_{jk}$$

$$=-\sum R_{a\bar{a}i\bar{j}}(\nabla_{\bar{b}}J_{ik})\nabla_b J_{\bar{j}k}$$

$$=-2\sum r^*_{ij}(\nabla_b J_{ik})\nabla_b J_{jk}$$

$$=4\sum r^*_{ij}(r_{ij}-r^*_{ij})$$

$$=\sum\{(r_{ij}+3r^*_{ij})-(r_{ij}-r^*_{ij})\}(r_{ij}-r^*_{ij})$$

$$=\sum\{(n+2)c\delta_{ij}(r_{ij}-r^*_{ij})\}-\|r-r^*\|^2$$

$$=(n+2)c(s-s^*)-\|r-r^*\|^2$$

$$=-\frac{(n+2)}{2}c\|\nabla J\|^2-\|r-r^*\|^2.$$

Lemma 3.4. $f_4=-\frac{3}{4}c\|\nabla J\|^2-\frac{1}{8}\|G\|^2.$

Proof. By (2.11),(2.18) and (2.22), we have

$$f_4=-\sum R_{abij}(\nabla_{\bar{b}}J_{ik})\nabla_{\bar{a}}J_{jk}$$

$$=-\sum R_{abij}(\nabla_b J_{i\bar{k}})\nabla_a J_{j\bar{k}}$$

$$=\sum R_{abij}(\nabla_a J_{ik})\nabla_b J_{jk}$$

$$=\frac{c}{4}\sum(\nabla_a J_{ik})\nabla_b J_{jk}(\delta_{ai}\delta_{bj}-\delta_{aj}\delta_{bi}+J_{ai}J_{bj}-J_{aj}J_{bi}+2J_{ab}J_{ij})$$

$$+\frac{1}{4}\sum(\nabla_a J_{ik})\nabla_b J_{jk}(2G_{abij}+G_{aibj}-G_{ajbi}). \tag{3.2}$$

We shall calculate the right hand side of (3.2) term by term. By (2.11), we get

$$\sum\nabla_a J_{ik}(\nabla_b J_{jk})\delta_{ai}\delta_{bj}=\sum(\nabla_a J_{ak})\nabla_b J_{bk}=0. \tag{3.3}$$

Taking account of

$$\|\nabla J\|^2=\sum(\nabla_i J_{jk})\nabla_i J_{jk}$$

$$=-\sum(\nabla_j J_{ki}+\nabla_k J_{ij})\nabla_i J_{jk}$$

$$=2\sum(\nabla_j J_{ik})\nabla_i J_{jk},$$

we have

$$\sum \nabla_a J_{ik}(\nabla_b J_{jk})\delta_{aj}\delta_{bi} = \sum (\nabla_j J_{ik})\nabla_i J_{jk}$$
$$= \frac{1}{2}\|\nabla J\|^2. \tag{3.4}$$

Similarly we have

$$\sum \nabla_a J_{ik}(\nabla_b J_{jk})J_{ai}J_{bj} = \sum (\nabla_a J_{\bar{a}k})\nabla_b J_{\bar{b}k}$$
$$= \sum (\nabla_a J_{a\bar{k}})\nabla_b J_{b\bar{k}}$$
$$=0. \tag{3.5}$$

$$\sum \nabla_a J_{ik}(\nabla_b J_{jk})J_{aj}J_{bi} = \sum (\nabla_a J_{\bar{b}k})\nabla_b J_{\bar{a}k}$$
$$= \sum (\nabla_a J_{bk})\nabla_b J_{ak}$$
$$= \frac{1}{2}\|\nabla J\|^2. \tag{3.6}$$

$$\sum \nabla_a J_{ik}(\nabla_b J_{jk})J_{ab}J_{ij} = \sum (\nabla_a J_{ik})\nabla_{\bar{a}} J_{\bar{i}k}$$
$$=- \|\nabla J\|^2. \tag{3.7}$$

By (2.4) and (2.10),

$$\sum \nabla_a J_{ik}(\nabla_b J_{jk})G_{abij} = \sum \nabla_a J_{\bar{i}k}(\nabla_b J_{\bar{j}k})G_{ab\bar{i}\bar{j}}$$
$$=- \sum \nabla_a J_{i\bar{k}}(\nabla_b J_{j\bar{k}})G_{abij}$$
$$=- \sum \nabla_a J_{ik}(\nabla_b J_{jk})G_{abij},$$

from which we have

$$\sum \nabla_a J_{ik}(\nabla_b J_{jk})G_{abij}=0. \tag{3.8}$$

By (2.3) and (2.15), we have

$$\sum \nabla_a J_{ik}(\nabla_b J_{jk})G_{aibj}$$
$$=\frac{1}{4} \sum (\nabla_a J_{ik}-\nabla_i J_{ak})(\nabla_b J_{jk}-\nabla_j J_{bk})G_{aibj}$$

$$=-\frac{1}{2}\sum G_{aibj}G_{aibj}$$

$$=-\frac{1}{2}\|G\|^2. \tag{3.9}$$

By the same reason as (3.8), we have

$$\sum \nabla_a J_{ik}(\nabla_b J_{jk})G_{ajbi}=0. \tag{3.10}$$

Substituting (3.3),...,(3.10) into (3.2), we obtain Lemma 3.4.

By virtue of above lemmas and Proposition 2.3, we have finally the following

$$\int_M \left(\frac{n-1}{2}c\,\|\nabla J\|^2+\|r-r^*\|^2+\frac{1}{4}\|G\|^2\right)\sigma=0. \tag{3.11}$$

From (3.11), we may easily show that if the constant c is non-negative, then ∇J vanishes identically on M, that is, M is a Kähler manifold. This completes the proof of Theorem B.

Taking account of (2.17) and (2.20), we can restate Theorem B as follows:

Corollary 3.5. Let M be a compact almost Kähler manifold of constant holomorphic sectional curvature. If M satisfies the condition (*) and the scalar curvature of M is non-negative, then M is a Kähler manifold.

REFERENCES

[1] E. Abbena: An example of an almost Kähler manifold which is not Kählerian, Boll. Un. Mat. Ital.(6) 3A (1984), 383-392.

[2] L. A. Cordero, M. Fernandez and M. de Leon: Examples of compact non-Kähler almost Kähler manifolds, Proc. Amer. Math. Soc. 95(1985), 280-286.

[3] S. I. Goldberg: Integrability of almost Kähler manifolds, Proc. Amer. Math. Soc. 21(1969), 96-100.

[4] A. Gray: Curvature identities for Hermitian and almost Hermitian manifolds, Tôhoku Math. J. 28(1976), 601-612.

[5] Z. Olszak: A note on almost Kaehler manifolds, Bull. Acad. Polon. Sci. 26(1978), 139-141.

[6] T. Sato: On some almost Hermitian manifolds with constant holomorphic sectional curvature, to appear.

[7] K. Sekigawa: On some 4-dimensional compact Einstein almost Kähler manifolds, Math. Ann. 271(1985), 333-337.

[8] K. Sekigawa: On some compact Einstein almost Kähler manifolds, J. Math. Soc. Japan 39(1987), 677-684.

[9] W. P. Thurston: Some simple examples of symplectic manifolds, Proc. Amer. Math. Soc. 55(1976), 467-468.

[10] F. Tricerri and L. Vanhecke: Curvature tensors on almost Hermitian manifolds, Trans. Amer. Math. Soc. 267(1981), 365-398.

[11] B. Watson: New examples of strictly almost Kähler manifolds, Proc. Amer. Math. Soc. 88(1983), 541-544.

[12] K. Yano: Differential geometry of complex and almost complex spaces, New York, Pergamon, 1965

Takuji SATO
Faculty of Technology
Kanazawa University
Kanazawa 920
Japan

On the rigidity and homogeneity for Kähler submanifolds of complex space forms

YOSHIHARU TANIGUCHI

0. Introduction

The main purpose of this note is to study conditions under which Kähler submanifolds of complex space forms become homogeneous. We use the so-called moving frame method which was originally treated by É. Cartan. We will introduce the S_c-structure for our study. It may be considered as a complex analogy of Sulanke and Švec's G,H-structure which interprets É. Cartan's method of moving frames in terms of fiber bundles and Lie algebra valued 1-forms on them. After preparing terminologies, we will give some basic properties of S_c-structures and a homogeneity theorem.

1. Preliminaries

In this section we briefly recall a few things about the method of moving frames for congruence problems for submanifolds of homogeneous spaces.

Let G be a Lie group with Lie algebra $\mathfrak{g}$, H a closed subgroup of G, and $S = G/H$ the quotient manifold. Two immersions f_1 and f_2 of a connected manifold M into S are said to be *G-congruent* if there exists $\tau \in G$ such that $f_1(x) = \tau f_2(x)$ $(x \in M)$. In general it is not easy to find conditions under which f_1 and f_2 become G-congruent. However in case $S = G$, we know a simple answer for the problem by using the

This work was supported by Grant-in-Aid for Scientific Research, The Ministry of Education, Science and Culture.

ISBN 0-12-640170-5

Maurer-Cartan form Φ of G which is a $\mathfrak{g}$-valued 1-form on G defined by $\Phi(X) = \tau^{-1}X$ $(X \in T_\tau G)$ under the usual identification $T_eG = \mathfrak{g}$ (e : the neutral element of G).

PROPOSITION 1. *In case $S = G$, two immersions f_1 and f_2 of a connected manifold M into S are G-congruent if and only if $f_1^*\Phi = f_2^*\Phi$ on M.*

This shows that the G-congruence class of immersion f of M into G is determined by the $\mathfrak{g}$-valued 1-form $f^*\Phi$ on M. On the other hand we have the following

PROPOSITION 2. *Suppose that M is simply connected. For a $\mathfrak{g}$-valued 1-form ω on M, there exists an immersion f of M into G such that $\omega = f^*\Phi$ if and only if*

$$(*) \qquad \begin{aligned} d\omega + [\,\omega, \omega\,] &= 0 \\ \operatorname{rank} \omega &= \dim M, \end{aligned}$$

where $[\,\cdot,\cdot\,]$ denotes the Lie bracket product of $\mathfrak{g}$.

From the above propositions we see that if M is simply connected, the set of G-congruence classes of immersions of M into $S = G$ is in one-to-one correspondence with the set of $\mathfrak{g}$-valued 1-forms on M with $(*)$.

In general case $S = G/H$ what we have to do is not so simple as in the case $S = G$. Let f be an immersion of a connected manifold M into $S = G/H$. Let $\pi_f : P_f \to M$ be the principal H-bundle over M induced by f from the principal H-bundle $\pi : G \to S$:

$$\begin{array}{ccc} P_f & \xrightarrow{F} & G \\ {\scriptstyle \pi_f}\downarrow & & \downarrow{\scriptstyle \pi} \\ M & \xrightarrow{f} & S, \end{array}$$

where F denotes the natural mapping covering f. We set $\omega_f = F^*\Phi$, a $\mathfrak{g}$-valued 1-form on P_f. According to Sulanke and Švec [9], one may take the pair (P_f, ω_f) as a candidate in the case $S = G/H$ for $f^*\Phi$ in the case $S = G$.

2. The structure equations of complex space forms and the Maurer-Cartan forms of their automorphism groups

Let $S_c(N)$ be the complex space form of complex dimension N with constant holomorphic sectional curvature $4c$, $\pi_c : U(S_c(N)) \to S_c(N)$ the unitary frame bundle of $S_c(N)$, and $G_c(N)$ the group of holomorphic isometries of $S_c(N)$ with Lie algebra $\mathfrak{g}_c(N)$. Fixing a frame $\mathbf{e}_0 \in U(S_c(N))$, we can identify $U(S_c(N))$ with $G_c(N)$ by virtue of complex space forms and, from now on, we often do so. Let H be the isotropy subgroup of $G_c(N)$ at $\pi_c(\mathbf{e}_0)$ with Lie algebra $\mathfrak{h}$. In the Cartan decomposition $\mathfrak{g} = \mathfrak{h} + \mathfrak{n}$, $\mathfrak{n}$ is naturally isomorphic to $T_o(S_c(N))$, and $T_o(S_c(N))$ can be identified with $\mathbf{C}^N$ by $\mathbf{e}_0$. So we identify $\mathfrak{n}$ with $\mathbf{C}^N$ in this way. In our case the linear isotropy representation ρ of H in $\mathfrak{n} = \mathbf{C}^N$ is an isomorphism of H onto the unitary group $U(N)$. By using these isomorphisms of $\mathfrak{h}$ and $\mathfrak{n}$ with $\mathfrak{u}(N)$ and $\mathbf{C}^N$, we will identify $\mathfrak{g}_c(N)$ with $\mathfrak{u}(N) + \mathbf{C}^N$ throughout the paper: so for any element $X \in \mathfrak{g}_c(N)$, we denote by $(X^A{}_B)_{A,B=1,\dots,N}$ and $(X^A)_{A=1,\dots,N}$ the $\mathfrak{u}(N)$-part and $\mathbf{C}^N$-part of X with respect to the decomposition $\mathfrak{g}_c(N) = \mathfrak{u}(N) + \mathbf{C}^N$ respectively. Now let Φ be the Maurer-Cartan form of $G_c(N)$. If we consider Φ as a $\mathfrak{g}_c(N)$-valued 1-form on $U(S_c(N))$, $(\Phi^A)_{A=1,\dots,N}$ and $(\Phi^A{}_B)_{A,B=1,\dots,N}$ turn out to be equal to the $\mathbf{C}^N$-valued canonical form $(\phi^A)_{A=1,\dots,N}$ and $\mathfrak{u}(N)$-valued Levi-Civita's connection form $(\phi^A{}_B)_{A,B=1,\dots,N}$ on $U(S_c(N))$ respectively. Now the structure equations of Kähler manifold $S_c(N)$ are expressed by

$$\text{(SE)} \qquad \begin{aligned} &d\phi^A + \phi^A{}_B \wedge \phi^B = 0 \\ &d\phi^A{}_B + \phi^A{}_C \wedge \phi^C{}_B = c(\phi^A \wedge \overline{\phi^B} + \delta^A{}_B \phi^C \wedge \overline{\phi^C}). \end{aligned}$$

We note here that the equations (SE) are equivalent to the Maurer-Cartan equation

$$\text{(MCE)} \qquad d\Phi + [\Phi, \Phi] = 0.$$

So $((\phi^A{}_B)_{A,B=1,\dots,N}, (\phi^A)_{A=1,\dots,N})$, regarded as a $\mathfrak{g}_c(N)$-valued 1-form on $U(S_c(N))$, can be considered as the Maurer-Cartan form on $G_c(N)$ under the identification $U(S_c(N)) = G_c(N)$.

3. The S_c-structures and their basic properties

In this section, we introduce the *S_c-structure* over a connected complex manifold M. It consists of a principal fiber bundle P and a $\mathfrak{g}_c(N)$-valued 1-form ω on it with some properties so that it would correspond to a $G_c(N)$-congruence class of full Kähler local immersions of M into $S_c(N)$[1].

We first consider a Kähler immersion $f : (M, g) \to S_c(N)$, where (M, g) is a Kähler manifold with Kähler metric g of complex dimension n. We denote by ∇ the induced connection in the induced bundle $f^*TS_c(N)$ over M. For the sake of simplicities, we assume that the immersion f is full. For $p = 1, 2, \dots$, we define the p-th *osculating* bundle $O^p(f) = \bigcup_{x\in M} O^p_x(f)$ of f, where $O^p_x(f) = \{(\nabla_{X_1} \cdots \nabla_{X_{p'-1}} f_*(X_{p'}))_x :$ $X_1, \dots, X_{p'-1}$, and $X_{p'}$ are vector fields over M $(p' \leq p)$, and $x \in M\}$. We assume that $O^p(f)$ forms a vector bundle over M, that is, $\dim O^p_x(f)$ is constant n_p in x. Then by virtue of complex space forms there exists a positive integer d such that $n = n_1 < \cdots < n_d = N$, although it is not so trivial. Let $B^p(f)$ be the orthogonal

[1] A full Kähler local immersion of M into $S_c(N)$ means a holomorphic isometric immersion of a non-empty open subset of M into $S_c(N)$ whose image does not lie in any proper totally geodesic submanifold of $S_c(N)$.

complement to $O^{p-1}(f)$ in $O^p(f)$ for $p > 1$ and q_p the rank of it ($q_p = n_p - n_{p-1}$). We have the following decomposition

$$f^*TS_c(N) = O^1(f) \oplus B^2(f) \oplus \cdots \oplus B^d(f).$$

Let P_f be the principal $U(q_1) \times \cdots \times U(q_d)$-bundle over M ($q_1 = n$) associated with the above decomposition: $P_f = \{(x, e_1, \ldots, e_N) : (e_1, \ldots, e_{n_1})$ forms a basis of $O^1_x(f)$, $(e_{n_1+1}, \ldots, e_{n_2})$ forms a basis of $B^2_x(f)$, ... , $(e_{n_{d-1}+1}, \ldots, e_{n_d})$ forms a basis of $B^d_x(f), x \in M\}$. Next we define a $\mathfrak{g}_c(N)$-valued 1-form $\omega_f = F^*\Phi$, where F denotes the natural mapping of P_f into $U(S_c(N))$ identified with $G_c(N)$. Then ω_f have the following properties:

(1) $\omega_f(E^*) = E \qquad (E \in \mathbf{Lie}(U(q_1) \times \cdots \times U(q_d))$;

(2) $R_a^*\omega_f = Ad\, a^{-1} \cdot \omega_f \qquad (a \in U(q_1) \times \cdots \times U(q_d))$;

(3) $d\omega_f + [\omega_f, \omega_f] = 0$;

(4) ${\omega_f}^r = 0 \qquad (n < r \leq n_d = N)$;

(5) ${\omega_f}^r{}_s = 0 \qquad (n_{p-1} < r \leq n_p,\ n_{p'-1} < s \leq n_{p'},\ |p - p'| \geq 2)$;

(6) for $i = 1, \ldots, n$, ${\omega_f}^i$ is "of type (1,0) in the directions of base manifold" and the ${\omega_f}^i$'s are linearly independent over $\mathbb{C}$;

(7) $\mathrm{rank}_{\mathbb{C}}({\omega_f}^r{}_s)_{n_{p-1} < r \leq n_p,\, n_{p-2} < s \leq n_{p-1}} = q_p \qquad (p = 2, \ldots, d)$.

The equation (3) actually consists of Gauss-Codazzi, Ricci-Minardi, and their higher order equations for f and the others directly come from the definition of P_f and the nature of f.

We make the following

DEFINITION. *Let M be an n-dimensional connected complex manifold. A pair (P, ω) is an S_c-structure over M of type $(n_1, \ldots, n_d)$ if it satisfies the following conditions:*

(1) *$n_1, \ldots, n_d$ are strictly increasing positive integers with $n = n_1$ and P is a*

principal $U(q_1) \times \cdots \times U(q_d)$*-bundle over* M, *where* $q_p = n_p - n_{p-1}$ *for* $p = 1, \ldots, d$ $(n_0 = 0)$;

(2) ω *is a* $\mathfrak{g}_c(N)$*-valued 1-form on* P *having all the properties (1)–(6) of* ω_f.

The previous (P_f, ω_f) will be call the *S_c-structure induced by* an immersion f.

Now let (P, ω) be an S_c-structure over M of type $(n_1, \ldots, n_d)$. We can define a Riemannian metric g_P on M such that $\pi^* g_P = \omega^i \otimes \overline{\omega^i} + \overline{\omega^i} \otimes \omega^i$ ($\pi : P \to M$ the projection). By the condition on ω, g_P turns out to be Kählerian. We will call this g_P the *Kähler metric defined by* an S_c-structure (P, ω). Two S_c-structures (P, ω) and (P', ω') over M will be said to be *isomorphic* and denoted by $(P, \omega) \cong (P', \omega')$ if there exists a principal bundle isomorphism $h : P \to P'$ such that $h^* \omega' = \omega$.

The following theorems give basic properties of S_c-structures.

THEOREM 1. *Let* (P, ω) *be an* S_c*-structure over a simply connected complex manifold* M *of type* $(n_1, \ldots, n_d)$ *and* g_P *the Kähler metric on* M *defined by* (P, ω). *Then there exists a full Kähler immersion* $f : (M, g_P) \to S_c(n_d)$ *such that* $(P, \omega) \cong (P_f, \omega_f)$. *Moreover such an* f *is unique up to* $G_c(n_d)$*-congruence.*

THEOREM 2. *Let* (P, ω) *and* (P', ω') *be* S_c*-structures over connected complex manifolds* M *and* M' *respectively. Let* f *be a holomorphic isometry of* (M, g_P) *onto* $(M', g_{P'})$, *where* g_P *and* $g_{P'}$ *denote the Kähler metrics on* M *and* M' *defined by* (P, ω) *and* (P', ω') *respectively. Then* f *gives rise to a unique isomorphism* $f_\# : (P, \omega) \to (P', \omega')$.

The above Theorem 2 might be considered as a reformulation of the rigidity theorem of E. Calabi.

4. Reduction of structure group of S_c-structure

We carry out a certain reduction procedure of structure group of P in a way which makes ω to a simple form as much as possible. After all, we can define a G-structure RF over a certain open subset M_0 of M and a G-connection θ on RF. Then RF have the following properties:

(1) G is a subgroup of $U(n)$ and RF is a principal G-subbundle of the unitary frame bundle of $(M_0, g_P|_{M_0})$; in addition there exists an injective homomorphism $\rho: G \to U(q_1) \times \cdots \times U(q_d)$ such that $RF \times_\rho (U(q_1) \times \cdots \times U(q_d)) \cong P|_{M_0}$;

(2) let $\mathfrak{g}$ denote the Lie algebra of G; taking a basis $\{E_\alpha\}_{\alpha\in\Lambda}$ of $\mathfrak{g}$, we set $\theta = \theta^\alpha \otimes E_\alpha$; let $\psi = \iota^*\omega$ where we denote by ι the natural inclusion $RF \to P|_{M_0}$; then all the coefficients of $\psi^A{}_B$ $(A, B = 1, \ldots, N)$ with respect to the coframe field $\{\psi^i, \overline{\psi^i}, \theta^\alpha : i = 1, \ldots, n, \alpha \in \Lambda\}$ of complex valued 1-forms on RF are constant along any fiber of $RF \to M_0$;

(3) the structure group of RF cannot be extended to any larger subgroup of $U(n)$ with the property (2) above.

We call the pair (RF, ψ) a *reduced S_c-structure* of (P, ω).

THEOREM 3. *We retain the notation above. Let M be a full, connected complete Kähler submanifold of $S_c(N)$ and (P, ω) the induced S_c-structure by the inclusion mapping of M into $S_c(N)$. Further let (RF, ψ) be a reduced S_c-structure over M_0 of (P, ω). Then (M, g_P) becomes a homogeneous Kähler manifold if and only if all the coefficients of $\psi^A{}_B$ $(A, B = 1, \ldots, N)$ with respect to $\{\psi^i, \overline{\psi^i}, \theta^\alpha : i = 1, \ldots, n, \alpha \in \Lambda\}$ are constant on RF. Moreover, if it is the case, $M_0 = M$ and RF is naturally diffeomorphic to the group of holomorphic isometries of (M, g_P). More precisely, as submanifolds of $U(S_c(N))$ RF coincides with $G_c(N) \cdot \mathbf{e}_0$, where $\mathbf{e}_0$ is an arbitrary element of RF.*

For the proof of this theorem we use the method of [**9**].

REFENCES

1. W. Ambrose and I. M. Singer, *On homogeneous Riemannian manifolds*, Duke Math. J. **25** (1958), 647–669.
2. G. E. Bredon, "Introduction to Compact Transformation Groups," Academic Press, New York, 1972.
3. E. Calabi, *Isometric imbedding of complex manifolds*, Ann. of Math. **58** (1953), 1–23.
4. É. Cartan, "Théorie des groupes finis et continus et la méthode du repère mobile," Gauthier-Villars, Paris, 1937.
5. P. A. Griffiths, *On Cartan's method of Lie groups and moving frames as applied to uniqueness and existence questions in differential geometry*, Duke Math. J. **41** (1974), 775–814.
6. K. Jänich, *Differenziarbare G-Mannigfaltigkeiten*, Lect. Notes in Math. **59** (1968), Springer-Verlag, Berlin-Heidelberg-New York.
7. R. Jensen, *Higher order contact of submanifolds of homogeneous spaces*, Lect. Notes in Math. **610** (1977), Springer-Verlag, Berlin-Heidelberg-New York.
8. R. Sulanke, *On É. Cartan's method of moving frames*, Proc. Colloq. Differential Geometry (1979), 681–704, Budapest.
9. R. Sulanke and A. Švec, *Zur Differentialgeometrie der Untermannigfaltigkeiten eines Kleinschen Raumes*, Beiträge zur Algebra und Geometrie **10** (1980), 63–85, Halle.
10. H. -S. Tai, *On Frenet frames of complex submanifolds in complex projective spaces*, Duke Math. J. **51** (1984), 163–183.

11. R. Takagi and M. Takeuchi, *Degree of symmetric Kähler submanifolds of a complex projective spaces*, Osaka J. Math. **14** (1977), 501–518.

Department of Mathematics, Osaka University, Toyonaka Osaka 560 JAPAN

Harmonic non-holomorphic maps of 2-tori into the 2-sphere

Dedicated to Professor Shingo Murakami on his 60th birthday.

Masaaki Umehara and Kotaro Yamada

0 Introduction.

The purpose of this paper is to investigate harmonic maps of Riemann surfaces of genus 1 into the unit sphere S^2. Holomorphic or anti-holomorphic maps of a Riemann surface Σ into S^2 are trivial harmonic maps.

Consider a Riemann surface Σ of genus γ. In [2], Eells and Wood proved that a harmonic map of Σ into S^2 with degree d must be holomorphic or anti-holomorphic if $|d| > \gamma - 1$. Hence the homotopy classes of harmonic maps which are neither holomorphic nor anti-holomorphic are restricted. So, it is meaningful to find harmonic maps of Riemann surfaces into S^2 which are neither holomorphic nor anti-holomorphic.

There are some examples of such surfaces. For a surface of genus $\gamma > 2$, Lemaire [5] showed that there exists a Riemann surface which admits a harmonic non-holomorphic and non anti-holomorphic map into S^2 with given degree $d \leq \gamma - 1$. In the case of genus one, that is torus, a harmonic map into S^2 is neither holomorphic nor anti-holomorphic if and only if its degree is 0. In this case, there exists a Riemann surface which admits surjective harmonic map into S^2 of degree 0 [6].

The notion of harmonic map of a Riemann surface Σ depends only on its conformal structure. In this paper, harmonic maps of tori with *arbitrary* conformal structure into S^2 is constructed. More precisely, the main theorem is the following.

Theorem. *There exist countably many "equivalent" classes of surjective harmonic maps of an arbitrary torus into S^2 which are neither holomorphic nor anti-holomorphic.*

The meaning of *equivalent* is described in Section 1.

ISBN 0-12-640170-5

Related to this problem, Eells and Wood [3] showed that there exist harmonic non-holomorphic, non anti-holomorphic maps with arbitrary degree of any conformal structure of torus into the complex projective space $\mathbf{C}P^n$ provided $n \geq 2$. Our main result is the case of $n = 1$ of their theorem.

A Gauss map of a conformal immersion of a Riemann surface into $\mathbf{E}^3$ with constant mean curvature is a harmonic map of the surface into S^2. Immersions of certain tori with constant mean curvature are constructed by Wente [8,9] and Abresch [1]. To prove our theorem, we apply their results and construct conformal immersions of the universal cover $\mathbf{C}$ of torus into $\mathbf{E}^3$ with constant mean curvature whose Gauss maps are invariant under the deck transformations. These immersions may not induce tori with constant mean curvature.

1 Harmonic maps of tori into the 2-sphere.

Let Σ be a compact Riemann surface and g its riemannian metric associated with the conformal structure of it. The energy of a smooth map f of Σ into the unit 2-sphere S^2 is defined as

$$E(f) = \frac{1}{2}\int_{\Sigma} |df|^2 \, dv_g, \tag{1.1}$$

where $|\cdot|$ and dv_g are the norm and the volume element of g respectively. Note that the definition of $E(\cdot)$ depends only on the conformal structure of Σ, not on the choice of a riemannian metric g.

A critical point of the functional $E(\cdot)$ is called a *harmonic map*. A map $f : \Sigma \to S^2$ is harmonic if and only if it satisfies the Euler-Lagrange equation

$$\frac{\partial^2 f}{\partial z \partial \bar{z}} - \frac{2\bar{f}}{1+|f|^2}\frac{\partial f}{\partial z}\frac{\partial f}{\partial \bar{z}} = 0 \tag{1.2}$$

with respect to the complex coordinate z of Σ. Here we identify S^2 with $\mathbf{C} \cup \{\infty\}$ by the stereographic projection at the north pole.

The automorphisms of Σ and the isometries of S^2 act on the set of harmonic maps of Σ into S^2.

Lemma 1.1 *Let $\varphi : \Sigma \to \Sigma$ be a holomorphic or anti-holomorphic map and $\tau : S^2 \to S^2$ an isometry. Then if $f : \Sigma \to S^2$ is harmonic, so is $\tau \circ f \circ \varphi$.*

From now on, we call $\tau \circ f \circ \varphi$ in the above lemma an *associated harmonic map* of f.

DEFINITION. Two harmonic maps $f_1, f_2 : \Sigma \to S^2$ are called *equivalent* if there exists a harmonic map $f : \Sigma \to S^2$ such that both f_1 and f_2 are associated harmonic maps of f.

In particular, if Σ is a torus, covering maps are holomorphic, and composing these with given harmonic map f, one can construct infinitely many associated harmonic maps of f. In this case, there are infinitely many equivalent harmonic maps with different energy value.

It is well-known that the Gauss map of a surface with constant mean curvature in the euclidean 3-space $\mathbf{E}^3$ is harmonic. In the rest of this section, we assume that Σ is a torus and study the relationships between harmonic maps and immersions of constant mean curvature.

Let $\Sigma = T$ be a torus. Then T is represented as a quotient $\mathbf{C}/\Gamma$, where Γ is a lattice of $\mathbf{C}$.

Consider a conformal immersion $\mathbf{x} : \mathbf{C} \to \mathbf{E}^3$. In the complex coordinate $z = u + \sqrt{-1}\,v$ of $\mathbf{C}$, the normal vector $\xi(p)$ of $\mathbf{x}$ at $p \in \mathbf{C}$ is written as

$$\xi(p) = \frac{\mathbf{x}_u(p) \times \mathbf{x}_v(p)}{|\mathbf{x}_u(p) \times \mathbf{x}_v(p)|}.$$

The immersion $\mathbf{x}$ is called *orientation preserving* (*resp. reversing*) if the orientation of $\mathbf{E}^3$ determined by the frame $(\mathbf{x}_u, \mathbf{x}_v, \xi)$ coincides (resp. does not coincide) with the canonical orientation of $\mathbf{E}^3$.

The Gauss map $\tilde{\nu}_{\mathbf{x}}$ of $\mathbf{x}$ is a map of $\mathbf{C}$ to S^2 which maps $p \in \mathbf{C}$ to the unit normal vector $\xi(p)$ at p. If $\tilde{\nu}_{\mathbf{x}}$ is invariant under the action of the lattice Γ, it induces a map $\nu_{\mathbf{x}} : T = \mathbf{C}/\Gamma \to S^2$. The following lemma is a criterion for the equivalence of such harmonic maps.

Lemma 1.2 *Let* $\mathbf{x}_i : \mathbf{C} \to \mathbf{E}^3\,(i = 1, 2)$ *be orientation preserving (resp. reversing) conformal immersions with constant mean curvature* $H = 1/2$ *whose Gauss maps are invariant under the action of a lattice* Γ. *Assume that the harmonic maps* $\nu_{\mathbf{x}_i} : T = \mathbf{C}/\Gamma \to S^2\,(i = 1, 2)$ *are equivalent. Then the image of* $\mathbf{x}_1$ *is congruent to that of* $\mathbf{x}_2$ *under the motions of* $\mathbf{E}^3$.

PROOF. Suppose Gauss maps $\nu_{\mathbf{x}_1}$ and $\nu_{\mathbf{x}_2}$ are equivalent. Then we can assume that there exist a holomorphic map $\varphi : T \to T$ and an isometry $\tau : S^2 \to S^2$ which satisfy

$$\nu_{\mathbf{x}_2} = \tau \circ \nu_{\mathbf{x}_1} \circ \varphi.$$

Lifting this to the universal cover $\mathbf{C}$ of T, we have

$$\tilde{\nu}_{\mathbf{x}_2} = \tau \circ \tilde{\nu}_{\mathbf{x}_1} \circ \tilde{\varphi},$$

where $\tilde{\varphi}$ is the lift of φ. Since $\mathbf{x}_1$ and $\mathbf{x}_2$ preserve (resp. reverse) the orientation, τ is extended to the motion $\tilde{\tau}$ of $\mathbf{E}^3$ as an element of $SO(3)$. Then,

$$\tilde{\nu}_{\mathbf{x}_2} = \tilde{\nu}_{\tilde{\tau} \circ \mathbf{x}_1} \circ \tilde{\varphi}.$$

On the other hand, the immersions $\mathbf{x}_1$ and $\mathbf{x}_2$ are determined by their Gauss maps and mean curvatures up to a translation [4]. Hence

$$\mathbf{x}_2 = \tilde{\tau} \circ \mathbf{x}_1 \circ \tilde{\varphi} + c,$$

where c is a vector in $\mathbf{E}^3$. □

REMARK. For a given harmonic non-holomorphic map $f : T \to S^2$, one can construct a conformal branched immersion $\mathbf{x}$ of the universal cover of T to $\mathbf{E}^3$ with constant mean curvature whose Gauss map $\nu_{\mathbf{x}} = f$ [4].

Moreover, if f is not anti-holomorphic, the degree of f is 0. In this case, $f_{\bar{z}}$ have no zeroes because of the index formula of Eells-Wood [2], and then, the immersion $\mathbf{x}$ have no branched points.

In particular, the set of harmonic maps of T to S^2 with degree 0 corresponds bijectively to the set of conformal immersions of $\mathbf{C}$ to $\mathbf{E}^3$ with non-zero constant mean curvature whose Gauss map is Γ-invariant.

2 Construction of harmonic maps.

In this section, we construct harmonic maps of tori into S^2 using the construction of twisted tori with constant mean curvature by Wente [9]. More precisely, we shall prove the following theorem mentioned in Section 0.

Theorem 2.1 *Let T be a torus with arbitrary conformal structure. Then there exist countably many equivalent classes of surjective harmonic maps of T into S^2 which are neither holomorphic nor anti-holomorphic.*

To begin with, we review the relationship between sinh-Gordon equation and surface theory.

Let $\Omega_{a_0 b_0} = (-a_0, a_0) \times (-b_0, b_0)$ be a rectangular domain of $\mathbf{R}^2$. Then the Dirichlet problem of the sinh-Gordon equation

$$\Delta\omega + \cosh\omega \, \sinh\omega = 0 \tag{2.1}$$

on $\Omega_{a_0 b_0}$ has the unique positive solution ω_0 when ${a_0}^{-2} + {b_0}^{-2} > 4\pi^{-2}$ [8,1,7]. This solution can be extended to the doubly periodic solution $\tilde{\omega}_0$ of (2.1) by the odd reflections about $\partial\Omega_{a_0 b_0}$. The fundamental domain of $\tilde{\omega}_0$ is a rectangle.

To construct harmonic maps, doubly periodic solutions of (2.1) with "skew" lattices which are constructed by Wente [9] are used.

Lemma 2.2 ([9], Theorem 1) *For sufficiently small positive a_0 and b_0, there exist a neighborhood U of $(a_0, b_0, 0) \in \mathbf{R}^3$ and a smooth function $\omega(u, v;\, a, b, c)$ on $\mathbf{R}^2 \times U$ which satisfy the following conditions.*

(1) *For each $(a, b, c) \in U$, $\omega(u, v;\, a, b, c)$ is a solution of (2.1) on $\mathbf{R}^2$.*

(2) *Let $\mathbf{p}_1 = (2a, 0)$ and $\mathbf{p}_2 = (2c, 2b)$. Then*

$$\omega(\mathbf{u} + \mathbf{p}_1;\, \mathbf{a}) = \omega(\mathbf{u} + \mathbf{p}_2;\, \mathbf{a}) = \omega(-\mathbf{u};\, \mathbf{a}) = -\omega(\mathbf{u};\, \mathbf{a}).$$

(3) $\omega(u, v;\ a_0, b_0, 0) = \omega_0(u, v)$.

Here $\mathbf{u}$ *(resp.* $\mathbf{a}$*) means* (u, v) *(resp.* (a, b, c)*).*

The solution $\omega(u, v) = \omega(u, v;\ a, b, c)$ of (2.1) determines a surface with constant mean curvature. That is, the Frenet equation

$$\begin{aligned} \mathbf{x}_{uu} &= \omega_u \mathbf{x}_u - \omega_v \mathbf{x}_v + L\xi \\ \mathbf{x}_{uv} &= \omega_u \mathbf{x}_v + \omega_v \mathbf{x}_u + M\xi \\ \mathbf{x}_{vv} &= \omega_v \mathbf{x}_v - \omega_u \mathbf{x}_u + N\xi \\ \xi_u &= -e^{-2\omega}(L\mathbf{x}_u + M\mathbf{x}_v) \\ \xi_v &= -e^{-2\omega}(M\mathbf{x}_u + N\mathbf{x}_v) \end{aligned} \tag{2.2}$$

is integrable and its solution $\mathbf{x} : \mathbf{R}^2 = \mathbf{C} \to \mathbf{E}^3$ determines an immersion with constant mean curvature $1/2$. Here, L, M and N are the coefficients of the second fundamental form of $\mathbf{x}$:

$$\begin{aligned} L &= L(\beta) &&= e^{\omega}(\sinh\omega \cos^2\beta + \cosh\omega \sin^2\beta), \\ M &= M(\beta) &&= -\sin\beta \cos\beta, \\ N &= N(\beta) &&= e^{\omega}(\cosh\omega \cos^2\beta + \sinh\omega \sin^2\beta). \end{aligned} \tag{2.3}$$

The fundamental forms of the immersion $\mathbf{x}$ with respect to the unit normal vector ξ are

$$\begin{aligned} |dx|^2 &= e^{2\omega}(du^2 + dv^2), \\ -dx \cdot d\xi &= L\,du^2 + 2M\,dudv + N\,dv^2. \end{aligned} \tag{2.4}$$

The principal curvatures of $\mathbf{x}$ are $\lambda_1 = \cosh\omega$ and $\lambda_2 = \sinh\omega$. The lines of curvature with respect to λ_1 and λ_2 are straight lines and β is the angle between λ_1-curvature lines and the u-axis.

Since the immersion $\mathbf{x}$ is determined by a, b, c, β up to a motion of $\mathbf{E}^3$, we write $\mathbf{x}(u, v) = \mathbf{x}(u, v;\ a, b, c, \beta)$.

The Gauss map $\tilde{\nu}$ of $\mathbf{x}$ determines a harmonic map of $\mathbf{R}^2 = \mathbf{C}$ into S^2. Using this, we construct harmonic maps of the torus of given conformal structure into S^2.

To construct harmonic maps of tori into S^2, it is enough to find (a, b, c, β) such that the normal vector ξ of $\mathbf{x}(\mathbf{u};\ \mathbf{a}, \beta)$ is doubly periodic with respect to $\mathbf{u}$.

By (2) in Lemma 2.2 and (2.4), there exist motions E_i $(i = 1, 2)$ of $\mathbf{E}^3$ which satisfies

$$\mathbf{x}(\mathbf{u} + 2\mathbf{p}_i;\ a, b, c, \beta) = E_i \circ \mathbf{x}(\mathbf{u};\ a, b, c, \beta).$$

Decompose E_i to their linear parts and parallel translations as

$$E_i = A_i + c_i \qquad (i = 1, 2),$$

where $A_i \in SO(3)$ and $c_i \in \mathbf{R}^3$ (cf. [9]). Then A_i $(i = 1, 2)$ are normalized by orthogonal matrices P_i as

$${}^tP_i A_i P_i = \begin{pmatrix} 1 & 0 & 0 \\ 0 & \cos\theta_i & -\sin\theta_i \\ 0 & \sin\theta_i & \cos\theta_i \end{pmatrix} \qquad (i = 1, 2).$$

The rotational angles θ_i $(i = 1, 2)$ depend on a, b, c and β. If both $\theta_i \in 2\pi\mathbf{Q}$ $(i = 1, 2)$, then the Gauss map $\tilde{\nu}_{\mathbf{x}}$ is invariant under a certain lattice.

Fix $(a, b, c) \in U$, where U is the neighborhood in Lemma 2.2. For this triplet, we define a map from a neighborhood of $(1, 0)$ to $\mathbf{R}^2$ as

$$\Phi_{abc} : (t, \beta) \mapsto (\theta_1(at, bt, ct, \beta), \theta_2(at, bt, ct, \beta)).$$

Then the following lemma holds.

Lemma 2.3 *For a certain* (a_0, b_0), *the following assertions hold.*

(1) *The derivative* $d\Phi_{a_0 b_0 0}$ *does not degenerate at* $(t, \beta) = (1, 0)$.

(2) *The image of the regular points of the Gauss map* $\tilde{\nu} : \mathbf{C} \to S^2$ *of* $\mathbf{x}(a_0, b_0, 0, 0)$ *is* S^2.

We prove this in the next section.

The proof of Theorem 2.1 follows this fact.

PROOF OF THEOREM 2.1. Take a_0 and b_0 as in Lemma 2.3. Then there exist a neighborhood V of $(a_0, b_0, 0)$ in $\mathbf{R}^3$ and a positive number ε which satisfy:

(1) If $(a, b, c) \in V$, then $d\Phi_{abc}$ does not degenerate at $(t, \beta) = (1, 0)$.

(2) If $(a, b, c) \in V$ and $|\beta| < \varepsilon$, the Gauss map of the immersion $\mathbf{x}(u, v;\ a, b, c, \beta)$ is surjective.

We denote the lattice generated by the periods $2\mathbf{p}_1 = 4(a, 0)$ and $2\mathbf{p}_2 = 4(b, c)$ of $\tilde{\omega}(a, b, c)$ as $\Gamma(2\mathbf{p}_1, 2\mathbf{p}_2)$ and define $T(a, b, c) = \mathbf{C}/\Gamma(2\mathbf{p}_1, 2\mathbf{p}_2)$.

Not that for a given torus T, the set of (a, b, c) for which there exists a conformal map of T to $T(a, b, c)$ is dense in V (see Appendix). Take (a, b, c) from this set and consider a holomorphic map

$$\varphi : T \to T(a, b, c). \tag{2.5}$$

Let $\mu_{0i} = \theta_i(a, b, c, 0)$ $(i = 1, 2)$. Then by (1), Φ_{abc} gives a diffeomorphism from a neighborhood of $(1, 0) \in \mathbf{R}^2$ to that of $(\mu_{01}, \mu_{02}) \in \mathbf{R}^2$. Take a pair (μ_1, μ_2) in the neighborhood of (μ_{01}, μ_{02}) as

$$\mu_i = \frac{n_i}{m_i} \cdot 2\pi \quad (i = 1, 2), \tag{2.6}$$

where n_i/m_i $(i = 1, 2)$ are irreducible fractions. Then there exists a pair (t, β) in the neighborhood of $(1, 0)$ such that

$$\theta_i(at, bt, ct, \beta) = \mu_i \quad (i = 1, 2). \tag{2.7}$$

The Gauss map $\tilde{\nu}(u, v;\ at, bt, ct, \beta)$ of the immersion $\mathbf{x}(u, v;\ at, bt, ct, \beta)$ is invariant under the action of $\Gamma(2m_1\mathbf{p}_1, 2m_2\mathbf{p}_2)$ at (t, β) in (2.7). Then this induces a harmonic map

$$\nu = \nu(u, v;\ at, bt, ct, \beta) : \mathbf{C}/\Gamma(2m_1\mathbf{p}_1, 2m_2\mathbf{p}_2) \to S^2.$$

Composing ν and the natural covering projection

$$\rho : T(a,b,c) = \mathbf{C}/\Gamma(2\mathbf{p}_1, 2\mathbf{p}_2) \to \mathbf{C}/\Gamma(\frac{2}{m_2}\mathbf{p}_1, \frac{2}{m_1}\mathbf{p}_2),$$

we have a harmonic map

$$\nu \circ \rho \circ \varphi : T \to S^2. \tag{2.8}$$

This map is surjective because of (2).

The Gauss map of a surface is holomorphic if and only if the surface is minimal, and anti-holomorphic if and only if the surface is totally umbilic. Hence the harmonic map (2.8) is neither holomorphic nor anti-holomorphic.

Since the set of (a,b,c) admitting a conformal map (2.5) is dense in V, one can construct countably many harmonic maps of T to S^2.

Now it is sufficient to show that there exist countably many equivalent classes of such harmonic maps.

Let (a_1, b_1, c_1) and (a_2, b_2, c_2) be two points of V admitting conformal maps as (2.5) which are sufficiently close in V. Then $\nu_i = \nu(u, v;\, a_i t_i, b_i t_i, c_i t_i, \beta_i)$ $(i = 1, 2)$ determine harmonic maps of T into S^2 for some (t_i, β_i). If $t_1 \neq t_2$, the functions $\omega_i = \omega(u, v;\, a_i t_i, b_i t_i, c_i t_i)$ $(i = 1, 2)$ are distinct each other because their fundamental periods are independent. Hence $(\mathbf{C}, e^{2\omega_1}|dz|^2)$ is not isometric to $(\mathbf{C}, e^{2\omega_2}|dz|^2)$. Then by Lemma 1.2, ν_1 and ν_2 are not equivalent. □

3 Proof of Lemma 2.3.

PROOF OF THE FIRST PART OF LEMMA 2.3.

In the case of $c = \beta = 0$, ω and $\mathbf{x}$ are the solutions explained by Abresch [1]. By his calculation, $\theta_1(ta_0, tb_0, 0, 0) \equiv 0$. Then $\partial\theta_1/\partial t = 0$ at $t = 1$ and $\beta = 0$. Take (a_0, b_0) such that $a_0 = b_0$. Observing the calculation in [1, Lemma 4.5], we have $\partial\theta_2/\partial t < 0$ at $t = 1$ and $\beta = 0$.

Now it is sufficient to prove $\partial\theta_1/\partial\beta \neq 0$. To prove this, we shall prepare some terminology. Let $e_1 = e^{-\omega}\mathbf{x}_u$, $e_2 = e^{-\omega}\mathbf{x}_v$, $e_3 = \xi$ be the Frenet frame of the immersion $\mathbf{x}(u, v;\, a_0, b_0, 0, \beta)$ and

$$X(u,\beta) = \begin{pmatrix} e_1 \\ e_2 \\ e_3 \end{pmatrix}.$$

for fixed $v = 0$. We normalize $X(0, \beta) = I$. So, $X(u, \beta)$ is expanded with respect to β as

$$X(u,\beta) = X_0(u) + \beta Y(u) + o(\beta),$$

where $X_0(u) = X(u, 0)$ and $Y(u) = \partial X/\partial\beta|_{\beta=0}$. Note that $X(4a_0, 0) = I$ because of [1]. So we have the following lemma by direct calculations.

Lemma 3.1

$$\left.\frac{\partial\theta_1}{\partial\beta}\right|_{\substack{t=1\\ \beta=0}} = 0 \textit{ if and only if } Y(4a_0) = 0.$$

PROOF. Since $X \in SO(3)$, there exists an orthogonal matrix $P = P(u,\beta) \in SO(3)$ which normalize X:

$$X(u,\beta) = {}^tPDP, \tag{3.1}$$

where

$$D = \begin{pmatrix} 1 & 0 & 0 \\ 0 & \cos\theta & -\sin\theta \\ 0 & \sin\theta & \cos\theta \end{pmatrix}.$$

Here, θ is a function of u such that $\theta(4a_0) = \theta_1$. Expand P and D with respect to β as

$$\begin{aligned} P &= P_0 + \beta T + o(\beta), \\ D &= I + \beta H + o(\beta), \end{aligned}$$

where

$$H = \left.\frac{\partial\theta_1}{\partial\beta}\right|_{\substack{t=1\\ \beta=0}} \times \begin{pmatrix} 0 & 0 & 0 \\ 0 & 0 & -1 \\ 0 & 1 & 0 \end{pmatrix}.$$

Since ${}^tPP = I$, we have

$${}^tP_0P_0 = I, \qquad {}^tP_0T - TP_0 = 0.$$

Then, by (3.1) and keeping in mind that $X_0(4a_0, 0) = I$,

$$\begin{aligned} X(4a_0,\beta) &= I + \beta\,{}^tPDP + o(\beta) \\ &= I + \beta\,{}^tP_0HP_0 + o(\beta). \end{aligned}$$

Hence the conclusion holds. □

To complete the proof, it is enough to show $Y(4a_0) \neq 0$.

The Frenet equation (2.2) shows that $e_2 \equiv (0,1,0)$ along the curve $\gamma(u) = \mathbf{x}(u, 0; a_0, b_0, 0, \beta)$. Then the curve $\gamma(u) = (x, 0, z)$ lies in xz-plane. Using this, we have

$$Y(4a_0) = X_0(4a_0)\begin{pmatrix} 0 & -z(4a_0) + z(0) & 0 \\ z(4a_0) - z(0) & 0 & -x(4a_0) + x(0) \\ 0 & x(4a_0) - x(0) & 0 \end{pmatrix}$$

by Wente's calculation [9, equation (3.6)].

Now we assume $a_0 = b_0$. In this case, the curve γ, the λ_1-curvature line of $x(a_0, b_0, 0, 0)$ never closes up [1]. Then $x(4a_0) - x(0)$ or $z(4a_0) - z(0)$ does not vanish. Hence $Y(4a_0) \neq 0$, and then $\partial\theta_1/\partial\beta \neq 0$ at $t = 1$ and $\beta = 0$. □

PROOF OF THE SECOND PART OF LEMMA 2.3.

Here, we shall prove the surjectivity of the Gauss map ν when $c = \beta = 0$. In this case, the solutions ω and $\mathbf{x}$ are those of Abresch [1]. Consider the curves

$$\begin{aligned}\gamma_1(u) &= \mathbf{x}(u, b_0;\ a_0, b_0, 0, 0)\\ \gamma_2(v) &= \mathbf{x}(a_0, v;\ a_0, b_0, 0, 0),\end{aligned}$$

which are the curvature lines of the immersion $\mathbf{x} = \mathbf{x}(u, v;\ a_0, b_0, 0, 0)$. Both γ_1 and γ_2 are plane curve in $\mathbf{E}^3$. Moreover the first normal vector of γ_i coincides with the normal vector ξ of $\mathbf{x}$ along γ_i. The behavior of normal vectors along γ_1 and γ_2 is analyzed by Abresch [1]. First, the Gauss image of γ_2 is a great circle in S^2. We assume this circle is the equator.

Then the Gauss image of γ_1 contains a meridian as a regular set of the Gauss map [1, page 185]. Hence, there exists an open strip F containing this meridian in the regular set of the Gauss image of the surface.

In case $\theta_2 = (n/m)2\pi$, the equator intersects m-times with copies of the Gauss image of γ_1.

Now we perturb a_0 and b_0 such that the denominator m of $\theta_2/(2\pi)$ is sufficiently large. Then $m$copies of the strip F cover whole S^2. Hence the image of the regular points of the Gauss map is S^2. □

A Appendix.

The purpose of this section is to prove the following proposition referred in Section 2.

Throughout this section, we denote a lattice of $\mathbf{C} = \mathbf{R}^2$ generated by $\mathbf{p}_1$ and $\mathbf{p}_2$ by $\Gamma(\mathbf{p}_1, \mathbf{p}_2)$.

Propostion A.1 *Let $T_0 = \mathbf{C}/\Gamma(\mathbf{p}_1, \mathbf{p}_2)$ be a given torus. Then for any torus T, the set of lattices Γ which admit holomorphic maps from T to $\mathbf{C}/\Gamma$ is dense in the neighborhood of $\Gamma(\mathbf{p}_1, \mathbf{p}_2)$.*

In the proof of this proposition, the following lemma is essential.

Lemma A.2 *Let $\Gamma(\mathbf{p}_1, \mathbf{p}_2)$ be a lattice of $\mathbf{C}$. Then for any rational numbers λ and μ ($\mu \neq 0$), there exists a holomorphic map from a torus $\Gamma(\mathbf{p}_1, \mathbf{p}_2)$ to $\mathbf{C}/\Gamma(p_1, \lambda\mathbf{p}_1 + \mu\mathbf{p}_2)$.*

PROOF. First, we claim that there exists a holomorphic map

$$\varphi : \mathbf{C}/\Gamma(\mathbf{p}_1, \mathbf{p}_2) \to \mathbf{C}/\Gamma(\mathbf{p}_1, \frac{n}{m}\mathbf{p}_2),$$

where m and n are integers. In fact, composing the natural projection

$$\varphi_1 : \mathbf{C}/\Gamma(\mathbf{p}_1, \mathbf{p}_2) \to \mathbf{C}/\Gamma(\frac{1}{n}\mathbf{p}_1, \frac{1}{m}\mathbf{p}_2)$$

and the homothety

$$\varphi_2 : \mathbf{C}/\Gamma(\frac{1}{n}\mathbf{p}_1, \frac{1}{m}\mathbf{p}_2) \to \mathbf{C}/\Gamma(\mathbf{p}_1, \frac{n}{m}\mathbf{p}_2),$$

we have a desired holomorphic map.

If $\lambda = 0$, the conclusion follows the above fact immediately.

Assume $\lambda \neq 0$. Since $\Gamma(\mathbf{p}_1, \mathbf{p}_1 + (\mu/\lambda)\mathbf{p}_2) = \Gamma(\mathbf{p}_1, (\mu/\lambda)\mathbf{p}_2)$, we have a holomorphic map

$$\psi_1 : \mathbf{C}/\Gamma(\mathbf{p}_1, \mathbf{p}_2) \to \mathbf{C}/\Gamma(\mathbf{p}_1, \mathbf{p}_1 + \frac{\mu}{\lambda}\mathbf{p}_2)$$

using the above fact. Moreover, there exists a holomorphic map

$$\psi_2 : \mathbf{C}/\Gamma(\mathbf{p}_1, \mathbf{p}_2) \to \mathbf{C}/\Gamma(\mathbf{p}_1, \lambda(\mathbf{p}_1 + \frac{\mu}{\lambda}\mathbf{p}_2)).$$

Then a holomorphic map $\psi_2 \circ \psi_1$ from $\mathbf{C}/\Gamma(\mathbf{p}_1, \mathbf{p}_2)$ to $\mathbf{C}/\Gamma(\mathbf{p}_1, \lambda\mathbf{p}_1 + \mu\mathbf{p}_2)$ is obtained.
□

PROOF OF PROPOSITION A.1. We can assume one of the generator of the lattice of T coincides with that of T_0 without any loss of generality. That is, we assume

$$T = \mathbf{C}/\Gamma(\mathbf{p}_1, \mathbf{r}).$$

Then, to prove the proposition, it is enough to show that the following claim.

For any $\varepsilon > 0$, there exists a vector $\mathbf{q}$ such that

(1) $|\mathbf{p}_2 - \mathbf{q}| < \varepsilon$, and

(2) the lattice $\Gamma(\mathbf{p}_1, \mathbf{q})$ admits a holomorphic map from $\mathbf{C}/\Gamma(\mathbf{p}_1, \mathbf{r})$ to $\mathbf{C}/\Gamma(\mathbf{p}_1, \mathbf{q})$.

Since $\mathbf{p}_1$ and $\mathbf{r}$ are linearly independent, there exist numbers α and β $(\beta \neq 0)$ such that

$$\mathbf{p}_2 = \alpha\mathbf{p}_1 + \beta\mathbf{r}.$$

Take rational numbers λ and μ which satisfy

$$|\lambda - \alpha| < \frac{\varepsilon}{2|\mathbf{p}_1|} \quad \text{and} \quad |\mu - \beta| < \frac{\varepsilon}{2|\mathbf{r}|}.$$

Then $\mathbf{q} = \lambda\mathbf{p}_1 + \mu\mathbf{r}$ satisfies the conditions (1) and (2). □

References

[1] U. Abresch, *Constant mean curvature tori in terms of elliptic functions,* J. Reine Angew. Math., **374**(1987),169–192.

[2] J. Eells and J. C. Wood, *Restrictions on harmonic maps of surfaces,* Topology, **15**(1976),263–266.

[3] J. Eells and J. C. Wood, *Harmonic maps from surfaces to complex projective space*, Adv. in Math., **49**(1983),217–263.

[4] K. Kenmotsu, *Weierstrass formula for surfaces of prescribed mean curvature*, Math. Ann., **245**(1979),89–99.

[5] L. Lemaire, *Applications harmoniques de surfaces riemanniennes*, J. Differential Geometry, **13**(1978),51–78.

[6] R. T. Smith, *Harmonic mappings of spheres*, PhD thesis, Univ. of Warwick, 1972.

[7] J. Spruck, *The elliptic sinh Gordon equation and the construction of troidal soap bubbles*, In Lect. Notes in Math. Vol. 1340, pages 275–301, Springer-Verlag, 1988.

[8] H. C. Wente, *Counterexample to a conjecture of H. Hopf*, Pacific J. of Math., **121** (1986),193–243.

[9] H. C. Wente. *Twisted tori in constant mean curvature in* R^3, Seminar on New Results in Nonlinear Partial Differential Equations, Aspects of Mathematics, pages 1–36, Max-Planck-Institut für Mathematik, Bonn, 1987.

Masaaki UMEHARA

Institute of Mathematics
University of Tsukuba
Tsukuba, Ibaraki 305
Japan

Kotaro YAMADA

Department of Mathematics
Faculty of Science and Technology
Keio University
Yokohama 223
Japan

Chapter III
Lie Sphere Geometry

WAVE EQUATION AND ISOPARAMETRIC HYPERSURFACES

Jyoichi Kaneko

Dedicated to Professor Toshihusa Kimura on his 60th birthday

§1. Introduction

A simple progressing wave solution, called a spw solution for short, of the (n+1)-dimensional wave equation

$$\Box u \equiv \frac{\partial^2 u}{\partial t^2} - \sum_{i=1}^{n} \frac{\partial^2 u}{\partial (x^i)^2} = 0 \tag{1.1}$$

is, by definition, a solution of the form

$$u = U(t,x)f(S(t,x)), \tag{1.2}$$

where U and S are specific C^∞ functions defined on some open set in $\mathbb{R}^{n+1}$ and f is an arbitrary distribution on $\mathbb{R}$. It is readily observed that (1.2) remains a solution of (1.1) for arbitrary f if and only if U and S satisfy the following system

$$\langle \nabla S, \nabla S \rangle = 0, \quad 2\langle \nabla S, \nabla U \rangle + U\Box S = 0, \quad \Box U = 0, \tag{1.3}$$

ISBN 0-12-640170-5

where $\langle\ ,\ \rangle$ denotes the Lorentzian inner product. For a spw solution of the form

$$u = V(x)f(t-\tau(x)), \tag{1.4}$$

the conditions (1.3) become

$$|\mathrm{grad}\tau(x)|^2 = 1,\ 2\mathrm{grad}\tau\cdot\mathrm{grad}V + V\Delta\tau = 0,\ \Delta V = 0, \tag{1.5}$$

where Δ denotes the Laplacian in $\mathbb{R}^n$.

As to the existence of a spw solution, it is known by Friedlander's procedure [4,pp.111] that the general case (1.3) can be reduced locally to the special case (1.5), in the sense that a solution of (1.5) can be constructed from that of (1.3). Moreover, the family of wave fronts of (1.2): $\{(t,x)|S(t,x) = \text{const.},\ t = \text{const.}\}$, coincides with that of (1.4): $\{(t,x)|\tau(x) = \text{const.},\ t = \text{const.}\}$.

In [3] Friedlander found that when $n = 3$ the necessary and sufficient condition for the existence of a spw solution of (1.1) of the form (1.4) is that the surfaces $\{x\in\mathbb{R}^3|\tau(x) = \text{const.}\}$ are cyclides of Dupin. This can be generalized to higher dimensions as follows:

Theorem([6]). *If the wave equation* (1.1) *has a spw solution of the form* (1.4), *then the wave fronts* $\{x\in\mathbb{R}^n|\tau(x) = \text{const.}\}$ *are Dupin hypersurfaces. Conversely, if* $|\mathrm{grad}\tau| = 1$ *and* $\{x\in\mathbb{R}^n|\tau(x) = 0\}$ *is a Dupin hypersurface having exactly two distinct principal curvatures at each point, then there exists a function* $V \not\equiv 0$ *which satisfies* (1.5).

Here by a Dupin hypersurface we mean a hypersurface in $\mathbb{R}^n$ such that each principal curvature is constant along its leaf of principal distribution [11].

In this paper we shall investigate the existence of nontrivial wave amplitude V for Dupin hypersurfaces with three or more distinct principal curvatures. In particular we shall be concerned with the cases where the Dupin hypersurfaces are Lie equivalent to isoparametric hypersurfaces in spheres, i.e., hypersurfaces having constant principal curvatures in spheres. For the definition of Lie equivalence, refer to [11] and [1].

Consider the spherical wave equation:

$$\frac{\partial^2 u}{\partial t^2} - \Delta_S u + \left(\frac{n-1}{2}\right)^2 u = 0, \tag{1.6}$$

where Δ_S denotes the Laplace-Beltrami operator on the n-dimensional unit sphere S^n. It is well known that by a conformal transformation of $\mathbb{R} \times S^n$ to $\mathbb{R}^{n+1}$ and multiplication of solution by a suitable function, the Euclidean wave equation (1.1) can be locally transformed into the spherical wave equation (1.6), cf.[7]. Moreover, the wave fronts of a spw solution of (1.1) are Lie equivalent to those of (1.6), and every Lie equivalence is induced in this manner, cf.[6],[11]. Hence we may work directly with the spherical wave equation and investigate the cases where the wave fronts are isoparametric hypersurfaces. Since Friedlander's procedure works also for the spherical wave equation, it suffices to consider the spw solution of the form (1.4). The conditions (1.5) are now transformed into

$$|\mathrm{grad}\tau| = 1,\ 2\,\mathrm{grad}\tau\cdot\mathrm{grad}V + V\Delta_S\tau = 0,$$
$$\Delta_S V - \left(\frac{n-1}{2}\right)^2 V = 0. \tag{1.7}$$

We now state our main result of this paper.

Theorem. *Suppose* $|\mathrm{grad}\tau| = 1$ *and that* $\{x\in S^n \mid \tau(x) = 0\}$ *is an isoparametric hypersurface in* S^n *having three or more distinct principal curvatures with the same multiplicity* $m = 1$ *or* 2. *Then the following hold.*

(1) *If* $m = 1$, *then there exists no nontrivial* V *which satisfies* (1.7).

(2) *If* $m = 2$, *then there exists a nontrivial* V *which satisfies* (1.7).

It should be noted that the number p of distinct principal curvatures is restricted: $p\in\{1,2,3,4,6\}$, [8]. Moreover, except possibly the case $p = 6$, $m = 2$, the pair (p,m) with $m = 1$ or 2 determines the isoparametric hypersurface uniquely, cf.[1].

The arguments in the paper will be purely local.

§2. Isoparametric hypersurfaces in spheres

First we review the basic results on isoparametric hypersurfaces in spheres, cf.[1],[8].

Let $M \subset S^n$ ($\subset \mathbb{R}^{n+1}$) be an isoparametric hypersurface with p distinct curvatures and ξ the unit normal vector field to M in S^n. Let $\cot\theta_i$ ($0 < \theta_0 <\ldots< \theta_{p-1} < \pi$) be the principal curvatures of M with respect to ξ, and m_i the multiplicities of $\cot\theta_i$. Münzner[8] showed that (1) $\theta_i = \theta_0 +$

$i\pi/p$ (2) $m_i = m_{i+2}$ (indices mod.p) (3) $p\in\{1,2,3,4,6\}$ (4) if $p = 3$, then $m_1 = m_2 = m_3$. For each $x \in M$, $t \in \mathbb{R}$, let ϕ_θ: $M \longrightarrow S^n$ be the normal exponential map defined by

$$\phi_\theta(x) = \cos\theta\, x + \sin\theta\, \xi_x.$$

It is known that if θ is not congruent to any θ_i mod.π, then ϕ_θ immerses M as an isoparametric hypersurface with principal curvatures $\cot(\theta_i - \theta)$ of multiplicity m_i, $i = 0,\dots,p-1$, cf. [1]. We denote $\phi_\theta(M)$ by M_θ. Münzner's key result states that there exists a homogeneous polynomial F of degree p in $\mathbb{R}^{n+1}$, called a Cartan polynomial, such that the pararell family $\{ M_\theta \}$ are level sets of the restriction of F to S^n:

$$M_\theta = \{x\in S^n | f(x) = \cos p(\theta_0 - \theta)\},$$

where $f = F|_{S^n}$. F satisfies

$$|\mathrm{grad}\, F|^2 = p^2 r^{2p-2}, \quad \Delta F = c r^{p-2}, \quad c = p^2(m_1 - m_2)/2,$$

where $r = ((x^1)^2 + \cdots + (x^{n+1})^2)^{1/2}$. It follows from this that

$$|\mathrm{grad}\, f| = p^2(1-f^2), \quad \Delta_S f = -p(p+n-1)f + c.$$

Put $\tau = \theta_0 - \theta = \frac{1}{p}\arccos f(x)$. Then we obtain

$$|\mathrm{grad}\tau| = 1, \quad \Delta_S\tau = (n-1)\cot p\tau - \frac{c}{p \sin p\tau}. \tag{2.1}$$

We are going to get a spw solution of (1.6) usig this τ. Substituting τ into the second equation of (1.7), we have

$$2\mathrm{grad}\tau\cdot\mathrm{grad}V + \{(n-1)\cot p\tau - \frac{c}{p\sin p\tau}\}V = 0. \qquad (2.2)$$

Since $\mathrm{grad}\tau\cdot\mathrm{grad}V = \partial V/\partial\tau$, it is easy to solve (2.2) to get

$$V = \frac{\psi}{(\sin p\tau)^{(n-1)/2p}}(\tan(p\tau/2))^{c/2p^2},$$

$$\partial\psi/\partial\tau = 0. \qquad (2.3)$$

Substituting this into the third equation of (1.7) and using (2.1), we obtain

$$\Delta_S\psi + \frac{1}{\sin^2 p\tau}\{(\frac{n-1}{2})(p-\frac{n-1}{2}) - \frac{c}{4p^2} + \frac{c}{2}(\frac{n-1}{p}-1)\cos p\tau\}\psi = 0.$$

As it seems hard to deal with the general case, we shall restrict ourselves to the case that the principal curvatures have the same multiplicity $m = 1$ or 2. Then, as $c = 0$, we get

$$\Delta_S\psi + \frac{(n-1)(p-(n-1)/2)}{2\sin^2 p\tau}\psi = 0. \qquad (2.4)$$

Now the part (2) of our theorem is evident: If $m = 2$, $\psi \equiv$ const. $\neq 0$ is a nontrivial solution of (2.3) and (2.4), for $2p = n-1$ in this case.

§3. Homogeneous hypersurfaces in spheres

It is known that the isoparametric hypersurfaces with $m = 1$ are homogeneous [1],[2],[10]. Moreover, it is shown by Hsiang and Lawson [5] that every homogeneous hypersurface in the unit sphere is represented as an orbit of a linear isotropy group of a Riemannian symmetric space of rank 2. follow the presentations in [9] and [12].

Let $(\mathfrak{u},\theta)$ be an effective orthogonal symmetric Lie algebra of compact type of rank 2 and (U,K) a symmetric pair associated to $(\mathfrak{u},\theta)$. Let $\mathfrak{k}$ be the Lie algebra of K and $\mathfrak{u} = \mathfrak{k} + \mathfrak{p}$ the orthogonal decomposition of $\mathfrak{u}$ with respect to a fixed $\mathrm{Ad}(U)$-invariant inner product $(\ ,\ )$ on $\mathfrak{u}$. We choose a maximal abelian subspace $\mathfrak{a}$ in $\mathfrak{p}$ and denote by Σ the set of all roots of $(\mathfrak{u},\theta)$ with respect to $\mathfrak{a}$. Let Σ_+ be the set of all positive elements in Σ with respect to a fixed linear order in the dual space of $\mathfrak{a}$. Then $\mathfrak{k}$ and $\mathfrak{p}$ have the following decomposition:

$$\mathfrak{k} = \mathfrak{k}_0 + \sum \mathfrak{k}_\lambda, \quad \mathfrak{p} = \mathfrak{a} + \sum \mathfrak{p}_\lambda, \quad \lambda \in \Sigma_+,$$

where

$$\mathfrak{k}_\lambda = \{X \in \mathfrak{k} \mid (\mathrm{ad}\, H)^2 X = -(\lambda(H))^2 X \text{ for all } H \in \mathfrak{a}\},$$

$$\mathfrak{p}_\lambda = \{X \in \mathfrak{p} \mid (\mathrm{ad}\, H)^2 X = -(\lambda(H))^2 X \text{ for all } H \in \mathfrak{a}\}.$$

Note that $\mathfrak{k}_\lambda = \mathfrak{k}_{-\lambda}$, $\mathfrak{p}_\lambda = \mathfrak{p}_{-\lambda}$. It is known that $\mathfrak{k}_0$ is the centralizer of $\mathfrak{a}$ in $\mathfrak{k}$ and $\dim \mathfrak{k}_\lambda = \dim \mathfrak{p}_\lambda$, which we denote by $m(\lambda)$.

Let $H \in \mathfrak{a}$ be a unit regular element, i.e. $\lambda(H) \neq 0$ for all $\lambda \in \Sigma_+$, and L the stabilizer of the adjoint action of K at H whose Lie algebra $\mathfrak{l}$ is $\{X \in \mathfrak{k} \mid ad(X)(H) = 0\} = \mathfrak{k}_0$. Define an embedding Φ_H of K/L into $\mathfrak{p}$ by $\Phi_H(kL) = Ad(k)H$. Since the adjoint action is an isometry and H is a unit regular element, the image N(H) of Φ_H is a hypersurface of the unit sphere S with respect the inner product (,). We identify the tangent space of $\mathfrak{p}$ with $\mathfrak{p}$ itself and give K/L the Riemannian metric g induced by the embedding Φ_H. Put $\mathfrak{m} = \sum_{\lambda \in \Sigma_+} \mathfrak{k}_\lambda$. Then we have the Ad(L)-invariant decomposition $\mathfrak{k} = \mathfrak{l} + \mathfrak{m}$. Identifying the tangent space $T_o(K/L)$ at the origin $o = \{L\}$ with $\mathfrak{m}$ by $\mathfrak{m} \ni X \longmapsto X_o \in T_o(K/L)$, we get an inner product < , > on $\mathfrak{m}$ defined by $\langle X,Y \rangle = g(X_o,Y_o) = ([X,H],[Y,H])$, $X,Y \in \mathfrak{m}$. Hence $\{|\lambda(H)|^{-1} X_{\lambda,i} \mid \lambda \in \Sigma_+, i=1,\ldots,m(\lambda)\}$ is an orthonormal basis of $\mathfrak{m}$ with respect < , >, where $\{X_{\lambda,i} \mid i=1,\ldots,m(\lambda)\}$ is an orthonormal basis of $\mathfrak{k}_\lambda$ with respect to (,) for each $\lambda \in \Sigma_+$. Now the Laplacian Δ_H of (K/L,g) is given by

$$\Delta_H = \sum_{\lambda \in \Sigma_+} \sum_{i=1}^{m(\lambda)} \lambda(H)^{-2} L^2_{X_{\lambda,i}},$$

where L_X denotes the Lie derivation on K with respect to the left invariant vector field X.

The homogeneous hypersurface N(H) in S is isoparametric and its principal curvatures and multiplicities are known as follows [12]. We choose $Z \in \mathfrak{a}$ in such a way that {H,Z} is an orthonormal basis of $\mathfrak{a}$. Let $\Sigma_+^* = \{\lambda \in \Sigma_+ \mid \lambda/2 \notin \Sigma_+\}$ and put $\Sigma_+^* = \{\lambda_0, \lambda_1, \ldots, \lambda_{p-1}\}$, $p = |\Sigma_+^*|$. Let H_i, $i=0,\ldots,p-1$, be unit vectors satisfying $\lambda_i(H_i) = 0$, and $0 < \theta_i < \pi$, the angles between H and H_i. We may choose λ_i, $i=0,\ldots,p-1$, in such a way that $0 < \theta_0 < \ldots < \theta_{p-1}$. Then the distinct principal curvatures k_i of $N(H)$ with respect to the normal vector Z and their multiplicities m_i are given by $k_i = -\lambda_i(Z)/\lambda_i(H)$, $i=0,\ldots,p-1$, and $m_i = m(\lambda_i) + m(2\lambda_i)$, $i=0,\ldots,p-1$. By the definition of θ_i, we have

$$-\lambda_i(Z)/\lambda_i(H) = \cot \theta_i, \quad i = 0,1,\ldots,p-1.$$

Let $\theta \in \mathbb{R}$ be not congruent to any θ_i mod.π, and put

$$H_\theta = \cos\theta\, H + \sin\theta\, Z.$$

Then H_θ is a unit regular element. By the definition of θ_i we have

$$(\lambda_i(H_\theta))^2 = (\lambda_i,\lambda_i)\sin^2(\theta_i - \theta), \quad i = 0,1,\ldots,p-1. \tag{3.1}$$

We identify $N(H)$ with M in §2. Then, choosing Z properly, we get $N(H_\theta) = M_\theta$. Hence the Laplacian Δ_θ of M_θ is given by

$$\Delta_\theta = \sum_{\lambda\in\Sigma_+} \sum_{i=1}^{m(\lambda)} \lambda(H_\theta)^{-2} L_{X_{\lambda,i}}. \tag{3.2}$$

Here we identify the functions on M_θ with those on K/L_θ by the embedding Φ_{H_θ}, L_θ being the stabilizer of $Ad(K)$ at H_θ.

§4. Case of multiplicity one

4.1. The list of effective orthogonal symmetric Lie algebras of compact type of rank 2 with $m_i = 1$ is the following [12]:

(1) $(\mathfrak{u},\mathfrak{k}) = (\mathfrak{su}(3),\mathfrak{so}(4))$, $p = 3$,

(2) $(\mathfrak{u},\mathfrak{k}) = (\mathfrak{so}(5),\mathfrak{so}(3)+\mathfrak{so}(2))$, $p = 4$,

(3) $(\mathfrak{u},\mathfrak{k}) = (G_2,\mathfrak{so}(4))$, $p = 6$.

Note that $\Sigma_+ = \Sigma_+^* = \{\lambda_0,\lambda_1,\dots,\lambda_{p-1}\}$ and that $\mathfrak{l} = \mathfrak{k}_0 = \{0\}$ in these cases. Since $\dim \mathfrak{k}_\lambda = 1$, $\lambda\in\Sigma_+$, we denote $X_{\lambda_i,1}$ by X_i, $i = 0,1,\dots,p-1$.

Let ψ be the common solution of (2.3) and (2.4) with $n-1 = p$. Since $\partial\psi/\partial\theta = 0$ by (2.3), it follows from (3.2) that the equation (2.4) reduces to

$$\sum_{i=0}^{p-1} \lambda(H_\theta)^{-2} L_{X_i}^2 \psi + \frac{p^2}{4\sin^2 p(\theta_0 - \theta)} \psi = 0. \tag{4.1}$$

In the following, we denote $L_X\psi$ by $X\psi$, $L_X^2\psi$ by $X^2\psi$ etc. for simplicity. We substitute (3.1) into (4.1) and multiply both sides of (4.1) by $\sin^2(\theta_i - \theta)$. Then, letting $\theta \longrightarrow \theta_i$, we get

$$\frac{4}{(\lambda_i,\lambda_i)} X_i^2 \psi + \psi = 0, \quad i = 0,1,\dots,p-1. \tag{4.2}$$

Conversely, if ψ satisfies (4.2), then (4.1) holds by virtue of the following identity:

$$\sum_{i=0}^{p-1} \frac{1}{\sin^2(\theta+i\pi/p)} = \frac{p^2}{\sin^2 p\theta}.$$

We shall prove that the equations (4.2), $i=0,1,\dots,p-1$, have no nontrivial common solution in each case (1),(2) and

(3) respectively.

4.2. Case of $(\mathfrak{su}(3), \mathfrak{so}(3))$. Let $(\mathfrak{u}, \mathfrak{k}) = (\mathfrak{su}(3), \mathfrak{so}(3))$, $\mathfrak{p} = \{\sqrt{-1}\, X \mid X \text{ real } 3 \times 3\text{-matrix}, X - {}^tX = 0, \mathrm{tr}X = 0\}$ and $(U,K) = (SU(3), SO(3))$, where ${}^tX$ (resp. $\mathrm{tr}X$) is the transpose (resp. the trace) of X. Put $\mathfrak{a} = \{\sqrt{-1}\, \mathrm{diag}(\xi_1, \xi_2, \xi_3) \mid \xi_i \in \mathbb{R}, \sum_{i=1}^{3} \xi_i = 0\}$ where $\mathrm{diag}(\xi_1, \xi_2, \xi_3)$ is the diagonal 3×3 matrix whose diagonal entries are ξ_1, ξ_2 and ξ_3. Let (,) be the Ad(U)-invariant inner product on $\mathfrak{su}(3)$ defined by $(X,Y) = -2^{-1}\mathrm{tr}XY$. Σ_+^* is given by $\Sigma_+^* = \{\xi_i - \xi_j \mid 1 \le i < j \le 3\}$. We put $\lambda_0 = \xi_2 - \xi_3$, $\lambda_1 = \xi_1 - \xi_2$, $\lambda_2 = \xi_1 - \xi_3$. Then the unit root vectors in $\mathfrak{k}$ are

$$X_0 = \begin{pmatrix} 0 & 0 & 0 \\ 0 & 0 & -1 \\ 0 & 1 & 0 \end{pmatrix}, \quad X_1 = \begin{pmatrix} 0 & -1 & 0 \\ 1 & 0 & 0 \\ 0 & 0 & 0 \end{pmatrix}, \quad X_2 = \begin{pmatrix} 0 & 0 & 1 \\ 0 & 0 & 0 \\ -1 & 0 & 0 \end{pmatrix},$$

so that we have

$$[X_i, X_j] = X_k, \tag{4.3}$$

where (i,j,k) is a cyclic permutation of (0,1,2). It is clear that $(\lambda_i, \lambda_i) = 4$, i=0,1,2. Thus in this case the equations (4.2) become

$$X_i^2\psi + \psi = 0, \quad i=0,1,2. \tag{4.4}$$

If we compute $X_i^2X_j^2\psi - X_j^2X_i^2\psi$ using (4.3) and (4.4), we get

$$X_iX_jX_k\psi = -2^{-1}\psi, \tag{4.5}$$

where (i,j,k) is a cyclic permutation of (0,1,2).

Let W_1 be the complex vector space defined by

$$W_1 = \mathbb{C}\psi + \sum_{i=0}^{2} \mathbb{C}X_i\psi + \sum_{0\le i<j\le 2} \mathbb{C}X_iX_j\psi.$$

Then W_1 is an irreducible $\mathfrak{k}$-module by virtue of (4.3),(4.4), (4.5) and complete reducibility. It is well known that if we put dim $W_1 = 2l+1$, then $(X_0^2+X_1^2+X_2^2)\varphi + l(l+1)\varphi = 0$ for any $\varphi\in W_1$. So if dim $W_1 \neq 0$, then it follows from (4.4) that $l(l+1) = 3$. This contradicts that l is an integer or a half integer.

4.3. Case of $(\mathfrak{so}(5),\mathfrak{so}(3)+\mathfrak{so}(2))$. Let $\mathfrak{k} = \mathfrak{so}(3)+\mathfrak{so}(2) = \{\begin{pmatrix} A & 0 \\ 0 & B \end{pmatrix} | A\in\mathfrak{so}(3),\ B\in\mathfrak{so}(2)\}$, $\mathfrak{p} = \{\begin{pmatrix} 0 & X \\ -{}^tX & 0 \end{pmatrix} | X$ real 3×2 matrix$\}$. We denote $\begin{pmatrix} A & 0 \\ 0 & B \end{pmatrix}\in\mathfrak{k}$ by $A + B$. Put $\mathfrak{a} = \{\begin{pmatrix} 0 & X \\ -{}^tX & 0 \end{pmatrix}\in\mathfrak{p} |\ {}^tX = \begin{pmatrix} \xi_1 & 0 & 0 \\ 0 & \xi_2 & 0 \end{pmatrix}\}$. The inner product is defined by $(X,Y) = -2^{-1}\mathrm{tr}XY$, $X,Y\in\mathfrak{so}(5)$. Then $\Sigma_+^* = \{\xi_1,\xi_2,\xi_1-\xi_2,\xi_1+\xi_2\}$. We put $\lambda_0 = \xi_1$, $\lambda_1 = \xi_2$, $\lambda_2 = \xi_1-\xi_2$, $\lambda_3 = \xi_1+\xi_2$. Then the root vectors are

$$X_0 = \begin{pmatrix} 0 & 0 & 1 \\ 0 & 0 & 0 \\ -1 & 0 & 0 \end{pmatrix} + \begin{pmatrix} 0 & 0 \\ 0 & 0 \end{pmatrix},\quad X_1 = \begin{pmatrix} 0 & 0 & 0 \\ 0 & 0 & -1 \\ 0 & 1 & 0 \end{pmatrix} + \begin{pmatrix} 0 & 0 \\ 0 & 0 \end{pmatrix},$$

$$X_2 = \begin{pmatrix} 0 & -1 & 0 \\ 1 & 0 & 0 \\ 0 & 0 & 0 \end{pmatrix} + \begin{pmatrix} 0 & -1 \\ 1 & 0 \end{pmatrix},\quad X_3 = \begin{pmatrix} 0 & -1 & 0 \\ 1 & 0 & 0 \\ 0 & 0 & 0 \end{pmatrix} + \begin{pmatrix} 0 & 1 \\ -1 & 0 \end{pmatrix}.$$

So $\{X_0, X_1, 2^{-1/2}X_2, 2^{-1/2}X_3\}$ is an orthonormal basis of $\mathfrak{k}$ with respect to (,). It is readily seen that $(\lambda_i,\lambda_i) = 1$, $i=0, 1$, $(\lambda_j,\lambda_j) = 2$, $j=2,3$. Hence the equations (4.2) become

$$4X_i^2\psi + \psi = 0, \quad i = 0,1 \tag{4.6}$$

$$X_j^2\psi + \psi = 0, \quad j = 2,3. \tag{4.7}$$

We put $Y_0 = X_0$, $Y_1 = X_1$, $Y_2 = 2^{-1}(X_2 + X_3)$, so that $[Y_i, Y_j] = Y_k$, where (i,j,k) is a cyclic permutation of $(0,1,2)$. It follows from (4.7) and $[X_2, X_3] = 0$ that

$$Y_2^3\psi + Y_2\psi = 0. \tag{4.8}$$

By computing $Y_0^2Y_1^2\psi - Y_1^2Y_0^2\psi$, we get

$$Y_0Y_1Y_2\psi = 2^{-1}Y_2^2\psi. \tag{4.9}$$

Let W_2 be the complex vector space defined by

$$W_2 = \mathbb{C}\psi + \sum_{i=0}^{2} \mathbb{C}Y_i\psi + \sum_{0\leq i<j\leq 2} \mathbb{C}Y_iY_j\psi + \mathbb{C}Y_2^2\psi + \mathbb{C}Y_0Y_2^2\psi + \mathbb{C}Y_1Y_2^2\psi.$$

By virtue of (4.7),(4.8),(4.9) and complete reducibility, W_2 is an irreducible $\mathfrak{so}(3)$-module, $\mathfrak{so}(3) = \sum_{i=0}^{2} \mathbb{C}Y_i$. Put $\varphi = Y_2\psi$. Then by (4.6) and (4.8) we obtain $(Y_0^2+Y_1^2+Y_2^2)\varphi + 3/2\ \varphi = 0$. Put $\dim W_2 = 2l+1$ and suppose $\varphi \neq 0$. Then, as in 4.2, we get $l(l+1) = 3/2$, a contradiction. Hence $\varphi = 0$ and we have $(Y_0^2+Y_1^2+Y_2^2)\psi + 2^{-1}\psi = 0$. So if $\psi \neq 0$, we get $l(l+1) = 2^{-1}$, which is also a contradiction.

4.4. Case of $(G_2, \mathfrak{so}(4))$. We follow the notation in [5,pp.27-29]. Let $\{E_{ij}\}$ be the standard basis of 7×7 matrices with coefficients in $\mathbb{R}$. Put $G_{ij} = E_{ij} - E_{ji}$, $i,j = 1,\ldots 7$, and

$$\mathfrak{G}_i = \{\eta_1 G_{i+1,i+3} + \eta_2 G_{i+2,i+6} + \eta_3 G_{i+4,i+5} | \eta_i \in \mathbb{R},\ \Sigma\eta_i = 0\}$$

Put $\mathfrak{G} = \sum_{i=1}^{7} \mathfrak{G}_i$. Then the following commutation relations hold

$$[\mathfrak{G}_i, \mathfrak{G}_i] = \{0\},\ [\mathfrak{G}_i, \mathfrak{G}_{i+1}] = \mathfrak{G}_{i+3},\ [\mathfrak{G}_{i+1}, \mathfrak{G}_{i+3}] = \mathfrak{G}_i,$$

$$[\mathfrak{G}_{i+3}, \mathfrak{G}_i] = \mathfrak{G}_{i+1}.$$

$\mathfrak{G}$ is a compact simple Lie algebra of type G_2. We put

$$\mathfrak{k} = \mathfrak{G}_3 + \mathfrak{G}_4 + \mathfrak{G}_6, \quad \mathfrak{p} = \mathfrak{G}_1 + \mathfrak{G}_2 + \mathfrak{G}_5 + \mathfrak{G}_7.$$

We take $\mathfrak{a} = \mathfrak{G}_1 = \{\xi_1 G_{2,4} + \xi_2 G_{3,7} + \xi_3 G_{5,6} | \xi_i \in \mathbb{R},\ \Sigma\xi_i = 0\}$. We define the inner product (,) on $\mathfrak{G}$ by $(X,Y) = -2^{-1}\mathrm{tr}XY$, $X,Y \in \mathfrak{G}$. Σ_+^* consists of 6 roots: $\lambda_0 = -\xi_2$, $\lambda_1 = \xi_1 - \xi_2$, $\lambda_2 = \xi_1$, $\lambda_3 = \xi_1 - \xi_3$, $\lambda_4 = -\xi_3$, $\lambda_5 = \xi_2 - \xi_3$. The root vectors are

$$X_0 = G_{4,6} + G_{5,2} - 2G_{7,3},\quad X_1 = G_{7,2} - G_{3,4},$$

$$X_2 = G_{5,7} + G_{6,3} - 2G_{1,2},\quad X_3 = G_{4,6} - G_{5,2},$$

$$X_4 = G_{7,2} - 2G_{1,5} + G_{3,4},\quad X_5 = G_{5,7} - G_{6,3}.$$

Then $\{6^{-1/2}X_i,\ 2^{-1/2}X_j,\ i=0,2,4,\ j=1,3,5\}$ is an orthonormal basis of $\mathfrak{k}$. It is easily verified that $(\lambda_i, \lambda_i) = 2/3$, $i=0,2,4$, $(\lambda_j, \lambda_j) = 2$, $j=1,3,5$. Hence the equations (4.2) become

$$X_i^2\psi + \psi = 0,\ i = 0,1,\ldots,5.$$

We note $[X_i, X_j] = X_k$, where (i,j,k) is a cyclic permutation of

(3,1,5). Hence the proof of the case ($\mathfrak{su}(3)$,$\mathfrak{so}(4)$) applies also to this case, and we get $\psi = 0$.

References

[1] T. Cecil and P. Ryan: Tight and taut immersions of manifolds, Res. Notes Math. 107, Pitman, London, 1985.

[2] J. Dorfmeister and E. Neher: Isoparametric hypersurfaces, case g = 6, m = 1, Comm. in Alg. 13 (1985), 2299-2368.

[3] F.G. Friedlander: Simple progressive wave solution of the wave equation, Proc. Camb. Phil. Soc. 43 (1946), 360-373.

[4] F.G. Friedlander: The wave equation on a curved space-time, Cambridge Univ. Press, Cambridge, 1975.

[5] W.Y. Hsiang and H.B. Lawson: Minimal submanifolds of low cohomogeneity, J. Differential Geometry 5 (1971), 1-38.

[6] J. Kaneko: Wave equation and Dupin hypersurface, Mem.Fac. Sci. Kyushu Univ. 40 (1986), 51-55.

[7] P.D. Lax and R.S. Phillips: An example of Huygens' principle, Comm. Pure Appl. Math. 31 (1978), 415-421.

[8] H.F. Münzner: Isoparametrische Hyperflächen in Sphären, I and II, Math. Ann. 251 (1980), 57-71 and 256 (1981), 215-232.

[9] H. Muto, Y. Ohnita and H. Urakawa: Homogeneous hypersurfaces in the unit spheres and the first eigenvalues of their Laplacian, Tôhoku Math. J. 36 (1984), 253-267.

[10] H. Ozeki and M. Takeuchi: On some types of isoparametric hypersurfaces in spheres II, Tôhoku Math. J. 28 (1976), 7-55.

[11] U. Pinkall: Dupin hypersurfaces, Math. Ann. 270 (1985),

427-440.

[12] R. Takagi and T. Takahashi: On the principal curvatures of homogeneous hypersurfaces in a sphere, Diff. Geom. in honor of K. Yano, Kinokuniya, Tokyo, 1972, 469-481.

Jyoichi KANEKO
Department of Mathematics
College of General Education
Kyushu University
Ropponmatsu, Fukuoka 810
Japan

CONSTRUCTION OF TAUT EMBEDDINGS AND CECIL-RYAN CONJECTURE

Reiko Miyaoka and Tetsuya Ozawa

§1. Introduction. Dupin hypersurfaces are, recently, studied in Lie sphere geometry (see, for example, [P], [M1], [M2], [M3], [CC1], [CC2]), and one interesting question concerning global theory of Dupin hypersurfaces, is whether the following is true or not;

Conjecture([CR 2],p.184). Every closed embedded Dupin hypersurface in a Euclidean space is Lie equivalent to an isoparametric hypersurface in a sphere.

Here, by "Dupin", we mean a hypersurface or more generally, a submanifold in a space form, whose principal curvatures have globally constant multiplicities and are constant along the corresponding curvature leaves, which lie in the unit normal bundle if the codimension is greater than one.

From the works of Münzner [Mü] and Thorbergsson [T], the possible number g of principal curvatures of a closed embedded Dupin hypersurface is 1, 2, 3, 4 or 6. The above conjecture is known to be true for $g = 1$ (trivial), 2 ([CR1], [P1]) and 3 ([M1]). But very recently, Pinkall and Thorbergsson [PT] have shown counterexamples in the case $g = 4$. Here we give counterexamples for $g = 4$ and 6, using the idea independent of [PT]. Our construction is based on a new

ISBN 0-12-640170-5

method of producing taut embeddings due to the idea of the second author. In order to know if a Dupin hypersurface is Lie equivalent to isoparametric hypersurfaces or not, we calculate "Lie curvature", a Lie invariant defined in [M2]. For the definitions of isoparametric hypersurfaces, tight and taut embeddings, Lie transformations, we refer [CR2] and [M2].

§2. Construction of taut embeddings. Let $h: S^7 \longrightarrow S^4$ be the Hopf fibering over the quaternion $\mathbb{H}$, which is written, in the canonical coordinate as follows;

$$h(u,v) = (2u\bar{v}\ ,\ \|u\|^2 - \|v\|^2).$$

Here we identify $S^7 = \{(u,v)\in\mathbb{H} \times \mathbb{H}; \|u\|^2+\|v\|^2=1\}$, and $S^4 = \{(w,t)\in\mathbb{H} \times \mathbb{R};\ \|w\|^2+t^2=1\}$.

Proposition 1. *Let* X *be a connected closed submanifold in* S^4. *If* X *is taut, then* $h^{-1}(X)$ *is taut in* S^7.

Proof. The fiber $h^{-1}(w,t)$ over a point $(w,t) \in S^4$ is equal to

$$h^{-1}(w,t) = \{(wz/\sqrt{2(1-t)},\ \sqrt{(1-t)/2}\cdot z)\ ;\ z \in S^3\}.$$

We write linear functions on $\mathbb{H} \times \mathbb{H}$ in the form $f_{a,b}(u,v) = \mathrm{Re}\{au + bv\}$, which is a height function with respect to the vector $(\bar{a},\bar{b})$. Then the function restricted to $h^{-1}(X)$ is equal to $\mathrm{Re}\{(aw/\sqrt{2(1-t)} + b\sqrt{(1-t)/2}\ )z\}$. On each fiber, this function has two critical points with critical values $\pm\|aw/\sqrt{2(1-t)} + b\sqrt{(1-t)/2}\|$, generically. Note that

$$\|aw/\sqrt{2(1-t)} + b\sqrt{(1-t)/2}\|^2 = \frac{1}{2}\ \mathrm{Re}\{2a\bar{b}w+(\|a\|^2-\|b\|^2)t\} + \frac{\|a\|^2+\|b\|^2}{2}$$

is a linear function in the variable (w,t), which we denote by $g_{a,b}(w,t)$. If X is a taut submanifold in S^4, then $g_{a,b}(w,t)$ restricted to X has, generically, the number of critical points equal to the sum of Betti numbers $\beta(X)$. Thus $f_{a,b}$ generically has the number of critical points on $h^{-1}(X)$ equal to $2\beta(X) = \beta(S^3 \times X) = \beta(h^{-1}(X))$. Therefore $h^{-1}(X)$ is tight. Since $h^{-1}(X)$ is spherical, it is taut. q.e.d.

The above is also true for the Cayley algebra. Higher dimensional analogues can be obtained by means of maps τ : $S^p(1) \longrightarrow SO(q+1;\mathbb{R})$ which satisfy the following:

For each $y \in \mathbb{R}^{q+1}(y \neq 0)$, the map : $x \in S^p(1) \longrightarrow \tau(x)y \in \mathbb{R}^{q+1}$ is a homothety onto a round p-sphere in $\mathbb{R}^{q+1}$.

We can construct such maps using Clifford algebra. Let $T : S^p(1) \times S^q(1/\sqrt{2}) \longrightarrow S^{2q+1}(1)$ be the map given by $T(x,y) = (\tau(x)y,y)$. The proof of the following is similar to that of Proposition 1 (see the end of §3).

Proposition 2. *For any taut submanifold* X *in* $S^p(1)$, *the image* $T(X \times S^q)$ *is taut in* S^{2q+1}.

We remark that if X is a totally geodesic sphere $S^n(1) \subset S^p(1)$ $(n=1,\ldots,p)$, then tubes with constant radius around the image $T(X \times S^q)$ are isoparametric hypersurfaces in S^{2q+1} (cf. [FKM], where $p = m-1$, $q = l-1$).

§3. Dupin submanifolds. A submanifold X in S^n is a Dupin submanifold if the r-tube around X is a Dupin hypersurface where r is a positive small number [CR 2, Theorem 3.2 on p.131]. A Dupin submanifold is taut [P2, Theorem 2].

Proposition 3. *Let* X *be a Dupin submanifold in* S^4. *Then* $h^{-1}(X)$ *is a Dupin submanifold in* S^7. *If* X *is a Dupin hypersurface in* S^4, *the principal curvatures of* $h^{-1}(X)$ *are given by* $\tilde{\lambda}_{\pm} = \lambda \pm \sqrt{1+\lambda^2}$, *where* λ *is a principal curvature of* X.

Proof. By Proposition 1, $\tilde{X} = h^{-1}(X)$ is taut. By arguments in [H] or [O], $\tilde{X}$ is Dupin if every connected component of critical sets of any linear function is nondegenerate or a sphere of dimension k, $1 \le k \le 6$. Obviously, if $\tilde{x} \in \tilde{X}$ is a critical point of $f_{a,b}|_{\tilde{X}}$, then $x = h(\tilde{x})$ is a critical point of $g_{a,b}|_X$. Conversely, if x is a critical point of $g_{a,b}|_X$ with non-zero critical value, there are two critical points of $f_{a,b}|_{\tilde{X}}$ on $h^{-1}(x)$. Thus a component of critical set of $\tilde{X}$ with non-zero critical value is homeomorphic to the corresponding critical component of X, which is a point or a sphere of some dimension since X is Dupin. Now, consider a critical level of $f_{a,b}|_{\tilde{X}}$ with critical value = 0. It is sufficient to consider the case $(a,b) \in S^7$. Then we have

$$g_{a,b}(w,t) = \tfrac{1}{2}\mathrm{Re}\{2a\bar{b}w + (\|a\|^2 - \|b\|^2)t\} + \tfrac{1}{2}$$

and $g_{a,b}(w,t) = 0$ if and only if $(w,t) = -(2\bar{a}b, \|a\|^2 - \|b\|^2)$. Thus a critical point of $g_{a,b}|_X$ with critical value = 0 is nondegenerate, and the critical set of $f_{a,b}|_{\tilde{X}}$ with critical value = 0 is a (3-k)-sphere where k = dim X. Finally, all components of critical sets of $f_{a,b}|_{\tilde{X}}$ are points or spheres and $\tilde{X}$ is Dupin.

Now, to show the latter part, put $f_{a,b}(\tilde{x}) = \cos\tilde{\theta}$ and $g_{a,b}(h(\tilde{x})) = \frac{1}{2}\cos\theta + \frac{1}{2}$, where $\tilde{x}$ is a critical point of $f_{a,b}|_{\tilde{X}}$, and $(\bar{a},\bar{b}) \in S^7$. Then we have

$$\cos 2\tilde{\theta} = \cos\theta, \tag{$*$}$$

since $g_{a,b}(h(\tilde{x})) = \{f_{a,b}(\tilde{x})\}^2$. This means that the normal geodesic spanned by $(\bar{a},\bar{b})$ and $\tilde{x}$ in S^7 covers the normal geodesic spanned by $h(\bar{a},\bar{b})$ and $h(\tilde{x})$ in S^4 twice by h. Now, if $\tilde{x}$ is a degenerate critical point of $f_{a,b}|_{\tilde{X}}$, or equivalently, $(\bar{a},\bar{b})$ is a focal point of $\tilde{x}$, where we assume $(\bar{a},\bar{b}) \neq -\tilde{x}$, then $h(\tilde{x})$ is a degenerate critical point of $g_{a,b}|_X$, i.e. $h(\bar{a},\bar{b})$ is a focal point of $h(\tilde{x})$, or $h(\tilde{x})$ is a nondegenerate critical point of $g_{a,b}|_X$ and codim X $\geq$ 2. In the latter case, $h(\bar{a},\bar{b}) = -h(\tilde{x})$ is in fact a focal point of $h(\tilde{x})$. Finally, a focal point $(\bar{a},\bar{b})(\neq -\tilde{x})$ of $\tilde{x}$ is mapped into a focal point $h(\bar{a},\bar{b})$ of $h(\tilde{x})$, and the correspondense is two by one by ($*$). Note that if $(\bar{a},\bar{b}) = -\tilde{x}$ is a focal point of $\tilde{x}$, i.e. if the codim $\tilde{X} \geq 2$, no corresponding focal point appears since $h(\bar{a},\bar{b}) = h(\tilde{x})$. In hypersurface case, when $(\bar{a},\bar{b})$ ($h(\bar{a},\bar{b})$, resp.) is a focal point of $\tilde{x}$ ($h(\tilde{x})$, resp.), the principal curvature $\tilde{\lambda}$ of $\tilde{X}$ (λ of X, resp.) is given by $\cot\tilde{\theta}$ ($\cot\theta$, resp.). Thus from $\tilde{\theta} = \frac{\theta}{2}$ or $\frac{\theta}{2} + \frac{\pi}{2}$ follows the last assertion. q.e.d.

Proposition 3 is also true for the Caylay algebra, but *not* true for the higher dimensional analogue $T(X \times S^q)$ except when $X \subset S^p$ is a round sphere. Generally, the critical level of the linear function

$$f_{a,b}(x,y) = \langle a, \tau(x)y \rangle + \langle b, y \rangle = \langle {}^t\tau(x)a + b, y \rangle,$$

$(a,b) \in S^{2q+1}$, $(x,y) \in X \times S^q(1/\sqrt{2})$, with critical value $= 0$ is a sphere bundle over a sphere.

§4. Counterexamples to Cecil-Ryan conjecture. Let X be a Dupin hypersurface in S^4 with g principal curvatures $\lambda_i = \cot \theta_i$, $1 \le i \le g$, $g = 1$, 2 or 3. Then $\tilde{X} = h^{-1}(X)$ is a Dupin hypersurface in S^7 with 2g principal curvatures $\tilde{\lambda}_{i\pm}$, $1 \le i \le g$. Therefore if $g \ge 2$, a Lie curvature Ψ ([M2,3]) of $\tilde{X}$ is computed as, for instance,

$$\Psi(\tilde{\lambda}_{1+}, \tilde{\lambda}_{2+}; \tilde{\lambda}_{1-}, \tilde{\lambda}_{2-}) = \frac{(\tilde{\lambda}_{1+} - \tilde{\lambda}_{1-})(\tilde{\lambda}_{2+} - \tilde{\lambda}_{2-})}{(\tilde{\lambda}_{1+} - \tilde{\lambda}_{2-})(\tilde{\lambda}_{2+} - \tilde{\lambda}_{1-})}$$

$$= \frac{2}{1 + \cos(\theta_1 - \theta_2)}.$$

Since Ψ is Lie invariant, Ψ is a constant function if $\tilde{X}$ is a Lie image of an isoparametric hypersurface. When X is, for instance, a non-isometric conformal image of an isoparametric hypersurface in S^4, Ψ is not a constant function on $\tilde{X}$. Thus we get:

Assertion 4. *Let X be a non-isoparametric Dupin hypersurface in S^4 with g principal curvatures, where $g = 2$ or 3. Then $h^{-1}(X)$ is a Dupin hypersurface in S^7 with 2g principal curvatures which is not Lie equivalent to any isoparametric hypersurface in S^7.*

When X is a Dupin submanifold in S^4 with codim $X \geq 2$, a small tube around $h^{-1}(X)$ is the inverse image by h of a small tube around X. Thus the simplest counterexample is given by taking a tube around the inverse image $h^{-1}(X)$ of a *small* circle X of S^4. This is the original version of counterexamples constructed by the second author. It can be generalized by using the maps $h : S^{15} \longrightarrow S^8$ and $T : S^p(1) \times S^q(1/\sqrt{2}) \longrightarrow S^{2q+1}(1)$, to obtain other examples for $g = 4$. A counterexample for $g = 6$ in S^7 is given, in this version, by taking a small tube around $h^{-1}(X)$ of a non-isometric conformal image X of the standard Veronese surface in S^4. But an analogous argument using Cayley algebra or map T can not produce counterexamples for $g = 6$ in a higher dimensional sphere, because of the restriction on multiplicities of principal curvatures [A], and of the remark in the end of §3.

References

[A] Abresch,U.; Isoparametric hypersurfaces with four or six principal curvatures, Math. Ann. 264 (1983) 283-302.

[CC1] Cecil,T and Chern,S.S.; Tautness and Lie sphere geometry, Math. Ann. 278 (1987) 381-399.

[CC2] ____________________; Dupin submanifolds in Lie sphere geometry, preprint (1988).

[CR1] Cecil,T and Ryan,P.; Focal sets, taut embeddings and the cyclides of Dupin, Math. Ann. 236 (1978) 177-190.

[CR2] ___________________; Tight and taut immersion of manifolds, Res. Notes in Math. 107 (1985) Pitman, London.

[FKM] Ferus,D., Karcher,H. and Münzner,H.F.; Cliffordalgebren und neue isoparametrische Hyperflächen, Math.Z. 177 (1981) 479-502.

[H] Hebda,J; The possible cohomology of certain types of taut submanifolds, preprint (1986), Saint Louis Univ.

[M1] Miyaoka,R; Dupin hypersurfaces with three principal curvatures, Math.Zeit. 187 (1984) 433-452.

[M2] _________; Dupin hypersurfaces and a Lie invariant, to appear in Kodai Math. J..

[M3] _________; Dupin hyperusurfaces with six principal curvatures, to appear in Kodai Math. J..

[Mü] Münzner,H.F; Isoparametrische Hyperflächen in Sphären I, Math. Ann. 251 (1980) 57-71.

[O] Ozawa,T; On the critical sets of distance functions to taut submanifolds, Math. Ann. (1986) 91-96.

[P1] Pinkall,U; Dupin hypersurfaces, Math. Ann. 270 (1985) 427-440.

[P2] _________; Curvature properties of taut submanifolds, Geom. Dedicata 20 (1986) 79-83.

[PT] _________ and Thorbergsson,G; Deformations of Dupin hypersurfaces, To appear in Proc. of A.M.S..

[T] Thorbergsson,G.; Dupin hypersurfaces, Bull.London Math. Soc. (1983) 493-498.

Department of Mathematics
Tokyo Institute of Technology
Meguroku, Tokyo 152
Japan

Department of Mathematics
Nagoya University
Chikusaku, Nagoya 464
Japan

LIE CONTACT MANIFOLDS

Hajime SATO and Keizo YAMAGUCHI

Dedicated to Professor Akio Hattori on his sixtieth birthday

§0. Introduction

Around 1870, **Sophus Lie** , with the collaboration of F. Klein, constructed the **sphere geometry**, which is a sythesis of the **Möbius geometry** and the **Laguerre geometry**. The set of transformations in the sphere geometry is a tractable subset of the contact transformation group. We call the subset the **spherical Lie contact tranformation group**. The spherical Lie contact transformation group is an example of today called a Lie group. The sphere geometry has given Klein an idea of Erlangen program and has made Lie develop the theory of continuous groups and differential equations.

The transformation group in the Möbius geometry (Möbius group) is equal to the group of motions of the special relativity. The Möbius geometry on general manifolds is called the conformal geometry and has been investigated extensively.

On the other hand, the transformation group in the sphere geometry (spherical Lie contact transformation group) is equal to the conformal transformation group of the special relativity. But there has been no theory of the sphere geometry on general contact manifolds. The purpose of this paper is to construct the sphere geometry on suitable contact manifolds including the unit tangent sphere bundles of all

ISBN 0-12-640170-5

Riemannian manifolds. We call such contact manifolds **Lie contact manifolds** and call the structure the **Lie contact structure**.

Let $G = PO(n+1,2)$ be the projective orthogonal group of signature $(n+1,2)$. Then G acts transitively on the unit tangent sphere bundle $T_1(S^n)$ of the unit sphere S^n (see §1). An element $g \in G$ keeps invariant the canonical contact structure of $T_1(S^n)$ and G is a subgroup of the contact transformation group of $T_1(S^n)$. This group G is called the spherical Lie contact transformation group. Let G' be the isotropy subgroup of G at a base point o of $T_1(S^n)$ and let $\rho : G' \to GL(2n-1,\mathbb{R})$ be the linear isotropy representation. Put $\tilde{G} = \rho(G') \subset GL(2n-1,\mathbb{R})$.

Let N be a $(2n-1)$-dimensional contact manifold. Then the set $F_c(N)$ of adapted frames to the contact structure is a principal C-bundle over N, where $C \subset GL(2n-1)$ is the linear contact group (see §3). Naturally the linear isotropy group $\tilde{G}$ is a subgroup of C. By a **Lie contact structure** on N, we mean a reduction of the structure group of $F_c(N)$ to the subgroup $\tilde{G}$. With this structure, N is called a **Lie contact manifold**.

With these definitions, we can apply the theory of Tanaka ([18], [19]). Tanaka has settled the eqivalence problems for linear group structures associated with simple graded Lie algebras by introducing the normal Cartan connection and the Weyl curvature tensor. Following his theory, we can define the Cartan connection and the Weyl tensor for Lie contact manifolds. Theoretically the equivalence problem for the Lie contact structure is completely solved in this manner.

Let (M,g) be an n-dimensional Riemannian manifold. Then the unit tangent sphere bundle $T_1(M)$ has a canonical Lie contact structure. The Lie contact structure depends only on the conformal class of the metric g on M. If $n = 3$, then the Lie contact structure on $T_1(M)$ is equivalent to the (integrable) CR -structure with non-degenerate and indefinite Levi form. This construction is equal to the twistor construction of Penrose [14] (cf. Atiyah - Hitchin - Singer [1]) and is also discovered by LeBrun [12] .

Now we explain the contents of each section. §1 is devoted to the brief review of the Lie sphere geometry (cf. Blaschke [4], Cecil-Chern [6]). In §2 we introduce a gradation of the (simple) Lie algebra $\mathfrak{g}$ of the spherical Lie contact transformations group G, which plays the fundamental role in the subsequent sections. The notion of Lie contact manifolds is introduced in §3. As an important example, we shall show that the unit tangent bundle of a Riemannian manidfold has a canonical Lie contact structure and discuss its relation with the conformal structure. In §4 we apply Tanaka's theory of Cartan connections to Lie contact manifolds. A normal Cartan connection (P,ω) is associated to a Lie contact manifold N, which settles the equivalence problem for Lie contact structure in the following sense; A Lie isomorphism between two Lie contact manifolds lifts uniquely to an isomorphism between corresponding normal connections and vice versa. The torsion and the curvature derived from the normal connection determines the structure of Lie contact manifolds. Because of the special isomorphisms among the simple Lie algebras in low dimension, a Lie contact structure on a (2n-1)- dimensional contact manifold behaves differently according to $n = 3$ or $n \geq 4$. §§ 5 and 6 are devoted to the case $n = 3$. In §5 we study the relation between Lie contact structures and CR-structures on the unit tangent bundle of 3-dimensional Riemannian manifolds. In §6 we consider the case where the geodesic flow is a Lie contact transformation.

The notion of Lie contact manifolds first appeared in the talks of the first named author in DG Colloquim in Mizusawa 1986 and Topology Symposium in Okinawa 1987 (see [16]).

In his master's thesis [21], T.Yamazaki studied local and global Lie contact geometry. Especially he classified surfaces in $\mathbb{R}^3$ by the Lie contact eqivalence. We are grateful to S.Bando for useful discussions.

§1. Lie Sphere Geometry

Our model geometry is the Lie geometry of oriented spheres in S^n. In this section, we shall briefly recall the bijective correspondence between the set K of kugels in S^n and the Lie quadric Q^{n+1} in $P^{n+2}(\mathbb{R})$, and recall the action of the group $PO(n+1,2)$ of Lie transformations on the unit tangent bundle $T_1(S^n)$ of S^n (cf. Blaschke[4] and Cecil-Chern[6]).

The set K of kugels of the unit sphere S^n in $\mathbb{R}^{n+1}$ consists of points in S^n and oriented hyperspheres in S^n. Each $k \in K$ is represented by a pair $(x, \rho) \in S^n \times [0, \pi)$ such that x is the center of k and ρ is the spherical distance from x ;

$$k = \{ y \in S^n \mid (x,y) = \cos \rho \},$$

where (x,y) denotes the inner product in $\mathbb{R}^{n+1}$. The orientation of k is determined by the inward normal vector. As the limit case $\rho \to 0$, $k = (x,0)$ represents a point sphere x in S^n . Then two kugels $k_1 = (x_1, \rho_1)$ and $k_2 = (x_2, \rho_2)$ are in oriented contact if and only if $(x_1,x_2) = \cos(\rho_1 - \rho_2)$.

S^n is embedded in $P^{n+1}(\mathbb{R})$ as the Möbius space by the embedding; $(x_1,\ldots,x_{n+1}) \to [(1,x_1,\ldots,x_{n+1})]$. $P^{n+1}(\mathbb{R})$ is further embedded into $P^{n+2}(\mathbb{R})$ as the hyperplane $x_{n+2} = 0$. Then the Lie quadric Q^{n+1} in $P^{n+2}(\mathbb{R})$ is defined in homogeneous coordinates by

$$\langle z,z \rangle = - z_0^2 + z_1^2 + \ldots + z_{n+1}^2 - z_{n+2}^2 = 0 .$$

The bijective correspondence between K and Q^{n+1} is given by

$$k = (x,\rho) \to \hat{k} = [(\cos \rho , x, \sin \rho)].$$

By this correspondence, two kugels k_1 and k_2 are in oriented contact if and only if $\langle \hat{k}_1,\hat{k}_2 \rangle = 0$. Namely two kugels in oriented contact determine a line on Q^{n+1} , and converesely a line on Q^{n+1} determines an S^1-family of kugels which are mutually in oriented contact. This S^1-family of kugels contains a unique point sphere $x \in S^n$, which is the common point of contact, and determines the common unit normal vector y at x . In this way the bijective correspondence is established between the unit tangent bundle

$$T_1(S^n) = \{(x,y) \in \mathbb{R}^{n+1}\times \mathbb{R}^{n+1} \mid (x,x) = (y,y) = 1,\ (x,y) = 0\}$$

of S^n and the space Λ of projective lines on Q^{n+1}, which is given by;

$$(x,y) \to \text{the line through } [(1,x,0)] \text{ and } [(0,y,1)].$$

The group G of Lie transformations is defined as the group of projective transformations of $P^{n+2}(\mathbb{R})$ which leave invariant Q^{n+1}. Thus $G = PO(n+1,2)$ acts on Λ, hence on $T_1(S^n)$ as a group of contact transformations. The Möbius group G_M in the Lie geometry is the subgroup of G which leaves invariant the hyperplane $z_{n+2} = 0$, or equivalently, which leaves invariant the point spheres $S^n = Q^{n+1} \cap P^{n+1}(\mathbb{R})$. The group G_M is isomorphic to $O(n+1,1)$.

Remark 1.1. The Möbius group $PO(n+1,1)$ in the conformal geometry is the subgroup of projective transformations in $P^{n+1}(\mathbb{R})$ leaving invariant the Möbius space S^n in $P^{n+1}(\mathbb{R})$. Each $\sigma \in PO(n+1,1)$ lifts in two ways to Lie transformations of $T_1(S^n)$. Namely we have two Lie transformations σ^+ and σ^- of $T_1(S^n)$ defined by

$$\sigma^+(v) = \frac{\sigma_*(v)}{\|\sigma_*(v)\|} \quad \text{and} \quad \sigma^-(v) = -\frac{\sigma_*(v)}{\|\sigma_*(v)\|}$$

for $v \in T_1(S^n)$ (cf. §3). Obviously we have an isomorphism κ of $PO(n+1,1)$ into G defined by $\kappa(\sigma) = \sigma^+$. We denote by G_M^+ the image of $PO(n+1,1)$ under κ, which is an open subgroup of G_M. In fact G_M consists of σ^+ and σ^- for all $\sigma \in PO(n+1,1)$.

Now we will give the above correspondence between $T_1(S^n)$

and Λ more explicitly. For this purpose, we start from $\mathbb{R}^{n+3}$ with the (indefinite) inner product $\langle\ ,\ \rangle$ as above. Then Λ is the space of isotropic subspaces of dimension 2 in $\mathbb{R}^{n+3}$. Hence we consider the following space;

$$F = \{(z,w) \in \mathbb{R}^{n+3}\times \mathbb{R}^{n+3} \mid \langle z,z\rangle=\langle z,w\rangle=\langle w,w\rangle= 0,\ z \wedge w \neq 0\ \}.$$

On this space, $GL(2,\mathbb{R})$ acts on the right as follows;

$$(z,w)\sigma = (\ az+cw,\ bz+dw\ ),\ \text{where}\ \ \sigma = \begin{pmatrix} a & b \\ c & d \end{pmatrix} \in GL(2,\mathbb{R})\ .$$

Moreover $O(n+1,2)$ acts on F in the obvious way. F is the total space of the universal 2-frame bundle over $\Lambda = F\ /\ GL(2,\mathbb{R})$. Let π be the projection of F onto Λ. We have an embedding φ of $T_1(S^n)$ into F given by;

$$(x,y) \rightarrow ((1,x,0),\ (0,y,1)) \in F\ .$$

Then $\Phi = \pi\circ\varphi$ gives a diffeomorphism of $T_1(S^n)$ onto Λ.

Furthermore the contact structure on Λ is given as follows: The 1-form $\Theta = \langle w,dz\rangle - \langle z,dw\rangle$ defines a linear symplectic structure on $\mathbb{R}^{n+3}\times \mathbb{R}^{n+3}$. Let θ be the restriction of $\frac{1}{2}\,\Theta$ on F. Then $\theta = \langle w,dz\rangle$ and $\sigma^*\theta = (\det \sigma)\theta$ for $\sigma \in GL(2,\mathbb{R})$. On the other hand, the canonical 1-form α on $T_1(S^n)$ is given by $\alpha = (y,dx)$. Thus θ and α coincides on the image of $T_1(S^n)$ under φ. Therefore $\theta = 0$ defines the contact distribution on Λ, which is preserved by the action of the group G of Lie transformations, and Φ is a contact diffeomorphism of $T_1(S^n)$ onto Λ.

§2. Gradation of the Lie algebra $\mathfrak{g}$ of G

The group G of Lie transformations acts transitively

on Λ , hence on $T_1(S^n)$. Let G' be the isotropy subgroup of G at $o = (e_1, e_{n+1}) \in T_1(S^n)$, or equivalently at $\hat{o} = \pi((e_0 + e_1, e_{n+1} + e_{n+2})) \in \Lambda$, where $\{e_0, e_1, \ldots, e_{n+2}\}$ is the natural basis of $\mathbb{R}^{n+3}$. In this section, we introduce a gradation of the Lie algebra $\mathfrak{g}$ of G , which plays the fundamental role in the subsequent sections, and study the linear isotropy representation of the homogeneous space $T_1(S^n) = G / G'$.

To study the homogeneous space G / G' , let us fix another basis $\{f_0, f_1, \ldots, f_{n+2}\}$ of $\mathbb{R}^{n+3}$ such that $f_0 = e_0 + e_1$, $f_1 = e_{n+1} + e_{n+2}$, $f_{n+1} = \frac{1}{2}(e_1 - e_0)$, $f_{n+2} = \frac{1}{2}(e_{n+1} - e_{n+2})$ and $f_i = e_i$ for $i = 2, \ldots, n$. Then the coefficient matrix $S = (\langle f_i, f_j \rangle)$ of the inner product $\langle\ ,\ \rangle$ takes the following form;

$$S = \begin{pmatrix} 0 & 0 & I_2 \\ 0 & I_{n-1} & 0 \\ I_2 & 0 & 0 \end{pmatrix},$$

where I_k denotes the unit matrix of degree k . The Lie algebra $\mathfrak{g}$ of G can be identified with $\mathfrak{o}(S)$, that is,

$$\mathfrak{g} = \{ X \in \mathfrak{gl}(n+3, \mathbb{R}) \mid {}^tX \cdot S + S \cdot X = 0 \}.$$

Explicitly each $X \in \mathfrak{g}$ can be written as a matrix of the following form;

$$X = \begin{pmatrix} a & -{}^tD & f \\ C & B & D \\ e & -{}^tC & -{}^ta \end{pmatrix},$$

where $a \in \mathfrak{gl}(2, \mathbb{R})$, $B \in \mathfrak{o}(n-1)$, $e, f \in \mathfrak{o}(2)$, and C , D are $(n-1) \times 2$ matrices. Thus $\mathfrak{g}$ can be decomposed as follows;

$$\mathfrak{g} = \Sigma_{p=-2}^{2} \ \mathfrak{g}_p \ ,$$

$$\mathfrak{g}_{-2} = \left\{ \begin{pmatrix} 0 & 0 & 0 \\ 0 & 0 & 0 \\ e & 0 & 0 \end{pmatrix} \right\} , \quad \mathfrak{g}_2 = \left\{ \begin{pmatrix} 0 & 0 & f \\ 0 & 0 & 0 \\ 0 & 0 & 0 \end{pmatrix} \right\},$$

$$\mathfrak{g}_0 = \left\{ \begin{pmatrix} a & 0 & 0 \\ 0 & B & 0 \\ 0 & 0 & -{}^t a \end{pmatrix} \right\} ,$$

$$\mathfrak{g}_{-1} = \left\{ \begin{pmatrix} 0 & 0 & 0 \\ C & 0 & 0 \\ 0 & -{}^t C & 0 \end{pmatrix} \right\} , \quad \mathfrak{g}_1 = \left\{ \begin{pmatrix} 0 & -{}^t D & 0 \\ 0 & 0 & D \\ 0 & 0 & 0 \end{pmatrix} \right\} .$$

In fact, this is the eigenspace decomposition of the semi simple endomorphism $ad(E)$ of $\mathfrak{g}$, where $E = \begin{pmatrix} I_2 & 0 & 0 \\ 0 & 0 & 0 \\ 0 & 0 & -I_2 \end{pmatrix}$ $\in \mathfrak{g}_0$. Furthermore $\mathfrak{g}_p$ is the eigenspace of $ad(E)$ for the eigenvalue p . Thus, putting $\mathfrak{g}_p = \{0\}$ for $|p| > 2$, we see that $\mathfrak{g} = \Sigma_{p\in\mathbb{Z}} \mathfrak{g}_p$ becomes a graded Lie algebra;

$$[\mathfrak{g}_p, \mathfrak{g}_q] \subset \mathfrak{g}_{p+q} \quad \text{for} \quad p, q \in \mathbb{Z} .$$

Moreover if we put

$$\mathfrak{m} = \mathfrak{g}_{-2} \oplus \mathfrak{g}_{-1} \quad \text{and} \quad \mathfrak{g}' = \mathfrak{g}_0 \oplus \mathfrak{g}_1 \oplus \mathfrak{g}_2 \ ,$$

then $\mathfrak{m}$ and $\mathfrak{g}'$ are subalgebras of $\mathfrak{g}$ and we have the decomposition;

$$\mathfrak{g} = \mathfrak{m} \oplus \mathfrak{g}' \ .$$

It is easy to see that $\mathfrak{g}'$ is the Lie algebra of G' . As usual, $\mathfrak{m}$ is identified with the tangent space $T_{\hat{o}}(\Lambda)$ of Λ at $\hat{o}$ as follows:

For $X = \begin{pmatrix} 0 & 0 & 0 \\ C & 0 & 0 \\ e & -{}^t C & 0 \end{pmatrix} \in \mathfrak{m}$, we take $\exp X = I_{n+3} + X + \frac{1}{2} X^2$

$\in O(S)$. If we represent $(z,w) \in F$ by $(n+3)\times 2$ matrices with respect to the basis $\{f_0, f_1, \ldots, f_{n+2}\}$ of $\mathbb{R}^{n+3}$, we have

$$\exp X \left(\begin{pmatrix} I_2 \\ 0 \\ 0 \end{pmatrix} \right) = \begin{pmatrix} I_2 \\ C \\ e - \frac{1}{2}{}^t C \cdot C \end{pmatrix} \in F \text{ , where } (f_0, f_1) = \begin{pmatrix} I_2 \\ 0 \\ 0 \end{pmatrix} .$$

Then a map ψ ; $\mathfrak{m} \to \Lambda$, defined by

$$\psi(X) = \pi(\exp X\,((f_0, f_1))) ,$$

gives a coordinate system of Λ with origin $\hat{o}$. Since $\theta = \langle w, dz \rangle = 0$ defines the contact distribution on Λ , we see that the contact structure on this coordinate (η, ξ_1, ξ_2) is given by

$$d\eta + \frac{1}{2}({}^t\xi_2 \cdot d\xi_1 - {}^t\xi_1 \cdot d\xi_2) = 0 ,$$

where $e = \begin{pmatrix} 0 & -\eta \\ \eta & 0 \end{pmatrix}$, $C = (\xi_1, \xi_2)$ and ξ_1, $\xi_2 \in \mathbb{R}^{n-1}$ are the column vectors. Especially $\mathfrak{m}$ is identified with the tangent space $T_{\hat{o}}(\Lambda)$ of Λ at $\hat{o}$ and, by this identification, $\mathfrak{g}_{-1}$ corresponds to the contact distribution of Λ at $\hat{o}$. Furthermore, by a direct calculation using the above bases of $\mathbb{R}^{n+3}$, we see that the (n-1)-dimensional subspace of $\mathfrak{g}_{-1}$ spanned by the vectors $(0,0,\xi_2)$, $\xi_2 \in \mathbb{R}^{n-1}$ corresponds to the vertical subspace $(\operatorname{Ker} p_*)(o)$ of $T_o(T_1(S^n))$ at $o = \Phi^{-1}(\hat{o})$, where p ; $T_1(S^n) \to S^n$ is the projection.

To study the linear isotropy representation of G / G' , we identify $G = PO(S)$ with the adjoint group $Ad(O(S))$ of $O(S)$. Thus each $g \in G'$ is written as $g = Ad(A)$ by a matrix $A \in O(S)$ of the following form;

$$A = \begin{pmatrix} a & D^* & f \\ 0 & B & D \\ 0 & 0 & {}^t a^{-1} \end{pmatrix},$$

where $a \in GL(2,\mathbb{R})$, $B \in O(n-1)$, $D^* = -a \cdot {}^tD \cdot B$, $a^{-1} \cdot f + \frac{1}{2}\,{}^tD \cdot D \in \mathfrak{o}(2)$ and D is a $(n-1) \times 2$ matrix.

Then the linear isotropy representation ρ of G' into $GL(\mathfrak{m})$ is given via the following commutative diagram;

$$\begin{array}{ccc} \mathfrak{g} & \xrightarrow{g} & \mathfrak{g} \\ p\downarrow & & \downarrow p \\ \mathfrak{m} & \xrightarrow{\rho(g)} & \mathfrak{m} \end{array} \qquad \text{for } g \in G',$$

where p is the projection of the decomposition $\mathfrak{g} = \mathfrak{m} \oplus \mathfrak{g}'$.

Explicitly, for $X = \begin{pmatrix} 0 & 0 & 0 \\ C & 0 & 0 \\ e & -{}^tC & 0 \end{pmatrix} \in \mathfrak{m}$, and $g = Ad(A) \in G'$,

we have

$$\rho(g)(X) = \begin{pmatrix} 0 & 0 & 0 \\ C' & 0 & 0 \\ e' & -{}^tC' & 0 \end{pmatrix} \in \mathfrak{m},$$

where $C' = B \cdot C \cdot a^{-1} + D \cdot e \cdot a^{-1}$ and $e' = (\det a)^{-1} \cdot e$. Thus, if we fix the bases of $\mathfrak{g}_{-1}$ and $\mathfrak{g}_{-2}$, and use the coordinate (η, ξ_1, ξ_2) of $\mathfrak{m}$ as above, each element σ of the linear isotropy subgroup $\tilde{G} = \rho(G')$ of $GL(\mathfrak{m})$ is given as a matrix of the following form;

$$\sigma = \begin{pmatrix} (\det a)^{-1} & 0 \\ \xi & B \otimes a \end{pmatrix} \in GL(\mathfrak{m}),$$

where $a \in GL(2,\mathbb{R})$, $B \in O(n-1)$, $\xi \in \mathbb{R}^{2(n-1)}$ and $B \otimes a((\xi_1, \xi_2)) = (B(\xi_1), B(\xi_2))\, a^{-1}$. Here the right action of $GL(2,\mathbb{R})$ on $\mathbb{R}^{n-1} \times \mathbb{R}^{n-1}$ is defined as in §1, or $B \otimes a$ is a matrix of the following form;

$$B \otimes a = \begin{pmatrix} \hat{\alpha} \cdot B & \hat{\gamma} \cdot B \\ \beta \cdot B & \delta \cdot B \end{pmatrix} \quad \text{for} \quad a^{-1} = \begin{pmatrix} \hat{\alpha} & \beta \\ \hat{\gamma} & \delta \end{pmatrix} .$$

Now we consider the action of the Möbius group G_M on Λ. Since G_M is the subgroup of G leaving invariant the point spheres, $g = Ad(A) \in G$ belongs to G_M if and only if $A(\langle e_{n+2} \rangle) = \langle e_{n+2} \rangle$, or equivalently $A(\langle f_1 - 2 \cdot f_{n+2} \rangle) = \langle f_1 - 2 \cdot f_{n+2} \rangle$.

G_M acts transitively on Λ. Let G_M' be the isotropy subgroup of G_M at $\hat{o}$. Then $g = Ad(A) \in G$ belongs to G_M' if and only if $A(\langle e_{n+2} \rangle) = \langle e_{n+2} \rangle$ and $A(\langle f_0, f_1 \rangle) = \langle f_0, f_1 \rangle$, or equivalently

$$A(f_0) = \alpha \cdot f_0 ,$$

$$A(f_1) \pm f_1 = 2(A(f_{n+2}) \pm f_{n+2}) = \beta \cdot f_0 ,$$

for some $\alpha, \beta \in \mathbb{R}$. Explicitly, each $g = Ad(A) \in G_M'$ is represented by a matrix of the following form;

$$A = \begin{pmatrix} a & D^* & f \\ 0 & B & D \\ 0 & 0 & {}^t a^{-1} \end{pmatrix} ,$$

where $a = \begin{pmatrix} \alpha & \beta \\ 0 & 1 \end{pmatrix} \in GL(2,\mathbb{R})$, $B \in O(n-1)$, $D = (\xi_1, 0)$, $\xi_1 \in \mathbb{R}^{n-1}$, $D^* = - a \cdot {}^t D \cdot B$, $f = \frac{1}{2} \begin{pmatrix} \hat{\alpha} & \beta \\ \hat{\gamma} & 0 \end{pmatrix}$, $\hat{\alpha} = - \alpha \cdot {}^t \xi_1 \cdot \xi_1 - \alpha^{-1} \cdot \beta^2$ and $\hat{\gamma} = - \alpha^{-1} \cdot \beta$. Thus the restriction to G_M' of the linear isotropy representation ρ is faithful. Moreover we see that the linear isotropy subgroup $\tilde{G}_M = \rho(G_M')$ of $T_1(S^n) = G_M / G_M'$ consists of elements in $GL(\mathfrak{m})$ of the following form;

$$A = \begin{pmatrix} \alpha & 0 & 0 \\ 0 & \alpha\cdot B & 0 \\ \xi & \beta\cdot B & B \end{pmatrix} \in GL(\mathfrak{m}),$$

where $B \in O(n-1)$, $\xi \in \mathbb{R}^{n-1}$, $\alpha \in \mathbb{R}^{\times}$ and $\beta \in \mathbb{R}$. One should note here that, if we consider the open subgroup G_M^+ of G_M , which is a natural lift of $PO(n+1,1)$ (see Remark 1.1), the corresponding linear isotropy subgroup $\tilde{G}_M^+$ of $T_1(S^n) = G_M^+ / (G_M')^+$ is an open subgroup of $\tilde{G}_M$ consisting of elements A such that $\alpha > 0$.

In order to apply Tanaka's theory ([18] and [19]) of Cartan connections in §4, we now mention how our model space G / G' is constructed from the simple graded Lie algebra $\mathfrak{g} = \Sigma_{p\in\mathbb{Z}}\, \mathfrak{g}_p$. We denote by $Aut(\mathfrak{g})$ the group of Lie algebra automorphisms of $\mathfrak{g}$. First we consider,

$$G_0 = \{\ g = Ad(A) \in G \mid A = \begin{pmatrix} a & 0 & 0 \\ 0 & B & 0 \\ 0 & 0 & {}^t a^{-1} \end{pmatrix},\ a \in GL(2,\mathbb{R})\ ,\ B \in O(n-1)\ \}.$$

Then G_0 is a subgroup of G with the Lie algebra $\mathfrak{g}_0$. In fact, it follows from the fact $Aut(\mathfrak{g}) = Ad(O(S))$ that , as a subgroup of $Aut(\mathfrak{g})$, G_0 coincides with the automorphism group of the graded Lie algebra $\mathfrak{g} = \Sigma_{p\in\mathbb{Z}}\, \mathfrak{g}_p$, that is, G_0 is the subgroup of $Aut(\mathfrak{g})$ consisting of all $g \in Aut(\mathfrak{g})$ such that $g(\mathfrak{g}_p) = \mathfrak{g}_p$ for $p \in \mathbb{Z}$. Thus, starting from the graded Lie algebra $\mathfrak{g} = \Sigma_{p\in\mathbb{Z}}\, \mathfrak{g}_p$, we first get G_0 and G is defined by $G = Int(\mathfrak{g})\cdot G_0$, where $Int(\mathfrak{g})$ is the adjoint group of $\mathfrak{g}$ or the identity component of $Aut(\mathfrak{g})$ (in our case, $G = Aut(\mathfrak{g})$). Moreover, if we consider the filtration $\{\mathfrak{g}^{\ell}\}_{\ell\in\mathbb{Z}}$ of $\mathfrak{g}$ by putting $\mathfrak{g}^{\ell} = \Sigma_{p\geq\ell}\, \mathfrak{g}_p$, G' is obtained as the automorphism group of this filtration. The contact structure on G / G' is given as follows; As a subspace of left invariant vector fields, $\mathfrak{g}^{-1}$ defines a distribution on G of codimension one, which is preserved by the left action of G and also by the right action of G' . Thus it induces the codimension one distribution D on G / G' , which is preserved by the action of G and is seen to define a contact structure from the structure equation of $\mathfrak{m}$.

The restriction to G_0 of the linear isotropy representation ρ ; $G' \to \tilde{G} \subset GL(\mathfrak{m})$ is faithful. Since each $g \in G_0$ preserves the gradation of $\mathfrak{g}$, $\rho(g) \in \rho(G_0)$ coincides with the restriction $g|_{\mathfrak{m}}$. Hence G_0 acts on $\mathfrak{m} = \mathfrak{g}_{-2} \oplus \mathfrak{g}_{-1}$ as a group of graded Lie algebra automorphisms. In this connection, $\tilde{G}$ is characterized as follows (§4 in [18]); For $\sigma \in GL(\mathfrak{m})$ such that $\sigma(\mathfrak{g}_{-1}) = \mathfrak{g}_{-1}$, the graded map $\hat{\sigma}$ of σ is the linear isomorphism of $\mathfrak{m}$ satisfying $\hat{\sigma}(\mathfrak{g}_p) = \mathfrak{g}_p$ for $p = -1,-2$ such that

$$\hat{\sigma}(Y) \equiv \sigma(Y) \pmod{\mathfrak{g}_{-1}} \quad \text{for } Y \in \mathfrak{g}_{-2} ,$$
$$\hat{\sigma}(X) = \sigma(X) \quad \text{for } X \in \mathfrak{g}_{-1} .$$

Then we have

$$\tilde{G} = \{ \sigma \in GL(\mathfrak{m}) \mid \sigma(\mathfrak{g}_{-1}) = \mathfrak{g}_{-1} \text{ and } \hat{\sigma} \in \rho(G_0) \} .$$

Namely $\tilde{G} = G_0^{\#}$ in the notation of §4 [18].

§3. Lie contact manifolds

In this section, we introduce the notion of Lie contact structure as a certain linear group structure on contact manifolds. As an important example, we shall show that the unit tangent bundle of a Riemannian manifold (M, g) has a Lie contact structure induced from g and also dicuss its relation with the conformal structure on M.

Let N be a contact manifold of dimension 2n-1. Namely N is a manifold of dimension $2n-1$ endowed with a distribution of codimension 1 on N, or equivalently a subbundle D of the tangent bundle $T(N)$ of rank $2(n-1)$, such that, on a neiborhood U of $x \in N$, each fibre $D(y)$ of D at $y \in U$ is defined by

$$D(y) = \{ X \in T_y(N) \mid \theta(X) = 0 \},$$

where θ is a 1-form on U such that $\theta \wedge (d\theta)^{n-1}$ gives a volume element. Then the 2-form $\theta_x = (d\theta)_x|_D$ is non-degenerate on $D(x)$, that is, $(D(x), \theta_x)$ is a symplectic vector space. Since θ is unique up to nowhere zero function, the contact structure on N defines a conformal symplectic structure on each fibre of D. Let N_1 and N_2 be contact manifolds with the contact distributions D_1 and D_2 respectively. A diffeomorphism $\Phi ; N_1 \to N_2$ is called a contact diffeomorphism if $\Phi_*(D_1) = D_2$.

According to Tanaka [18], this conformal symplectic structure can be described more precisely as in the followig: First we introduce the graded Lie algebra $\mathfrak{m}(x)$ associated with D at each $x \in N$. We put $\mathfrak{m}(x) = \mathfrak{g}_{-2}(x) \oplus \mathfrak{g}_{-1}(x)$, where $\mathfrak{g}_{-2}(x) = T_x(N) / D(x)$ and $\mathfrak{g}_{-1}(x) = D(x)$. The Lie algebra structure of $\mathfrak{m}(x)$ is given as follows; For any cross sections $\tilde{X}$ and $\tilde{Y}$ of D, we have $[f\cdot\tilde{X}, g\cdot\tilde{Y}] = f\cdot g\,[\tilde{X},\tilde{Y}] + f\cdot(\tilde{X}g)\,\tilde{Y} - g\cdot(\tilde{Y}f)\,\tilde{X}$. Hence, for X and $Y \in D(x) = \mathfrak{g}_{-1}(x)$, we can define $[X,Y] \in \mathfrak{g}_{-2}(x)$ by

$$[X,Y] = q([\tilde{X},\tilde{Y}]_x),$$

where q is the projection of $T_x(N)$ onto $\mathfrak{g}_{-2}(x)$ and $\tilde{X}$, $\tilde{Y}$ are sections of D such that $X = \tilde{X}_x$ and $Y = \tilde{Y}_x$. Moreover we define $[\mathfrak{g}_{-2}(x), \mathfrak{m}(x)] = \{0\}$. Thus $\mathfrak{m}(x)$ becomes a nilpotent graded Lie algebra which is isomorphic with $\mathfrak{m}$ in §2. It is easy to see that the automorphism group of the graded Lie algebra $\mathfrak{m} = \mathfrak{g}_{-2} \oplus \mathfrak{g}_{-1}$ is isomorphic with the conformal symplectic group $CSp(n-1)$, where

$$CSp(n-1) = \{\, A \in GL(2(n-1),\mathbb{R}) \mid {}^tA\cdot J\cdot A = c\cdot J \,\},$$

and J is the skew-symmetric matrix given by

$$J = \begin{pmatrix} 0 & -I_{n-1} \\ I_{n-1} & 0 \end{pmatrix}.$$

The contact structure on N gives rise to the canonical reduction of the structure group of the frame bundle $F(N)$ of N: By a frame z at $x \in N$, we mean a linear isomorphism z of $\mathfrak{m} \simeq \mathbb{R}^{2n-1}$ onto $T_x(N)$. A frame z ; $\mathfrak{m} \to T_x(N)$ is called adapted if $z(\mathfrak{g}_{-1}) = \mathfrak{g}_{-1}(x)$ and the graded map $\hat{z}$; $\mathfrak{m} \to \mathfrak{m}(x)$ is a Lie algebra isomorphism. By fixing the basis of $\mathfrak{m}$ as in §2, an adapted frame z is given, as an ordered basis of $T_x(N)$, as follows ; First X_0 is a vector transversal to $D(x)$. Let θ be a defining 1-form of D around x such that $\theta(X_0) = 1$. We take a sympletic basis $\{X_1,\ldots,X_{2(n-1)}\}$ of $(D(x), \theta_x)$, that is,

$$d\theta(X_{n+i-1}, X_i) = -d\theta(X_i, X_{n+i-1}) = 1 \quad \text{for} \quad i = 1, \ldots, n-1,$$

$$d\theta(X_i, X_j) = 0 \quad \text{otherwise}.$$

Then $\{X_0, X_1, \ldots, X_{2(n-1)}\}$ forms an adapted frame at x. In fact, putting $\bar{X}_0 = q(X_0)$, we see that $\bar{X}_0$ and $\{X_1, \ldots, X_{2(n-1)}\}$ form the bases of $\mathfrak{g}_{-2}(x)$ and $\mathfrak{g}_{-1}(x)$ respectively such that

$$\bar{X}_0 = [X_i, X_{n+i-1}] = -[X_{n+i-1}, X_i] \quad \text{for} \quad i = 1, \ldots, n-1,$$

$$[X_i, X_j] = 0 \quad \text{otherwise}.$$

Then the set $F_c(N)$ of adapted frames on N forms a subbundle of $F(N)$ with structure group C. Here C is the subgroup of $GL(\mathfrak{m})$ consisting of all $\sigma \in GL(\mathfrak{m})$ such that $\sigma(\mathfrak{g}_{-1}) = \mathfrak{g}_{-1}$ and the graded map $\hat{\sigma}$ is a graded Lie algebra automorphism of $\mathfrak{m}$ ($C = G_0(\mathfrak{M})^{\#}$ in the notation of §4 [18]). Explicitly, by using the coordinate (η, ξ_1, ξ_2) of $\mathfrak{m}$ in §2, each $\sigma \in C$ is represented as a matrix of the following form;

$$\begin{pmatrix} c & 0 \\ \xi & A \end{pmatrix},$$

where ${}^tA \cdot J \cdot A = c \cdot J$ and $\xi \in \mathbb{R}^{2(n-1)}$. Obviously, $\tilde{G}$ in §2 is a subgroup of C.

Now we introduce the following notion:

Definition 3.1. Let N be a contact manifold of dimension $2n-1$. A Lie contact structure on N is a subbundle $\tilde{P}$ of $F_c(N)$ with structure group $\tilde{G}$. With this structure, N is called a Lie contact manifold.

Let N_1 and N_2 be contact manifolds with Lie contact structures $\tilde{P}_1$ and $\tilde{P}_2$ respectively. A contact diffeomorphism Φ of N_1 onto N_2 is called a Lie isomorphism if $\Phi_*(\tilde{P}_1) = \tilde{P}_2$.

Remark 3.2. In the terminology in [18], a Lie contact structure on N is a $\tilde{G}$-structure of type $\mathfrak{M}$ on N associated with the simple graded Lie algebra $\mathfrak{g} = \Sigma_{p\in\mathbb{Z}}\, \mathfrak{g}_p$.

Let (M,g) be an n-dimensional Riemannian manifold. We shall show that the unit tangent bundle $T_1(M)$ of (M,g) has a Lie contact structure induced from g . In order to see this, we recall several notions on the tangent bundle p ; $T(M) \to M$ (cf. Besse[3]): First, for $v \in T(M)$ and $x = p(v)$, we have the notion of the vertical lift $w^V \in (\mathrm{Ker}\, p_*)(v)$ of a vector $w \in T_x(M)$. The subspace $(\mathrm{Ker}\, p_*)(v)$ of $T_v(T(M))$ is a tangent space at v to the fibre $T_x(M)$ of p ; $T(M) \to M$. Then w^V is defined as the vector tangent to a curve $c(t) = v + t\cdot w$ of $T_x(M)$ at $c(0) = v$.

A canonical 1-form α on $T(M)$ is given by

$$\alpha_v(X) = g(v, p_*(X)) \quad \text{for} \quad X \in T_v(T(M)) \ .$$

Then $(T(M), d\alpha)$ is a symplectic manifold and $\alpha = 0$ defines a contact distribution D on $T_1(M)$. A geodesic vector field (or a geodesic flow) Z_g is a Hamiltonian vector field on $T(M)$ given by

$$i_{Z_g} d\alpha = - dE_g \ ,$$

where $E_g(v) = \frac{1}{2} g(v,v)$ is the energy function on $T(M)$. Z_g is tangent to $T_1(M)$ and $\alpha(Z_g) = 1$ on $T_1(M)$. Furthermore the Riemannian connection on M defines the horizontal subspace $H(v)$ of $T_v(T(M))$ at each $v \in T(M)$ (cf. Besse[3, p.37]) such that $H(v)$ is tangent to $T_1(M)$, $d\alpha|_{H(v)} = 0$ and that

$$T_v(T(M)) = H(v) \oplus (\text{Ker } p_*)(v) .$$

Thus we have the notion of the horizontal lift $w^H \in H(v)$ of a vector $w \in T_x(M)$, where $x = p(v)$. Here we note that, for $w_1, w_2 \in T_x(M)$, we have

$$d\alpha(w_1^V, w_2^H) = g(w_1, w_2) ,$$

where w_1^V and w_2^H are the vertical lift of w_1 and the horizontal lift of w_2 respectively. Moreover we have

$$Z_g(v) = v^H \quad \text{and} \quad E(v) = v^V \quad \text{at} \quad v \in T(M) ,$$

where E is the Liouville vector field on $T(M)$.

We put $H_1(v) = D(v) \cap H(v)$ at $v \in T_1(M)$ and denote the projection of $T_1(M)$ onto M also by p. Then, we have the following decomposition at $v \in T_1(M)$;

$$T_v(T_1(M)) = \langle Z_g(v) \rangle \oplus D(v) ,$$
$$D(v) = H_1(v) \oplus (\text{Ker } p_*)(v) ,$$
$$H(v) = \langle Z_g(v) \rangle \oplus H_1(v) ,$$

where $\langle Z_g(v) \rangle$ is the 1-dimensional subspace spanned by $Z_g(v)$. Here $H_1(v)$ and $(\text{Ker } p_*)(v)$ are both Legendrian subspaces of $D(v)$. Then, putting $x = p(v) \in M$, we see that $p_*(Z_g(v)) = v$ and p_* is a linear isomorphism of $H(v)$ onto $T_x(M)$ such that

$$T_x(M) = \langle v \rangle \oplus p_*(H_1(v)) ,$$

is an orthogonal decomposition. Moreover it is easy to see that $D(v)$ is spanned by the vectors w^H, w^V for $w \in T_x(M)$ satisfying $g(v,w) = 0$.

Starting from an orthonormal frame $\{e_0,e_1,\ldots,e_{n-1}\}$ at $x \in M$, we can construct an adapted frame $\{X_0,X_1,\ldots,X_{2(n-1)}\}$ of $T_1(M)$ at $v = e_0$ as follows ; First we take $X_i = e_i^H$ for $i = 0,1,\ldots,n-1$. Then $X_0 = Z_g(v)$, $\alpha(X_0) = 1$ and $\{X_1,\ldots,X_{n-1}\}$ forms a basis of $H_1(v)$. The tangent space $T_v(T_1(M))$ of $T_1(M)$ at v is spanned by the vectors e_0^H, $e_1^H,\ldots, e_{n-1}^H$ and $e_1^V,\ldots, e_{n-1}^V$. Since $d\alpha(e_i^V, e_j^H) = \delta_{ij}$, $\{e_1^H,\ldots, e_{n-1}^H, e_1^V,\ldots, e_{n-1}^V\}$ forms a symplectic basis of $(D(v), (d\alpha)_v)$. Thus, putting $X_{n+i-1} = e_i^V$, we obtain an adapted frame $\{X_0,X_1,\ldots,X_{2(n-1)}\}$ at v .

In this way we obtain an embedding of the orthonormal frame bundle $O(M)$ of (M,g) into $F_c(T_1(M))$. It is easy to see that the image $F_0(T_1(M))$ of this embedding forms a subbundle of $F_c(T_1(M))$ with structure group $\tilde{G}_0$, which is isomorphic with $O(n-1)$. As a subgroup of $GL(\mathfrak{m})$, each $\sigma \in \tilde{G}_0$ is represented, in the notation of §2, as a matrix of the following form;

$$\begin{pmatrix} 1 & 0 \\ 0 & B\otimes I_2 \end{pmatrix}$$

where $B \in O(n-1)$. In particular $\tilde{G}_0$ is a subgroup of $\tilde{G}$. Then, by enlarging the structure group,

$$\tilde{P}(T_1(M)) = F_0(T_1(M)) \times_{\tilde{G}_0} \tilde{G}$$

gives a Lie contact structure on $T_1(M)$.

Now we shall show that a conformal structure on an n-dimensional manifold M induces a Lie contact structure on the tangent sphere bundle $S(M)$ of M . More precisely, a conformal structure gives rise to the canonical reduction of the structure group of $F_c(S(M))$ to the subgroup $\tilde{G}^+_M$ of $\tilde{G}$, where $\tilde{G}^+_M$ is the linear isotropy subgroup of the Möbius group G^+_M at $o \in T_1(S^n)$ (see §3 and Remark 1.1). We show this by exhibiting how an adapted frame in $F_0(T_1(M))$ changes under a conformal chage of the metric g .

Let M be an n-dimensional manifold with a conformal structure and let $\mathcal{G}_M$ be the set of Riemannian metrics on M which induce the given conformal structure. Let κ be the projection of $\mathring{T}(M)$ onto the tangent sphere bundle $S(M) = \mathring{T}(M) / \mathbb{R}^+$ of M , where $\mathring{T}(M) = T(M) / \{\text{zero section}\}$. In the following, we denote by $T^g_1(M)$ and α_g the unit tangent bundle and the canonical 1-form on $T(M)$ respectively, which are defined by a Riemannian metric $g \in \mathcal{G}_M$. Moreover we denote by κ_g the restriction of κ to $T^g_1(M)$. Then κ_g is a fibre-preserving diffeomorphism of $T^g_1(M)$ onto $S(M)$.

Let g and $\hat{g}$ be two elements of $\mathcal{G}_M$. Then g and $\hat{g}$ are conformally equivalent, that is, there exists a positive function f on M such that $\hat{g} = f \cdot g$. Since $\alpha_{\hat{g}} = f \cdot \alpha_g$ and $(R_\rho)^* \alpha_g = \rho \cdot \alpha_g$ for $\rho \in \mathbb{R}^+$, $\alpha_g = 0$ defines a contact distribution D on $S(M)$ which is independent of a conformal change of the metric,where R_ρ ; $T(M) \to T(M)$ is the multiplication by $\rho \in \mathbb{R}^+$. Namely the contact structure on $S(M)$ is uniquely determined by the conformal structure on M . κ_g

and $\kappa_{\hat{g}}$ are contact diffeomorphisms onto $S(M)$. In particular, the vertical subspace $(\mathrm{Ker}\ q_*)(u)$ is a Legendrian subspace at each $u \in S(M)$, where $q; S(M) \to M$ is the projection. Moreover the image $(\kappa_g)_*(F_0(T_1^g(M)))$ of $F_0(T_1^g(M))$ under $(\kappa_g)_* ; F_c(T_1^g(M)) \to F_c(S(M))$ is a subbundle of $F_c(S(M))$ with structure group $\tilde{G}_0$.

We consider now the image of horizontal subspaces under κ_g. Let Ψ_f be the bundle automorphism of $T(M)$ defined by

$$\Psi_f(v) = \lambda(x)\cdot v \quad \text{for} \quad v \in T(M) ,$$

where $x = p(v)$ and $\lambda(x) = (f(x))^{1/2}$. Then $\kappa \circ \Psi_f = \kappa$ and Ψ_f sends $T_1^{\hat{g}}(M)$ diffeomorphically onto $T_1^g(M)$. Let us fix a point $u_o \in S(M)$ and put $x_o = q(u_o)$. Let v_o and $\hat{v}_o$ be unique vectors in $T_{x_o}(M)$ satisfying $\kappa(v_o) = \kappa(\hat{v}_o) = u_o$ and $\|v_o\|_g = \|\hat{v}_o\|_{\hat{g}} = 1$. Then $v_o \in T_1^g(M)$, $\hat{v}_o \in T_1^{\hat{g}}(M)$ and $\Psi_f(\hat{v}_o) = v_o$. For each $w \in T_{x_o}(M)$, we denote by $w^H \in T_{v_o}(T(M))$ and $\hat{w}^H \in T_{\hat{v}_o}(T(M))$ the horizontal lifts of w with respect to the metrics g and $\hat{g}$ respectively.

We take a coordinate system $(x_1,\ldots,x_n)$ of M around x_o. Let $(x_1,\ldots,x_n,v_1,\ldots,v_n)$ be the coordinate system of $T(M)$ induced from $(x_1,\ldots,x_n)$. In this coordinate, for $w = w_i \frac{\partial}{\partial x_i} \in T_{x_o}(M)$, w^H and $\hat{w}^H$ are given by

$$w^H = w_i \frac{\partial}{\partial x_i} - \Gamma^i_{jk}(x)\cdot v_j \cdot w_k \frac{\partial}{\partial v_i} ,$$

$$\hat{w}^H = w_i \frac{\partial}{\partial x_i} - \hat{\Gamma}^i_{jk}(x)\cdot \hat{v}_j \cdot w_k \frac{\partial}{\partial v_i} ,$$

where $v_o = v_i \frac{\partial}{\partial x_i}$, $\hat{v}_o = \hat{v}_i \frac{\partial}{\partial x_i}$ and $\Gamma^i_{jk} = \frac{1}{2} g^{im}(\partial g_{mk}/\partial x_j$

$+ \partial g_{jm}/\partial x_k - \partial g_{jk}/\partial x_m)$ is the Christoffel symbol associated with g (cf. Besse [3, p.33]). Here we adopt the Einstein's convention for indices. Then, by a direct calculation, we obtain

$$(\Psi_f)_*(\hat{w}^H) = w^H - \gamma\left(w_i \frac{\partial}{\partial v_i}\right) + g(v,w)\cdot Y ,$$

where $\gamma = \frac{1}{2}\lambda(x_o)^{-1} v_j \frac{\partial f}{\partial x_j}(x_o)$ and

$$Y = \frac{1}{2}\lambda(x_o)^{-1} g^{ij}(x_o) \frac{\partial f}{\partial x_j}(x_o) \frac{\partial}{\partial v_i} .$$

Let $\{e_0, e_1, \ldots, e_{n-1}\}$ be an orthonormal frame of (M,g) at x_o such that $e_0 = v_o$. Then, by putting $\hat{e}_i = \lambda(x_o)^{-1}\cdot e_i$ for $i = 0,1,\ldots,n-1$, $\{\hat{e}_0, \hat{e}_1, \ldots, \hat{e}_{n-1}\}$ is an orthonormal frame of $(M,\hat{g})$ at x_o such that $\hat{e}_0 = \hat{v}_o$. Let $z \in F_c(T_1^g(M))$ and $\hat{z} \in F_c(T_1^{\hat{g}}(M))$ be the adapted frames at v_o and $\hat{v}_o$ respectively which are constructed from these orthonormal frames at x_o . Since Ψ_f gives a contact diffeomorphism of $T_1^{\hat{g}}(M)$ onto $T_1^g(M)$, $(\Psi_f)_*(\hat{z})$ is an adapted frame at v_o . In order to compare $(\Psi_f)_*(\hat{z})$ and z , we choose the normal coordinate system $(x_1,\ldots,x_n)$ of (M,g) determined by the frame $\{e_0, e_1, \ldots, e_{n-1}\}$. In this coordinate, we have $\Gamma^i_{jk}(x_o) = 0$, $(d\alpha)_{v_o} = \Sigma_{i=1}^n dv_i \wedge dx_i$ and $\frac{\partial}{\partial x_i} = e_{i-1}$ for $i = 1,\ldots,n$ at x_o . Then it is easy to see that, as an ordered basis of $T_{v_o}(T_1^g(M))$, adapted frames z and $(\Psi_f)_*(\hat{z})$ is given, in this coordinate, by

$$z = \left\{ \frac{\partial}{\partial x_1}, \frac{\partial}{\partial x_2}, \ldots, \frac{\partial}{\partial x_n}, \frac{\partial}{\partial v_2}, \ldots, \frac{\partial}{\partial v_n} \right\} ,$$

$$(\Psi_f)_*(\hat{z}) = \{ \hat{X}_0, \hat{X}_1, \ldots, \hat{X}_{n-1}, \frac{\partial}{\partial v_2}, \ldots, \frac{\partial}{\partial v_n} \} ,$$

where $\quad \hat{X}_0 = \lambda(x_o)^{-1}(\frac{\partial}{\partial x_1} + \frac{1}{2} \lambda(x_o)^{-1} \Sigma_{i=2}^{n} \frac{\partial f}{\partial x_i}(x_o) \frac{\partial}{\partial v_i}) ,$

and $\quad \hat{X}_i = \lambda(x_o)^{-1}(\frac{\partial}{\partial x_{i+1}} - \gamma \frac{\partial}{\partial v_{i+1}})$ for $i = 1, \ldots, n-1$.

Namely $(\Psi_f)_*(\hat{z}) = z \cdot \sigma$ for an element $\sigma \in GL(\mathfrak{m})$ given by

$$\sigma = \begin{pmatrix} \alpha & 0 & 0 \\ 0 & \alpha \cdot I & 0 \\ \xi & \beta \cdot I & I \end{pmatrix} ,$$

where $\alpha = \lambda(x_o)^{-1}$, $\beta = -\frac{1}{2} \lambda(x_o)^{-2} \frac{\partial f}{\partial x_1}(x_o)$, $\xi_{i-1} = \frac{1}{2} \lambda(x_o)^{-2} \frac{\partial f}{\partial x_i}(x_o)$, $\xi = (\xi_1, \ldots, \xi_{n-1}) \in \mathbb{R}^{n-1}$ and I is the unit matrix of degree n-1.

Now we consider the union $F_M(S(M))$ of all adapted frames of $S(M)$ obtained from $F_0(T_1^g(M))$ for all $g \in \mathcal{G}_M$. Namely we consider

$$F_M(S(M)) = \cup_{g \in \mathcal{G}_M} (\kappa_g)_*(F_0(T_1^g(M))) .$$

Then, by the above argument and the explicit matrix representation of elements of $\tilde{G}_M$ in §2 , we see that $F_M(S(M))$ is a subbundle of $F_c(S(M))$ with structure group $\tilde{G}_M^+$. Hence, by enlarging the structure group,

$$\tilde{P}(S(M)) = F_M(S(M)) \times_{\tilde{G}_M^+} \tilde{G}$$

gives a Lie contact structure on $S(M)$. It follows from the definition that κ_g is a Lie isomorphism of $T_1^g(M)$ onto $S(M)$ for $g \in \mathcal{G}$.

Let M_1 and M_2 be n-dimensional manifolds with

conformal structures. For a conformal diffeomorphism φ of M_1 onto M_2 , φ_* induces diffeomorphisms $\tilde{\varphi}^+$ and $\tilde{\varphi}^-$ of $S(M_1)$ onto $S(M_2)$ by

$$\tilde{\varphi}^+(u) = q_2(\varphi_*(v)) \quad \text{and} \quad \tilde{\varphi}^-(u) = q_2(-\varphi_*(v)) ,$$

for $u \in S(M_1)$, $q_1(v) = u$ and $v \in T(M_1)$, where q_i ; $T(M_i) \to S(M_i)$ is the projection. Let g_1 and g_2 be representative metrics of M_1 and M_2 respectively. Since $\varphi^* g_2 = f \cdot g_1$ for a positive function f on M_1 , it is easy to see that $\tilde{\varphi}^+$ and $\tilde{\varphi}^-$ are Lie isomorphisms of $S(M_1)$ onto $S(M_2)$. In particular a conformal transformation on M lifts to Lie transformations on $S(M)$ in two ways.

§4. Normal Cartan connection and Tanaka's Theorem.

In this section we apply Tanaka's theory ([18] and [19]) of Cartan connections associated with simple graded Lie algebras to Lie contact manifolds.

In order to apply Tanaka's theory to Lie contact manifolds, we first check that, for $n \geq 3$, $\mathfrak{g} = \Sigma_{p\in\mathbb{Z}}\, \mathfrak{g}_p$ is the prolongation of $(\mathfrak{m}, \mathfrak{g}_0)$. Let $\mathfrak{h}_0$ be the ideal of $\mathfrak{g}_0$ defined by

$$\mathfrak{h}_0 = \{ X \in \mathfrak{g}_0 \mid [X, \mathfrak{g}_{-2}] = 0 \} .$$

Since ρ ; $G_0 \to \tilde{G} \subset GL(\mathfrak{m})$ is faithful, $\mathfrak{h}_0$ can be regarded as a subalgebra of $\mathfrak{gl}(\mathfrak{g}_{-1})$. Explicitly, by fixing the basis of $\mathfrak{m}$ as in §2 , each $X \in \mathfrak{h}_0 \subset \mathfrak{gl}(\mathfrak{g}_{-1})$ is given as a matrix of the following form;

$$X = \begin{pmatrix} B - \alpha \cdot I & \gamma \cdot I \\ \beta \cdot I & B + \alpha \cdot I \end{pmatrix},$$

where $B \in \mathfrak{o}(n-1)$, $\alpha, \beta, \gamma \in \mathbb{R}$ and I is the unit matrix of degree n-1. Thus $\mathfrak{h}_0$ is a direct sum of $\mathfrak{o}(n-1)$ and $\mathfrak{sl}(2,\mathbb{R})$. Then we have

Lemma 4.1. For $n \geq 3$, the first prolongation $\mathfrak{h}_0^{(1)}$ of $\mathfrak{h}_0$ vanishes.

Hence, by Lemma 3.4 [18], $\mathfrak{g} = \Sigma_{p \in \mathbb{Z}} \mathfrak{g}_p$ is the prolongation of $(\mathfrak{m}, \mathfrak{g}_0)$ for $n \geq 3$.

Proof. We give here an elementary proof of this fact. Let u be an arbitrary element of $\mathfrak{h}_0^{(1)}$. Namely u is a linear map of $\mathfrak{g}_{-1}$ into $\mathfrak{h}_0$ satisfying $u(v)(w) = u(w)(v)$ for any $v, w \in \mathfrak{g}_{-1}$. For $v = (\xi_1, \xi_2) \in \mathfrak{g}_{-1}$, we write

$$u(\xi_1, \xi_2) = \begin{pmatrix} B(\xi_1,\xi_2) - \alpha(\xi_1,\xi_2) \cdot I & \gamma(\xi_1,\xi_2) \cdot I \\ \beta(\xi_1,\xi_2) \cdot I & B(\xi_1,\xi_2) + \alpha(\xi_1,\xi_2) \cdot I \end{pmatrix}.$$

Then, for $v = (\xi_1, \xi_2)$, $w = (\eta_1, \eta_2) \in \mathfrak{g}_{-1}$, $u(v)(w) = u(w)(v)$ is equivalent to the following two equalities;

$$(4.1) \qquad \begin{aligned} &B(\xi_1,\xi_2)(\eta_1) - \alpha(\xi_1,\xi_2)\eta_1 - \gamma(\xi_1,\xi_2)\eta_2 \\ &= B(\eta_1,\eta_2)(\xi_1) - \alpha(\eta_1,\eta_2)\xi_1 - \gamma(\eta_1,\eta_2)\xi_2 \quad . \end{aligned}$$

$$(4.2) \qquad \begin{aligned} &B(\xi_1,\xi_2)(\eta_2) - \beta(\xi_1,\xi_2)\eta_1 + \alpha(\xi_1,\xi_2)\eta_2 \\ &= B(\eta_1,\eta_2)(\xi_2) - \beta(\eta_1,\eta_2)\xi_1 + \alpha(\eta_1,\eta_2)\xi_2 \quad . \end{aligned}$$

Putting $\xi_1 = \eta_1 = 0$ in (4.1), we get

$$\gamma(0,\xi)\eta = \gamma(0,\eta)\xi$$

for any ξ , $\eta \in \mathbb{R}^{n-1}$. Since $n-1 \geq 2$, we conclude $\gamma(0,\xi) = 0$ for any $\xi \in \mathbb{R}^{n-1}$. Similarly, putting $\xi_2 = \eta_2 = 0$ in (4.2), we obtain $\beta(\xi,0) = 0$ for any $\xi \in \mathbb{R}^{n-1}$. Moreover, putting $\xi_1 = \eta_2 = 0$ in (4.1), we have

$$(4.3) \qquad B(0,\xi)(\eta) - \alpha(0,\xi)\eta + \gamma(\eta,0)\xi = 0 \quad ,$$

for any ξ , $\eta \in \mathbb{R}^{n-1}$. For any $\xi \in \mathbb{R}^{n-1}$, we can take non-zero vector $\eta \in \mathbb{R}^{n-1}$ such that $(\xi,\eta) = 0$. Since $B(0,\xi) \in \mathfrak{o}(n-1)$, we have $(B(0,\xi)(\eta),\eta) = 0$. Then, by taking the inner product of η and both sides of (4.3), we obtain $\alpha(0,\xi) = 0$ and $B(0,\xi)(\xi) + \gamma(\xi,0)\xi = 0$ for any $\xi \in \mathbb{R}^{n-1}$. Taking the inner product of ξ and both sides of the last equality again, we conclude $\gamma(\xi,0) = 0$, which implies $B(0,\xi) = 0$ for any $\xi \in \mathbb{R}^{n-1}$. Similarly, from (4.2), we obtain $\alpha(\xi,0) = 0$, $\beta(0,\xi) = 0$ and $B(\xi,0) = 0$ for any $\xi \in \mathbb{R}^{n-1}$. Thus we obtain $u = 0$, which completes the proof .

In the case $n = 2$, we have $\mathfrak{h}_0 = \mathfrak{sl}(\mathfrak{g}_{-1})$. Hence $\mathfrak{h}_0$ is of infinite type. Thus $\mathfrak{g} = \Sigma_{p\in\mathbb{Z}}\, \mathfrak{g}_p$ is not the prolongation of $(\mathfrak{m},\mathfrak{g}_0)$, when $n = 2$.

Remark 4.2. Since $\mathfrak{h}_0$ is an irreducible subalgebra of $\mathfrak{gl}(\mathfrak{g}_{-1})$, Lemma 4.1 follows also from Lemma 3.4 [18] and the classification of the irreducible linear Lie algebras of infinite type (cf. Kobayashi - Nagano [9]). Furthermore, by Lemma 1.14 [19], Lemma 4.1 is equivalent to the vanishing of the cohomology groups $H^{p,1}(\mathfrak{m},\mathfrak{g})$ for $p \geq 1$, which is associated with the representation $\mathrm{ad} ; \mathfrak{m} \to \mathfrak{gl}(\mathfrak{g})$ (see below). Then Lemma 4.1 can be shown by calculating the cohomology groups $H^{p,1}(\mathfrak{m},\mathfrak{g})$ by the method of Kostant [11].. In fact, by Kostant's method, one can check which simple graded Lie algebra $\mathfrak{g} = \Sigma_{p\in\mathbb{Z}} \mathfrak{g}_p$ is the prolongation of $(\mathfrak{m},\mathfrak{g}_0)$ or $\mathfrak{m}$ among all simple graded Lie algebras over $\mathbb{R}$. We shall report on this in another occasion.

Now we recall the notion of Cartan connections of type G / G'. Let N be a manifold of dimension $2n-1$. The pair (P,ω) is called a Cartan connection of type G / G' on N if P and ω satisfy the following:

(1) P is a principal fibre bundle over N with structure group G'.

(2) ω is a $\mathfrak{g}$-valued 1-form on P satisfying the following conditions;

(a) $\omega(X) \neq 0$ for every nonzero vector X of P.

(b) $(R_a)^*\omega = \mathrm{Ad}(a^{-1})\,\omega$ for every $a \in G'$.

(c) $\omega(A^*) = A$ for every $A \in \mathfrak{g}'$,

where A^* is the fundamental vector field on P corresponding to $A \in \mathfrak{g}'$.

Let ω_G be the Maurer-Cartan form on G. Then the pair (G,ω_G) gives a Cartan connection of type G / G' on $T_1(S^n) = G / G'$, which is called the standard connection of type G / G'. The curvature form Ω of (P,ω) is defined by the following structure equation:

$$d\omega + \frac{1}{2}[\omega \wedge \omega] = \Omega .$$

(P,ω) induces a $\tilde{G}$ -structure on N as follows [18]; Let θ be the $\mathfrak{m}$-component of ω with respect to the decomposition $\mathfrak{g} = \mathfrak{m} \oplus \mathfrak{g}'$. Then θ satisfies (Lemma 1.1 [18]);

(1) For $X \in T(P)$, $\theta(X) = 0$ if and only if X is a vertical vector.

(2) $(R_a)^* \theta = \rho(a^{-1}) \theta$ for every $a \in G'$,

where $\rho ; G' \to \tilde{G}$ is the linear isotropy representation.

For a point $z \in P$, we put $x = \pi(z)$, where $\pi ; P \to N$ is the projection. It is easy to see that θ_z defines a linear isomorphism $\hat{\rho}(z) ; T_x(N) \to \mathfrak{m}$ by $\hat{\rho}(z)(Y) = \theta_z(X)$ for $Y \in T_x(N)$, where X is a vector in $T_z(P)$ satisfying $\pi_*(X) = Y$. Thus we have a map ρ of P into the frame bundle $F(N)$ of N defined by $\rho(z) = (\hat{\rho}(z))^{-1} ; \mathfrak{m} \to T_x(N)$ for $z \in P$. Then the image $\tilde{P}$ of ρ is a subbundle of $F(N)$ with structure group $\tilde{G}$, and $\rho ; P \to \tilde{P}$ is a bundle homomorphism corresponding to $\rho ; G' \to \tilde{G}$.

To state the normality condition for the curvature Ω of a Cartan connection (P,ω) , we now mention about the harmonic theory for the cohomology associated with the representation $\mathrm{ad} ; \mathfrak{m} \to \mathfrak{gl}(\mathfrak{g})$ following Tanaka [19] (cf. Kostant [11]).

Let $(C(\mathfrak{m},\mathfrak{g}), \partial)$ be the cochain complex associated with the representation $\mathrm{ad} ; \mathfrak{m} \to \mathfrak{gl}(\mathfrak{g})$. Namely $C(\mathfrak{m},\mathfrak{g}) = \Sigma_q C^q(\mathfrak{m},\mathfrak{g})$, where

$$C^q(\mathfrak{m},\mathfrak{g}) = \mathrm{Hom}(\Lambda^q \mathfrak{m} , \mathfrak{g}) ,$$

and the coboundary operator $\partial : C^q(\mathfrak{m},\mathfrak{g}) \to C^{q+1}(\mathfrak{m},\mathfrak{g})$ is defined by

$$(\partial c)(X_1,\ldots,X_{q+1}) = \Sigma_i \ (-1)^{i+1}[X_i, c(X_1,\ldots,\hat{X}_i,\ldots,X_{q+1})]$$
$$+ \Sigma_{i<j} \ (-1)^{i+j} c([X_i,X_j],X_1,\ldots,\hat{X}_i,\ldots,\hat{X}_j,\ldots,X_{q+1}) \ ,$$

for $c \in C^q(\mathfrak{m},\mathfrak{g})$ and $X_1,\ldots,X_{q+1} \in \mathfrak{m}$.

Let σ be an involutive automorphism of $\mathfrak{g}$ defined by $\sigma(X) = -{}^tX$ for $X \in \mathfrak{g}$. Then it follows that $\sigma(\mathfrak{g}_p) = \mathfrak{g}_{-p}$ for $p \in \mathbb{Z}$, $B(\mathfrak{g}_p,\mathfrak{g}_q) = 0$ if $p + q \neq 0$ and $B(X,\sigma(X))$ is negative definite on $\mathfrak{g}$, where B is the Killing form of $\mathfrak{g}$. Utilizing σ , we define an inner product $(\ ,\ )$ in $\mathfrak{g}$ by $(X,Y) = -B(X,\sigma(Y))$ for $X, Y \in \mathfrak{g}$. Explicitly $(X,Y) = (n+1)\cdot \mathrm{tr}\, X\cdot {}^tY$ in our case. Since B gives a pairing of $\mathfrak{g}_p$ and $\mathfrak{g}_{-p}$, $\Sigma_{p>0}\, \mathfrak{g}_p$ may be regarded as the dual space $\mathfrak{m}^*$ of $\mathfrak{m} = \Sigma_{p<0}\, \mathfrak{g}_p$. Hence, for a basis $\{e_0,e_1,\ldots,e_{2(n-1)}\}$ of $\mathfrak{m}$, we have the dual basis $\{e_0^*,e_1^*,\ldots,e_{2(n-1)}^*\}$ of $\mathfrak{m}^* = \Sigma_{p>0}\, \mathfrak{g}_p$ such that $B(e_i,e_j^*) = \delta_{ij}$. The above inner product in $\mathfrak{g}$ induces the inner product in $C^q(\mathfrak{m},\mathfrak{g})$ in the usual manner.

The adjoint operator $\partial^* : C^{q+1}(\mathfrak{m},\mathfrak{g}) \to C^q(\mathfrak{m},\mathfrak{g})$ with respect to this inner product is given by the following formula (§1 [19]);

$$(\partial^* c)(x_1,\ldots,X_q) = \Sigma_j \ [e_j^*,\ c(e_j,X_1,\ldots,X_q)]$$
$$+ \frac{1}{2} \Sigma_{ij} \ (-1)^{i+1} c([e_j^*, X_i]_-, e_j, X_1,\ldots,\hat{X}_i,\ldots,X_q) \ ,$$

where $c \in C^q(\mathfrak{m},\mathfrak{g})$, $X_1,\ldots,X_q \in \mathfrak{m}$ and $[e_j^*,X_i]_-$ denotes the $\mathfrak{m}$-component of $[e_j^*,X_i]$ with respect to the decomposition $\mathfrak{g} = \mathfrak{m} \oplus \mathfrak{g}'$.

According to [19], a bigradation of $C(\mathfrak{m},\mathfrak{g})$ is introduced from the gradation of $\mathfrak{g}$ and $\mathfrak{m}$ as follows; First, from $\mathfrak{m} = \mathfrak{g}_{-2} \oplus \mathfrak{g}_{-1}$ and $\dim \mathfrak{g}_{-2} = 1$, we have

$$(\Lambda^q \mathfrak{m})^* = \Lambda^q \mathfrak{m}^* = \mathfrak{g}^*_{-2} \otimes \Lambda^{q-1} \mathfrak{g}^*_{-1} \oplus \Lambda^q \mathfrak{g}^*_{-1} .$$

By the gradation of $\mathfrak{g}$, we have

$$C^q(\mathfrak{m},\mathfrak{g}) = \mathfrak{g} \otimes \Lambda^q \mathfrak{m}^* = \Sigma_j \mathfrak{g}_j \otimes \Lambda^q \mathfrak{m}^* .$$

Then the bigradation of $C(\mathfrak{m},\mathfrak{g})$ is introduced by

$$C^{p,q}(\mathfrak{m},\mathfrak{g}) = \mathfrak{g}_{p-2} \otimes (\mathfrak{g}^*_{-2} \otimes \Lambda^{q-1} \mathfrak{g}^*_{-1}) \oplus \mathfrak{g}_{p-1} \otimes \Lambda^q \mathfrak{g}^*_{-1}$$

We have (Lemma 1.13 [19]),

$$\partial\, C^{p,q}(\mathfrak{m},\mathfrak{g}) \subset C^{p-1,q+1}(\mathfrak{m},\mathfrak{g}) ,$$
$$\partial^*\, C^{p,q}(\mathfrak{m},\mathfrak{g}) \subset C^{p+1,q-1}(\mathfrak{m},\mathfrak{g}) .$$

Since the curvature form Ω is a tensorial form, Ω can be regarded as a $C^2(\mathfrak{m},\mathfrak{g})$-valued function K on P (Lemma 2.2 [19]) by putting

$$\Omega = \frac{1}{2} K(\theta \wedge \theta) ,$$

where θ is the $\mathfrak{m}$-component of ω. We denote by K^p the $C^{p,2}(\mathfrak{m},\mathfrak{g})$ -component of K. Moreover we denote by K_j the $\mathfrak{g}_j \otimes \Lambda^2 \mathfrak{m}^*$-component of the decomposition $C^2(\mathfrak{m},\mathfrak{g}) = \Sigma_j \mathfrak{g}_j \otimes \Lambda^2 \mathfrak{m}^*$. Then a Cartan connection (P,ω) of type G / G' is called normal if the curvature K satisfies the following conditions:

(1) $K^{-1} = 0$,

(2) $\partial^* K^p = 0$ for $p \geq 0$.

Now we have the following theorem due to Tanaka

(Theorem 2.7 [19]) which is fundamental to the equivalence problem for Lie contact manifolds.

Theorem 4.3. Assume that $n \geq 3$.

(1) Every normal connection (P,ω) of type G / G' on a manifold N induces a Lie contact structure $\tilde{P}$ on N in a natural manner. Conversely if $\tilde{P}$ is a Lie contact structure on N , there is a normal connection (P,ω) of type G / G' on N which induces the given $\tilde{P}$.

(2) Let (P,ω) and (P',ω') be normal connections of type G / G' on manifolds N and N' respectively. Let $\tilde{P}$ and $\tilde{P}'$ be the Lie contact structures on N and N' induced from (P,ω) and (P',ω') respectively. Then every isomorphism Φ ; $(P,\omega) \to (P',\omega')$ induces a Lie isomorphism φ ; $N \to N'$ in a natural manner. Conversely if φ ; $N \to N'$ is a Lie isomorphism, there is a unique isomorphism Φ ; $(P,\omega) \to (P',\omega')$ which induces the given φ .

Remark 4.4. In [19], it is further shown (Theorem 2.9) that the harmonic part $H(K)$ of the curvature K gives a fundamental system of invariants, that is, K vanishes if and only if $H(K)$ vanishes. Concernig with this, it can be shown by the Kostant's method [11] that if $n \geq 4$, then $H^{p,2}(\mathfrak{m},\mathfrak{g})$ vanishes except for $p = 0$, whereas $H^{p,2}(\mathfrak{m},\mathfrak{g})$ vanishes except for $p = 0$ and 1 when $n = 3$. Hence we see that if $n \geq 4$, K vanishes if and only if $H(K^0)$ vanishes. Thus $H(K^0)$ should be called the Weyl curvature of the Lie contact structure. We shall discuss about this in another occasion. The case $n = 3$ is quite different from the other cases and will be discussed in the subsequent sections. This difference comes from the fact that $\mathfrak{g}$ is a simple Lie algebra of type B_ℓ or D_ℓ for $n \geq 4$, whereas $\mathfrak{g}$ is of type A_3 when $n = 3$.

As is well known, it follows from Theorem 4.3 that the group L(N) of Lie transformations on a Lie contact manifold N is a Lie group of $\dim L(N) \leq \dim G = \frac{1}{2}(n+3)(n+2)$. As a simple application of Theorem 4.3, we have

Theorem 4.5. Let N be a connected Lie contact manifold of dimension 2n-1 ($n \geq 3$). If $\dim L(N) = \frac{1}{2}(n+3)(n+2)$, then N is Lie isomorphic to $T_1(S^n)$.

We give here a sketch of the proof (cf. Kobayashi [8], Yamaguchi [20] and Yamazaki [21]). Let (P,ω) be the normal connection on N. By Theorem 4.3, L(N) lifts to the group L(P) of isomorphisms of (P,ω). Let us fix a point $u \in P$. We consider an embedding ι_u of L(P) into P defined by $\iota_u(\Phi) = \Phi(u)$ for $\Phi \in L(P)$. Since $\dim L(P) = \dim P$, ι_u is an open embedding. Thus $\iota_u(L^o(P))$ coincides with one of the connected component of P, where $L^o(P)$ is the identity component of L(P). Then it follows that $L^o(N)$ acts transitively on N. Furthermore, utilizing the defining element E of the gradation of $\mathfrak{g}$, we see that N is flat, that is, $\Omega = 0$, by a standard argument (cf. Theorem 3.2 [8] or Proposition 5.6 [20]). Then $(\iota_u)^*\omega$ is a $\mathfrak{g}$-valued left invariant 1-form on $L^o(P)$ and gives a Lie algebra isomorphism of the Lie algebra $\mathfrak{l}(P)$ of L(P) onto $\mathfrak{g}$. Let

$L^o_x(N)$ be the isotropy subgroup of $L^o(N)$ at $x = \pi(u)$. By identifying $L^o(P)$ with $L^o(N)$, $\iota_u(L^o_x(N))$ is an open submanifold of the fibre $\pi^{-1}(x)$. Then, by $\iota_u(\Phi) = u\cdot\rho_u(\Phi)$ for $\Phi \in L^o_x(N)$, we have an isomorphism ρ_u of $L^o_x(N)$ into G' such that $(\rho_u)_* = (\iota_u)^*\omega \mid_{\mathfrak{l}'(N)}$ as a Lie algebra isomorphism, where $\mathfrak{l}'(N)$ is the Lie algebra of $L^o_x(N)$. Namely $\rho_u(L^o_x(N))$ is an open subgroup of G' . Thus $L^o(N) \to N = L^o(N) / L^o_x(N)$ can be regarded as an open subbundle of $P \to N$.

On the other hand, we have the standard connection (G,ω_G) on the model space $T_1(S^n) = G / G'$. Let G^o and $(G')^o$ be the identity component of G and G' respectively. Then $G^o \to T_1(S^n) = G^o / (G')^o$ is an open subbundle of $G \to T_1(S^n)$. Here we note that $T_1(S^n)$ is simply connected and $N_{G^o}((G')^o) = (G')^o$, where $N_{G^o}((G')^o)$ is the normalizer of $(G')^o$ in G^o .

Since the Lie algebra $\mathfrak{l}(N)$ of $L^o(N)$ is isomorphic with $\mathfrak{g}$, we have a covering homomorphism μ of $L^o(N)$ onto $G^o = \mathrm{Int}(\mathfrak{g})$. Here we may assume that $\mu_* = (\iota_u)^*\omega$ as a Lie algebra isomorphism. By $N_{G^o}((G')^o) = (G')^o$, we have $\mu(L^o_x(N)) = (G')^o$. Then a simple homotopy argument shows that μ is an isomorphism of $L^o(N)$ onto G^o such that $\mu(L^o_x(N)) = (G')^o$ (cf. the proof of Proposition 7.1 [20]). Namely μ is a bundle isomorphism of $L^o(N) \to N$ onto $G^o \to T_1(S^n)$, which implies that N is Lie isomorphic with $T_1(S^n)$.

§5. CR-structure of tangent sphere bundle of 3-manifolds.

Let M be a 3-dimensional oriented Riemannian manifold.

In this section, we shall show that the Lie contact structure of the unit tangent bundle $T_1(M)$ of M is equal to the CR-structure with the non-degenerate and indefinite Levi form. Thus we obtain that the unit tangent bundle $T_1(M)$ of M has the canonical CR-structure. This result is originally due to LeBrun [12], but the first named author found that independently in the course of the study of Lie structures. This CR-structure is equal to the structure given by the twistor method of Penrose [14] . If M is hyperbolic, this is called mini-twistor by Atiyah [2] . For the definition of CR-structure, we follow Tanaka [18], although he called the structure PC-structure (pseudo-complex structure).

Let N be a (2n-1)-dimensional manifold. **An almost CR-structure** (D,I) on N consists of a 2(n-1)-dimensional subbundle D of the tangent bundle $T(N)$ of N and a bundle isomorphism $I : D \to D$ with $I^2 = -id$ satisfying the following condition;

$$\text{(A.1)} \quad [IX,IY] - [X,Y] \in \Gamma(D) , \quad \text{for} \quad X, Y \in \Gamma(D) .$$

We say that an almost CR-structure (D,I) is integrable or (D,I) is a **CR-structure** if the following conditions are further satisfied;

$$\text{(A.2)} \quad [IX,IY] - [X,Y] = I([IX,Y] + [X,IY]) ,$$

for $X,Y \in \Gamma(D)$.

For a real vector bundle V , we denote by $\mathbb{C}V$ the complexification $\mathbb{C} \otimes_{\mathbb{R}} V$. If (D,I) is an almost CR-structure, we have the natural extension $I : \mathbb{C}D \to \mathbb{C}D$. Put

$$S = \{ X \in \mathbb{C}D \mid IX = \sqrt{-1}\cdot X \} \quad .$$

Then an almost CR-structure (D,I) is a CR-structure if and only if

$$\text{(C.1)} \qquad [\Gamma(S),\Gamma(S)] \subset \Gamma(S) \ .$$

Let (D,I) be an almost CR-structure on N. Choose a local 1-form θ of N such that locally $D = \{ X \in T(N) \mid \theta(X) = 0 \}$. Then the Levi form L is defined by

$$L(X,Y) = \sqrt{-1}\cdot d\theta(X,\bar{Y}) \qquad \text{for} \quad X,\ Y \in S \ .$$

Then L is a hermitian form. We say that the Levi form L is non-degenerate if L has no zero eigenvalue and L is indefinite if L has both positive and negative eigenvalues. The condition that L is non-degenerate and indefinite is independent of the choice of θ.

Now let (M,g) be a 3-dimensional oriented Riemannian manifold and let $N = T_1(M)$ be the unit tangent bundle. For a point $v \in N$, we denote by w^H and $w^V \in T_v(N)$ the horizontal lift and the vertical lift of $w \in T_x(M)$, where $x = p(v)$ (see §3). Then we recall that the contact distribution D on N is given by

$$D(v) = \{ w^H,\ w^V \mid g(v,w) = 0 \} \ .$$

For each $v \in N = T_1(M)$, choose an oriented orthonormal frame $\{v,a,b\}$ of $T_x(M)$. We define a bundle isomorphism I ; $D \to D$ by

$$I(a^H) = b^H \ , \qquad I(b^H) = -\, a^H$$

$$I(a^V) = b^V , \qquad I(b^V) = - a^V .$$

Then, from $d\alpha(w_1^V, w_2^H) = g(w_1, w_2)$ for $w_1, w_2 \in T_x(M)$, it follows that the pair (D, I) is an almost CR-structure on N.

Proposition 5.1. The almost CR-structure on the unit tangent bundle $T_1(M)$ of an oriented 3-dimensional Riemannian manifold (M, g) is integrable. The Levi form is non-degenerate and indefinite at every point of N.

Proof. Let $\pi ; O^o(M) \to M$ be the oriented orthonormal frame bundle of M. Then $P = O^o(M)$ is a principal $SO(3)$-bundle over M. Each $u \in P$ is a linear isometry of $\mathbb{R}^3$ onto $T_x(M)$, where $x = \pi(u)$. Let $\{e_1, e_2, e_3\}$ be the natural basis of $\mathbb{R}^3$. We take a basis $\{A_1, A_2, A_3\}$ of $\mathfrak{o}(3)$ given by the following matrices;

$$A_1 = \begin{pmatrix} 0 & 0 & 0 \\ 0 & 0 & -1 \\ 0 & 1 & 0 \end{pmatrix}, \quad A_2 = \begin{pmatrix} 0 & -1 & 0 \\ 1 & 0 & 0 \\ 0 & 0 & 0 \end{pmatrix}, \quad A_3 = \begin{pmatrix} 0 & 0 & -1 \\ 0 & 0 & 0 \\ 1 & 0 & 0 \end{pmatrix}.$$

Then the standard horizontal vector fields $B(e_1)$, $B(e_2)$, $B(e_3)$ and the fundamental vector fields A_1^*, A_2^*, A_3^* define an absolute parallerism in P, that is, these vetor fields form a basis of $T_u(P)$ at each $u \in P$ (cf. Kobayashi-Nomizu [10]).

We consider a map $q ; P \to T_1(M)$ defined by

$$q(u) = u(e_1) .$$

By this map, $T_1(M)$ is identified with the quotient space of P by the right action of a maximal torus $\exp t \cdot A_1$ of $SO(3)$. In particular, we have $q_*(A_1^*) = 0$. By the definition of horizontal subspaces of $T(T_1(M))$, q_* sends a horizontal subspace at $u \in P$ isomorphically onto the horizontal subspace at $q(u) = v$. More precisely we have

$$q_*((B(e_i))_u) = u(e_i)^H \qquad \text{for } i = 1,2,3,$$

where $u(e_i)^H \in T_v(T_1(M))$ is the horizontal lift of $u(e_i) \in T_x(M)$, $x = \pi(u)$. In particular, we get

$$q_*(B(e_1)) = Z_g ,$$

where Z_g is the geodesic vector field. Furthermore it is easy to see that we have

$$q_*((A_2^*)_u) = u(e_2)^V \quad \text{and} \quad q_*((A_3^*)_u) = u(e_3)^V ,$$

where $u(e_i)^V \in T_v(T_1(M))$ is the vertical lift of $u(e_i) \in T_x(M)$.

Let $\tilde{D}$ be the 4-dimensional subbundle of $T(P)$ spanned by the vector fields $B(e_2)$, $B(e_3)$, A_2^* and A_3^* at each $u \in P$. We define a bundle isomorphism $\tilde{I} ; \tilde{D} \to \tilde{D}$ by

$$\tilde{I}(B(e_2)) = B(e_3) , \qquad \tilde{I}(B(e_3)) = - B(e_2) ,$$
$$\tilde{I}(A_2^*) = A_3^* , \qquad \tilde{I}(A_3^*) = - A_2^* .$$

Then $\tilde{I}^2 = -\mathrm{id}$ on $\tilde{D}$. Since $(R_a)_*B(\xi) = B(a^{-1}\xi)$ and $(R_a)_*A^* = (Ad(a^{-1})A)^*$ for $a \in SO(3)$, $\xi \in \mathbb{R}^3$ and $A \in \mathfrak{o}(3)$, it follows easily that $(R_a)_*\tilde{D} = \tilde{D}$ for $a = \exp t\cdot A_1$. By the above argument, we see that q_* induces a bundle map $\hat{q}_* ; \tilde{D} \to D$. Namely $\tilde{D}$ is isomorphic with the induced bundle $q^{-1}(D)$. Hence $\mathbb{C}\tilde{D}$ is isomorphic with $q^{-1}(\mathbb{C}D)$. Obviously we have $\hat{q}_*\circ\tilde{I} = I\circ\hat{q}_*$. Thus, in order to show that (D,I) is integrable, it suffices to show

$$[\Gamma(\tilde{S}),\Gamma(\tilde{S})] \subset \Gamma(\tilde{S}) \ ,$$

where $\tilde{S} = \{ X \in \mathbb{C}\tilde{D} \mid \tilde{I}X = \sqrt{-1}\cdot X \}$. However this is easily checked by using a global generator $X = B(e_2) - \sqrt{-1}\cdot B(e_3)$, $Y = A_2^* - \sqrt{-1}\cdot A_3^*$ of $\Gamma(\tilde{S})$. In fact we have $[X,Y] = 0$.

The Levi form L is non-degenerate since D is a contact distribution, and L is indefinte since horizontal or vertical subspaces are complex isotropic subspaces. This completes the proof.

According to Tanaka [18], an almost CR-structure on a 5-dimensional manifold with non-degenerate and indefinite Levi form is equivalent to a $\tilde{G}$ -structure of type $\mathfrak{M}$ associated with the simple graded Lie algebra $\mathfrak{su}(2,2) = \Sigma_{p\in\mathbb{Z}}\, \mathfrak{s}_p$ (see §10 [18] or I.3 [20], for the gradation of $\mathfrak{su}(2,2)$). There is a (graded) Lie algebra isomorphism of $\mathfrak{su}(2,2)$ onto $\mathfrak{o}(4,2) \simeq \mathfrak{g}$ (cf. Helgason [7,p.522]). By this isomorphism, the Lie contact structure on a 5-dimensional contact manifold is equal to a non-degenerate almost CR-structure with indefinite Levi form (see Remark 3.2). In fact, by the notation in §2, a complex structure in $\mathfrak{g}_{-1}$ is given by the following element $I \in \mathfrak{g}_0$;

$$I = \begin{pmatrix} 0 & 0 & 0 \\ 0 & R & 0 \\ 0 & 0 & 0 \end{pmatrix} \in \mathfrak{g}_0 \ , \quad R = \begin{pmatrix} 0 & -1 \\ 1 & 0 \end{pmatrix} \in \mathfrak{o}(2) \ .$$

Then, identifying $\mathfrak{g}_{-1}$ with $\mathbb{R}^2\times \mathbb{R}^2$ as in §2 , we see that $[I,(\xi_1,\xi_2)] = (R(\xi_1),R(\xi_2))$ for $(\xi_1,\xi_2) \in \mathfrak{g}_{-1}$. Furthermore it is not difficult to see that $\mathfrak{g}_0$ coincides with the derivation algebra of $(\mathfrak{M},I)$. Namely $\mathfrak{g}_0$ coincides with the set of derivations of $\mathfrak{m} = \Sigma_{p<0}\, \mathfrak{g}_p$ preserving the complex structure I in $\mathfrak{g}_{-1}$. Then, by Lemma 4.1, $\mathfrak{g} = \Sigma_{p\in\mathbb{Z}} \mathfrak{g}_p$ is the prolongation of $(\mathfrak{M}, I)$. This, combined with Lemma 10.1 [18], establishes the fact that $\mathfrak{g} = \Sigma_{p\in\mathbb{Z}} \mathfrak{g}_p$ is isomorphic with $\mathfrak{su}(2,2) = \Sigma_{p\in\mathbb{Z}} \mathfrak{s}_p$.

Now, by the definition of the complex structure in $\mathfrak{g}_{-1}$, it is obvious that the CR-structure constructed by the proposition is equal to the Lie contact structure on the unit tangent bundle studied in the previous sections.

Moerover, by Theorem 11.1 [18], a non-degenerate almost CR-structure is integrable if and only if the K_{-1}-curvature (the $\mathfrak{g}_{-1}\otimes \Lambda^2 \mathfrak{m}^*$-component of the curvature K) of the normal connection vanishes. Thus we obtain

Corollary. Let M be a 3-dimensional manifold with a conformal structure. Then K_{-1}-curvature of the Lie contact structure on the tangent sphere bundle of M vanishes.

§6. Geodesic flow on twistor CR-manifold

In the previous section, we have shown that the tangent S^2 -bundle $T_1(M)$ of an oriented 3-dimensional Riemannian manifold M has the natural structure of a CR-manifold. In this section, we show that the geodesic flow vector field on the 5-dimensional manifold $T_1(M)$ is a Lie contact vector field if and only if M is of constant curvature. In such case, we can naturally give a complex structure on the 6-dimensional manofold $T_1(M) \times S^1$. This complex structure is equal to the twistor holomorphic projective structure on the twistor S^2-bundle over the conformally flat manifold

$M \times S^1$ (Atiyah-Hitchin- Singer[1]). Then we obtain that, if M is of constant curvature, the geodesic flow vector field on $T_1(M)$ defines a (transversely) holomorphic foliation of codimension 2 . The exotic characteristic classes of the foliations may be considered as invariants of the 3-manifold M .

We have the following theorem.

Theorem 6.1. Let (M,g) be a 3-dimensional Riemannian manifold. Then the geodesic vector field Z_g is a Lie contact vector field of the Lie contact manifold $T_1(M)$ if and only if M is of constant sectional curvature.

Proof. Let (D,I) be the CR-structure given in §5 , which is equivalent to the Lie contact structure on $T_1(M)$. Then the geodesic vector field Z_g is a Lie contact vector field if and only if it is an infinitesimal CR -automorphism.

Now we utilize the situation in the proof of Proposition 5.1. In particular Z_g lifts to the standard horizontal vector field $B(e_1)$ on $P = O^o(M)$. Since $(\tilde{D},\tilde{I})$ is a pull back of (D,I) by $q ; P \to T_1(M)$, Z_g is a Lie contact vector field if and only if $B(e_1)$ is an infinitesimal automorphism of $(\tilde{D},\tilde{I})$, that is,

$$[B(e_1),\tilde{I}Y] = \tilde{I}[B(e_1),Y] \quad \text{for} \quad Y \in \Gamma(\tilde{D}) .$$

From $[A^*,B(e_1)] = B(A(e_1))$ for $A \in \mathfrak{o}(3)$ and the definition of $\tilde{I}$, it follows easily that

$$[B(e_1),\tilde{I}A_2^*] = \tilde{I}[B(e_1),A_2^*] = - B(e_3) ,$$

$$[B(e_1), \tilde{I}A_3^*] = \tilde{I}[B(e_1), A_3^*] = B(e_2) .$$

Moreover, we have

$$[B(e_1), B(\xi)]_u = 2\ (\Omega_u(B(\xi), B(e_1)))_u^* \quad \text{for} \quad \xi \in \mathbb{R}^3,$$

where Ω is the curvature form on P . Thus $B(e_1)$ is an infinitesimal automorphism of $(\tilde{D}, \tilde{I})$ if and only if

$$\tilde{I}(\Omega_u(B(\xi), B(e_1))) = \Omega_u(B(I_R(\xi)), B(e_1)) \quad \text{for} \quad \xi \in \langle e_2, e_3\rangle ,$$

where $I_R = \begin{pmatrix} 0 & -1 \\ 1 & 0 \end{pmatrix}$.

On the other hand, for $v = q(u) \in T_1(M)$, we put $W(v) = \{ w \in T_x(M) \mid g(v,w) = 0 \}$, where $x = p(v)$. Then we have $w = \pi_*(B(\xi)_u) \in W(v)$ for $\xi \in \langle e_2, e_3\rangle$ and

$$R(w,v)v = u(\ 2\ \Omega_u(B(\xi), B(e_1))(e_1))$$

where R denotes the curvature of the Levi-Civita connection (cf. Kobayashi-Nomizu [10]). Then, since $(\tilde{I}A)(e_1) = I_R(A(e_1))$ for $A \in \langle A_2, A_3\rangle$, the condition above is equivalent to

$$J(R(w,v)v) = R(Jw,v)v \quad \text{for} \quad v \in T_1(M) \quad \text{and} \quad w \in W(v) ,$$

where $J ; W(v) \to W(v)$ is the rotation by the angle $\frac{\pi}{2}$. Then, putting $\varphi_v(w) = R(w,v)v$, we see that Z_g is a Lie contact vector field if and only if $\varphi_v ; W(v) \to W(v)$ commutes with J for every $v \in T_1(M)$. Since φ_v satisfies $g(\varphi_v(w_1), w_2) = g(w_1, \varphi_v(w_2))$ for $w_1, w_2 \in W(v)$, it follows easily that φ_v commutes with J if and only if there exsists a constant $c \in \mathbb{R}$ such that

$$R(w,v)v = c \cdot w \quad \text{for} \quad v \in T_1(M) \quad \text{and} \quad w \in W(v) .$$

This shows that Z_g is a Lie contact vetor field if and only if M is of constant sectional curvature.

Finally we refer to Cecil-Chern [6,p.393] , for the explicit description of Φ_t as Lie transformations, when M is a space form, where Φ_t is the geodesic flow generated by Z_g .

According to Sasaki [15] (cf. Tanaka [17, III]) , a CR -manifold N with non-degenerate Levi form is called normal if there exists a nowhere vanishing vector field ξ on N such that

(N.1) $$[\xi,\Gamma(S)] \subset \Gamma(S) ,$$

(N.2) $$\{\xi\} \cap D = 0 .$$

The condition (N.1) is equivalent to the conditions $[\xi,\Gamma(D)] \subset \Gamma(D)$ and $[\xi,IY] = I[\xi,Y]$ for $Y \in \Gamma(D)$. If N is a normal CR -manifold, then we define an almost complex structure J on $N \times S^1$ or $N \times \mathbb{R}$ by

$$J(X,0) = (IX,0) \quad \text{for} \quad X \in D ,$$

$$J(\xi,0) = (0,d/dt) ,$$

$$J(0,d/dt) = (-\xi,0) .$$

Then the following is easy to see.

Proposition 6.2. If N is a normal CR-manifold, then the almost complex structure J on $N \times S^1$ or $N \times \mathbb{R}$ is integrable.

By the theorem above, if M is a 3-dimensional orientable Riemannian manifold of constant sectional curvature, then the geodesic vector field Z_g on $N = T_1(M)$ defines a normal

CR-structure on N . Thus we obtain a complex structure on $M \times S^2 \times S^1$ or $M \times S^2 \times \mathbb{R}$. On the other hand, if M is of constant cruvature, then we can give a product metric on $W = M \times S^1$ or $M \times \mathbb{R}$ so that it is conformally flat. Especially if M is orientable, the 4-manifold W is self dual or half conformally flat (see Atiyah-Hitchin-Singer [1]). The projective self dual bundle $P(V_-)$ is diffeomorphic to the product $W \times S^2$. The Penrose twistor theory gives a natural complex structure on $P(V_-)$. The following is easy to see.

Proposition 6.3. The complex structure on $M \times S^2 \times S^1$ or $M \times S^2 \times \mathbb{R}$ defined from the normal CR-structure on M coincides with the complex structure defined from the self dual structure on $M \times S^1$ or $M \times \mathbb{R}$.

Now we will show that the geodesic flow on $T_1(M)$ defines a (transeversely) holomorphic complex structure of codimension 2 if M is of constant curvature.

Let N be a real smooth manifold. By a foliation $\mathfrak{F}$ on M , we mean an involutive subbundle E of the tangent bundle $T(M)$ of M . Each integrable submanifold is called a leaf. The quotient bundle $Q = T(M)/E$ is called the normal bundle of the foliation $\mathfrak{F}$.

We say that $\mathfrak{F}$ is a transverse holomorphic foliation, or simply, holomorphic foliation, if the following two conditions are satisfied;

(F.1) There exists an almost complex structure $J : Q \rightarrow$

Q with $J^2 = -1$ such that, for any smooth section X of E

$$L_X J = 0 ,$$

where L_X denotes the Lie derivative.

(F.2) Let X and Y be two smooth sections of Q and let $s : Q \to T(M)$ be a cross section of the projection $q : T(M) \to Q$. Then the following torsion tensor $N(X,Y)$ vanishes, that is,

$$N(X,Y) = p[sX,sY]-p[sJX,sJY]+Jp[sX,sJY]+Jp[sJX,sY] = 0 .$$

If $\mathfrak{F}$ is a holomorphic foliation on M , by (F.1), the normal bundle Q is a complex vector bundle.

Theorem 6.4. Let M be an orientable 3-dimensional Riemannian manifold of constant sectional curvature and let E be the 1-dimensional subbundle of $N = T_1(M)$ consisting of geodesic flow vectors. Then E is a holomorphic foliation on $N = T_1(M)$ such that the complex normal bundle is isomorphic to the trivial normal bundle.

Proof. The contact distribution D on $T(N)$ is isomorphic to the quotient bundle $Q = T(N)/E$. We define an almost complex structure J on Q by I . Then by Proposition 5.1 , J satisfies the condition $N(X,Y) = 0$ of (F.2). By Thorem 6.1 , we have $L_\xi J = 0$, where $\xi = Z_g$. This shows that $\mathfrak{F}$ is a holomorphic foliation. The bundle Q is isomorphic to the direct sum of two complex line bundles $Q_1 = \{w^H\}$ and $Q_2 = \{w^V\}$. The manifold $T_1(M)$ is homeomorphic to $M \times S^2$ and $H^2(T_1(M);\mathbb{Z}) \cong H^2(M;\mathbb{Z}) \oplus$

$H^2(S^2;\mathbb{Z})$. The first Chern classes are given by $c_1(Q_1) = 2a$ and $c_1(Q_2) = -2a$, where a is a generator of $H^2(S^2;\mathbb{Z})$ regarded as an element of $H^2(T_1(M);\mathbb{Z})$. Thus $c_1(Q) = 0$ and Q is isomorphic to the trivial bundle, which completes the proof.

If M is of constant negative curvature, then stable and unstable vectors to the geodesic flow give codimension 2 foliations on N respectively. Our complex codimension 2 foliation is the intersection of these stable and unstable foliations. This remark is due to Morita [13] .

According to Bott [5], there are exotic characteristic classes $h_1c_1^2$ and h_1c_2 on complex codimension 2 foliation on a 5-dimensional manifold with trivial normal bundle. These characteristic classes are invariants of 3-manifolds M . By the remark above $h_1c_1^2 = h_1c_2$ if M is of constant negative curvature.

References

[1] M.F.Atiyah, N.J.Hitchin and I.M.Singer, Self-duality in four dimensional Riemannian geometry, Proc.R.Soc.Lond. A 362 (1978), 425-461.

[2] M.F.Atiyah, Magnetic monopoles in hyperbolic space, in Proc. of Bombay Colloq., 1984, on Vector bundles on algebraic varieties, Oxford Univ. Press(1987), 1-34.

[3] A.Besse, Manifolds all of whose Geodesics are Closed,

Springer-Verlag, Berlin-Heidelberg-New York, (1978).

[4] W.Blaschke, Vorlesungen über Differentialgeometrie, Vol.3, Springer, Berlin (1929).

[5] R.Bott, On the Lefshetz formula and exotic characteristic classes, Symposia Math. X (1972), 95-105.

[6] T.Cecil and S.Chern, Tautness and Lie Sphere Geometry, Math. Ann. 278(1987), 381-399.

[7] S.Helgason, Differential Geometry, Lie groups and Symmetric Spaces, Academic Press, New York-San Francisco-London (1978).

[8] S.Kobayashi, Transformation Groups in Differential Geometry, Springer, Berlin-Heiderberg-New York (1972).

[9] S.Kobayashi and T.Nagano, On fitered Lie algebras and geometric structures, III, J. Math. Mech. 14(1965), 679-706.

[10] S.Kobayashi and K.Nomizu, Foundations of Differential Geometry, Vol.1. Interscience Publishers (1969).

[11] B.Kostant, Lie algebra cohomology and the generalized Borel-Weil theorem, Ann. of Math. 74(1961), 329-387.

[12] C.R.LeBrun, Twistor CR manifolds and three-dimensional conformal geometry, Trans.Amer.Math.Soc. 284(1984), 601-617.

[13] S.Morita, Discontinuous Invariants of Foliations, Advanced Study in Pure Math. 5(1985) Foliations, 160-193.

[14] R.Penrose, The twistor programme, Reports on Math.Phys. 12(1977), 65-76.

[15] S.Sasaki, On differential manifolds with certain structures which are closely related to almost contact structures I, Tôhoku Math. J. 12(1960), 456-476; II(with Y.Hatakeyama), ibid. 13(1961), 281-294.

[16] H.Sato, Lie's contact geometry and Lie manifolds (in

Japanese). Proc. of 35th Top. Symp.(1988), Ryukyu Univ., 41-63.

[17] N.Tanaka, A Differential Geometric Study on Strongly Pseudo-Convex Manifolds, Lect. in Math. Kyoto Univ. 9. Kinokuniya Book-Store, Tokyo(1975).

[18] N.Tanaka, On non-degenerate real hypersurfaces, graded Lie algebras and Cartan connections, Japanese J. Math. 2 (1976), 131-190.

[19] N.Tanaka, On the equivalence problems associated with simple graded Lie algebras, Hokkaido Math. J. 8(1979), 23-84.

[20] K.Yamaguchi, Non-degenerate real hypersurfaces in complex manifolds admitting large groups of pseudo-conformal transformations I, Nagoya Math. J. 62(1976), 55-96.

[21] T.Yamazaki, On Dupin hypersurfaces and Lie geometry, Master's Thesis (1988), Tohoku Univ.

Department of Mathematics
Nagoya University
Nagoya 464-01, JAPAN

Department of Mathematics
Hokkaido University
Sapporo 060, JAPAN

On simple graded Lie algebras of finite depth

Tomoaki Yatsui

A graded Lie algebra (GLA) $\mathfrak{g} = \bigoplus_{p\in\mathbb{Z}} \mathfrak{g}_p$ is called of finite depth if $\dim \mathfrak{g}_- < \infty$, where $\mathfrak{g}_- := \bigoplus_{p<0} \mathfrak{g}_p$. In this note, the ground field is assumed to be the field of complex numbers. We will study the class $\mathscr{C}$ of infinite dimensional GLAs of finite depth satisfying the following conditions:

(G.1) $\mathfrak{g}_0$ contains the defining element E of $\mathfrak{g} = \bigoplus_{p\in\mathbb{Z}} \mathfrak{g}_p$ (We call the element $E \in \mathrm{Der\,gr}(\mathfrak{g})$ the defining element of $\mathfrak{g} = \bigoplus_{p\in\mathbb{Z}} \mathfrak{g}_p$ if $E(x) = px$ for $x \in \mathfrak{g}_p$).

(G.2) Every non-trivial ideal in $\mathfrak{g}$ contains $\mathfrak{g}_-$.

We remark that, by (G.2), $\mathfrak{g} = \bigoplus_{p\in\mathbb{Z}} \mathfrak{g}_p \in \mathscr{C}$ is a transitive GLA (i.e., for $x \in \mathfrak{g}_p (p\geq 0)$, $[x, \mathfrak{g}_-] = \{0\}$ implies $x = 0$).

The purpose of this note is to give the classification of the class $\mathscr{C}$ of GLAs of finite depth and to give the classification of simple GLAs of finite depth as a corollary. Details of the proof will be published elsewhere.

The typical examples of GLAs of the class $\mathscr{C}$ appear in GLAs of Cartan type. First we will formulate Lie algebras of Cartan type and their gradations. Let $\mathfrak{A}(\mathrm{m}) = \mathbb{C}[x_1, \ldots, x_{\mathrm{m}}]$, $W(\mathrm{m}) = \{ \sum_{i=1}^{\mathrm{m}} f_i \partial/\partial x_i : f_i \in \mathfrak{A}(\mathrm{m}) \}$. Namely $W(\mathrm{m})$ is the Lie algebra of polynomial vector fields. We consider the following differential forms in $dx_1, \ldots, dx_{\mathrm{m}}$. Let

$\omega_S = dx_1 \wedge \ldots \wedge dx_{\mathrm{m}}$,

$\omega_H = \sum_{i=1}^{\mathrm{n}} dx_i \wedge dx_{i+\mathrm{n}}$,

$\omega_K = dx_{2n+1} - \sum_{i=1}^{n} x_{n+i} dx_i$.

ISBN 0-12-640170-5

Then we define subalgebras of W(m) as follows:

$S(m) = \{ D \in W(m): D\omega_S = 0 \}$,

$CS(m) = \{ D \in W(m): D\omega_S \in \mathbb{C}\omega_S \}$,

$H(n) = \{ D \in W(2n): D\omega_H = 0 \}$,

$CH(n) = \{ D \in W(2n): D\omega_H \in \mathbb{C}\omega_H \}$,

$K(n) = \{ D \in W(2n+1): D\omega_K \in \mathfrak{U}(m)\omega_K \}$,

(Here the action of D on differential forms is defined by Lie derivative). These Lie algebras are called Lie algebras of Cartan type, which is the classes of infinite Lie algebras appeared in the Cartan's classification of Lie pseudogroups.

Now, following Kac[Kac70], we can define the gradation of Lie algebras of Cartan type as follows: For each m-tuple $s=(s_1,\ldots,s_m)$ of positive integers we set $W(m:s)_p = \sum_i \mathfrak{U}(m:s)_{p-s_i} \partial/\partial x_i$, where $\mathfrak{U}(m:s)_p = \{ x^a: \sum_i \alpha_i s_i = p \}$. Then $W(m:s) := \oplus_{p\in\mathbb{Z}} W(m:s)_p$ gives the structure of GLAs on W(m). Moreover if for X = S or CS we set $X(m:s)_p = W(m:s)_p \cap X(m)$, then $X(m:s) := \oplus_{p\in\mathbb{Z}} X(m:s)_p$ gives the structure of GLA on X(m). For an n-tuple $\mathbf{t}=(t_1,\ldots,t_n)$ of positive integers and an integer $\mu\geq 2$ such that $t_i < \mu$, we set $H(n:\mathbf{t}:\mu)_p = W(2n:s)_p \cap H(n)$, $CH(n:\mathbf{t}:\mu)_p = W(2n:s)_p \cap CH(n)$(where $s=(t_1,\ldots,t_n,\mu-t_1,\ldots,\mu-t_n))$ and $K(n:\mathbf{t}:\mu)_p = W(2n+1:s)_p \cap K(n)$ (where $s=(t_1,\ldots,t_n,\mu-t_1,\ldots,\mu-t_n,\mu))$. Then for X= H,CH or K, $X(n:\mathbf{t}:\mu) := \oplus_{p\in\mathbb{Z}} X(n:\mathbf{t}:\mu)_p$ gives the structure of a GLA on X(n). Here we remark the following facts: For X = W,S or CS if $s=(s_1,\ldots,s_m)$ and $s'=(s'_1,\ldots,s'_m)$ coincide as a set, X(m:s) is isomorphic to X(m:s') as a GLA. Hence, in what follows, we will assume that $s_1\leq\ldots\leq s_m$. For X = H,CH or K if $s=(t_1,\ldots,t_n,\mu-t_1,\ldots,\mu-t_n)$ and $s'=(t_1',\ldots,t_n',\mu-t_1',\ldots,\mu-t_n')$ coincide as a set, then $X(n:\mathbf{t}:\mu)$ is isomorphic to $X(n:\mathbf{t}':\mu)$ as a GLA. Hence, in what follows, we will assume that $t_1\leq\ldots\leq t_n$ and $t_i\leq[\mu/2]$. For $\mathfrak{g}$ = $W(m:\mathbb{1})$, $S(m:\mathbb{1})$, $CS(m:\mathbb{1})$, $H(n:\mathbb{1}:2)$, $CH(n:\mathbb{1}:2)$, $K(n:\mathbb{1}:2)$(where $\mathbb{1}=(1,\ldots,1)$), they appear in the classification of the infinite primitive Lie algebras. Moreover, for $W(m:(1,\ldots,1,\mu,\ldots,\mu))(\mu\geq2)$, $K(n:\mathbb{1}:\mu)(\mu\geq3)$, they appear in higher order contact geometry (cf.[Yam82] and [Yam83]).

This note gives the following main theorem:

Theorem 1. Let $\mathfrak{g}=\underset{p\in\mathbb{Z}}{\oplus}\mathfrak{g}_p\in\mathcal{G}$. Then $\mathfrak{g}=\underset{p\in\mathbb{Z}}{\oplus}\mathfrak{g}_p$ is isomorphic to one isomorphic to one of $W(m:s)$, $CS(m:s)$, $CH(n:\mathbf{t}:\mu)$, $K(n:\mathbf{t}:\mu)$.

We describe an outline of our proof. Let L^0 be a maximal subalgebra containing $\underset{p\geq0}{\oplus}\mathfrak{g}_p$ and L^{-1} be an $\mathrm{ad}L^0$-invariant subspace containing L^0 such that L^{-1}/L^0 is $\mathrm{ad}L^0$-irreducible. Define L^i inductively (following [Wei68]) by

$$L^{-i-1}=[L^{-i},L^{-1}]+L^{-i},\quad i\geq1$$

and $$L^{i+1}=\{\,x\in L^i:\ [x,L^{-1}]\subset L^i\,\},\quad i\geq0.$$

Then we get a filtration of L, and corresponding associated GLA $\mathrm{gr}(L)=\underset{p\in\mathbb{Z}}{\oplus}\mathrm{gr}(L)_p$ is an irreducible transitive GLA such that $\mathrm{gr}(L)_-$ is generated by $\mathrm{gr}(L)_{-1}$. As is well-known (cf. [KN65] and [MT70]), $\mathrm{gr}(L)=\underset{p\in\mathbb{Z}}{\oplus}\mathrm{gr}(L)_p$ is isomorphic to one of $W(m:\mathbb{1})$, $S(m:\mathbb{1})$, $CS(m:\mathbb{1})$, $H(n:\mathbb{1}:2)$, $CH(n:\mathbb{1}:2)$, $K(n:\mathbb{1}:2)$. Let $\mathfrak{t}$ be a commutative subalgebra of $\mathfrak{g}_0$ such that the $\mathfrak{t}$-module $\mathfrak{g}$ is completely reducible and $E\in\mathfrak{t}$. Then, by an analogous method of [KN66], we have

Lemma 1. There exist the $\mathfrak{t}$-invariant subspaces G_p of

L^p such that $L^p = G_p \oplus L^{p+1}$ and $[G_p, G_q] \subset G_{p+q}$, whence $\mathfrak{g} = \oplus_{p\in\mathbb{Z}} G_p \simeq \mathrm{gr}(L)$.

It follows from the Lemma and (G.1) that $\mathfrak{g}$ is isomorphic to one of W(m),CS(m),CH(n),and K(n), and the element E is explicitly described by an element of W(m). This establishs the theorem. Since S(m) (resp.H(n)) is an ideal in CS(m) (resp.CH(n)) of codimension one, we have

Corollary. Let $\mathfrak{g} = \oplus_{p\in\mathbb{Z}} \mathfrak{g}_p$ be an infinite dimensional simple GLA of finite depth over $\mathbb{C}$. Then $\mathfrak{g} = \oplus_{p\in\mathbb{Z}} \mathfrak{g}_p$ is isomorphic to one of W(m:s), S(m:s), H(n:**t**:μ), K(n:**t**:μ).

This result is a partial solution of the problem conjectured by Kac in [Kac70].

Let $\mathfrak{g} = \oplus_{p\in\mathbb{Z}} \mathfrak{g}_p$ be a transitive GLA of finite depth. We put $\mathrm{Trun}_k(\mathfrak{g}) = \oplus_{p\le k} \mathfrak{g}_p$. A GLA $\check{\mathfrak{g}} = \oplus_{p\in\mathbb{Z}} \check{\mathfrak{g}}_p$ is called the prolongation of $\mathrm{Trun}_k(\mathfrak{g})$ ($k\ge -1$) if $\check{\mathfrak{g}} = \oplus_{p\in\mathbb{Z}} \check{\mathfrak{g}}_p$ is the maximum transitive GLA such that $\mathrm{Trun}_k(\mathfrak{g}) = \mathrm{Trun}_k(\check{\mathfrak{g}})$. In particular, $\mathfrak{g} = \oplus_{p\in\mathbb{Z}} \mathfrak{g}_p$ is called the prolongation of $\mathrm{Trun}_k(\mathfrak{g})$ if $\mathfrak{g} = \check{\mathfrak{g}}$. For $\mathfrak{g}$ = W(n:**t**) and K(n:**t**:μ), we have

Theorem 2. (1) If $\mathfrak{g} = \oplus_{p\in\mathbb{Z}} \mathfrak{g}_p =$ K(n:**t**:μ), then $\mathfrak{g} = \oplus_{p\in\mathbb{Z}} \mathfrak{g}_p$ is the prolongation of $\mathrm{Trun}_k(\mathfrak{g})$ for all $k\ge -1$.

(2) Let $\mathfrak{g} = \oplus_{p\in\mathbb{Z}} \mathfrak{g}_p =$ W(n:**t**). Then $\mathfrak{g} = \oplus_{p\in\mathbb{Z}} \mathfrak{g}_p$ is the prolongation of $\mathrm{Trun}_k(\mathfrak{g})$ if one of the following conditions holds: (α) $t_n = t_{n-1}$; (β) $t_n > t_{n-1}$, $t_n < 2t_{n-1} + k + 1$.

(3) Let $\mathfrak{g} = \oplus_{p\in\mathbb{Z}} \mathfrak{g}_p =$ W(n:**t**). If $t_n > t_{n-1}$, $t_n \ge 2t_{n-1} + k + 1$, then the prolongation of $\mathrm{Trun}_k(\mathfrak{g})$ is isomorphic to K(n-1:s:t_n), where $s_i = \min\{t_i, t_n - t_i\}$.

For the remaining GLAs of Cartan type, we obtain similar similar results; we will state them in a forthcoming paper.

References

[Kac70] V.G.Kac, The classification of the simple Lie algebras over a field with nonzero characteristic. Izv.Acad.Nauk SSSR Ser Mat34.(1970) 358-408 [Russian]; Math.USSR-Izv 4(1970),391-413. [English transl.].

[Wei68] B.J.Weisfeiler, Infinite dimensional filtered Lie algebras and their connection with graded Lie algebras, Funktsional.Anal.i Prilozhen.2 (1968), 94-95[Russian];Functional Anal.Appl.2(1968), 88-89 [English transl.].

[KN65] S.Kobayashi and T.Nagano, On filtered Lie algebras and geometric structure III.J.Math Mech.14(1965), 697-706.

[KN66] S.Kobayashi and T.Nagano, On filtered Lie algebras and geometric structure IV,J.Math.Mech.15(1966), 163-175.

[MT70] T.Morimoto and N.Tanaka, The classification of the real primitive infinite Lie algebras, J.of Math Kyoto Univ.10 (1970),207-243.

[Yam82] K.Yamaguchi, Contact geometry of higher order, Japan J.Math.8 (1982),109-176.

[Yam83] K.Yamaguchi, Geometrization of Jet bundle, Hokkaido Math.8 (1983)27-40

Department of Mathematics
Hokkaido University
Sapporo 060, Japan

Chapter IV
Riemannian Geometry

On Decompositions of Riemannian Manifolds with Nonnegative Ricci Curvature

Ryosuke Ichida

§0. Introduction.

Let M be a complete connected Riemannian manifold. We can assign to each connected submanifold N embedded in M a certain nonnegative value $\beta(N)$, which is possible to be $+\infty$. The definition will be given in §1. In the case where codim N > 1, $\beta(N)$ is positive. On the other hand, if N is a hypersurface in M, then $\beta(N)$ can take the value 0. In that case, the condition $\beta(N) = 0$ means that N is minimal.

The purpose of this paper is to investigate the geometric structures of complete connected Riemannian manifolds which contain embedded compact connected submanifolds N with $\beta(N) < +\infty$ under a certain Ricci curvature condition.

Let N be a connected submanifold embedded in a complete connected Riemannian manifold M. We will say that the pair (M,N) satisfies the condition (R_0) if for any unit speed geodesic $c:[0,+\infty) \to M$ emanating orthogonally from N $\mathrm{Ric}(c'(t)) \geqq 0$ holds for every $t \geqq 0$ where $\mathrm{Ric}(c'(t))$ denotes the Ricci curvature for the tangent vector $c'(t)$ to c. We denote by i(N) the injectivity radius of the normal exponential map.

Let M be an m-dimensional $(m > 2)$ compact connected Riemannian manifold and let N be an embedded compact connected

ISBN 0-12-640170-5

submanifold in M, $0 \leqq \dim N \leqq m-1$. We suppose that $\pi_1(M,N) \neq 0$ and that (M,N) satisfies the condition (R_0) and $\beta(N) \leqq i(N)$. Under the assumption stated above, we show in §3 that if codim $N > 1$, then the cut locus C_N of N is an embedded compact connected minimal hypersurface in M and $\pi_1(M,C_N) = 0$. Moreover $M = B_r(N) \cup C_N$ (disjoint union) where $r = i(N)$ and $B_r(N)$ is the open tubular r-neighborhood of N. To consider the case where N is a hypersurface, we study in §4 the geometric structures of compact connected Riemannian manifolds with boundary under certain conditions. Using the results in §4, we show in §5 that if N is a hypersurface, then M is one of Riemannian manifolds of certain four types.

In §6 we consider an m-dimensional $(m > 2)$ complete noncompact connected Riemannian manifold M with nonnegative Ricci curvature which contains an embedded compact connected submanifold N such that $\beta(N) < +\infty$. We first show that if codim $N > 1$, then the normal exponential map $\mathrm{Exp}_N : \nu(N) \to M$ is a diffeomorphism and $M \setminus B_r(N)$, $r = \beta(N)$, is isometric to the Riemannian product manifold $\partial B_r(N) \times [0,+\infty)$. We next show that if N is a hypersurface, then M is one of Riemannian manifolds of certain three types.

We explain in §1 the notations and prepare in §2 some lemmas, which will be used in the later sections.

Throughout this paper we always assume that manifolds and all apparatus on them are of class C^∞ unless otherwise stated.

§1. Preliminaries.

In this section let M denote a complete connected Riemannian manifold of dimension $m > 2$ and let N be a connected submanifold embedded in M, $0 \leqq \dim N \leqq m-1$. We denote by $\nu(N)$ the total space of the normal bundle of N. If N is a point p of M, then $\nu(p)$ denotes the tangent vector space T_pM to M at p. Throughout this paper we use the following notations: $\nu_r(N) = \{\xi \in \nu(N);\ \|\xi\| < r\}$, $\partial\nu_r(N) = \{\xi \in \nu(N);\ \|\xi\| = r\}$, $\bar{\nu}_r(N) = \nu_r(N) \cup \partial\nu_r(N)$, where r is a positive and $\|\xi\|$ denotes the length of ξ. In the following, we put $U(N) = \nu_1(N)$. Let $\mathrm{Exp}_N : \nu(N) \to M$ be the normal exponential map, which is the restriction of the exponential map $\mathrm{Exp}: TM \to M$ to $\nu(N)$ where TM is the total space of the tangent bundle of M. Let $i(N)$ be the injectivity radius of Exp_N, which is defined by $i(N) = \sup\{r \geqq 0;\ \mathrm{Exp}_N | \nu_r(N)$ is an embedding$\}$. In case of $i(N) > 0$, for an r such that $0 \leqq r \leqq i(N)$, we put $B_r(N) = \mathrm{Exp}_N(\nu_r(N))$, $\partial B_r(N) = \mathrm{Exp}_N(\partial\nu_r(N))$ and $\bar{B}_r(N) = \mathrm{Exp}_N(\bar{\nu}_r(N))$. We identify N with the zero section in $\nu(N)$. For each $\xi \in U(N)$ we define $\delta(\xi) \in \mathbf{R} \cup \{+\infty\}$, $\mathbf{R}$ denotes the set of all real numbers, as follows. If $t\xi$, $t > 0$, is the first focal point of N in $\nu(N)$, then we let $\delta(\xi) = t$, and if there are no focal points of N along the ray $t\xi$ $(t \geqq 0)$ in $\nu(N)$, then we put $\delta(\xi) = +\infty$. Put $D = \{t\xi \in \nu(N);\ \xi \in U(N),\ 0 \leqq t < \delta(\xi)\}$. Let g be the Riemannian metric on D induced from M by Exp_N. We put $L_t = D \cap \partial\nu_t(N)$ for $t \geqq 0$ and define $\rho : D \to \mathbf{R}$ by $\rho(\xi) = \|\xi\|$. For each $t > 0$ such that $L_t \neq \emptyset$, let H_t denotes the mean curvature of L_t with respect to $\mathrm{grad}\,\rho$. We have $\Delta\rho(\xi) = -(m-1)H_t(\xi)$ at $\xi \in L_t$ where Δ is the Laplace-Beltrami operator on (D,g) defined as $\Delta\rho = \mathrm{div}(\mathrm{grad}\,\rho)$.

Now we shall give the definition of $\beta(N)$ stated in the introduction. To do it, for $\xi = (p,\xi_p) \in U(N)$, we put $H_0(0\xi)$

$= -\infty$ if codim $N > 1$ and denote by $H_0(0\xi)$ the mean curvature of N with respect to ξ_p if N is a hypersurface. We define β_N: $U(N) \to \mathbf{R} \cup \{+\infty\}$ as follows: $\beta_N(\xi) = \inf\{t \geqq 0;\ H_t(t\xi) \geqq 0\}$ if there exists a t $(0 \leqq t < \delta(\xi))$ such that $H_t(t\xi) \geqq 0$, and $\beta_N(\xi) = \delta(\xi)$ otherwise. We put $\beta(N) = \sup\{\beta_N(\xi);\ \xi \in U(N)\}$. It should be noted that $\beta_N(\xi) > 0$ holds for any $\xi \in U(N)$ if codim $N > 1$. In the case where N is a hypersurface, it is possible to be $\beta(N) = 0$. In this case, $\beta(N) = 0$ means that N is minimal.

We now suppose $\pi_1(M,N) \neq 0$, i.e., the homomorphism $i_{\#}$: $\pi_1(N) \to \pi_1(M)$ induced from the inclusion $i: N \to M$ is not surjective. Let M' be the universal Riemannian covering manifold of M with covering map π. Then $\pi^{-1}(N)$ has at least two connected components. Let N' be a connected component of $\pi^{-1}(N)$. Since π is locally isometric, we have $\beta(N') = \beta(N)$.

§2. Lemmas.

Throughout this section let M denote an m-dimensional $(m > 2)$ complete connected Riemannian manifold and let S be a connected hypersurface embedded in M with a nonvanishing normal vector field. In the following we assume that there exists an $r > 0$ such that $\mathrm{Exp}_N: \nu_r(S) \to M$ is an embedding. We choose a unit normal vector field η to S. Let d be the distance function on M induced from the Riemannian metric. Let $\rho(x) = \pm d(x,S)$ be the signed distance function on $B_r(S)$ such that ρ takes positive values in the same side as the direction of η. We denote by H_t the mean curvature of the level hypersurface $L_t = \rho^{-1}(t)$ with respect to grad ρ, $|t| < r$. Let x be a point of S. Along the geodesic $c_x(t) = \mathrm{Exp}_x t\eta_x$, $|t| < r$, we have

$$(m-1)\frac{dH_t}{dt}(c_x(t)) = \mathrm{Ric}(c_x'(t)) + \mathrm{trace}(A_t{}^2)$$

where A_t denotes the second fundamental form of L_t at $c_x(t)$, $|t| < r$. This formula yields the following.

Lemma 2.1. Let M, S and r be as above. Suppose that (M,S) satisfies the condition (R_0). Then for each $x \in S$ $H_t(c_x(t))$ is non-decreasing for t. For some a, b ($a < b$), if $H_a(c_x(a)) = H_b(c_x(b))$ holds for all $x \in S$, then L_t is totally geodesic for all t such that $a \leqq t \leqq b$.

Let u be a real-valued C^∞ function on S such that $0 \leqq u < r$. We put $L(u) = \{ \mathrm{Exp}_S u(x)\eta_x;\ x \in S\}$. We choose a unit normal vector field ξ to L(u) so that $\langle\xi, \mathrm{grad}\,\rho\rangle > 0$ on L(u) where $\langle\ ,\ \rangle$ denotes the Riemannian metric of M. Let Λ be the mean curvature of L(u) with respect to ξ. Let V be a local coordinate neighborhood in S and let $x_1,\ldots,x_{m-1}$ be coordinate functions on V. Let $X_j = \frac{\partial}{\partial x_j} + \frac{\partial u}{\partial x_j}\frac{\partial}{\partial \rho}$, $1 \leqq j \leqq m-1$. Then Λ is expressed in the form:

$$(m-1)\Lambda = -\sum_{i,j=1}^{m-1} g^{ij}\langle\nabla_{X_i}\xi, X_j\rangle$$

where the matrix (g^{ij}) is the inverse matrix of $(\langle X_i, X_j\rangle)$ and ∇ denotes the Riemannian connection of M. By rewriting the above equality, we obtain a quasilinear elliptic partial differential equation of second order on V:

$$\sum_{i,j=1}^{m-1} A_{ij}(x,u,Du)u_{ij} = B(x,u,Du)$$

where $Du = (\frac{\partial u}{\partial x_1},\ldots,\frac{\partial u}{\partial x_{m-1}})$, $u_{ij} = \frac{\partial^2 u}{\partial x_i \partial x_j}$ and $B(x,u(x),0) = (m-1)(\Lambda(c_x(u(x)))-H_{u(x)}(c_x(u(x))))$. Suppose that (M,S) satisfies the condition (R_0) and that $H_0 \geqq 0$ and $\Lambda \leqq 0$ hold everywhere. Then, using Lemma 2.1, we have $B(x,u(x),0) \leqq 0$ in V. By making use of E. Hopf's method in [8], we can show the following.

Lemma 2.2. Let M, S, r, and L(u) be as above where $u \in C^\infty(S)$ and $0 \leqq u < r$. Suppose that (M,S) satisfies the condition (R_0) and that $H_0 \geqq 0$ and $\Lambda \leqq 0$ hold everywhere. If u takes its minimum at a point of S, then u is constant.

Using the above lemma, we shall show the following.

Lemma 2.3. Let M be an m-dimensional (m > 2) complete connected Riemannian manifold. Let N_1 and N_2 be disjoint embedded submanifolds in M, $0 \leqq \dim N_1, \dim N_2 \leqq m-1$. Suppose that (M,N_1) and (M,N_2) satisfy the condition (R_0). Assume that there exists a minimizing unit speed geodesic $c:[0,r] \to M$, $r = d(N_1,N_2) > 0$, from N_1 to N_2 and assume that $\beta(N_1) + \beta(N_2) \leqq d(N_1,N_2)$. Then there exists an open neighborhood V of $\eta_1 = (c(0),c'(0))$ in $U(N_1)$ such that $\mathrm{Exp}_{N_1} r\xi \in N_2$ for all $\xi \in V$.

<u>Proof</u>. Let $a_j = \beta(N_j) + a$ $(j = 1,2)$ where $2a = r - \beta(N_1) - \beta(N_2) \geqq 0$. By minimality of c, c'(0) and c'(r) are unit normal vectors to N_1 and N_2 respectively and each point c(t),

$0 < t < r$, is not focal point to N_1 and N_2 along c. We can take connected open neighborhoods V_1 of η_1 in $U(N_1)$ and V_2 of $\eta_2 = (c(r), -c'(r))$ in $U(N_2)$ so that $Exp_{N_1}|CV_1$ and $Exp_{N_2}|CV_2$ are embeddings, where $CV_j = \{t\xi \in \nu(N_j);\ 0 \leqq t \leqq a_j,\ \xi \in V_j\}$, $j = 1,2$. We put $S_j = Exp_{N_1}(a_j V_j)$ where $a_j V_j = \{a_j\xi \in \nu(N_j);\ \xi \in V_j\}$, $j = 1,2$. Let ξ_1 and ξ_2 be unit normal vector fields to S_1 and S_2 such that ξ_1 directs to the outside relative to N_1 and ξ_2 directs to the inside relative to N_2. Since $a_1 + a_2 = d(N_1, N_2)$, S_2 lies in the same side as the direction of ξ_1 relative to S_1. ξ_1 coincides with ξ_2 at $c(a_1)$, which is a common point of S_1 and S_2. Let H_1 and H_2 be the mean curvatures of S_1 and S_2 with respect to ξ_1 and ξ_2 respectively. By Lemma 2.1 and the hypothesis on Ricci curvature, we obtain that $H_1 \geqq 0$ on S_1 and $H_2 \leqq 0$ on S_2. By taking V_1 and V_2 sufficiently small, S_2 can be expressed as $S_2 = \{Exp_x u(x)\xi_1(x);\ x \in S_1\}$, where $u \in C^\infty(S_1)$, $u \geqq 0$ and $u(c(a_1)) = 0$. Then, by Lemma 2.2, we have $S_1 = S_2$. This implies that $Exp_{N_1} r\xi \in N_2$ for all $\xi \in V_1$. We complete the proof.

§3. Decompositions of Riemannian manifolds.

In this section we generalize the results in the author's paper [10].

Theorem 3.1. Let M be an m-dimensional $(m > 2)$ complete connected Riemannian manifold. Let N_1 and N_2 be disjoint compact connected submanifolds embedded in M such that $0 \leqq \dim N_1$, $\dim N_2 \leqq m-2$. Suppose that (M, N_1) and (M, N_2) satisfy the condition (R_0) and that $\beta(N_1) + \beta(N_2) \leqq d(N_1, N_2)$. Then we have $M = B_r(N_1) \cup N_2 = N_1 \cup B_r(N_2)$ (disjoint union), where r =

$d(N_1, N_2)$. Moreover $S = \mathrm{Exp}_{N_1}(\partial\nu_a(N_1))$ is a connected compact minimal hypersurface embedded in M where $2a = r + \beta(N_1) - \beta(N_2)$.

<u>Proof</u>. Let $A = \{\xi \in U(N_1);\ \mathrm{Exp}_{N_1} r\xi \in N_2\}$. Then A is a non-empty closed set in $U(N_1)$. By Lemma 2.3, A is open in $U(N_1)$. Since $U(N_1)$ is connected , we have $A = U(N_1)$. This implies that $i(N_1) = r$ and $\mathrm{Exp}_{N_1}(\partial\nu_r(N_1)) = N_2$. Similarly, $i(N_2) = r$ and $\mathrm{Exp}_{N_2}(\partial\nu_r(N_2)) = N_1$. We put $2a = r + \beta(N_1) - \beta(N_2)$. Then we have $M = B_a(N_1) \cup S \cup B_{r-a}(N_2)$ (disjoint union), where $S = \partial B_a(N_1) = \partial B_{r-a}(N_2)$. By the definition of $\beta(N_j)$ and Lemma 2.1, S is a minimal hypersurface in M. We complete the proof.

The above theorem yields the following.

Corollary 3.1. Let M be an m-dimensional $(m > 2)$ complete connected Riemannian manifold. Suppose that M contains distinct points p and q with the following properties: (M,p) and (M,q) satisfy the condition (R_0), and $\beta(p) + \beta(q) \leqq d(p,q)$. Then M is homeomorphic to a standard m-sphere.

As an application of Theorem 3.1 we shall show the following.

Theorem 3.2. Let M be an m-dimensional $(m > 2)$ complete connected Riemannian manifold with nonnegative Ricci curvature. Suppose that M contains an embedded compact connected submanifold N such that $0 \leqq \dim N \leqq m-2$ and $\beta(N)$ is finite. Assume that $\pi_1(N)$ is finite and $\pi_1(M,N)$ is non trivial. Then $\pi_1(M)$ is finite.

<u>Proof</u>. Let M' be the universal Riemannian covering manifold

of M with covering map π. Since $\pi_1(M,N) \neq 0$ and $\pi_1(N)$ is finite, $\pi^{-1}(N)$ has at least two connected components and each connected component is a compact covering manifold of N. Suppose $\pi_1(M)$ is infinite. Then there exist two connected components N_1 and N_2 of $\pi^{-1}(N)$ such that $2\beta(N) \leq d'(N_1,N_2)$ where d' is the distance function on M' induced from the Riemannian metric. Note $\beta(N) = \beta(N_j)$, $j = 1,2$. By Theorem 3.1, M' is compact, which is a contradiction.

Theorem 3.3. Let M be an m-dimensional ($m > 2$) compact connected Riemannian manifold and N a compact connected submanifold embedded in M, $0 \leq \dim N \leq m-2$. Suppose that $\pi_1(M,N) \neq 0$ and that (M,N) satisfies the condition (R_0) and $\beta(N) \leq i(N)$. Then the cut locus C_N of N is an embedded compact connected minimal hypersurface in M. Moreover, $\pi_1(M,C_N) = 0$ and $M = B_r(N) \cup C_N$ (disjoint union) where $r = i(N)$.

Proof. Let C be the family of all continuous curves in M parametrized on [0,1] whose both end points belong to N. Let C' be the subfamily of C consisting of all piecewise smooth curves which can not be deformed to a curve in N through curves in C. By compactness of N and $\pi_1(M,N) \neq 0$, $a = \inf\{L(c); c \in C'\}$ is positive where L(c) denotes the length of c. There exists a geodesic $\gamma:[0,1] \to M$ such that $\gamma \in C'$ and $L(\gamma) = a$. By minimality of γ, the tangent vectors $\gamma'(0)$ and $\gamma'(1)$ are pendicurar to N. Let M' be the universal Riemannian covering manifold of M with covering map π. Since $\pi_1(M,N) \neq 0$, $\pi^{-1}(N)$ is disconnected. Let $\sigma:[0,1] \to M'$ be a lift of γ. Then both end points $\sigma(0)$ and $\sigma(1)$ belong to distinct connected components of $\pi^{-1}(N)$. Let N_j, $j = 0,1$, be the connected component of $\pi^{-1}(N)$ containing $\sigma(j)$. Then σ is a minimizing geodesic between N_0 and N_1. Hence we have $d'(N_0,N_1) = a$, where d' is the distance function on M' induced from the Riemannian metric. From the assumption $\beta(N) \leqq i(N)$, we have $\beta(N_0) + \beta(N_1) \leqq d'(N_0,N_1)$. By the same argument as in the proof of Theorem 3.1, $M' = B_a(N_0) \cup N_1 = N_0 \cup B_a(N_1)$ (disjoint union). We put $L = \{x \in M';\ d'(x,N_0) = d'(x,N_1)\}$ and $S = \pi(L)$. From Lemma 2.1 and the definition of $\beta(N)$, L is a connected minimal hypersurface embedded in M'. Since L is invariant by the deck transformation group, S is an embedded compact connected minimal hypersurface in M and $\pi_1(M,S) = 0$. Moreover, we have $M = B_r(N) \cup S = N \cup B_r(S)$ (disjoint union) where $a = 2r$. This implies that $S = C_N$ and $r = i(N)$. We complete the proof.

As a consequence of the above theorem we have

Corollary 3.2. Under the same assumption as in Theorem 3.3, let N be a point of M. Then M is doubly covered by a standard m-sphere.

Remark. Let M and N be as in Theorem 3.3. Suppose that (M,N) satisfies the condition (R_0) and $\pi_1(M,N) \neq 0$. Moreover, we assume $\beta(N) < i(N)$. Then, by the above theorem and Lemma 2.1, $L := \partial B_s(N)$, $s = i(N) - \beta(N)$, and C_N are totally geodesic and $M \setminus (B_s(N) \cup C_N)$ is isometric to the Riemannian product manifold $L\times[0,s)$. Furthermore, $M \setminus B_s(N)$ is doubly covered by the Riemannian product manifold $L\times[0,2s]$ (see §4).

From the above remark we have the following.

Theorem 3.4. Let M be an m-dimensional $(m > 2)$ compact connected Riemannian manifold and N an embedded compact connected submanifold in M such that $0 \leqq \dim N \leqq m-2$. Suppose that for any unit speed geodesic $c:[0,+\infty) \to M$ emanating perpendicularly from N $\mathrm{Ric}(c'(t)) > 0$ holds for any $t \geqq 0$ and that $\pi_1(M,N) \neq 0$ holds. Then we have $i(N) \leqq \beta(N)$. If the equality $i(N) = \beta(N)$ holds, then the cut locus of N is an embedded compact connected minimal hypersurface in M.

§4. Riemannian manifolds with boundary.

In this section let W denote an m-dimensional ($m > 2$) connected Riemannian manifold with boundary ∂W. Let d be the distance function on W induced from the Riemannian metric. We say that W is complete if (W,d) is complete as a metric space. In the following we assume that W is complete. Let ν denote the inner unit normal vector field to ∂W. For each $p \in \partial W$ let c_p denote the geodesic in W emanating from p with the initial tangent vector ν_p. For each $p \in \partial W$ let t_p be the supremum of the set consisting of all positive t for which $c_p(s) \in \mathrm{Int}\, W$, $0 < s \leqq t$, where Int W denotes the interior of W. By completeness of W, $c_p(t_p)$ is a boundary point of W if t_p is finite. In what follows we always assume that for any $p \in \partial W$ the parameter values of c_p belong to the interval $[0,t_p]$, where $[0,t_p] = [0,+\infty)$ if $t_p = +\infty$. For each $p \in \partial W$ we define $\delta(p) \in \mathbf{R} \cup \{+\infty\}$ as follows. If $c_p(t)$, $0 < t \leqq t_p$, is the first focal point to ∂W along c_p, then we put $\delta(p) = t$, and if c_p has no focal points to ∂W along c_p, then we let $\delta(p) = t_p$. Now let N be a connected component of ∂W. We put $D_N = \{(p,t\nu_p);\ p \in N,\ 0 \leqq t < \delta(p)\}$. Then $\mathrm{Exp}:D_N \to W$, $\mathrm{Exp}((p,t\nu_p)) = c_p(t)$, is regular. Let g be the Riemannian metric on D_N induced from W by Exp. We define $\rho:(D_N,g) \to \mathbf{R}$ by $\rho((p,t\nu_p)) =$

t. For each $t \geqq 0$ such that $L_t = \rho^{-1}(t)$ is nonempty, let H_t be the mean curvature of L_t with respect to $\operatorname{grad}\rho$ in (D_N, g). We define $\beta_+ : N \to \mathbf{R} \cup \{+\infty\}$ as follows: $\beta_+(p) = \inf\{t \geqq 0;\ H_t \geqq 0$ at $(p, t\nu_p)\}$ if there exists a t $(0 \leqq t < \delta(p))$ such that $H_t \geqq 0$ at $(p, t\nu_p)$, and we put $\beta_+(p) = \delta(p)$ otherwise. Let $\beta_+(N) = \sup\{\beta_+(p);\ p \in N\}$ and $\beta_+(\partial W) = \sup\{\beta_+(N);\ N$ is a connected component of $\partial W\}$. The condition $\beta_+(N) = 0$ means that the mean curvature of N with respect to ν is everywhere nonnegative.

We now suppose that ∂W is connected and $\pi_1(W, \partial W) \neq 0$. Let X be the universal Riemannian covering manifold of W with covering map π. Then the boundary ∂X of X is disconnected. Since π is locally isometric, for each connected component N of ∂X we have $\beta_+(N) = \beta_+(\partial W)$.

We say that $(W, \partial W)$ satisfies the condition (R_0) if for any $p \in \partial W$ $\operatorname{Ric}(c_p'(t)) \geqq 0$ holds, $0 \leqq t \leqq t_p$. Let $i_W(\partial W) = \sup\{r \geqq 0;\ \operatorname{Exp}|\nu_{+r}(\partial W)$ is an embedding$\}$ where $\nu_{+r}(\partial W) = \{(p, t\nu_p);\ p \in \partial W,\ 0 \leqq t \leqq r\}$.

Lemma 4.1. Let W be an m-dimensional $(m > 2)$ complete connected Riemannian manifold with compact boundary ∂W. Assume that ∂W has exactly two connected components N_1 and N_2. Suppose that $(W, \partial W)$ satisfies the condition (R_0).

(1) If $\beta_+(N_1) + \beta_+(N_2) \leqq d(N_1, N_2)$, then M is diffeomorphic to the product manifold $N_1 \times [0, r]$, $r = d(N_1, N_2)$. The level set $L_a = \{x \in \partial W;\ d(x, N_1) = a\}$ is a connected minimal hypersurface embedded in W, where $2a = r + \beta_+(N_1) - \beta_+(N_2)$.

(2) If $\beta_+(\partial W) = 0$, then W is isometric to the Riemannian product manifold $N_1 \times [0, r]$, $d(N_1, N_2) = r$.

<u>Proof</u>. Note that there exists a minimizing geodesic in W be-

tween N_1 and N_2. Such a geodesic is perpendicular to ∂W at both end points. Let $A = \{p \in N_1;\ c_p(r) \in N_2\}$. It is a non-empty closed set in N_1. We can apply Lemma 2.3 to the present situation. Hence A is open in N_1. Thus we have $A = N_1$. Then the map $f: N_1 \times [0,1] \to W$ defined by $f(p,t) = c_p(t)$ is a diffeomorphism. By the definition of β_+ and Lemma 2.1, L_a is minimal. We next assume $\beta_+(\partial W) = 0$. Then Lemma 2.1 implies that $f(N_1 \times \{s\})$ is totally geodesic for any s, $0 \leqq s \leqq r$. Thus f is an isometry.

Theorem 4.1. Let W be an m-dimensional ($m > 2$) compact connected Riemanian manifold with connected boundary ∂W. Suppose that $\pi_1(W, \partial W) \neq 0$ and that $(W, \partial W)$ satisfies the condition (R_0) and $\beta_+(\partial W) \leqq i_W(\partial W)$. Then Int W contains an embedded compact connected minimal hypersurface S and $\mathrm{Exp}_S : \nu_r(S) \to W$ is a diffeomorphism where $r = d(S, W) = \max\{d(x, \partial W);\ x \in W\}$. Moreover, $\pi_1(W, S) = 0$ and W is doubly covered by the product manifold $\partial W \times [0, 2r]$.

Proof. Let X be the universal Riemannian covering manifold of W with covering map π. The boundary ∂X of X is disconnected because of $\pi_1(W, \partial W) \neq 0$. Since π is locally isometric, X is complete and $(X, \partial X)$ satisfies the condition (R_0). Take a connected component N_0 of ∂X. We can show that there exists a geodesic $\sigma : [0,1] \to X$ with the following properties: (1) $\sigma(0) \in N_0$ and $\sigma(1)$ belongs to the other connected component N_1 of ∂X, (2) $\sigma((0,1)) \subset \mathrm{Int}\, X$ and (3) the length of σ is equal to $a = \inf\{d_X(N_0, N);\ N$ is a connected component of ∂X such that $N_0 \cap N = \emptyset\}$, where d_X is the distance function on X induced from the Riemannian metric. Hence σ is a minimizing geodesic between N_0 and N_1. Note that $2i_W(\partial W) \leqq a$ and $\beta(N_j) = \beta(\partial W)$, $j = 0,1$. Thus we have $\beta(N_0) + \beta(N_0) \leqq d_X(N_0,N_1)$. We put $A = \{x \in N_0;\ \gamma_x(a) \in N_1\}$, where $\gamma_x:[0,a] \to X$ denotes the geodesic such that $\gamma_x(0) = x$ and the tangent vector $\gamma_x'(0)$ is the inner unit normal vector to N_0. A is a nonempty closed set in N_0. By applying Lemma 2.3 to the present case, we see that A is open in N_0. Hence, $A = N_0$. Then the map $h:N_0\times[0,a] \to X$ defined by $h(x,t) = \gamma_x(t)$ is a diffeomorphism. Hence we have $\partial X = N_0 \cup N_1$. We put $L = \{x \in X;\ d_X(x,N_0) = d_X(x,N_1)\}$. Then X can be decomposed as $X = N_0 \cup B_r(L) \cup N_1$ (disjoint union). Since $\beta_+(N_j) \leqq r$, by the definition of $\beta_+(N_j)$ and Lemma 2.1, L is a connected minimal hypersurface in X. We put $S = \pi(L)$. Since L is invariant by the deck transformation group, S is an embedded minimal hypersurface in Int W and $\pi_1(M,S) = 0$. Moreover, we have $W = \partial W \cup B_r(S) = B_r(\partial W) \cup S$ (disjoint union) where $2r = a$. This implies that $\mathrm{Exp}_S:\nu_r(S) \to W$ is a diffeomorphism. The map $f:\partial W\times[0,2r] \to W$ defined by $f(p,t) = c_p(t)$ is a double covering map.

Theorem 4.2. Let W be an m-dimensional ($m > 2$) compact connected Riemannian manifold with connected boundary ∂W such that $\pi_1(W,\partial W) \neq 0$. Suppose that $(W,\partial W)$ satisfies the condition (R_0) and that the mean curvature of ∂W with respect to the inner normal direction is everywhere nonnegative. Then ∂W is totally geodesic and admits an isometric involution ϕ without fixed points. Furthermore, W is isometric to the Riemannian manifold $\partial W\times[0,r]/\phi$, $r = \max\{d(x,\partial W);\ x \in W\}$, which is obtained from the Riemannian product manifold $\partial W\times[0,r]$ by identifying each (x,r) with $(\phi(x),r)$.

Proof. By the previous theorem, Int W contains an embeded compact connected minimal hypersurface S and $W = \bar{B}_r(S)$, where $r = d(S,\partial W) = \max\{d(x,\partial W);\ x \in W\}$. The map $\phi:\partial W \to \partial W$ defined by $\phi(p) = c_p(2r)$ is a diffeomorphic involution without fixed points. Since $\beta_+(\partial W) = 0$, by Lemma 2.1 the parallel hypersurface $\partial B_t(S)$ of S is totally geodesic, $0 \leqq t \leqq r$. Hence ϕ is an isometry on ∂W. Define $f:\partial W\times[0,r] \to W$ by $f(p,t) = c_p(t)$. From the above facts, f is locally isometric and $f:\partial W\times[0,r) \to W \setminus S$ is an isometry. Since $f(p) = f(\phi(p))$ holds for any $p \in \partial W$, $f:\partial W\times\{r\} \to S$ is a double Riemannian covering map. It is now easy to show the last assertion.

§5. Classification of certain Riemannian manifolds.

In this section we consider compact connected Riemannian manifolds which contain embedded compact connected hypersurfaces N with $\beta(N) < +\infty$ under certain conditions. To do it, we need to describe certain compact connected Riemannian manifolds, which are the model spaces in our consideration.

Type A. Let W be an m-dimensional $(m > 2)$ compact connected Riemannian manifold with connected boundary ∂W. We say that W is of type A if ∂W admits an isometric involution ϕ without fixed points and W is isometric to the Riemannian manifold $\partial W\times[0,r]/\phi$, $r = \max\{d(x,\partial W);\ x \in W\}$, which is obtained from the Riemannian product manifold $\partial W\times[0,r]$ by identifying each (p,r) with $(\phi(p),r)$.

Type B. Let W be an m-dimensional $(m > 2)$ compact connected Riemannian manifold with connected boundary ∂W. Suppose that Int W contains an embedded compact connected minimal hypersurface S. We say that W is of type B if $\mathrm{Exp}_S:\nu_r(S) \to W$, $r = \max\{d(x,\partial W);\ x \in W\}$, is a diffeomorphism and W is not isometric to a Riemannian manifold of type A.

Let W be a Riemannian manifold of type A or type B. We note that $\pi_1(W,\partial W) \neq 0$ and W is doubly covered by a product manifold $\partial W\times[0,a]$.

<u>Type I</u>. Let N be an n-dimensional ($n \geq 2$) compact connected Riemannian manifold without boundary which admits an isometry ϕ and let r be a positive. Let $N\times[0,r]/\phi$ denote the quotient Riemannian manifold obtained from the Riemannian product manifold $N\times[0,r]$ by identifying $N\times\{0\}$ with $N\times\{r\}$ by ϕ. We say that such a Riemannian manifold is of type I.

<u>Type II</u>. Let W_1 and W_2 be Riemannian manifolds of type A. We assume that there exists an isometry $f:\partial W_1 \to \partial W_2$. Let $W_1 \cup_f W_2$ be a Riemannian manifold obtained from the disjoint sum $W_1 \cup W_2$ by identifying each $x \in \partial W_1$ and $f(x) \in \partial W_2$. We say that a Riemannian manifold constructed in the way stated above is of type II.

<u>Type III</u>. Let M be an m-dimensional ($m > 2$) compact connected Riemannian manifold without boundary. M is called a Riemannian manifold of type III if M can be decomposed in the form: $M = W_1 \cup W_2$, $\mathrm{Int}\, W_1 \cap \mathrm{Int}\, W_2 = \emptyset$ and $\partial W_1 = \partial W_2$, where W_1 and W_2 are compact domains in M with smooth boundaries ∂W_1 and ∂W_2 respectively such that W_1 is of type A and $\pi_1(W_2,\partial W_2) = 0$.

<u>Type IV</u>. Let M be an m-dimensional ($m > 2$) compact connected Riemannian manifold without boundary. M is called a

Riemannian manifold of type IV if M can be decomposed in the form: $M = W_1 \cup W_2$, $\text{Int } W_1 \cap \text{Int } W_2 = \emptyset$ and $\partial W_1 = \partial W_2$, where W_1 and W_2 are compact domains in M with smooth boundaries ∂W_1 and ∂W_2 respectively such that W_1 is of type B and $\pi_1(W_2, \partial W_2) = 0$.

Theorem 5.1. Let M be an m-dimensional ($m > 2$) compact connected Riemannian manifold and N an embedded compact connected hypersurface in M. Suppose that (M,N) satisfies the condition (R_0) and $\beta(N) \leq i(N)$ and that $\pi_1(M,N) \neq 0$. Then M is one of Riemannian manifolds of four types described above.

To give the proof, we shall prepare two lemmas.

Lemma 5.1. Let M be an m-dimensional ($m > 2$) compact connected Riemannian manifold containing an embedded compact connected minimal hypersurface N. Let r be a positive such that $\text{Exp}_N : \nu_{2r}(N) \to M$ is an embedding. Suppose that $M \setminus B_r(N)$ is connected and that N admits a nonvanishing normal vector field. Moreover assume that (M,N) satisfies the condition (R_0). Then N is totally geodesic and M is of type I.

Proof. Let $W = M \setminus B_r(N)$. By assumption, the boundary ∂W of W consists of just two connected components N_1 and N_2 such $\partial B_r(N) = N_1 \cup N_2$. Since (M,N) satisfies the condition (R_0) and N is minimal, by Lemma 2.1 the mean curvature of ∂W with respect to the inner normal direction is nonnegative everywhere. Then, by Lemma 4.1, W is isometric to the Riemannian product manifold $N_1 \times [0, r_1]$ where $r_1 = d_W(N_1, N_2)$, here d_W is the distance function on W induced from the Riemannian metric. It follows from Lemma 4.1 that N is totally geodesic and

$\bar{B}_r(N)$ is isometric to the Riemannian product manifold $N \times [-r, r]$. Let η be a unit normal vector field to N and let $a = 2r + r_1$. Define $\phi : N \to N$ by $\phi(x) = \mathrm{Exp}_x a\eta_x$, $x \in N$. Then ϕ is an isometry on N. It is now easy to see that M is of type I.

Lemma 5.2. Let M be an m-dimensional ($m > 2$) compact connected Riemannian manifold and N an embedded compact connected hypersurface in M. Suppose that N is a one-sided hypersurface and $\pi_1(M,N) \neq 0$. Moreover assume that (M,N) satisfies the condition (R_0) and $\beta(N) \leqq i(N)$. Then N is totally geodesic and M is of type II.

Proof. Take an $r > 0$ so that $\mathrm{Exp}_N : \nu_{2r}(N) \to M$ is an embedding. In case of $\beta(N) > 0$, we may assume $r < \beta(N)$. Let $W = M \setminus B_r(N)$. Since N is a one-sided hypersurface in M, W and ∂W are connected. Since N is a deformation retract of $B_r(N)$, the assumption $\pi_1(M,N) \neq 0$ implies $\pi_1(W,\partial W) \neq 0$. From the assumption $\beta(N) \leqq i(N)$, we obtain $\beta_+(\partial W) \leqq i_W(\partial W)$ because $\beta_+(\partial W) \leqq \beta(N) - r$ and $i_W(\partial W) = i(N) - r$. Since $(W,\partial W)$ satisfies the condition (R_0), by Theorem 4.1, Int W contains an embedded compact connected minimal hypersurface S and $W = \bar{B}_s(S)$, where $s = d(\partial W, S)$. Then we have $\partial B_t(N) = \partial B_{a-t}(S)$, where $a = r + s = d(N,S)$ and $0 \leqq t \leqq a$. It follows from this fact that (M,S) satisfies the condition (R_0). Since S is minimal, by Lemma 2.1, N, S and the parallel hypersurfaces $\partial B_t(N)$ of N ($0 < t < a$) are totally geodesic. Hence both $B_r(N)$ and W are of type A. Therefore, M is of type II.

Proof of Theorem 5.1. Take an $r > 0$ so that $\mathrm{Exp}_N : \nu_{2r}(N) \to M$ is an embedding. In case of $\beta(N) > 0$, we may assume $r < \beta(N)$. Let $W = M \setminus B_r(N)$. The boundary ∂W of W has at most two connected components. If ∂W is connected, then by Lemma 5.2 M is of type II. From now on we assume that ∂W consists of two connected components N_1 and N_2. Then W has at most two connected components. Suppose first W is connected. On W we have $\beta_+(N_j) = 0$ ($j = 1,2$) if $\beta(N) = 0$, and $\beta_+(N_j) \leq \beta(N) - r$ ($j = 1,2$) if $\beta(N) > 0$. Moreover, we have $2(i(N) - r) \leq d_W(N_1,N_2)$ on W, where d_W is the distance function on W. From the above facts and the assumption $\beta(N) \leq i(N)$, we obtain $\beta_+(N_1) + \beta_+(N_2) \leq d_W(N_1,N_2)$ on W. By Lemma 4.1, W is diffeomorphic to the product manifold $N_1 \times [0,a]$, $a = d_W(N_1,N_2)$, and some parallel hypersurface of N_1 in W is minimal in M. It follows from Lemma 5.1 that M is of type I. We next assume that W consists of two connected components. Let W_1 and W_2 be compact domains in M such that $\mathrm{Int}\, W_1 \cap \mathrm{Int}\, W_2 = \emptyset$, $\partial W_1 = \partial W_2 = N$, $N_j \subset W_j$, $j = 1,2$, and $M = W_1 \cup W_2$. On each W_j we have $i(N) \leq i_{W_j}(\partial W_j)$ and $\beta_+(\partial W_j) \leq \beta(N)$. By the assumption $\beta(N) \leq i(N)$, we have $\beta_+(\partial W_j) \leq i_{W_j}(\partial W_j)$, $j = 1,2$. Since $\pi_1(M,N) \neq 0$, so is at least one of $\pi_1(W_1,\partial W_1)$ and $\pi_1(W_2,\partial W_2)$. Each $(W_j,\partial W_j)$ satisfies the condition (R_0). If $\pi_1(W_j,\partial W_j) \neq 0$, then W_j is of type A or type B (see Theorem 4.1 and 4.2). Assume now that both $\pi_1(W_1,\partial W_1)$ and $\pi_1(W_2,\partial W_2)$ are nontrivial. By Theorem 4.1, each $\mathrm{Int}\, W_j$ contains an embedded compact connected minimal hypersurface S_j and $W_j = \overline{B}_{r_j}(S_j)$, where $r_j = d(N,S_j)$, $j = 1,2$. By applying Lemma 2.1, we see that N is minimal. By Theorem 4.2, both W_1 and W_2 are of type A. Therefore, M is of type II. From what we have just proved in the above, in the case where W consists of two connected components, we see that M is one of Riemannian manifolds of three types II, III and IV. We complete the proof.

As a corollary of Theorem 5.1, we have the following.

Corollary 5.1. Let M be an m-dimensional ($m > 2$) compact connected Riemannian manifold containing an embedded compact connected minimal hypersurface N. Suppose that (M,N) satisfies the condition (R_0) and $\pi_1(M,N) \neq 0$. Then M is one of Riemannian manifolds of types I, II and III described above. In particular, N is totally geodesic.

§6. Noncompact manifolds.

In this section we study the geometric structure of a complete noncompact connected Riemannian manifold with nonnegative Ricci curvature which contains an embedded compact connected submanifold N such that $\beta(N)$ is finite.

Let M be a complete noncompact connected Riemannian manifold with $\dim M \geqq 2$. A unit speed geodesic $c:[0,+\infty) \to M$ is called a ray if $d(c(0),c(t)) = t$ for all $t > 0$, where d is the distance function on M induced from the Riemannian metric. Let $c:[0,+\infty) \to M$ be a ray. The Busemann function $F_c:M \to R$ for c is defined by $F_c(x) = \lim_{t\to+\infty} (t - d(x,c(t)))$, $x \in M$. F_c is Lipschitz continuous. Suppose now M is of nonnegative Ricci curvature. Then ΔF_c(weakly) $\geqq 0$ holds on M, i.e., for each $x \in M$ and for any $\varepsilon > 0$ F_c has a C^∞ support function $f_{x,\varepsilon}$ in an open neighborhood V of x such that $\Delta f_{x,\varepsilon} \geqq -\varepsilon$ in V. It follows from this property that F_c is subharmonic ([1], [3], [4], [15]).

Making use of the way employed in [3], [4] and [9], we shall show the following.

Theorem 6.1. Let M be an m-dimensional ($m > 2$) complete noncompact connected Riemannian manifold with nonnegative Ricci curvature. Suppose that M contains an embedded compact connected submanifold N such that $0 \leqq \dim N \leqq m-2$ and $\beta(N)$ is finite. Then $\mathrm{Exp}_N : \nu(N) \to M$ is a diffeomorphism. Moreover $M \setminus B_r(N)$, $r = \beta(N)$, is isometric to the Riemannian product manifold $\partial B_r(N) \times [0, +\infty)$.

<u>Proof</u>. Let A be the subset of U(N) which consists of all $\xi \in U(N)$ such that the geodesic $c_\xi(t) = \mathrm{Exp}_N t\xi$ $(t \geqq 0)$ satisfies the condition that $d(c(t), N) = t$ holds for any $t > 0$. It is easy to see that A is a nonempty closed set in U(N). We shall show that A is open in U(N). Let $\eta = (p, \eta_p)$ be an arbitrary element of A and $c : [0, +\infty) \to M$ the geodesic emanating from p such that $c'(0) = \eta_p$ and $d(c(t), N) = t$ for any $t > 0$. Take an r so that $r > \beta(N)$. Since c has no focal points to N along c, we can take an open connected neighborhood V of η in U(N) so that $\mathrm{Exp}_N : C(V) \to M$ is an embedding, where $C(V) = \{t\xi \in \nu(N);\ 0 \leqq t \leqq 3r,\ \xi \in V\}$. Taking V sufficiently small if necessary, we may assume that $\mathrm{Exp}_N(C(V))$ does not meet the cut locus of N. Let $D = \{\mathrm{Exp}_N t\xi;\ r < t < 3r,\ \xi \in V\}$. Define $G : D \to \mathbf{R}$ by $G(x) = F_c(x) - d_N(x)$, $x \in D$, where $d_N(x) = d(x, N)$. By the definition of a Busemann function, we have $G \leqq 0$ in D and $G(c(t)) = 0$, $r < t < 3r$. On the other hand, by the definition of $\beta(N)$, we obtain $\Delta d_N \leqq 0$ in D because $\beta(N) < d_N$ in D. Thus, $\Delta G(\text{weakly}) \geqq 0$ in D. By the maximum principle ([1], [4]), we have $F_c = d_N$ in D. Hence d_N is harmonic in D. Then each parallel hypersurface $L_t = \{\mathrm{Exp}_N t\xi;\ \xi \in V\}$, $r < t < 3r$, is minimal. By Lemma 2.1, L_t is totally geodesic, $r < t < 3r$. Thus D is isometric to the Riemannian product manifold $L_{2r} \times (-r, r)$. It follows from this fact together with $F_c = d_N$ in D

that for any $\xi \in V$ the geodesic $c_\xi(t) = \mathrm{Exp}_N t\xi$ $(t \geq 0)$ is a ray asymptotic to c and satisfies $d(c_\xi(t),N) = t$ for all $t > 0$ (see Theorem 1.1 in [14]). Hence we have $V \subset A$. Therefore A is open in $U(N)$. By connectedness of $U(N)$, we have $A = U(N)$. Hence $\mathrm{Exp}_N : \nu(N) \to M$ is is a diffeomorphism. From the above argument done by using a Busemann function and Lemma 2.1, we see that $\partial B_t(N)$ is a totally geodesic hypersurface in M for any $t \geq \beta(N)$. Hence $M \setminus B_r(N)$, $r = \beta(N)$, is isometric to the Riemannian product manifold $\partial B_r(N) \times [0,+\infty)$. We complete the proof.

Corollary 6.1. Let M be an m-dimensional $(m > 2)$ complete connected Riemannian manifold with nonnegative Ricci curvature. Suppose that M contains an embedded compact connected submanifold N such that $0 \leq \dim N \leq m-2$ and $\beta(N)$ is finite. If $\pi_1(M,N) \neq 0$, then M is compact.

As an application of Theorem 6.1 we have the following.

Theorem 6.2. Let M be an m-dimensional $(m > 2)$ complete connected Riemannian manifold with nonnegative Ricci curvature. If there exist distinct points p and q of M such that both $\beta(p)$ and $\beta(q)$ are finite, then M is compact.

Proof. Suppose M is noncompact. Then by the above theorem $\mathrm{Exp}_p : T_pM \to M$ is a diffeomorphism and for any $r > \beta(p)$ the geodesic sphere $\partial B_r(p)$ is a connected totally geodesic hypersurface in M. The same properties also hold for q. Let $c:[0,+\infty) \to M$ be the ray which starts at p and passes through q. Take an $r > 0$ so that $r > \beta(p) + \beta(q) + d(p,q)$. Then the geodesic spheres $\partial B_r(p)$ and $\partial B_t(q)$, $t = r - d(p,q)$, have the same tangent space at $c(t)$. Since both $\partial B_r(p)$ and $\partial B_t(q)$ are totally geodesic, it must be $\partial B_r(p) = \partial B_t(q)$, which is a contradiction. Thus M is compact.

In the following we shall classify complete noncompact connected Riemannian manifolds with nonnegative Ricci curvature which contain embedded compact connected hypersurfaces N such that $\beta(N)$ is finite.

Theorem 6.3. Let M be an m-dimensional $(m > 2)$ complete noncompact connected Riemannian manifold with nonnegative Ricci curvature which contains an embedded compact connected hypersurface N such that $\beta(N)$ is finite. Then one of the following three cases holds:

(1) N is totally geodesic and M is isometric to a Riemannian manifold $L\times[0,+\infty)/\phi$, where L is isometric to either N or a double Riemannian covering manifold of N and ϕ is an isometric involution on L without fixed points, and where $L\times[0,+\infty)/\phi$ is the quotient Riemannian manifold obtained from the Riemannian product manifold $L\times[0,+\infty)$ by identifying each $(x,0)$ with $(\phi(x),0)$.

(2) M is isometric to the Riemannian product manifold $N\times\mathbf{R}$.

(3) M contains two closed domains D_1 and D_2 with the following properties: (i) $M = D_1 \cup D_2$, $\operatorname{Int} D_1 \cap \operatorname{Int} D_2 = \emptyset$ and $\partial D_1 = \partial D_2 = N$, (ii) D_1 is diffeomorphic to the product manifold $N\times[0,+\infty)$, and (iii) D_2 is compact and $\pi_1(D_2,N) = 0$.

To prove this theorem, we prepare the following lemma. Since the proof is similar to that of Theorem 6.1, we omit it.

Lemma 6.1. Let M and N be as in Theorem 6.3. Suppose that N admits a unit normal vector field η and that there exists a unit speed geodesic $c:[0,+\infty) \to M$ such that $c'(0) = \eta_{c(0)}$ and $d(c(t),N) = t$ for any $t > 0$. Then $\mathrm{Exp}_N:\nu_+(N) \to M$ is an embedding, where $\nu_+(N) = \{(x,t\eta_x);\ x \in N,\ t \geq 0\}$. Furthermore, $\mathrm{Exp}_N(\nu_+(N)) \setminus B_{+r}(N)$, $r > \beta(N)$, is isometric to the Riemannian product manifold $L_{+r}\times[0,+\infty)$, where $B_{+r}(N) = \mathrm{Exp}_N(\nu_{+r}(N))$, $\nu_{+r}(N) = \{(x,t\eta_x);\ x \in N,\ 0 \leq t < r\}$ and $L_{+r} = \{\mathrm{Exp}_x r\eta_x;\ x \in N\}$.

Proof of Theorem 6.3. We first suppose that N is a one-sided hypersurface embedded in M, i.e., U(N) is connected. Then, by the same argument as in the proof of Theorem 6.1, we can show that $\mathrm{Exp}_N:\nu(N) \to M$ is a diffeomorphism and $M \setminus B_r(N)$, $r > \beta(N)$, is isometric to the Riemannian product manifold $\partial B_r(N)\times[0,+\infty)$. Take an $r > \partial(N)$, and let $L = \partial B_r(N)$. Then we have $\pi_1(B_r(N),L) \neq 0$. Since L is totally geodesic, by Theorem 4.2, $\overline{B}_r(N)$ is of type A. Hence (1) holds. We next suppose that N is a two-sided hypersurface embedded in M, i.e., U(N) is disconnected. Since M is noncompact and N is compact, there exists a unit speed geodesic $c:[0,+\infty) \to M$ emanating from N with the property that $c'(0)$ is perpendicular to N and $d(c(t),N) = t$ for all $t > 0$. Let η be the unit normal vector field to N which agrees with $c'(0)$ at $c(0)$. For η and $t > 0$ we define $\nu_+(N)$, $B_{+t}(N)$ and $L_{+t}(N)$ by the same way as in Lemma 6.1. It follows from Lemma 6.1 that $\mathrm{Exp}_N:\nu_+(N) \to M$ is an embedding. We put $D_1 = \mathrm{Exp}_N(\nu_+(N))$ and take a $s > \beta(N)$. By Lemma 6.1, $D_1 \setminus B_{+s}(N)$ is isometric to the Riemannian product manifold $L_{+s}(N)\times[0,+\infty)$. We put $D_2 = M \setminus \mathrm{Int}\, D_1$. Assume now D_2 is noncompact. Then there exists a unit speed geodesic $\gamma:(-\infty,+\infty) \to$

M such that $\gamma'(0)$ is perpendicular to N and $d(\gamma(t_1),\gamma(t_2)) = |t_1 - t_2|$ for any $t_1, t_2 \in \mathbf{R}$. Then, by the splitting theorem ([3]), M is isometric to the Riemannian product manifold $N\times\mathbf{R}$. We next consider the case where D_2 is compact. We suppose $\pi_1(M,N) \neq 0$. Put $D = \{x \in M;\ d(x,D_2) \leq s\}$. Since $\pi_1(M,N) \neq 0$, so is $\pi_1(D,\partial D)$. By Theorem 4.2, D is of type A. Therefore N is totally geodesic and is isometric to ∂D, and hence N admits an isometric involution without fixed points. It is now easy to show that (1) holds. We next assume $\pi_1(M,N) = 0$. Then it must be $\pi_1(D_2,N) = 0$ because D_1 is diffeomorphic to the product manifold $N\times[0,+\infty)$. Hence (3) holds. We complete the proof.

From Theorem 6.3 we have the following.

Corollary 6.2. Let M be an m-dimensional $(m > 2)$ complete connected Riemannian manifold with nonnegative Ricci curvature which contains an embedded compact connected hypersurface N such that $0 < \beta(N) < +\infty$. If either N is a one-sided hypersurface or $\pi_1(M,N) \neq 0$, then M is compact.

Remark. After the author had finished the manuscript, he found Galloway's paper [6]. Galloway obtained there similar results to Theorems 5.1 and 6.3 in this paper. However, in [6] Galloway considered only minimal hypersurfaces.

References

[1] E. Calabi, An extension of E. Hopf's maximum principle with an application to Riemannian geometry, Duke Math. J., 25(1957), 45-56.

[2] J. Cheeger and D. G. Ebin, Comparison Theorems in Riemannian Geometry, North Holland, Amsterdam, 1975.

[3] J. Cheeger and D. Gromoll, The splitting theorem for manifolds of nonnegative Ricci curvature, J. Diff. Geom., 6(1971), 119-128.

[4] J. H. Eschenburg and E. Heintze, An elementary proof of the Cheeger-Gromoll Splitting Theorem, Ann. Glob. Analysis and Geometry, 2(1984), 141-151.

[5] T. Frankel, On the fundamental group of a compact minimal submanifold, Ann. of Math., 83(1966), 68-73.

[6] G. J. Galloway, A note on the fundamental group of a compact minimal hypersurface, Pacific J. Math., 126 (1987), 243-251.

[7] R. S. Hmilton, Four-manifolds with positive curvature operator, J. Diff. Geom., 24(1986), 153-179.

[8] E. Hopf, Elementare Bemerkungen über die Lösungen partieller Differentialgleichungen zweiter Ordnung vom elliptischen Typus, Sitzungsberichte, Preussische Akademie der Wissenschaften, 19(1927), 147-152.

[9] R. Ichida, On Riemannian manifolds of nonnegative Ricci curvature containing compact minimal hypersurfaces, Advanced Studies in Pure Mathematics, 3(1984), 473-485.

[10] R. Ichida, On manifolds of nonnegative Ricci curvature and an extension of Myers'theorem, Yokohama Math. J., 34 (1986), 73-81.

[11] A. Kasue, Ricci curvature, geodesics and some geometric properties of Riemannian manifolds with boundary, J. Math. Soc. Japan, 35(1983), 117-131.

[12] B. Lawson, The unknottedness of minimal embeddings,

Invent. Math., 11(1970), 183-187.

[13] W. Meeks III, L. Simon and S. T. Yau, Embedded minimal surface, exotic spheres, and manifolds with positive Ricci curvature, Ann. of Math., 116(1982), 621-659.

[14] K. Shiohama, Topology of complete noncompact manifolds, Advance Studies in Pure Mathematics, 3(1984), 423-450.

[15] H. Wu, An elementary method in the study of nonnegative curvature, Acta Math., 142(1979), 57-78.

Ryosuke Ichida

Department of Mathematics
Yokohama City University
22-2 Seto Kanazawa-ku
Yokohama, 236 Japan

p-convex domains in $\mathbf{R}^n$

Jin-ichi ITOH

1. Introduction

Let D be a compact domain in $\mathbf{R}^n$ with smooth boundary. It is well known that if D is convex, then D is a disk. On the other hand it follows from the works of J.Cheeger, D.Gromoll, W.Meyer (cf.[1,2]) that a compact positively curved manifold with convex boundary is a disk.

Recentry J.Sha defined p-convex boundary by the sum of any p principal curvatures at each point of boundary is positive and he showed that if n-dimensional manifold M carries a riemannian metric with nonnegative sectional curvature and p-convex boundary $(1 \leq p \leq (n-1))$, then M has the homotopy type of CW-complex of dimension $\leq (p-1)$ (cf.[4]). Although this result is sharp, it seems that for a long time p-convexity generally means that at least $(n-p)$ eigenvalues of the second fundamental form (or the Levi form) are positive.

In this paper we study only compact domains in $\mathbf{R}^n$ and generalize the above well known result.

Definition. *For any integer p, $1 \leq p \leq (n-1)$, D is called a p-convex domain if at least $(n-p)$ principal curvatures (with respect to the inward normal vector) at each point of ∂D are positive.*

Note that 1-convex domain usually means convex domain and this definition of p-convex is weaker than J.Sha's.

Theorem. *If D is a p-convex domain, then D has the homotopy type of CW-complex of dimension $\leq (p-1)$.*

We will first consider an example, a solid torus. Put $ST(k) := S^k \times B^{n-k}$, where $1 \leq k \leq (n-1)$, S^k is a k-dimensional canonical sphere and B^{n-k} is a $(n-k)$-dimensional canonical disk. In this case $ST(k)$ is $(k+1)$-convex and $ST(k)$ is homotopic to S^k (i.e. has the homotopy type of CW-complex of dimension $\leq k$). In view of the above example our result is optimal.

To prove Theorem we will use some kind of Morse theory.

ISBN 0-12-640170-5

2. Proof of Theorem

For any point $y \in \mathbf{R}^n \setminus D$, define the function $L_y : D \to \mathbf{R}^n$ by $L_y(x) := |x-y|^2$. We can take $y \in \mathbf{R}^n \setminus D$ such that $L_y|_{\partial D} : \partial D \to \mathbf{R}^n$ has only nondegenerate critical points (See p.36 in [3].). Let v be the vector field on D defined by

$$v(x) := grad(L_y)(x)/|grad(L_y)(x)|.$$

At each critical point q of $L_y|_{\partial D}$, the inward unit normal vector n_q coinsides with $v(q)$ or $-v(q)$. We call a critical point q of $L_y|_{\partial D}$ *a positive critical point* (resp. *a negative critical point*) if $n_q = v(q)$ (resp. $n_q = -v(q)$). From easy consideration it follows that the index of a positive critical point of $L_y|_{\partial D}$ is less than or equal to the number of negative principal curvatures. Hence under the assumption of Theorem the index of a positive critical point of $L_y|_{\partial D}$ is less than or equal to $(p-1)$. Morse functions form an open dense subset of $C^\infty(M)$ with C^2-topology. Then without loss of generality we can assume that each critical value of $L_y|_{\partial D}$ is different.

From now on we simply denote L_y by L. Put

$$D^a := L^{-1}[0,a](= \{x \in D | L(x) \leq a\}).$$

Lemma. (i) *If there is no critical point of $L|_{\partial D}$ in $L^{-1}[a,b]$ ($a < b$), then D^a is a deformation retract of D^b.*

(ii) *Let q be a negative critical point of $L|_{\partial D}$ with $L(q) = c$. For a sufficiently small number $\varepsilon > 0$, $D^{c-\varepsilon}$ is a deformation retract of $D^{c+\varepsilon}$.*

(iii) *Let q be a positive critical point of $L|_{\partial D}$ with $L(q) = c$. If the index of q is equal to i, then for a sufficiently small number $\varepsilon > 0, D^{c-\varepsilon} \cup \{$ i-dimensional cell $\}$ is a deformation retract of $D^{c+\varepsilon}$.*

Theorem now follows from Lemma immediately.

Proof of Lemma. (i) We first define a smooth vector field on D as follows. Let g_α be a C^∞-function such that

$$g_\alpha(t) = 1 \quad t \in (-\infty, 0], \qquad 0 < g_\alpha(t) < 1 \quad t \in (0, \alpha),$$

$$g_\alpha(t) = 0 \quad t \in [\alpha, \infty), \qquad g'_\alpha(t) < 0 \quad t \in (0, \alpha).$$

Put

$$N_\delta := \{\ \exp_u(tn_u) \in D \mid 0 \leq t \leq \delta, u \in \partial D\},$$

where n_u is the inward unit normal vector at u. Define the map $\varphi : \partial D \times [0,\delta] \to N_\delta$ by $\varphi(u,t) = \exp(tn_u)$. We can take $\delta > 0$ sufficiently small such that φ is a diffeomorphism. For any $x \in N_\delta$, let u_x be the point on ∂D such that $x = \exp(tn_{u_x})$

for some $0 \le t \le \delta$. We define a vector field $X_\delta(x)$ by

$$X_\delta(x) := \begin{cases} v(= grad(L)/|grad(L)|) & x \in D \setminus N_\delta \\ grad(L|_{\partial D}) & x \in \partial D \\ g_\delta(|x - u_x|) \cdot P^x_{u_x}(grad(L|_{\partial D})(u_x)) \\ \qquad +(1 - g_\delta(|x - u_x|) \cdot v & x \in N_\delta, \end{cases}$$

where $P^x_{u_x}$ is the parallel translation along the segment $u_x x$.

Note that $\{x \in D | X_\delta(x) = 0\}$ coincides with the set of all critical points of $L|_{\partial D}$ and for any $x \in D \setminus \{$ *critical points* $\}$, $\angle(X_\delta(x), v(x)) < \pi/2$. Then X_δ is transverse to $L^{-1}(a)$ for all a's but critical values of $L|_{\partial D}$. Let $\Phi_{X_\delta}(x,t)$ be a 1-parameter group of diffeomorphisms generated by $-X_\delta$. For any $x \in D^b \setminus D^a$, we take $t_1(x)$ as the minimum value of t such that $\Phi_{X_\delta}(x,t) \in D^a$. Then the deformation retraction $\Psi_1 : D^b \times [0,1] \to D^b$ is defined by

$$\Psi_1(x,t) := \begin{cases} \Phi_{X_\delta}(x, t_1(x)t) & x \in D^b \setminus D^a \\ id & x \in D^a. \end{cases}$$

(ii) Take $\varepsilon > 0$ small enough such that there is exactly one critical point q in $D^{c+\varepsilon} \setminus D^{c-\varepsilon}$. For any critical point q of $L|_{\partial D}$ whose index is i, we can take local coordinates $(u^1, u^2, \cdots, u^{n-1})$ in a neighborhood U_q of q on ∂D such that

$$u^j(q) = 0 \qquad\qquad 1 \le j \le n-1,$$

$$L|_{\partial D}(u) = c - \sum_{j=1}^{i}(u^j(u))^2 + \sum_{j=i+1}^{n-1}(u^j(u))^2 \qquad u \in U_q$$

(See p.15 in [3].). Take a small neighborhood $U^1(\subset U_q)$ of q such that $\angle(n_u, -v(u)) < \pi/2$ for any $u \in U^1$, where n_u is the inward unit normal vector at u. For a sufficiently small number $l > 0$, we take two neighborhoods U^2, U^3 as follows;

$$U^2 := \{u \in U^1 | \ |u| < 2l\}, \qquad U^3 := \{u \in U^1 | \ |u| < l\},$$

where $|u| = \sqrt{\sum_{j=1}^{n-1}(u^j(u))^2}$. For any subset V in ∂D, put

$$V \times [0,\delta] := \{\exp_u(tn_u) \in D \mid 0 \le t \le \delta, u \in V\}.$$

For any $x \in U^2 \times [0,\delta]$, let u_x be the point on U^2 as in (i), and put $|x|_1 := |u_x| (= \sqrt{\sum_{j=1}^{n-1}(u^j(u_x))^2})$.

Now we define a new vector field Y on D by

$$Y(x) := \begin{cases} v & x \in U^3 \times [0,\delta] \\ g_l(|x|_1 - l) \cdot v + (1 - g_l(|x|_1 - l)) \cdot X_\delta & x \in (U^2 \setminus U^3) \times [0,\delta] \\ X_\delta & \text{otherwise.} \end{cases}$$

Note that for any $x \in D^{c+\varepsilon} \setminus D^{c-\varepsilon}$, Y is transverse to level surfaces of L. Let $\Phi_Y(x,t)$ be a 1-parameter group of diffeomorphisms generated by $-Y$. For any $x \in D^{c+\varepsilon} \setminus D^{c-\varepsilon}$, we take $t_2(x)$ as the minimum value of t such that $\Phi_Y(x,t) \in D^{c-\varepsilon}$. Note that $t_2(x)$ is a continuous function. Then there is a deformation retraction $\Psi_2 : D^{c+\varepsilon} \times [0,1] \to D^{c+\varepsilon}$ defined by

$$\Psi_2(x,t) := \begin{cases} \Phi_Y(x, t_2(x)t) & x \in D^{c+\varepsilon} \setminus D^{c-\varepsilon} \\ \\ id & x \in D^{c-\varepsilon}. \end{cases}$$

Hence $D^{c-\varepsilon}$ is a deformation retract of $D^{c+\varepsilon}$.

(iii) Take local coordinates $(u^1, \cdots, u^{n-1})$ in a neighborhood U_q of q on ∂D as in (ii). Take a small neighborhood U^4 of q in U_q such that for any $u \in U^4$, $\angle(n_u, v(u)) < \pi/2$. We can take a sufficiently small number $\varepsilon > 0$ such that at each point on $\{u \in U^4 \setminus D^{c-\varepsilon} \mid u^j(u) = 0, i+1 \le j \le n-1\}$, (Hessian of $L|_{\partial D}$)$(\partial u^j, \partial u^j) > 0$ for any $i+1 \le j \le n-1$. For positive numbers m, m', set

$$U_{m,m'} := \left\{ u \in U^4 \;\middle|\; \begin{array}{c} \sqrt{\sum_{j=i+1}^{n-1}(u^j(u))^2} < m, \\ \\ \sqrt{\sum_{j=1}^{i}(u^j(u))^2} < m' + \sqrt{m^2 - \sum_{j=i+1}^{n-1}(u^j(u))^2} \end{array} \right\}.$$

We can find positive numbers m, m' such that

(1) $$U_{m,m'} \subset D^{c+\varepsilon}$$

(2) $$\{u \in U_{m,m'} | \sqrt{\textstyle\sum_{j=1}^{i}(u^j(u))^2} > m'\} \subset D^{c-\varepsilon}$$

(3) for any $i+1 \le j \le n-1$, (Hessian of $L|_{\partial D}$)$(\partial u^j, \partial u^j) > 0$ on $U_{m,m'}$.

For the above m and m', put $U^5 := U_{m,m'}$. Note that $grad(L|_{\partial D})$ is transverse to $\partial(U^5 \setminus D^{c-\varepsilon})$. We can take a sufficiently small number $\delta > 0$ such that

(1) $$U^5 \times [0,\delta] \subset D^{c+\varepsilon}$$

(2) $$\{u \in U^5 | \sqrt{\textstyle\sum_{j=1}^{i}(u^j(u))^2} > m'\} \times [0,\delta] \subset D^{c-\varepsilon},$$

where $U^5 \times [0,\delta]$ is defined as in (ii). For such δ, we again define a vector field X_δ on D and a 1-parameter group of diffeomorphisms Φ_{X_δ} as in (i). Note that v is transverse to $\partial(U^5 \times [0,\delta]) \setminus D^{c-\varepsilon}$. Then X_δ is transverse to $\partial((U^5 \times [0,\delta]) \cup D^{c-\varepsilon})$. Put $D_1^c := D^{c-\varepsilon} \cup (U^5 \times [0,\delta])$. For any $x \in D^{c+\varepsilon} \setminus D_1^c$, we take $t_3(x)$ as the minimum value of t such that $\Phi_{X_\delta}(x,t) \in D_1^c$. $t_3(x)$ is a continuous function from the transversalities. Then there is a deformation retraction $\Psi_3 : D^{c+\varepsilon} \times [0,1] \to D^{c+\varepsilon}$ defined by

$$\Psi_3(x,t) = \begin{cases} \Phi_{X_\delta}(x, t_3(x)t) & x \in D^{c+\varepsilon} \setminus D_1^c \\ \\ id & x \in D_1^c. \end{cases}$$

Thus D_1^c is a deformation retract of $D^{c+\varepsilon}$. Put $D_2^c := D^{c-\varepsilon} \cup U^5$. Then it is obvious that D_2^c is a deformation retract of D_1^c. Moreover it holds that $D^{c-\varepsilon} \cup \{$ *i-dimensional cell* $\}$ is a deformation retract of D_2^c (See p.18-19 in [3].). Hence the proof of Lemma is completed.

References

1. Cheeger, J., Gromoll, D.: On the structure of complete open manifolds of nonnegative curvature, Ann. Math. 96, 413-443 (1972)
2. Gromoll, D., Meyer, W.: On complete open manifolds of positive curvature, Ann. Math. 90, 75-90 (1969)
3. Milnor, J.: Morse theory, Princeton Univ. Press, Princeton, 1975
4. Sha, J.: p-convex Riemannian manifolds, Invent. Math. 83, 437-447 (1986)

Jin-ichi Itoh

Kyushu Institute of Technology
Iizuka 820, Japan

GEODESIC SPHERES AND POLES

MASAO MAEDA

0. Introduction.

Let M be a 2-dimensional complete non-compact Riemannian manifold with non-negative Gaussian curvature K. Then in [6], we showed that the subset $P \subset M$ of points which are poles in M is compact if $\int_M K\,dv > 0$. Here a point $p \in M$ is called a pole when the exponential mapping from the tangent space T_pM of M at p to M is of maximal rank at every point of T_pM. dv denotes the volume element of M and hence $\int_M K\,dv =: C(M)$ is the total curvature of M.

If M satisfy further strong condition that $C(M) = 2\pi$ and $K \equiv 0$ outside some compact set, then it was proved that the number of elements of P is at most one, i.e. there exists at most one pole on M.

In [7], we generalized this result as follows ;

Fact. Let M be a 2-dimensional complete non-compact Riemannian manifold with non-negative Gaussian curvature. If there exist a constant L such that for some point $p \in M$

diameter of $S_t(p) \leqq L$ for all $t \geqq 0$,

then there exists at most one pole on M.

Here $S_t(p) := \{q \in M : d(p,q)=t\}$ is the geodesic circle with radius t

ISBN 0-12-640170-5

centerd at p. d denotes the distance function on M induced from the Riemannian metric on M.

As is easily seen, the condition mentioned in the above fact is satisfied for a manifold satisfying the conditions that $C(M) = 2\pi$ and $K \equiv 0$ outside some compact set.

The purpose of this note is to give a further generalization of the Fact mentioned above.

1. Diameter of geodesic spheres.

Let M be a complete non-compact Riemannian manifold. For a point $p \in M$ and a non-negative number t, geodesic ball with radius t centerd at p is denoted by $B_t(p) := \{q \in M : d(p,q) \leqq t\}$. So, $\partial B_t(p) = S_t(p)$ is the geodesic sphere with radius t centerd at p. For each t, let

$$D_t(p) := \text{diameter of } S_t(p) \; (= \sup \{d(q,q') : q,q' \in S_t(p)\}).$$

We now consider the following case that for some point $p_o \in M$

$$\overline{\lim_{t\to\infty}} \, (D_t(p_o)^2/t) < \infty$$

In this case, we have

Proposition 1. For all point $p \in M$,

$$\overline{\lim_{t\to\infty}} \, (D_t(p)^2/t) = \overline{\lim_{t\to\infty}} \, (D_t(p_o)^2/t).$$

Proof. This is essentially proved in [7;see Prop.1]. Fix another point $p \in M$ and let $\{t_i\}$, $t_i \uparrow \infty$ be a sequence such that $\lim_{i\to\infty} (D_{t_i}(p)^2/t_i) = \overline{\lim_{t\to\infty}} \, (D_t(p)^2/t)$. For each i, let $m_i, n_i \in S_{t_i}(p)$ are points such that

$$d(m_i,n_i) = D_{t_i}(p).$$

Put $t_o = d(p_o, p)$. Without loss of generality, we can assume that

$$d(p_o, n_i) \leqq d(p_o, m_i), \quad i=1,2,\cdots .$$

By triangle inequality,

$$t_i' := d(p_o, m_i) \leqq t_i + t_o ,$$

$$t_i'' := d(p_o, n_i) \geqq t_i - t_o .$$

So

$$t_i - t_o \leqq t_i'' \leqq t_i' \leqq t_i + t_o$$

and

$$t_i' - t_i'' \geqq 2t_o .$$

Since $t_i' > t_i''$, there is a point $m_i' \varepsilon S_{t_i''}(p_o)$ on a minimizing geodesic joining p_o to m_i such that $d(m_i, m_i') = t_i' - t_i''$. Since n_i is on $B_{t_i''}(p_o)$, we see that $d(m_i, n_i) \leqq 2t_o + D_{t_i''}(p_o)$, and this proves

$$D_{t_i}(p) \leqq 2t_o + D_{t_i''}(p_o).$$

Making use of the above inequality and $t_i'' \leqq t_i + t_o$, we have $D_{t_i}(p)^2/t_i \leqq 4t_o^2/t_i + 4t_o(D_{t_i''}(p_o)/t_i'')(1 + t_o/t_i) + (D_{t_i''}(p_o)^2/t_i'')(1 + t_o/t_i)$. Since $t_i'' \to \infty$ as $i \to \infty$, we observe

$$\lim_{i\to\infty} (D_{t_i}(p)^2/t_i) \leqq 4t_o \cdot \overline{\lim_{i\to\infty}} (D_{t_i''}(p_o)/t_i'') + \overline{\lim_{i\to\infty}} (D_{t_i''}(p_o)^2/t_i'')$$

$$\leqq 4t_o \cdot \overline{\lim_{t\to\infty}} (D_t(p_o)/t) + \overline{\lim_{t\to\infty}} (D_t(p_o)^2/t).$$

Suppose that $\overline{\lim}_{t\to\infty} (D_t(p_o)/t)>0$. Then there exists a monotone divergent sequence $\{t_i\}$ such that $\overline{\lim}_{t\to\infty} (D_t(p_o)/t) = \lim_{i\to\infty} (D_{t_i}(p_o)/t_i)$ and $D_{t_i}(p_o) \uparrow \infty$ as $i \to \infty$. For any small positive ε, there exists a number $i_o = i_o(\varepsilon)$ such that for all $i \geqq i_o$, $1/D_{t_i}(p_o)<\varepsilon$. Thus

$$\overline{\lim}_{t\to\infty} (D_t(p_o)/t) = \lim_{i\to\infty} (D_{t_i}(p_o)/t_i) = \lim_{i\to\infty} (D_{t_i}(p_o)^2/t_i)(1/D_{t_i}(p_o))$$

$$< \varepsilon\, \overline{\lim}_{t\to\infty} (D_t(p_o)^2/t).$$

Since ε is arbitrary small and since we have assumed that $\overline{\lim}_{t\to\infty} (D_t(p_o)^2/t)$ is bounded, a contradiction is derived by the above inequality. Therefore we have

$$\overline{\lim}_{t\to\infty} (D_t(p_o)/t) = 0.$$

The above computations show that

$$\overline{\lim}_{t\to\infty} (D_t(p)^2/t) \leqq \overline{\lim}_{t\to\infty} (D_t(p_o)^2/t) < \infty .$$

By replacing p to p_o in the above argument, we conclude the proof of Proposition 1. Q.E.D.

This Proposition says that the number

$$d_o := \overline{\lim}_{t\to\infty} (D_t(p)^2/t) \leqq \infty$$

does not depend on the choice of the point which was used to define it and thus $d_o = d_o(M)$ is a constant determined by the Riemannian structure of M.

2. Relationship between d_o and the set of poles.

In this section, we will give a meaning of the constant d_o defined in section 1.

Let M be a complete simply connected non-compact Riemannian manifold with non-negative sectional curvature. Let $p \in M$ be a pole. Then the exponential mapping $\exp_p : T_p(M) \to M$ is a diffeomorphism. Thus, M is diffeomorphic to an n-dimensional Euclidean space E^n and each geodesic $c : [0,\infty) \to M$ starting from p is a ray. Here geodesic $c : [0,\infty) \to M$ is called a ray if any subarc of c is a shortest connection between its end points. In this note, let all geodesics have arc-length as their parameter.

Now, we consider the case $d_o < \infty$. Let $c : [0,\infty) \to M$ be a ray from p. Let $\varepsilon > 0$ be an arbitraly fixed small number and $\{t_i\} \uparrow \infty$ be a sequence such that $d_o = \overline{\lim_{t\to\infty}} (D_t(p)^2/t) = \lim_{i\to\infty} (D_{t_i}(p)^2/t_i)$.
Then there exists a number i_o such that

$$(D_{t_i}(p)^2/t_i) - d_o < \varepsilon \quad \text{for all } i \geq i_o.$$

Fix a number $i \geq i_o$ and a point $q \in S_{t_i}(p) = \partial B_{t_i}(p)$, $q \neq c(t_i)$. Let $b : [0, d(c(t_i),q)] \to M$ be a shortest geodesic from $c(t_i)$ to q and $u := \measuredangle(\dot{b}(0), -\dot{c}(t_i))$, the angle between the vectors $\dot{b}(0)$ and $-\dot{c}(t_i)$. Let $\Delta_i = \Delta_i(b, c|[0,t_i],a)$ be the geodesic triangle with sides $b, c|[0,t_i], a$ where $a : [0,d(p,q)] \to M$ is the shortest geodesic from p to q. Then, by Toponogov's Comparison Theorem(see [5;pp 183∿]), there exists a triangle $\tilde{\Delta}_i(A,B,C)$ in a Euclidean plane E^2 such that

$$AB = d(p,c(t_i)) = t_i, \ AC = d(c(t_i),q), \ BC = d(p,q) = t_i$$

and

$$u \geq \angle A.$$

We then have

$$\cos\angle A = AC/(2AB) \leq D_{t_i}(p)/(2t_i).$$

<u>Lemma 1</u>. For all point $p\varepsilon M$ and for all $t>0$,

$$D_t(p)/t \leq 2.$$

<u>Proof</u>. For $q', q'' \varepsilon S_t(p)$ such that $D_t(p)$ = diameter of $S_t(p) = d(q', q'')$, we have

$$D_t(p) \leq d(p,q') + d(p,q'') = 2t. \qquad \text{q.e.d.}$$

Thus, from Lemma 1,

$$\cos^{-1}(D_{t_i}(p)/(2t_i)) \leq \angle A \leq u.$$

So, we have

$$(*) \qquad \pi - u \leq \pi - \cos^{-1}(D_{t_i}(p)/(2t_i)).$$

Now, for a fixed point $c(t'), t' > t_i$, consider the geodesic triangle $\Delta_{t'}(b, c|[t_i, t'], e)$ where $e : [0, d(q, c(t')] \to M$ is a shortest geodesic from q to $c(t')$. Then, by Toponogov's Comparison Theorem, there exists a triangle $\tilde{\Delta}_{t'}(A,B,C)$ in a Euclidean plane E^2 such that

$$AB = d(c(t_i), c(t')) = t' - t_i,\ AC = d(q, c(t_i)),\ BC = d(q, c(t'))$$

and

$$\measuredangle(\dot{c}(t_i), \dot{b}(0)) \geq \angle A.$$

From (*) and $\measuredangle(\dot{c}(t_i),\dot{b}(0)) = \pi - u$, we have

$$\pi - \cos^{-1}(D_{t_i}(p)/(2t_i) \geq \angle A.$$

Cosine rule implies that

$$BC \leq (AB^2 + AC^2 + AB\cdot AC\cdot D_{t_i}(p)/t_i)^{1/2}.$$

<u>Lemma 2</u>. For any t,t' with $t<t'$ and for $q \in S_t(p)$, we have

$$d(q,c(t')) \geq d(c(t_i),c(t')) = t' - t_i$$

where equality holds if and only if $q=c(t')$.

<u>Proof.</u> If $d(q,c(t')) \leq d(c(t_i),c(t'))$, then

$$t' = d(p,c(t')) \leq d(p,q) + d(q,c(t')) \leq t_i + d(c(t_i),c(t')) = t'.$$

Thus

$$d(p,c(t')) = d(p,q) + d(q,c(t')).$$

This equality holds if and only if $q \in c([0,\infty))$, because c is a ray.

q.e.d.

An upper bound for BC - AB is obtained as follows ;

$$BC - AB \leq [AC^2 + AB\cdot AC\cdot D_{t_i}(p)/t_i][(AB^2 + AC^2 + AB\cdot AC\cdot D_{t_i}(p)/t_i)^{1/2} + AB]^{-1}$$

$$\leq [AC^2 + AB\cdot AC\cdot D_{t_i}(p)/t_i](2AB)^{-1} \leq AC^2/(2AB) + D_{t_i}(p)^2/(2t_i)$$

$$\leq AC^2/(2AB) + (d_o + \varepsilon)/2.$$

By choosing t' sufficiently large, we may assume that

$$AC^2/(2AB) = d(q,c(t_i))/[2(t_i'- t_i)] < \varepsilon/2,$$

and hence we have $BC - AB < d_o/2 + \varepsilon$. This inequality means that

$$0 < d(q,c(t')) - d(c(t_i),c(t')) < d_o/2 + \varepsilon.$$

From above inequality we observe that q is contained entirely in the open ball with radius $t'- t_i + d_o/2 + \varepsilon$ centered at c(t'). Therefore by setting $C_t := \overline{\bigcap_{s \geq o} \{M - B_s(c(t+s))\}}$, we observe that $q \notin C_{t_i - d_o/2-\varepsilon}$, i.e. $S_{t_i}(p) \cap C_{t_i - d_o/2-\varepsilon} = \emptyset$. Thus we have

$$(**) \quad C_{t_i - d_o/2-\varepsilon} \subset B_{t_i}(p).$$

As was stated in [1], the family of subsets $\{C_t\}_{t \geq o}$ has the following properties ;

$$\text{(i)} \quad p \in \partial C_o \; , \; \text{(ii) if } t < t', \text{ then } C_t \subset C_{t'}$$

$$\text{(iii)} \quad C_t = \{q \in C_{t'} \mid d(q,\partial C_{t'}) \geq t'- t\} \quad \text{for } t < t'.$$

The family of subsets $\{B_t(p)\}_{t \geq o}$ also satisfies the same properties as (i) and (ii). And since p is a pole, the same property as (iii) is satisfied for $\{B_t\}_{t \geq o}$. That is

$$B_t(p) = \{q \in B_{t'}(p) \mid d(q,\partial B_{t'}(p)) \geq t'- t\} \quad \text{for } t < t'.$$

This can be checked easily. It follows from (**) that

$$C_o \subset B_{d_o/2+\varepsilon}.$$

Since ε is arbitraly, we have

Proposition 3.

$$C_o \subset B_{d_o/2}(p).$$

As a corollary to Proposition 3, we have

Corollary 4.

diameter of $C_o \leqq d_o$.

With these preparations, we prove the

Theorem. Let M be a complete non-compact Riemannian manifold with non-negative sectional curvature. Then

diameter of $P \leqq d_o$

where P is the set of all poles in M and $d_o = d_o(M^*)$ is the number determined by the universal covering manifold M* of M.

Proof. Let $f : M^* \to M$ be the universal covering of M and put $P^* := f^{-1}(P)$.

Lemma 5. P* coinside with the set of all poles in M*.

Proof. By definition, $p \in M$ is a pole when and only when on each geodesic $c : [0,\infty) \to M$ starting from p, there exists no conjugate points of p along c. $c(t)$ $(t > 0)$ is called a conjugate point of p along c if there exists a Jacobi field $Y \not\equiv 0$ along c such that $Y(0) = 0$ and $Y(t) = 0$. Here a vector field Y along c is called a Jacobi field if it satisfy the Jacobi equation $Y'' + R \circ Y = 0$ ($''$ is the covariant derivative along c and R is the curvature tensor of M). And, since f is a local isometry two solutions of the Jacobi equation along c and the Jacobi equation along a lifted geodesic $c^*: [0,\infty) \to M^*$ of c with same initial values coinside. Thus p is a pole in M if and only if each point $p^* \in f^{-1}(p)$ is a pole in M^*. q.e.d.

Lemma 6. For any subset $A \subset M$,

$$\text{diameter of } A \leqq \text{diameter of } A^* := f^{-1}(A).$$

Proof. Let $p,q \in A$ are points such that $d(p,q)$ = diameter of A and $c : [0,d(p,q)] \to M$ be a shortest geodesic from p to q. Then a lift $c^*: [0, d(p,q)] \to M^*$ of c is a shortest connection from $c^*(0) \in A^*$ to $c^*(d(p,q)) \in A^*$. q.e.d.

From Lemma 5,6, it sufficies to prove that

$$\text{diameter of } P^* \leqq d_o.$$

Since M^* is simply connected, we can use the result of Corollary 4. Let $p^*,q^* \in P^*$ and $c^*: [o,\infty) \to M^*$ be the ray such that $c^*(0) = p^*$ and $c^*(d(p^*,q^*)) = q^*$. Then the geodesic $c_1^* : [0,\infty) \to M^*$ defined by $c_1^*(t) = c^*(t+d(p^*,q^*))$ is also a ray from q^*. Then from Corollary 4, we have

(***) diameter of $C_o \leqq d_o$

By definition, $C_o = \overline{\bigcap_{t\geq o} (M^* - B_t(c_1^*(t))} = \overline{\bigcap_{t\geq o} (M^* - B_t(c^*(t+d(p^*,q^*)))}$.

Thus

$$c^*([0,d(p^*,q^*)]) \subset C_o.$$

Since c* is a shortest geodesic, this shows that

$$d(p^*,q^*) \leqq \text{diameter of } C_o \quad \text{for all points } p^*,q^* \in P^*.$$

Thus by (***), we have

$$\text{diameter of } P^* \leqq d_o. \qquad \text{Q.E.D.}$$

3. Examples.

Let M be a complete non-compact rotation surface in a 3-dimensional Euclidean space E^3 with non-negative Gaussian curvature K whose axis of rotation is z-axis and whose generating line in xz-plane is given by $z = f(x)$ for $x \geq 0$. We assume $f(0) = 0$. Since Gaussian curvature of M is non-negative $f' \geqq 0$ and $f'' \geqq 0$. Obviously original point $p_o := (0,0,0) \in M$ is a pole. Using this point p_o we compute the constant d_o. Let t be the distance from p_o to $q = (x,0,f(x))$. Then $t = \int_o^x (1 + f'(x)^2)^{1/2} dx$ and $D_t(p_o) \leqq 2\pi x$. Thus

$$D_t(p_o)^2/t \leqq 4\pi^2 x^2 / \int_o^x f'(x)\, dx = 4\pi^2 x^2/f(x).$$

Thus letting $t \to \infty$(then $x \to \infty$), we have

Proposition 7.

$$d_o = \lim_{t\to\infty} (D_t(p_o)^2/t) \leqq 4\pi^2 \cdot \lim_{x\to\infty} (x^2/f(x)).$$

Example 1 (elliptic paraboloid of revolution).

If M is given by the equation $z = a(x^2 + y^2)$, ($a>0$ is a constant), then the function of generating line is $z = ax^2$. Thus

$$d_o \leqq \lim_{x\to\infty} 4\pi^2 x^2/(ax^2) = 4\pi^2/a$$

and hence all poles of elliptic paraboloid of revolution are contained in $B_{4\pi^2/a}(p_o)$. Note that for a elliptic paraboloid of revolution, sharper result is obtained in [4;pp 258 example 6].

Example 2. If the function $f(x)$ of generating line of M coinside with the function $z = kx^n$, (k and $n>2$ are constants) on some interval $[x_o, \infty)$, $x_o>0$. Then

$$d_o \leqq \lim_{x\to\infty} 4\pi^2 x^2/f(x) = 0$$

for any choice of f on $[0, x_o]$ under the restriction $K \geqq 0$. Thus $p_o = (0, 0, 0) \in M$ is only one pole on M.

REFERENCES

[1] J.Cheeger and D.Gromoll : On the Structure of Complete Manifolds of Non-negative Curvature, Ann. of Math.,96(1972),413-443.

[2] ________ : The Splitting Theorem for Manifolds of Non-negative Ricci Curvature, Jour.Diff.Geom.,6(1971),119-128.

[3] S.Cohn-Vossen : Kürzeste Wege und Total Krümmung auf Flächen, Compositio Math.,2(1935),63-133.

[4] P.Do Carmo : Differential Geometry of Curves and Surfaces, Prentice-Hall Inc.,1976.

[5] D.Gromoll,W.Klingenberg and W.Meyer : Riemannsche Geometrie im Grossen, Springer-Verlag,1968.

[6] M.Maeda : A Note on the Set of Points which are Poles, Science Rep. of Yokohama National University, Sec.I,No.32(1985),1-5.

[7] ______ : Geodesic Circles and Total Curvature, to appear in Yokohama Math. Jour..

Masao MAEDA

Department of Mathematics

Faculty of Education

Yokohama National University

156 Tokiwadai,Hodogaya-ku,

Yokohama 240 Japan

Topology of Complete Open Manifolds with non-negative Ricci curvature

Yukio OTSU

1. Introduction

In some cases, restrictions of differential geometric quantities of Riemannian manifolds give restriction on topological structure of the manifolds. For example the classical sphere theorem states that a connected, complete and simply connected Riemannian n-manifold M is homeomorphic to the n-dimensional unit sphere $S^n(1)$, if its sectional curvature K_M satisfies $\frac{1}{4}<K_M\leq 1$. Recently in [7] Shiohama, Yamaguchi and the author present a new sphere theorem which states that a connected complete Riemannian manifold M is diffeomorphic to $S^n(1)$ if $K_M\geq 1$ and the volume of M is nearly equal to that of $S^n(1)$. We have also its real projective analogue. The results are strengthened to replace $K_M\geq 1$ by Ricci curvature $Ric_M \geq n-1$ and $K_M\geq -\Delta$ for any constant Δ in [9](see also [8]). In this article we introduce a class of complete open Riemannian manifolds as an analogue of the above theorems, and show that they are diffeomorphic to the Euclidean n-space $\mathbb{R}^n$.

Let M be a complete open Riemannian manifold with Ricci curvature $Ric_M\geq(n-1)\kappa$ for some constant $\kappa\leq 0$. By Bishop-Gromov's inequality there exists a non-negative constant

$$\lim_{r\to\infty}\frac{Vol(B(p,r))}{b_\kappa^n(r)}\leq 1$$

for $p\in M$, where $B(p,r)$ is the metric r-ball around p and $b_\kappa^n(r)$ is the volume of an r-ball in the n-dimensional space form with constant curvature κ. We denote it by $V_\kappa(p,M)$. If $\kappa=0$, $V_0(p,M)$ dose not depend on the choice of $p\in M$, so we denote it

ISBN 0-12-640170-5

briefly by $V_0(M)$.

By the Bishop-Gromov comparison theorem we have the following characterization in the maximal case:

Theorem 1.1. *Let* M *be a complete open manifold with Ricci curvature* $Ric_M \geq (n-1)\kappa$. *If* $V_\kappa(p,M)=1$, *then* M *is isometric to the* n*-dimensional space form with constant curvature* κ.

We say M is *an asymptotically non-negatively curved manifold by* λ *around* p_0 *with* C_0, if there is a non-negatively non-increasing function $\lambda:[0,\infty)\to\mathbb{R}$ such that

$$C_0:=\int_0^\infty r\lambda(r)dr<\infty,$$

and that the sectional curvature K_M is bounded from below by

$$K_M|_q \geq -\lambda(d(p_0,q)),$$

for any $q\in M$ (Abresch [1]). When the situation is clear, we simply say asymptotically non-negatively curved manifold with C_0, and so on.

For a point $p\in M$ the contractibility radius at p is the supremum of r such that the function $d_p(q):=d(p,q):B(r,p)\to\mathbb{R}$ is non-critical, that is, for any $q\in B(r,p)\setminus\{p\}$ there exists a vector $u\in T_qM$ such that for the initial vector v of any minimal geodesics from q to p the angle $\sphericalangle(u,v)$ is greater than $\frac{\pi}{2}$.

We extend the above theorem as follows:

Theorem 1.2. *There exists a positive constant* $\varepsilon=\varepsilon(n,C_0)$ *such that if* M *is a Ricci non-negative, asymptotically non-negatively curved manifold around* p_0 *with* C_0 *and*

$$V_0(M)\geq 1-\varepsilon,$$

then the contractibility radius at p_0 *is infinite. In particular,* M *is diffeomorphic to* $\mathbb{R}^n$.

Remark. (1) Since there is no curvature bound from above, the injectivity radius of M is not estimated in Theorem 1.2.

(2) If M is a complete open manifold with non-negative sectional curvature, more explicit results are known as follows: if $K_M>0$, then M is diffeomorphic to $\mathbb{R}^n$([5]). If $K_M\geq 0$ and $V_0(M)>0$, then M is diffeomorphic to $\mathbb{R}^n$([4]).

(3) Recently Abresch-Gromoll [2] give another criterion to restrict the differential structures on some class of complete open manifolds.

Acknowledgement. The author thanks Prof. K. Shiohama earnestly for his constant help and encouragement. The author is also indebted to K. Cho, T. Shioya and T. Yamaguchi.

2. Proof of Theorem

We first consider the measure of rays. A ray is, by definition, a unit speed geodesic $\gamma;[0,\infty)\to M$ satisfying $d(\gamma(t),\gamma(s))=|t-s|$ for all $s,t\in[0,\infty)$. For a fixed point $p\in M$ let A_p be the set of all unit initial vectors of rays emanating from p. We denote by $m(\cdot)$ the measure on the unit sphere U_pM induced from the metric on M. The set A_p is measurable. From an easy volume comparison argument [6] we can estimate the measure of rays of a manifold M with $Ric_M\geq(n-1)\kappa$ for a constant $\kappa\leq 0$ as follows:

$$m(A_p)\geq V_\kappa(M)\times vol(S^{n-1}) \qquad (1)$$

for any $p\in M$.

Remark. 1. If $K_M\geq 0$ and $V_0(M)>\frac{1}{2}$, then for any point p in M we have $m(A_p)>\frac{1}{2}vol(S^{n-1})$. For any point $q\neq p$ let γ be a minimal geodesic from p to q. It follows from $m(A_p)>\frac{1}{2}vol(S^{n-1})$ that there is a ray σ emanating from p with the angle $\measuredangle(\dot\gamma(0),\dot\sigma(0))<\frac{\pi}{2}$. From Toponogov's comparison theorem the angle between any minimal geodesic γ' from q to p and a ray σ' from q which is asymptotic to the ray σ is bounded from below by $\frac{\pi}{2}$ (more precise description is found in the proof of Theorem 1.2.) Hence for any p the contractibility radius at p is infinite. Note that if $K_M \geq 0$ and $V_0(M) < \frac{1}{2}$, then there is no estimate of contractibility radius. In fact a series of surfaces constructed from a cone with vertical angle $<\pi$ gives us an example. We extend this argument to the proof of Theorem 1.2.

We recall the generalized Toponogov comparison theorem for asymptotically non-negatively curved manifolds (Abresch [1]). Let M be an asymptotically non-negatively curved manifold by λ around p_0 with C_0, where $\lambda = \lambda(r)$ is a non-increasing non-negative function such that $C_0 = \int_0^\infty r\lambda(r)dr < \infty$. Then there is the unique solution $g;[0,\infty) \to [0,\infty)$ of the following differential equation:

$$g''(r) = \lambda(r)g(r) \quad \text{with } g(0)=0 \text{ and } g'(0)=1. \tag{2}$$

Let us denote by $M(-\lambda)$ the revolution surface with the metric

$$ds^2 = dr^2 + g^2(r)d\theta^2$$

in polar coordinates (r,θ) around a base point $\bar{p}_0$ in $M(-\lambda)$. Let $\Delta = (p_0, p_1, p_2)$ be a geodesic triangle in M with vertices p_0, p_1, p_2 and edges $\gamma_0, \gamma_1, \gamma_2$, where γ_i is opposed to the vertex p_i for i=0, 1, 2 respectively. Let ℓ_i be the length of the edge γ_i and let $\sphericalangle$ *at* p_i be the angle determined by two geodesics γ_{i-1} and γ_{i+1} at p_i, where indices are taken modulo 3. Overhead bar refers to an object in $M(-\lambda)$. Let $\bar{\Delta} = (\bar{p}_0, \bar{p}_1, \bar{p}_2)$ be a geodesic triangle in $M(-\lambda)$ with $\ell_i = \bar{\ell}_i$ for i=0,1,2. Note that p_0 and $\bar{p}_0$ are base points of M and $M(-\lambda)$ respectively. Assume that γ_1 and γ_2 are minimal geodesics and $\ell_0 < \ell_1 + \ell_2$. Then the theorem states that

$$\sphericalangle \textit{ at } \bar{p}_1 \leq \sphericalangle \textit{ at } p_1 \quad \text{and} \quad \sphericalangle \textit{ at } \bar{p}_2 \leq \sphericalangle \textit{ at } p_2.$$

Note that if $\ell_0 \leq \bar{\ell}_0$, $\ell_1 = \bar{\ell}_1$, $\ell_2 = \bar{\ell}_2$ and $\sphericalangle$ *at* $\bar{p}_1 \geq \frac{\pi}{2}$, then $\sphericalangle$ *at* $\bar{p}_1 \leq \sphericalangle$ *at* p_1. In fact, if we deform $\bar{\Delta} = (\bar{p}_0, \bar{p}_1, \bar{p}_2)$ to $\tilde{\Delta} = (\bar{p}_0, \tilde{p}_1, \bar{p}_2)$ in $M(-\lambda)$ so that $d(\bar{p}_0, \tilde{p}_1) = d(\bar{p}_0, \bar{p}_1) = \ell_2$ and $d(\tilde{p}_1, \bar{p}_2) = \ell_0 \leq d(\bar{p}_1, \bar{p}_2)$, then we have

$\measuredangle$ *at* $\bar{p}_1 \le \measuredangle$ *at* $\tilde{p}_1 \le \measuredangle$ *at* p_1,

because $\measuredangle$ *at* $\bar{p}_1 \ge \frac{\pi}{2}$ and $M(-\lambda)$ is a Hadamard manifold.

To estimate the total curvature of $M(-\lambda)$ we need the following:

Lemma. 1. *Let* $g;[0,\infty)\to[0,\infty)$ *be the solution of* (2). *Then we have*

$$\int_0^\infty g(r)\lambda(r)dr \le \exp C_0 - 1.$$

Proof . Integrate the above equation (2) twice in the domain $\{(r,s)\in[0,t]\times[0,t]: r\le s\}$. Then we have

$$g(t)=t+\int_0^t (t-r)\lambda(r)g(r)dr \le t\{1+\int_0^t \lambda(r)g(r)dr\}.$$

Hence

$$\frac{\lambda(t)g(t)}{1+\int_0^t \lambda(r)g(r)dr} \le t\lambda(t).$$

Integrating the above inequality, we have

$$\log\{1+\int_0^t \lambda(r)g(r)dr\} \le \int_0^t r\lambda(r)dr \le C_0,$$

which gives us the desired estimate.

Proof of Theorem 1.2. Assume that M satisfies the conditions in Theorem 1.2. Setting $\Phi:=\pi/2\exp C_0$ we define the constant $\varepsilon=\varepsilon(n,C_0)$ in Theorem 1.2 by

$$\varepsilon(n,C_0):=\frac{b_1^{n-1}(\Phi)}{b_1^{n-1}(\pi)},$$

where $b_1^{n-1}(r)$ is the volume of an r-ball on the n−1-dimensional unit sphere S^{n-1}. If $V_0(M)>1-\varepsilon$, then there is a positive constant δ_1 such that for any point p and $u\in S_pM$ there exists a ray σ from p with $\measuredangle(u,\dot{\sigma}(0))<\Phi-2\delta_1$ because of (1).

Let p be a point of M which differs from the base point p_0 and let γ be a minimal geodesic from p_0 to p with length $\ell:=d(p_0,p)$. From the above there exists a ray σ emanating from p_0 with $\measuredangle(\dot{\gamma}(0),\dot{\sigma}(0))<\Phi-2\delta_1=:\bar{\varphi}_0$. We take two rays $\bar{\gamma}$

and $\bar{\sigma}$ from $\bar{p}_0$ with angle $\bar{\varphi}_0$ in the model space $M(-\lambda)$. Set $\bar{p}_t:=\bar{\gamma}(t)$ and $\delta:=\delta_1\exp C_o>0$. Then there exists a positive constant L such that for $\bar{q}:=\bar{\sigma}(L)$ the angle $\sphericalangle$ *at* $\bar{q}$ of triangle $\bar{\Delta}_t:=(\bar{p}_0,\bar{q},\bar{p}_t)$ is smaller than δ for any $t\in[0,\ell)$. Let $\bar{\varphi}_t:=\pi-\sphericalangle$ at $\bar{p}_t$. By the Gauss-Bonnet Theorem and Lemma 1 we have

$$\bar{\varphi}_0-\bar{\varphi}_t+\delta \geq \bar{\varphi}_0\int_0^{\infty} -\lambda(r)g(r)dr \geq -\bar{\varphi}_0(\exp C_o-1),$$

that is,

$$\bar{\varphi}_t \leq \bar{\varphi}_0 \exp C_o+\delta < \frac{\pi}{2}-\delta,$$

for any $t\in[0,\ell]$. Let σ_t be a minimal geodesic from $p_t:=\gamma(t)$ to $q:=\sigma(L)$ and $\varphi_t:=\sphericalangle(\dot{\gamma}(t),\dot{\sigma}_t(0))$ for any $t\in[0,\ell]$. Then we have:

Claim. $\qquad \varphi_t \leq \bar{\varphi}_t \qquad$ *for any* $t\in[0,\ell]$.

In particular, $\qquad \varphi_\ell < \frac{\pi}{2}-\delta.$

Proof. Put $T:=\sup\{\tau: L(\sigma_t)<L(\bar{\sigma}_t)$ for any $t\in(0,\tau)\}$, where $L(*)$ indicates the length of a curve. It follows from the first variation formula that $T>0$. Suppose that $T\leq\ell=d(p_0,p)$. Then $L(\sigma_t)<L(\bar{\sigma}_t)$ for any $t\in(0,T)$ and

$$L(\sigma_T)=L(\bar{\sigma}_T).$$

From the note of the generalized Toponogov theorem we have $\varphi_t \leq \bar{\varphi}_t$ for any $t\in(0,T)$. It follows from the first variation formula that

$$L(\sigma_T)=L-\int_0^T \cos\varphi_t dt<L-\int_0^T \cos\bar{\varphi}_t dt=L(\bar{\sigma}_T),$$

which contradicts the above equation. Thus $T>\ell$ and therefore $\varphi_t \leq \bar{\varphi}_t < \frac{\pi}{2}-\delta$ for any $t\in[0,\ell]$.

For any minimal geodesic γ' from p_0 to p we consider a geodesic triangle $\Delta_{\gamma'}$ determined by three geodesics γ',σ_ℓ and σ. Since the length of each side of $\Delta_{\gamma'}$ is equal to the corresponding side of Δ_γ, we can use the same model triangle in Claim to estimate the angles of $\Delta_{\gamma'}$. Hence we have also

$$\sphericalangle(\dot{\gamma}'(\ell),\dot{\sigma}_\ell(0)) < \frac{\pi}{2}-\delta.$$

This means that for the vector $u = \dot{\sigma}_{\ell}(0) \in S_pM$ the angle between u and the initial vector of any minimal geodesic from p to p_0 is not smaller than $\frac{\pi}{2}+\delta$ and therefore contractibility radius at p_0 is infinite. Then from the above vector $u \in S_pM$ for $p \in M \setminus \{p_0\}$ we can construct a unit C^∞-vector field on $M \setminus \{p_0\}$ whose flow curve c(t) satisfies

$$d(c(t),p_0) \geq d(c(s),p_0)+(t-s)\sin\delta$$

for any $t \geq s \geq 0$. Hence the flow curves give a diffeomorphism from M to $\mathbb{R}^n$.

Finally we give a corollary which is clear from the proof of theorem 1.2.

Corollary (to Theorem 1.2) For any positive constant $\varepsilon<\frac{1}{2}$, there exists a positive constant $\delta=\delta(n,\varepsilon)$ such that if a Ricci non-negative, asymptotically non-negatively curved manifold M with C_0 satisfies $V_0(M) > \frac{1}{2} + \varepsilon$ and $C_0 < \delta$, then M is diffeomorphic to $\mathbb{R}^n$

References

[1] U. Abresch, Lower curvature bounds, Toponogov's theorem, and bounded topology I, Ann. scient. Ec. Norm. Sup. 18(1985), 651-670.

[2] U. Abresch and D. Gromoll, On complete manifolds with nonnegative Ricci curvature, preprint.

[3] J.Cheeger and D. Ebin, Comparison Theorems in Riemannian Geometry, North-Holland Math. Library 9,1975.

[4] J.Cheeger and D. Gromoll, On the structure of complete manifolds of non-negative curvature, Ann. of Math. 96(1972) 413-443.

[5] D. Gromoll and W. Meyer, On complete manifolds of positive curvature, Ann. of Math. 90(1969)75-90.

[6] A. Kasue, On manifolds of asymptotically nonnegative curvature.

[7] Y.Otsu, K. Shiohama and T. Yamaguchi, A new version of differentiable sphere theorem, to appear in Invent. Math.

[8] K. Shiohama, A sphere theorem for manifolds of positive Ricci curvature, Trans. Amer. Math. Soc. 273(1983), 811-819.

[9] T.Yamaguchi, Lipschitz convergence of manifolds of positive Ricci curvature with Large volume, preprint.

Yukio Otsu
Department of Mathematics
Faculty of Science
Kyushu University
Fukuoka 812
Japan

ON THE ISODIAMETRIC INEQUALITY FOR THE 2-SPHERE

Takashi SAKAI

Introduction

There are many important inequalities with respect to the riemannian invariants on manifolds ,e.g., isoperimetric inequality, isosystolic inequality, isoembolic inequality (see the recent Book [Bu-Za] on the subject). Especially for closed surfaces there are many attempts and results on these inequalities ([B-2],[Gr],[Cr],[Bu-Za]).Here we will be concerned with the following type of the inequality; Let (M,g) be a compact connected riemannian manifold. We denote by Vol(M,g) the volume of M with respect to the canonical measure v_g induced from g and by Diam(M,g) the diameter of (M,g). We consider the riemannian invariant

$$F(g) := \mathrm{Vol}(M,g)/\mathrm{Diam}^m(M,g),$$

where m is the dimension of the manifold. Clearly F(g) is a homothety invariant. In the present note we will be concerned with the simplest case , namely when M is diffeomorphic to the 2-sphere S^2. Then we consider the area and $F(g) := \mathrm{Area}(S^2,g)/\mathrm{Diam}^2(S^2,g)$ may be considered as the functional on the space Ω of the riemannian structures on S^2. We ask the behavior of F. What is the upper and the lower bound of F? If we consider the very long and thin

ISBN 0-12-640170-5

spheres $F(g)$ may be arbitraly small and we have $\inf F(g) = 0$. As for the upper bound ,in general when the Gauss curvature takes the negative value ,$F(g)$ may take arbitrary large value (see section 1).Thus we should retrict ourself to the subclass of Ω to get the finite upper bound. Let $\Omega^+ := \{g \in \Omega;\ K_g \geq 0\}$ be the space of riemannian structures on S^2 of non-negative Gauss curvature.

Then by Bishop volume comparison theorem we get easily

$$(*) \quad F(g) < \pi \quad \text{for} \quad g \in \Omega^+.$$

On the other hand A.D.Alexandrov conjectured that

$$(**) \quad F(g) \leq \pi/2 \quad \text{for} \quad g \in \Omega^+,$$

where the equality does hold exactly for the singular metric obtained as the double of the flat euclidean disk ([A],[Bu-Za]). Since very little seems to be known about the conjecture, we first try to improve (*) and show that $\sup\{F(g), g \in \Omega^+\}$ is strictly less than π. Namely in the present note we show the following:

THEOREM Let (S^2,g) be a riemannian structure with non-negative curvature on the 2-sphere. Then we have

$$\mathrm{Area}(S^2,g)/\mathrm{Diam}^2(S^2,g) < 0.985\pi.$$

In the first section we will give some observations on the isodiametric inequality on S^2 and in the second section we give the proof of the theorem. The author would like to thank B.C.Croke for comments and A.Katsuda for useful discussions.

1. Motivation and elementary results

Here we give some considerations on the isodiametric inequality on S^2. These seem to be known to people who considered the above problem ,but we present them here to show our motivation and for the sake of completeness.

(1.1) For the canonical riemannian structure g_o of positive constant curvature on S^2, we have $F(g_o) = 4/\pi$. If we consider the Zoll's metrics g on S^2, which are realized as the surfaces of revolution whose geodesics are all closed geodesics of the same length L, then we have again $F(g) = 4/\pi$, because $\mathrm{Area}(S^2,g)$ is equal to L^2/π in this case ([Be]). Thus we can not characterize the canonical round sphere in terms of the functional F.

(1.2) We consider the metric ball $B_r(p)$ in the hyperbolic plane H^2 of constant curvature -1. Then we get

$$\mathrm{Area}\, B_r(p) = 2\pi \int_0^r \sinh t\, dt = 4\pi \sinh^2(r/2).$$

Take the double of $\bar{B}_r(p) := \{q \in H^2 ; d(p, q) \leqslant r\}$ and we get the (singular) metric g_r^- on S^2 for which $\mathrm{Diam}(S^2, g_r^-)$ is equal to 2r (see the argument in the proof of (1.5)). Namely we have $F(g_r^-) = \pi/2(\sinh^2(r/2)/(r/2)^2)$ which tends to $+\infty$ when r goes to $+\infty$. Since it is possible to approximate g_r^- by C^∞-metrics with $K_g \geqslant -1$, where K_g denotes the Gauss curvature of g, we see that $\sup\{F(g); K_g \geqslant -1\} = +\infty$. Thus if we want to get the finite upper bound for F we should restrict ourself to the subclass $\Omega^+ := \{ g \in \Omega; K_g \geqslant 0\}$ of Ω.

(1.3) As in (1.2) we consider the metric ball $B_r(p)$ in the euclidean plane (resp. round sphere of constant curvature 1). We get as before the metric (S^2, g_r^o) (resp. (S^2, g_r^+)) by taking the double of $\bar{B}_r(p)$. Then we have easily

$$F(g_r^o) = \pi/2 \quad (\text{resp. } F(g_r^+) = \pi/2 \sin^2(r/2)/(r/2)^2).$$

Now Alexandrov conjectured $F(g) \leqslant \pi/2$ if $K_g \geqslant 0$, where the equality holds exactly when g is isometric to g_r^o ([A], [Bu-Za]). Since $F(g_r^+)$ converges to $\pi/2$ when r goes to zero, we may only conjecture that

$$F(g) < \pi/2 \quad \text{if } K_g \geqslant 1.$$

(1.4) Let (S^2, g) be a riemannian structure on S^2 with $K_g \geqslant 0$ (resp. $K_g \geqslant 1$) and put $d := \mathrm{Diam}(S^2, g)$. Then by Bishop's volume comparison theorem ([Bi-Cr]) we have

$$\mathrm{Area}\,(S^2, g) = \mathrm{Area}(B_d(p)) \leqslant d^2 \ (\text{resp. } 4\pi \sin^2 d/2).$$

and $F(g) < \pi$ (resp. $\leqslant \pi (\sin^2(d/2)/(d/2)^2)$. See also the arguments in [Bu-Za], p.42.

(1.5) If we restrict ourself to the surfaces of revolution in $\mathbf{R}^3$ which are diffeomorphic to the 2-sphere and of non-negative Gauss curvature, then we get $F(g) \leqslant \pi/2$, where the equality holds if and only if g is isometric to g_r^o.

Proof. Let c be a meridian line which intersects the axis at the two points p and q. Then c defines a minimal geodesic from p to q. All geodesics emanating from p are meridian lines, go through q and form closed geodesics. We show that the diameter is given by the distance d(p,q). In fact take points p_1, p_2 which realizes the diameter. Then p_1 and p_2 should lie on the same meridian curve. To see this we may assume p_i (i = 1,2) are different from p, q. Take meridian lines c_1 through p_1 and c_2 through p_2. Choose the point p_3 on c_2 with $d(p,p_3) = d(p,p_1)$ so that p_3 lies on the opposite side to p_2 with respect to p. Then we have $d(q,p_1) = d(q,p_3)$. Since p_2 and p_3 lie on the same meridian curve c_2 we have the following :

$$\mathrm{Diam}(S^2,g) \geqslant d(p,q) = 1/2\ L(c_2) \geqslant \min\{d(p,p_3) + d(p,p_2),\ d(q,p_3) + d(q,p_2)\} = \min\{d(p,p_1) + d(p,p_2),\ d(q,p_1) + d(q,p_2)\} \geqslant d(p_1,p_2) = \mathrm{Diam}(S^2,g).$$

Thus we see that the equalities hold in the above and d(p, q) is equal to the diameter. Now since every point of S^2 lies on a meridian curve which is a minimal geodesic from p to q, we see that $S^2 = \bar{B}_{d/2}(p) \cup \bar{B}_{d/2}(q)$ and $B_{d/2}(p) \cap B_{d/2}(q) = \phi$. Then by Bishop comparison theorem ([Bi-Cr]) we get

$$\mathrm{Area}(S^2,g) = \mathrm{Area}\ \bar{B}_{d/2}(p) + \mathrm{Area}\ \bar{B}_{d/2}(q) \leqslant \pi d^2/2 ,$$

and $F(g) \leqslant \pi/2$, where the equality holds if and only if $\bar{B}_{d/2}(p)$ and $\bar{B}_{d/2}(q)$ are flat euclidean disks.

(1.6) The above argument shows that if we have $\bar{B}_{d/2}(p) \cup \bar{B}_{d/2}(q) = S^2$ for p,q with $d(p,q) = \mathrm{Diam}(S^2,g)$ we can prove the desired inequality. But in general this does not hold. Treibergs pointed out that the Reuleaux triangle glued to itself yields a counterexample (C.B.Croke kindly taught me this fact, to whom I am very grateful). On the other hand if there exists a closed geodesic γ of length 2d which passes through p, q, then we may estimate the area by Heintze-Karcher volume comparison argument ([H-K]). In fact since γ is a closed geodesic , by considering the family of geodesics emanating from the points of γ perpendicularly to the both sides of γ we get

$$\mathrm{Area}(S^2,g) \leqslant d\ L_g(\gamma) = 2d^2$$

and consequently $F(g) \leqslant 2\ (< \pi)$. Again in general we can not expect to have such a closed geodesic.

2. Proof of the theorem

In this section we shall give the proof of our theorem. Let g be a riemannian metric on S^2 of non-negative curvature and p,q be points with d(p,q)

$= \mathrm{Diam}(S^2,g)$, which will be also denoted simply by d. Firstly we recall the following resut due to M.Berger.

Lemma 2.1.([B-1]) Let M be a compact connected riemannian manifold and p,q be points of M with $d(p,q) = \mathrm{Diam}(M,g)$. Then for any vector $v \in U_pM$ (:= unit sphere in the tangent space to M at p) there exists a minimizing geodesic τ from p to q such that the angle $\measuredangle(\dot{\tau}(0),v) \leq \pi/2$.

Applying the above to our case we have a sequence of minimal geodesics $\tau_1, \tau_2, \ldots, \tau_N$ from p to q such that $0 < \alpha_i, \beta_i \leq \pi$, where we put $\alpha_i := \measuredangle(\dot{\tau}_i(0), \dot{\tau}_{i+1}(0))$ and $\beta_i := \measuredangle(-\dot{\tau}_i(d), -\dot{\tau}_{i+1}(d))$ ($i = 1,\ldots, N$, $\tau_{N+1} = \tau_1$). Thus S^2 may be covered by sucsessive N biangles $H_i(\tau_i,\tau_{i+1})$ ($i = 1,\ldots, N+1; \tau_{N+1} = \tau_1$). We also denote the above biangles by $H(\alpha_i, \beta_i)$. Secondly we show that if $H(\alpha_i, \beta_i)$ is not "symmetric" in the following sense , then F(g) becomes small.

Lemma 2.2 Let biangles $H(\alpha_i,\beta_i)$ ($i = 1,\ldots,N$) cover S^2. Then we have

$$\mathrm{Area}(S^2,g) \leq \Sigma_{i=1}^{N} \min(\alpha_i,\beta_i)\, d^2/2.$$

Especially if there exists i with $|\alpha_i - \beta_i| > \varepsilon\pi$, then we get

$$F(g) < (1 - \varepsilon/2)\pi.$$

Proof. Let $\widetilde{H}_p(\alpha_i)$ be the convex sector in T_pS^2 bounded by $\dot{\tau}_i(0)$ and $\dot{\tau}_{i+1}(0)$. Then $\exp_p\{tu\ ;\ 0 \leqslant t \leqslant d,\ u \in U_pM \cap \widetilde{H}_p(\alpha_i)\}$ contains $H(\alpha_i, \beta_i)$. In fact for any point x in the interior of $H(\alpha_i, \beta_i)$, minimal geodesic from p to x can not intersect τ_i and τ_{i+1} and then we see that $\dot{\tau}(0) \in U_pM \cap \widetilde{H}_p(\alpha_i)$. Then by Bishop volume comparison theorem we have

$$\text{Area } H(\alpha_i, \beta_i) \leqslant \text{Area } \exp_p\{tu;\ 0 \leqslant t \leqslant d,\ u \in U_pM \cap \widetilde{H}_p(\alpha_i)\}$$
$$\leqslant \alpha_i\, d^2/2.$$

Similar fact also holds for $\widetilde{H}_q(\beta_i)$ and we get Area $H(\alpha_i, \beta_i) \leqslant \min(\alpha_i, \beta_i) d^2/2$. To show the last assertion we assume ,for instance, that $\alpha_1 < \beta_1 - \varepsilon\pi$. Then we get

$$\text{Area}(S^2, g) \leqslant (\beta_1 - \varepsilon\pi + \beta_2 + \ldots + \beta_N)\, d^2/2 \leqslant$$
$$(1 - \varepsilon/2)\, \pi\, d^2,$$

and $F(g) < (1 - \varepsilon/2)\pi$.

ε will be determined later and in the following we may assume that all biangles $H(\alpha, \beta)$ considered, including biangles obtained by joining together successive biangles as long as they form a convex angle at p ,q , satisfy $|\alpha - \beta| \leqslant \varepsilon\pi$.

Lemma 2.3. Suppose that there exists a biangle $H(\alpha, \beta)$ such that $\pi \geqslant \alpha \geqslant 2\pi/3$ and $|\alpha - \beta| \leqslant \varepsilon\pi$ hold. Then we have

$$F(g) \leqslant (2/\pi + 1/3 + \varepsilon/2)\pi.$$

Proof. Let τ_1, τ_2 be the minimal geodesics from p to q which bound $H(\alpha,\beta)$. Then the simple closed curve $c := \tau_1 \cup \tau_2^{-1}$ divides S^2 into two disks $S_1 := H(\alpha,\beta)$ and $S_2 := S^2 \setminus S_1$. Put $\eta_i := \max\{d(y,c);\ y \in S_i\}$ ($i = 1,2$).Then we see that $\eta_1 + \eta_2 \leqslant d$. In fact take points $p_i \in S_i$ such that $d(p_i,c) = \eta_i$ ($i = 1,2$) and take a minimal geodesic joining p_1 and p_2 which meets c at a point q'. Then we have

$$d \geqslant d(p_1,p_2) = d(p_1,q') + d(p_2,q') \geqslant \eta_1 + \eta_2.$$

Now $S_1 = H(\alpha,\beta)$ is convex at p (i.e., $\sphericalangle_{S^1}(\dot{\tau}_1(0), \dot{\tau}_2(0)) \leqslant \pi$) and first assume that $H(\alpha,\beta)$ is also convex at q. Then in this case we have

$S_1 \subset \{ \exp_c tn\ ;\ 0 \leqslant t \leqslant \eta_1$, where n denotes the unit normal vector field to c pointing to the inside of $S_1\}$.

This follows from the fact that if γ is a distance minimizing geodesic from a point $q_1 \in S_1$ to c ,then $-\dot{\gamma}(d(q_1, c))$ should make the angle greater than or equal to $\pi/2$ with both branches of c at the foot point. Then by Heintze-Karcher volume comparison theorem ([H-K]) we get

$$\mathrm{Area}(S_1,g) \leqslant 2d\eta_1,$$

since τ_1 and τ_2 are geodesics. Next we estimate the area of S_2. We set

$A_p := \{ \exp_p tu \ ; \ 0 \leqslant t \leqslant \eta_2$, where u are unit vectors at p belonging to the convex sector s_p bounded by the two vectors which are perpendicular to $\dot{\tau}_1(0)$ and $\dot{\tau}_2(0)$ respectively and pointing to the inside of $S_2 \}$.

Since $\alpha \leqslant 2\pi/3$ we see that the angle of the above sector s_p at p is not greater than $\pi/3$. We may define A_q by the same way for q and in this case the angle of s_q is not greaer than $(1/3 + \varepsilon)\pi$, because $\beta \geqslant \alpha - \varepsilon \geqslant (2/3 - \varepsilon)\pi$. Then S_2 is contained in $A_p \cup A_q \cup \{ \exp_c tn'; \ 0 \leqslant t \leqslant \eta_2$, where n' are unit normal vector fields to c pointing to the inside of S_2 }. Thus we have again by Bishop and Heintze-Karcher volume comparison theorem that

$$\mathrm{Area}(S_2,g) \leqslant 2d\,\eta_2 + \pi/3 \cdot \eta_2^2/2 + (1/3 + \varepsilon)\pi\,\eta_2^2/2$$

and consequently

$$\mathrm{Area}(S^2,g) \leqslant 2d(\eta_1 + \eta_2) + (1/3 + \varepsilon/2)\pi\eta_2^2 \leqslant (2/\pi + 1/3 + \varepsilon/2)\pi d^2.$$

Namely we get, $F(g) \leqslant (2/\pi + 1/3 + \varepsilon/2)\pi$.
The remaining case when S_2 is convex at q may be treated by the same argument and we get the same estimate for F(g).This completes the proof of the lemma.

Now we turn to the proof of the theorem. We take $\varepsilon =$ 0.03 in lemma(2.2). We cover S^2 by successive biangles H(

α_i, β_i) adjoinig each other and assume that $|\alpha_i - \beta_i| > 0.03\pi$ for some biangle $H(\alpha_i, \beta_i)$. Then we have

$$F(g) < (\pi - 0.03\ \pi/2) = 0.985\pi$$

by lemma(2.2).

Thus we may assume that $|\alpha_i - \beta_i| \leqslant 0.03\pi$ for all biangles $H(\alpha_i, \beta_i)$ which cover S^2. Now if there exist a biangle $H(\alpha_i, \beta_i)$ with $\alpha_i \geqslant 2\pi/3$, then we get by lemma(2.3)

$$F(g) \leqslant (2/\pi + 1/3 + 0.03/2)\pi < 0.985\pi.$$

Thus we may reduce to the case when all biangles satisfy the conditions $\alpha_i, \beta_i \leqslant 2\pi/3$ and $|\alpha_i - \beta_i| \leqslant 0.03\pi$. In this case starting from $H(\tau_1, \tau_2)$, $H(\tau_2, \tau_3), \ldots$ and adjusting $H(\tau_1, \tau_2)$ and $H(\tau_2, \tau_3)$ to get the biangle $H(\tau_1, \tau_3)$ if $\alpha_1 + \alpha_2 \leqslant 2\pi/3$ and continue the process. By this if we get a biangle $H(\alpha, \beta)$ with $\alpha \geqslant 2\pi/3$ then we are done by lemma(2.3). By the similar process reversing the order $H(\tau_N, \tau_1)$, $H(\tau_{N-1}, \tau_N), \ldots.$ we may reduce to the case when we have three minimal geodesics τ_1, τ_2, τ_3 from p to q such that τ_i and τ_{i+1} make the angle $2\pi/3$ at p and q. Then by the above argument we get $F(g) \leqslant (2/\pi + 1/3)\pi < 0.97\pi$. This completes the proof of the theorem.

Remark. To estimate the area of the biangle $H(\alpha, \beta)$ we may also use the isoperimetric inequality. We may show that

$$\text{Area}(H(\alpha,\beta)) \leqslant (\pi - (\alpha+\beta)/2)^{-1} d^2.$$

In fact let $L = 2d$ be the perimeter. Since $K_g \geqslant 0$ we have from the isoperimetric inequality ([Bu-Za])

$$L^2 + 2\left(\int_{H(\alpha,\beta)} K_g \, dv_g - 2\pi\right) \text{Area} \geqslant 0.$$

By Gauss-Bonnet we have $\int_{H(\alpha,\beta)} K_g \, dv_g = \alpha+\beta$ and we get the desired estimate. This gives a nice estimate only when α and β are small. For instance if we have three biangles $H_i(2\pi/3, 2\pi/3)$ $(i = 1, 2, 3)$ which cover S^2, we get

$$\text{Area}(S^2,g) \leqslant 9/\pi \, d^2 < 0.92\,\pi\, d^2.$$

REFERENCES

[A] Alexandrov, A.D.; Die innere Geometrie der konvexen Flächen. Akad.Verl.Berlin, 1955.

[B-1] Berger, M.; Sur quelques variétés riemanniennes suffisament pincées, Bull.Soc.Math.France 88, 57-71 (1960).

[B-2] Berger, M.; Lectures on geodesics in Riemannian Geometry. Tata Institute, Bombay. 1965.

[Be] Besse.A.L.; Manifolds all of whose geodesics are closed. Springer, Berlin. 1978.

[Bi-Cr] Bishop, R.L.-Crittenden, R.J.; Geometry of Manifolds. Academic Press, New York. 1964.

[Bu-Za] Burago, Yu.D.-Zalgaller, V.A.; Geometric

Inequalities. Springer, Berlin. 1988.

[Cro] Croke,B.C,; Area and the length of the shortest closed geodesic, J.Diff.Geo. **27**, 1-21 (1988).

[Gr] Gromov,M.; Filling Riemannian manifolds,J.Diff.Geo. **18**, 1-148 (1983).

[H-K] Heintze,E-Karcher.H; A general comparison theorem with applications to volume estimates for submanifolds, Ann.Sci.Ecole.Norm.Sup. **11** ,451-470 (1978).

Takashi Sakai

Department of Mathematics
Faculty of Science
Okayama University
Okayama 700
Japan

An Isoperimetric Problem for Infinitely Connected Complete Open Surfaces

(Dedicated to Professor T.Otsuki on his 70th Birthday)

K.Shiohama and M.Tanaka

Introduction. An isoperimetric problem for a Riemannian plane (e.g., a complete Riemannian manifold homeomorphic to R^2) was first investigated by Fiala in [3] and later by Hartman in [4]. Their works will be summarized as follows. For a simply closed smooth curve $\mathfrak{C}$ in a Riemannian plane M and for $t \geqq 0$ let $S(t) := \{x \in M \,;\, d(x,\mathfrak{C}) = t\}$ and $B(t) := \{x \in M \,;\, d(x,\mathfrak{C}) \leqq t\}$, where d is the distance function induced by the Riemannian metric of M. First of all they proved that $S(t)$ for almost all $t \geqq 0$ becomes a finite union of piecewise smooth simple closed curves (see Proposition 6.1 in [4]). Secondly, if $L(t)$ and $A(t)$ are the length of $S(t)$ and the area of $B(t)$ and if $\int_M |G|\, dM < \infty$, then

$$\lim_{t\to\infty} \frac{L(t)}{t} = \lim_{t\to\infty} \frac{2A(t)}{t^2} = 2\pi - c(M), \tag{1}$$

and

$$\lim_{t\to\infty} \frac{L^2(t)}{A(t)} = 2(2\pi - c(M)) \tag{2}$$

where G is the Gaussian curvature and $c(M)$ is the total curvature of M (e.g., an improper integral of G with respect to the area element dM of M).

ISBN 0-12-640170-5

Recently (1) and (2) were extended to finitely connected complete open surfaces in [7] as follows : Let M be a finitely connected complete open surface and $p_o \in M$ a fixed point. Let $L(t)$ and $A(t)$ for $t \geqq 0$ be the length of $S(t) := \{x \in M ; d(p_o,x) = t\}$ and the area of $B(t) := \{x \in M ; d(p_o,x) \leqq t\}$. If M admits total curvature $c(M)$ in $[-\infty, 2\pi\chi(M)]$, then

$$\lim_{t\to\infty} \frac{L(t)}{t} = \lim_{t\to\infty} \frac{2A(t)}{t^2} = 2\pi\chi(M) - c(M) \quad (1)'$$

and

$$\lim_{t\to\infty} \frac{L^2(t)}{A(t)} = 2(2\pi\chi(M) - c(M)). \quad (2)'$$

Here we note that a well known theorem due to Cohn-Vossen [2] states that the right hand sides of (1)' and (2)' are non-negative. We also note that p_o in the above relations can be replaced by an arbitrary fixed simply closed curve on M.

Our attempt of obtaining the relations (1)',(2)' corresponding to infinitely connected complete open surfaces gives raise to serious observation on the topology of M and on the analysis of the local structure of cut locus to $\mathfrak{C}$.

The first observation is concerned with the topology of M. Recall that M is by definition finitely connected iff it is homeomorphic to a closed 2-manifold from which finitely many points are removed. M is by definition <u>infinitely connected</u> iff it is not finitely connected. A famous result by Huber [5] states that if an infinitely connected complete M admits total curvature, then $c(M) = -\infty$. However the Euler characteristic of M is not well defined, and hence the right hand sides of (1)' and (2)' do not exist. The crucial point of our topological observation is to interpret the value $2\pi\chi(M) - c(M)$

so as to make a natural sense for an infinitely connected M. This value should depend only on the Riemannian metric which defines the total curvature. For this purpose we decompose M into a sequence $\{M_j\}$ of submanifolds of M, each M_j of which satisfies the following properties : (1) M_j is complete and finitely connected with nonempty compact boundary, (2) each component of ∂M_j is a nonnull homotopic simply closed geodesic, (3) $\{M_j\}$ is strictly monotone increasing with j and $\bigcup M_j = M$, (4) the sequence $\{\chi(M_j)\}$ of Euler characteristic of M_j is strictly monotone decreasing. We shall call such an $\{M_j\}$ a <u>filtration of finitely connected submanifolds of M</u>. In view of the original proof of the Cohn-Vossen theorem we see that for a filtration of finitely connected submanifolds of M the sequence $\{2\pi\chi(M_j) - c(M_j)\}$ is nonnegative and monotone non-decreasing with j. Furthermore it will be proved that the limit of this sequence

$$s(M) := \lim_{j\to\infty} [2\pi\chi(M_j) - c(M_j)]$$

is independent of the choice of filtration. These facts are proved in Theorem 1.1, §1. The $s(M)$ takes value in $[0,\infty]$ and depends only on the Riemannian metric around endpoints.

The second observation is concerned with the local structure of cut locus to an arbitrary fixed simple closed curve $\mathfrak{C}$ in an infinitely connected complete open surface M. In their earlier works by Fiala-Hartman the assumption for M being homeomorphic to R^2 was behind the discussion to establish the local structure of cut locus to $\mathfrak{C}$. We emphasize that their discussion is purely analytic. Whatever M is infinitely connected or not and whatever $\mathfrak{C}$ bounds a domain of M or not, we recognize that the local structure of cut locus to $\mathfrak{C}$ is established. This is achieved by using our notion of degenerate anormal points (see §3) and by observing cut locus in a small convex ball around a cut point. The crucial point is to verify that there exists a measure zero set E on $[0,\infty)$ such that if $t \in [0,\infty) - E$, then $S(t)$ consists of a finite union of piecewise smooth simple closed curves. This property was essentially used without proof to establish relations (1)' and (2)' (see Theorem A, [7]). Here we consider a filtration $\{M_j\}$ of finitely connected submanifolds of an infinitely connected M. Our Main Theorem is established by applying the above property to each M_j and by showing the right continuity of $L(t)$.

Main Theorem. Let M be an infinitely connected, complete, oriented and noncompact Riemannian 2-manifold admitting total curvature. For a simply closed curve $\mathfrak{C}$ in M and for $t \geqq 0$ let $L(t)$ and $A(t)$ be the length of $S(t) := \{x \in M; d(x,\mathfrak{C}) = t\}$ and the area of $B(t) := \{x \in M ; d(x,\mathfrak{C}) \leqq t\}$. Then we have

$$\liminf_{t \to \infty} \frac{L(t)}{t} \geqq s(M), \tag{3}$$

$$\liminf_{t \to \infty} \frac{2A(t)}{t^2} \geqq s(M) \tag{4}$$

and

$$\limsup_{t \to \infty} \frac{L^2(t)}{A(t)} \geqq 2s(M). \tag{5}$$

Note that the inequality (5) replaced by $\liminf_{t \to \infty} \frac{L^2(t)}{A(t)}$ instead of $\limsup_{t \to \infty} \frac{L^2(t)}{A(t)}$ does not hold in general. Such an example is provided in §5. Also examples constructed in §5 show that all inequalities in our Main Theorem are optimal.

Now it seems to us that the proof of Theorem 6.2 in [4] is somewhat unclear. This theorem is purely analytic. We extend it to an arbitrary fixed simply closed curve in an infinitely connected complete open surface M and publish it separately in [8].

At the end of the introduction we note that the Huber theorem is a direct consequence of the Cohn-Vossen theorem. In fact, $c(M) = \lim_{j\to\infty} c(M_j)$ and $c(M_j) \leqq 2\pi\chi(M_j)$ holds for any filtration $\{M_j\}$ of finitely connected submanifolds of M. Since $\{\chi(M_j)\}$ is strictly decreasing with j, we have $c(M) = -\infty$.

1. Filtration of finitely connected submanifolds of M.

In this section let M be a connected, oriented, infinitely connected complete and noncompact Riemannian 2-manifold admitting total curvature. A filtration of finitely connected submanifolds of M will be constructed as follows. Let $f : M \to R$ be a Morse function such that $f^{-1}((-\infty,a])$ is compact for all $a \in R$. For a regular value a of f the level set $f^{-1}(a)$ consists of a finite disjoint union of circles. If $U_1,\ldots,U_n$ and $V_1,\ldots,V_m$ are all components of $M - f^{-1}((-\infty,a))$ with the properties that U_i for each $i = 1,\ldots, n$ is homeomorphic to a tube $S^1 \times [0,\infty)$ and that V_k for each $k = 1,\ldots, m$ is compact, then we denote by $\hat{M}(a)$ a component of $f^{-1}((-\infty,a]) \cup \bigcup_{i=1}^{n} U_i \cup \bigcup_{k=1}^{m} V_k$. This $\hat{M}(a)$ is a finitely connected submanifold of M having nonempty boundary. Since each component σ of $\partial\hat{M}(a)$ does not bound a tube, there exists for σ a non-null homotopic closed curve in M which intersects all closed curves freely homotopic to σ. Therefore we find a simply closed geodesic which is freely homotopic to σ and has minimum length among all closed curves belonging to the homotopy class $[\sigma]$ of σ. Thus for all sufficiently large regular value a of f each component of $\partial\hat{M}(a)$ may be replaced by a simply closed geodesic just obtained. If $M(a)$ denotes a submanifold of M whose boundary consists of simply closed geodesics as just obtained now, then $M(a)$ is homeomorphic to $\hat{M}(a)$. To obtain a filtration of finitely connected submanifolds of M we only choose a monotone increasing divergent sequence $\{a_j\}$ of regular values of f such that $\{\chi(M(a_j))\}$ is strictly monotone decreasing. Setting $M(a_j) := M_j$ we see that $\{M_j\}$ is a filtration of finitely connected submanifolds of M.

For a filtration $\{M_j\}$ of finitely connected submanifolds of M and for a fixed $j = 1,\ldots,$ let n_j be the number of endpoints of M_j. By means of the discussion developed by Cohn-Vossen [2] it is possible to choose a sequence $\{K^j{}_\ell\}_{\ell=1,\ldots}$ of compact domains of M_j with the following properties : $\partial M_j \subset K^j{}_1$ and $\{K^j{}_\ell\}$ is monotone increasing and $\bigcup_{\ell=1}^{\infty} K^j{}_\ell = M_j$, (2) if W is a component of $M_j - \mathrm{Int}(K^j{}_1)$ and if $Q_\ell = \partial K^j{}_\ell \cap W$, then W is a tube and Q_ℓ is freely homotopic to ∂W in W and Q_ℓ is either a geodesic loop and $W - \mathrm{Int}(K^j{}_\ell)$ is locally convex, or else a closed curve consisting of broken geodesics and $K^j{}_\ell \cap W$ is locally convex around Q_ℓ. According to Busemann (see §43, [1]), W in the first case is called a contracting tube and in the second case an expanding tube. $M_j - K^j{}_\ell$ has n_j components and $K^j{}_1$ may be chosen so large that each component of $M_j - \mathrm{Int}(K^j{}_1)$ is either contracting or expanding. It was proved by Cohn-Vossen that if W is contracting, then $\{K^j{}_\ell\}$ can be chosen in such a way that the outer angle $\Theta_\ell(W)$ of the geodesic loop Q_ℓ measured with respect to $K^j{}_\ell$ is negative and $\lim_{\ell\to\infty} \Theta_\ell(W) = 0$, and that if W is expanding, then the sum $\Theta_\ell(W)$ of all outer angles of Q_ℓ measured with respect to $K^j{}_\ell$ is nonnegative for all ℓ and $\Theta_\ell(W)$ has a limit

$$\lim_{\ell\to\infty} \Theta_\ell(W) = \Theta_1(W) - c(W) \geqq 0.$$

Note that the right hand side of the above equation is independent of the choice of Q_1 in W. Therefore $\lim_{\ell\to\infty} \Theta_\ell(W)$ depends only on the Riemannian metric around endpoint in W of M_j. If $W^1,\ldots,W^{n_j}$ are all the components of M_j - $\mathrm{Int}(K^j{}_\ell)$, then they are tubes which are contracting or expanding and we have

$$s_j := 2\pi\chi(M_j) - c(M_j) = \lim_{\ell\to\infty} \sum_{i=1}^{n_j} \Theta_\ell(W^i) \geqq 0.$$

This fact means that s_j depends only on the Riemannian metric around endpoints of M_j. If $M_j - \mathrm{Int}(M_{j-1})$ has tubes $\tilde{W}^1,\ldots,\tilde{W}^{n_j-n_{j-1}}$, then

$$s_j - s_{j-1} = \lim_{\ell\to\infty} \sum_{i=1}^{n_j-n_{j-1}} \Theta_\ell(\tilde{W}^i) \geqq 0. \qquad (*)$$

The above relation shows that $\{s_j\}$ is monotone non-decreasing.

Let $\{\tilde{M}_h\}$ be another filtration of finitely connected submanifolds of M and set $\tilde{s}_h := 2\pi\chi(\tilde{M}_h) - c(\tilde{M}_h)$. The $\{\tilde{s}_h\}$ is monotone non-decreasing. It follows from (*) that if $\tilde{M}_h \subset M_j$, then $\tilde{s}_h \leqq s_j$. Since $M = \bigcup_h \tilde{M}_h = \bigcup_j M_j$, there exists for each h a number j such that $\tilde{M}_h \subset M_j$. Thus $\tilde{s}_h \leqq s_j \leqq \lim_{j\to\infty} s_j$ holds for each h, and hence $\lim_{h\to\infty} \tilde{s}_h \leqq \lim_{j\to\infty} s_j$. This shows $\lim_{h\to\infty} \tilde{s}_h = \lim_{j\to\infty} s_j$, and we have proved

Theorem 1.1. Let $\{M_j\}$ be a filtration of finitely connected submanifolds of M and set

$$s_j := 2\pi\chi(M_j) - c(M_j).$$

Then $\{s_j\}$ is nonnegative monotone non-decreasing and has the following properties:

(1) $s_j = s_{j-1}$ if and only if every tube W in $M_j - \mathrm{Int}(M_{j-1})$ is either contracting or else expanding with $\lim_{\ell\to\infty} \Theta_\ell(W) = 0$,

(2) $s_j > s_{j-1}$ if and only if $M_j - \mathrm{Int}(M_{j-1})$ contains an expanding tube W with $\lim_{\ell\to\infty} \Theta_\ell(W) > 0$,

(3) if $s(M) = \lim_{j\to\infty} s_j$, then $s(M)$ is independent of the choice of filtration of M.

Proof of Main Theorem by assuming the right continuity of L(t).

Choose a filtration $\{M_j\}$ of finitely connected submanifolds of M such that $\mathfrak{C} \subset M_1$. For each $j = 1,\ldots,$ let $A_j(t)$ and $L_j(t)$ be the area of $B_j(t) := B(t) \cap M_j$ and the length of $S_j(t) := S(t) \cap M_j$. It is clear that $L(t) \geqq L_j(t)$ and $A(t) \geqq A_j(t)$ for all $t \geqq 0$. (For the definition of L for all $t \geqq 0$, see §2.)

Assume for a moment that $s(M) < \infty$. For an arbitrary given $\varepsilon > 0$ there is a $j(\varepsilon)$ such that $s_j > s(M) - \varepsilon/2$ for all $j > j(\varepsilon)$. Since ∂M_j consists of a finite union of simply closed geodesics, Theorem A in [7] together with Lemma 3.5 implies that

$$\text{(a)} \qquad \lim_{t\to\infty} \frac{L_j(t)}{t} = \lim_{t\to\infty} \frac{2A_j(t)}{t^2} = 2\pi\chi(M_j) - c(M_j) = s_j$$

and

$$\text{(b)} \qquad \lim_{t\to\infty} \frac{L_j(t)^2}{A_j(t)} = 2s_j.$$

Here we note that L'Hospital's rule applies to prove (b) if $A_j(t)$ diverges to ∞ as $t \to \infty$. If $A_j(\infty) < \infty$, then we have $s_j = 0$ and the uniform continuity of $L_j(t)^2$ around ∞ implies that $\lim\sup_{t\to\infty} L_j(t) = 0$.

There exists for each $j > j(\varepsilon)$ a $t_j > 0$ such that for all $t > t_j$ we have

$$\frac{L(t)}{t} \geqq \frac{L_j(t)}{t} \geqq s(M) - \varepsilon$$

and

$$\frac{2A(t)}{t^2} \geqq \frac{2A_j(t)}{t^2} \geqq s(M) - \varepsilon.$$

Thus the proof of (3) and (4) in this case is immediate since $\varepsilon > 0$ is arbitrary.

In the case where $s(M) = \infty$, (3) and (4) are proved by a similar discussion and the proof is omitted.

We need the following Lemma 1.2 for the proof of (5).

<u>Lemma 1.2</u>. By setting $\overline{\alpha} := \limsup_{t \to \infty} \frac{L(t)}{t}$, $\underline{\alpha} := \liminf_{t \to \infty} \frac{L(t)}{t}$, $\overline{\beta} := \limsup_{t \to \infty} \frac{2A(t)}{t^2}$ and $\underline{\beta} := \liminf_{t \to \infty} \frac{2A(t)}{t^2}$, we have

$$\overline{\alpha} \geqq \overline{\beta} \geqq \underline{\beta} \geqq \underline{\alpha} \geqq s(M).$$

The proof of the above Lemma is omitted because it is easily obtained by using the following relation :

$$A(t) = \int_0^t L(u)\, du.$$

The above relation was proved by Hartman in the case where M is a Riemannian plane. In our case where M is not finitely connected and $\mathfrak{C} \subset M$ a simply closed curve, we shall prove the above relation in Lemma 2.4 (when $\mathfrak{C}$ bounds a domain) and in §4 (when $\mathfrak{C}$ does not bound any domain).

The proof of (5) proceeds as follows. If $s(M) = 0$, then it is obvious. In view of Lemma 1.2 we may assume that $\overline{\beta} > 0$. Suppose $\overline{\alpha}$ is finite. Let $\{t_n\}$ be a monotone divergent sequence such that $\lim_{n\to\infty} \frac{L(t_n)}{t_n} = \overline{\alpha}$. Then Lemma 1.2 implies

$$\liminf_{n \to \infty} \frac{L(t_n)^2}{2A(t_n)} \geqq \overline{\beta}^{-1} \cdot \lim_{n\to\infty} \left(\frac{L(t_n)}{t_n}\right)^2 \geqq (\overline{\beta}^{-1} \cdot \overline{\alpha})\overline{\alpha} \geqq \overline{\alpha}.$$

In particular we have

$$\limsup_{t \to \infty} \frac{L(t)^2}{2A(t)} \geqq \bar{\alpha} \geqq s(M).$$

If $\bar{\alpha}$ is infinite, then we choose for every positive integer n a number t_n such that $t_n := \inf \{t > 0 ; \frac{L(t)}{t} \geqq n \}$. This $\{t_n\}$ is monotone divergent. By the choice of t_n we see that $\frac{L(t)}{t} < n$ for all $t < t_n$. Since we have assumed that L is right continuous for every $t > 0$, we have

$$\frac{L(t_n)}{t_n} \geqq n.$$

Therefore the above discussion yields

$$A(t_n) = \int_0^{t_n} L(t)\,dt < n \cdot \int_0^{t_n} t\,dt \leqq \frac{n}{2} \cdot \left(\frac{L(t_n)}{n}\right)^2 = \frac{L(t_n)^2}{2n}.$$

This inequality implies that

$$\limsup_{t \to \infty} \frac{L(t)^2}{2A(t)} \geqq \limsup_{n \to \infty} \frac{L(t_n)^2}{2A(t_n)} = \infty = \bar{\alpha}.$$

This completes the proof of Main Theorem.

<u>2. Right continuity of $L(t)$</u>. In this and the next sections we discuss the local structure of cut locus to $\mathfrak{C}$ which bounds a domain in a complete M which is not necessarily finitely connected. We do not assume the existence of total curvature of M. Since our discussion proceeds in the same manner as developed by Hartman, we employ the same terminologies as used in [4].

Let L_0 be the length of $\mathfrak{C}$. A point on $\mathfrak{C}$ is expressed as $z_0(s)$ with respect to the arclength parameter $s \in [0,L_0]$. $z_0(s)$ and other functions of s will be considered periodic of period L_0 for convenience. Let g be the Riemannian metric on M and N the unit normal field along $\mathfrak{C}$ with $N_0 = N_{L_0}$. A map $z : R \times [0,L_0] \to M$ is defined by

$$z(t,s) := \exp_{z_0(s)} t \cdot N_s,$$

where $\exp_p$ is the exponential map of M at p. If t is sufficiently small, then z gives a coordinate system (t,s) and $g(\frac{\partial z}{\partial t},\frac{\partial z}{\partial t}) = 1$ holds around $\mathfrak{C}$ and $g(\frac{\partial z}{\partial t},\frac{\partial z}{\partial s}) = 0$ follows from Gauss Lemma. For every $s \in [0,L_0]$ let $\gamma_s : R \to M$ be a geodesic with $\gamma_s(t) = z(t,s)$ and $e_s(t)$ a unit parallel vector field along γ_s with $e_s(0) = \frac{\partial z}{\partial s}(0,s)$. For each s let $Y_s(t)$ denote the Jacobi field along γ_s with $Y_s(0) = e_s(0)$, $g(Y_s(t),\gamma_s'(t)) = 0$. Setting $f(t,s) := g(Y_s(t),e_s(t))$, we have $f(0,s) = 1$, $f_t(0,s) = \kappa(s)$ and $g(\frac{\partial z}{\partial s},\frac{\partial z}{\partial s}) = f(s,t)^2$, where $\kappa(s)$ is the geodesic curvature of $\mathfrak{C}$ at $z_0(s)$ and $f_t = \frac{\partial f}{\partial t}$. Since Y_s is a Jacobi field we have

$$f_{tt}(t,s) + G(z(t,s))\, f(t,s) = 0,$$

where we set $f_{tt} = \dfrac{\partial^2 f}{\partial t^2}$.

Let $P(s)$ (respectively $N(s)$) denote the least positive (respectively largest negative) t with $f(t,s) = 0$, or $P(s) = \infty$ (respectively $N(s) = -\infty$) if there is no such zero. If $P(s_0) < \infty$ (respectively $N(s_0) > -\infty$), then P (respectively N) is smooth around s_0 and $z(P(s_0),s_0)$ (respectively $z(N(s_0),s_0)$) is called the first positive (respectively negative) focal point to $\mathfrak{C}$ along γ_{s_0}. For any $t > 0$ the function $P_t(s) := \mathrm{Min}\,\{P(s), t\}$ (respectively $N_t(s) := \mathrm{Max}\,\{N(s), -t\}$) is uniformly Lipschitz continuous and hence for almost all $t > 0$ the number of $P^{-1}(t)$ (respectively $N^{-1}(-t)$) is finite (see Lemma 4.1, Corollary 4.1 in [4] for detail). It follows from the Sard theorem that the set

$$Z_0 := \{t > 0;\ P'(s)=0\ \&\ P(s)=t\} \cup \{t > 0;\ N'(s)=0\ \&\ N(s)=-t\}$$

is of Lebesgue measure zero.

A unit speed geodesic $\sigma : [0,\ell] \to M$ is called a $\mathfrak{C}$-segment iff $\sigma(0) \in \mathfrak{C}$ and $d(\sigma(t),\mathfrak{C}) = t$ holds for all $t \in [0,\ell]$. Every $\mathfrak{C}$-segment is a subarc of some γ_s. Let $\rho(s) := \sup\{t; d(\gamma_s(t),\mathfrak{C}) = t\}$ and $\nu(s) := \inf\{t < 0; d(\gamma_s(t),\mathfrak{C}) = -t\}$. $\rho(s)$ (respectively $\nu(s)$) is the cut point distance to $\mathfrak{C}$ along $\gamma_s|[0,\infty)$ (respectively $\gamma_s|(-\infty,0]$). $z(\rho(s),s)$ is called a cut point to $\mathfrak{C}$ along γ_s and $\gamma_s|[0,\rho(s)]$ is a maximal $\mathfrak{C}$-segment contained in $\gamma_s|[0,\infty)$. A cut point is a first focal point on a $\mathfrak{C}$-segment or the intersection of at least two distinct $\mathfrak{C}$-segments.

If we note that $\frac{d}{ds}z(P(s),s) = z_t(P(s),s)\cdot P'(s)$ and $\frac{d}{ds}z(N(s),s) = z_t(N(s),s)\cdot N'(s)$ (see Proposition 4.1 in [4]), then we observe that the following Proposition 2.1 holds in our case.

Proposition 2.1. ρ (respectively ν) is continuous on $[0,L_0]$ and $\rho(s) \leqq P(s)$ (respectively $\nu(s) \geqq N(s)$) for all s, while $\rho(s_0) = P(s_0) < \infty$ (respectively $\nu(s_0) = N(s_0) > -\infty$) implies that $P'(s_0) = 0$ (respectively $N'(s_0) = 0$).

A cut point to $\mathfrak{C}$ is called normal iff it is the endpoint of exactly two distinct $\mathfrak{C}$-segments and is not a first focal point to $\mathfrak{C}$ along either of them. A cut point to $\mathfrak{C}$ which is not normal is called anormal.

Since we do not assume that $\mathfrak{C}$ bounds a domain, $M - \mathfrak{C}$ has either one or two connected components. We need to develop different treatment between the two cases. Our assertions from Proposition 2.2 to Lemma 3.8 are for such a curve that $M - \mathfrak{C}$ has two comonents. We need a slightly modified discussion corresponding to the above stated assertions in the case where $M - \mathfrak{C}$ has exactly one component. This case is treated in §4.

Suppose that the connected components of $M - \mathfrak{C}$ are two and let M_1 be the component containing $\{z(\rho(s),s);\ \rho(s) < \infty\}$. Note that both sets $\{z(\rho(s),s);\ \rho(s) < \infty\}$ and $\{z(\nu(s),s)\ ;\ \nu(s) > -\infty\}$ have no common point. The proof of the following proposition is a consequence of the inverse function theorem (for detail of the proof see Proposition 5.6 in [4]).

<u>Proposition 2.2</u>. Let $p \in M_1$ be a normal point such that $p = z(\rho(s),s)$ has exactly two solutions $s = s_1, s_2 \pmod{L_0}$, $s_1 < s_2 < s_1 + L_0$ and $\rho(s_j) < P(s_j)$ for $j = 1,2$. Then (i) $\rho(s)$ is smooth on a neighborhood of $s = s_1, s_2$; (ii) there is a strictly monotone smooth function $v(s)$ on a neighborhood of s_2 with the properties $v(s_2) = s_1$, $\rho(v(s)) = \rho(s)$ and $z(\rho(v(s)),v(s)) = z(\rho(s),s)$ for s near s_2 ; finally (iii) the cut locus to $\mathfrak{C}$ is a smooth curve near s_2 which bisects the angle between the $\mathfrak{C}$-segments γ_s and $\gamma_{v(s)}$ meeting at $z(\rho(s),s)$.

It follows from the first variational formula that in the above proposition $\rho'(s) = 0$ holds if and only if the two $\mathfrak{C}$-segments $\gamma_{s_1}, \gamma_{s_2}$ meeting at $z(\rho(s_1),s_1) = z(\rho(s_2),s_2)$ makes an angle π at that point (see Proposition 5.7 in [4]).

Let $D := \{(t,s)\ ;\ 0 \leqq t < \rho(s),\ 0 \leqq s \leqq L_0\}$ and $\chi(t,s)$ the characteristic function of D such that $\chi(t,s) = 1$ or 0 according as $(t,s) \in D$ or not. For any $t \geqq 0$ set

$$L(t) := \int_0^{L_0} \chi(t,s)\cdot f(t,s)\, ds.$$

This $L(t)$ is the length of $S(t) = \{x \in M_1\ ;\ d(x,\mathfrak{C}) = t\}$ if t is a non-exceptional value. (For the definition of a non-exceptional value, see the paragraph after Corollary 3.4). However we do not know the geometric meaning of $L(t)$ for an exceptional t-value. For every $s \in [0,L_0]$ and for every $r \geqq 0$ we have $\lim_{t\to r+o} \chi(t,s)\cdot f(t,s) = \chi(r,s)\cdot f(r,s)$, and $\chi(t,s)\cdot f(t,s)$ is bounded on $[r-1,r+1]\times[0,L_0]$. Therefore $\lim_{t\to r+o} L(t) = L(r)$ follows from the Lebesgue dominated convergence theorem. Thus we have

Lemma 2.3. If $M - \mathfrak{C}$ is not connected, then $L(t) = L(t+0)$ holds for all $t \geqq 0$.

Furthermore we have for every non-exceptional value $r \geqq 0$,

$$A(r) = \int_{\{z(t,s);(t,s)\in D,t\leqq r\}} dM = \int_0^r \int_0^{L_0} \chi(t,s)f(t,s)dsdt = \int_0^r L(t)dt,$$

where $A(r)$ is the area of $B(r) = \{p \in M_1 \; ; \; d(p,\mathfrak{C}) \leqq r\}$. Thus we obtain the following

Lemma 2.4. If $M - \mathfrak{C}$ is not connected, then we have for every $t > 0$

$$A(t) = \int_0^t L(u)\, du.$$

3. Degenerate and totally nondegenerate anormal points.

To investigate the cut locus of $\mathfrak{C}$ which bounds a domain in a complete open M we introduce notions of totally nondegenerate and degenerate anomal points. An anormal cut point $z(\rho(s),s)$ (or $z(\nu(s),s)$) is called totally nondegenerate iff $z(\rho(s),s)$ (or $z(\nu(s),s)$) is not a first focal point to $\mathfrak{C}$ along any $\mathfrak{C}$-segment ending at $z(\rho(s),s)$ (or $z(\nu(s),s)$). An anormal cut point is called degenerate iff it is not totally nondegenerate. Degenerate cut points are not discussed in [4] (for instance, in Proposition 5.3 and Corollary 5.2 a cut point under consideration may be degenerate).

We restrict to consider the cut locus to $\mathfrak{C}$ in M_1, where $M_1 \subset M$ is the domain bounded by $\mathfrak{C}$ and containing $\{z(t,s) ; 0 < t < \rho(s), 0 \leqq s \leqq L_0\}$.

Proposition 3.1. Let $p \in M_1$ be a totally nondegenerate anormal point. Then ; (i) there exists only a finite number $n \geqq 3$ of s-values (mod L_0), say, $s = s_1,\ldots,s_n$, such that $p = z(\rho(s),s)$; (ii) $\rho(s)$ is smooth on all small closed intervals ending at s_j ; finally (iii) the cut locus in a neighborhood of p consists of n smooth curves ending at p ; at each point q on the cut locus these curve bisects the angle between a pair of $\mathfrak{C}$-segments intersecting at q. In particular, totally nondegenerate cut points are isolated.

Proof. By restricting our attention to a small convex ball around p we observe that the proof is obtained by a similar manner to that of Proposition 5.8 in [4]. Since p is not a focal point to $\mathfrak{C}$ along any $\mathfrak{C}$-segment ending at q, the assertion (i) is clear. Let $B_\varepsilon \subset M_1$ be a small convex ε-ball centered at p. For each s_j, $j = 1,\ldots,n$ let $s_j' \in M_1$ be the point of intersection of ∂B_ε (the boundary of B_ε) and $\gamma_{s_j}([0,\rho(s_j)])$. Then for each fixed s_j and for a small interval $I_1 \subset [0,L_0]$ abutting at s_j there exists some s_i, $i \neq j$, and an interval $I_2 \subset [0,L_0]$ abutting at s_i such that the open subarc (s_i',s_j') of ∂B_ε with endpoints s_i', s_j' has no s_k's in it and such that there exists a smooth monotone function $v : I_2 \to I_1$ satisfying $\rho(v(s)) = \rho(s)$ and $z(\rho(v(s)),v(s)) = z(\rho(s),s)$ for $s \in I_2$. This is possible because z is locally diffeomorphic in neighborhoods of $\rho(s_j)$ $N(s_j)$, $j = 1,\ldots,n$, and B_ε may be contained in the intersection of images of these local diffeomorphisms.

Thus (ii) and (iii) follow from the above discussion. Because every cut point to $\mathfrak{C}$ in $B_\varepsilon - \{p\}$ is normal, totally nondegenerate cut points are isolated.

Corollary 3.2. Let $\{p_n\}_{n=1,\ldots}$ be a sequence of distinct anormal points in M_1 with $\lim_{n\to\infty} p_n = q$. Then q is a degenerate cut point.

Proof. Since q is a cut point to $\mathfrak{C}$, it is either degenerate or else totally nondegenerate. Suppose that q is totally nondegenerate. Then Proposition 3.1 implies that q is an isolated anormal point, a contradiction.

Corollary 3.3. The set of all anormal points is closed.

A number $t > 0$ is called anormal iff there exists a value $s \in \rho^{-1}(t)$ (or $s \in \nu^{-1}(-t)$) such that $z(t,s)$ (or $z(-t,s)$) is anormal. If t is not anormal, then t is called normal.

Corollary 3.4. The set of all anormal values is closed and of Lebesgue measure zero.

The proof is the same as that of Corollary 5.3 in [4] and omitted here.

A positive number t is called exceptional iff it is either anormal or if it is normal but there exists an s such that $\rho(s) = t$ (or $\nu(s) = -t$) and $\rho' = 0$ (or $\nu' = 0$) at s. A positive number t is called non-exceptional iff it is not exceptional. The following Lemma 3.5 will be derived from Proposition 2.2, Corollaries 3.3 and 3.4, and the proof is omitted here.

Lemma 3.5. The set $E \subset [0,\infty)$ of all exceptional values is closed and of Lebesgue measure zero.

Remark. For a finitely connected M admitting total curvature (and this is the case where M_j is an element of a filtration $\{M_j\}$ of M in our Main Theorem), we observe from the discussion in the proof of Theorem A in [7] that there exists a $t_0 > 0$ such that there is no normal exceptional value on $[t_0,\infty)$.

The proof of the following Lemma 3.6 is easy and left to the reader.

Lemma 3.6. Let $t > 0$ be non-exceptional. Then the equation $\rho(s) = t$ has at most a finite number $n_\rho(t)$ of solutions s (mod L_0). The integer $n_\rho(t)$ is even and constant on every interval of non-exceptional values.

Proposition 3.7. If $t > 0$ is non-exceptional, then $\rho(s) = t$ has an even number of solutions $2m$(mod L_0) which (if $m > 0$) can be enumerated as $\alpha_1 < \beta_1 < \alpha_2, \ldots, < \alpha_m < \beta_m$ $(< \alpha_1 + L_0)$ such that $\rho > t$ on (α_k,β_k) and $\rho < t$ on (β_k,α_{k+1}) for $k = 1,\ldots,m$ with $\alpha_{m+1} = \alpha_1 + L_0$ and forms a set of simply closed piecewise smooth curves whose corners are at $z(t,\alpha_k)$, $z(t,\beta_k)$, $k = 1,\ldots,m$. The length of $S(t)$ is

$$L(t) = \sum_{k=1}^{m} \int_{\alpha_k}^{\beta_k} f(t,s)\, ds.$$

In particular, α_k, β_k and L are smooth on the set $(0,\infty) - E$. If θ_k is the angle between the two vectors tangent to $\mathfrak{C}$-segments at $z(t,\alpha_k)$, then

$$L'(t) = \sum_{k=1}^{m} \{\int_{\alpha_k}^{\beta_k} f(t,s)ds + f(t,\beta_k)\frac{d\beta_k}{dt} - f(t,\alpha_k)\frac{d\alpha_k}{dt}\}$$

$$= 2\pi\chi(B(t)) + \int_0^{L_0} \kappa(s)ds - c(B(t)) - \sum_{k=1}^{m} [2\tan\frac{\theta_k}{2} - \theta_k].$$

Remark. It is not known in general how many components of $S(t)$ exists for a non-exceptional t-value. If M is finitely connected with total curvature (and this is the case where M_j is a member of a filtration $\{M_j\}$ of M in our Main Theorem) and if M has n endpoints, then there exists a large number T such that $S(t)$ for all $t > T$ has exactly n components each of which is homeomorphic to a circle (see [6],[7]).

We also remark that there exists for $\varepsilon > 0$ a large number $t(\varepsilon)$ such that if $t \in [0,\infty) - E$ satisfies $t > t(\varepsilon)$ then

$$\sum_{k=1}^{m} \theta_k < \varepsilon.$$

This is obtained as a slight modification of Theorem C in [6], and the proof is omitted. Since the integral of geodesic curvature of $\mathfrak{C}$ cancels along ∂M_1 and $\partial(M - M_j)$, we have

$$\lim_{t\to\infty} L'(t) = 2\pi\chi(M) - c(M).$$

The above relation holds in the case where $\mathfrak{C}$ does not bound any domain (see Proposition 4.3 in the next section).

4. Simply closed curve bounding no domain. We deal with the case where a closed curve $\mathfrak{C}$ does not bound any domain of M. Our situation means that there exists a cut point p to $\mathfrak{C}$ such that $p = z(\rho(s_1),s_1) = z(\nu(s_2),s_2)$ for some s_1, s_2 on $[0,L_0]$. Three types of cut points to $\mathfrak{C}$ appear. A cut point p to $\mathfrak{C}$ is of <u>ρ-type</u> (respectively <u>ν-type</u>) iff all $\mathfrak{C}$-segments to p are tangent to N (respectively to $-N$) at their starting points. A cut point p to $\mathfrak{C}$ is of <u>mixed type</u> iff $p = z(\rho(s_1),s_1) = z(\nu(s_2),s_2)$ for some s_1, $s_2 \in [0,L_0]$. The structure of cut locus belonging to the first two types has already been established in §§2 and 3. For a mixed type cut point to $\mathfrak{C}$ the normality, anormality, degeneracy and all the

other properties are well defined by the same manner as before. These properties are defined for t-value where $S(t)$ contains a mixed type cut point having the corresponding properties. In view of the previous discussion the structure of cut locus was studied by restricting our attention to a small convex ball around a cut point. Therefore we observe that the same conclusions for the structure of mixed type cut points as obtained in §§2 and 3 are valid. The following propositions are easy to prove.

Proposition 4.1. Let $p = z(\rho(s_1),s_1) = z(\nu(s_2),s_2)$ be a mixed type normal cut point. Then we have ; (i) ρ and ν are smooth on neighborhoods of s_1 and s_2 ; (ii) there exists a strictly monotone smooth function v on a neighborhood of s_2 such that $v(s_2) = s_1$, $\rho(v(s)) = \nu(s)$ and $z(\rho(v(s)),v(s)) = z(\nu(s),s)$ for s near s_2 ; (iii) the cut locus to $\mathfrak{C}$ is a smooth curve near s_2 which bisects the angle between the $\mathfrak{C}$-segments corresponding to $\nu(s)$ and $\rho(v(s))$ meeting at $z(\nu(s),s)$; finally (iv) $\rho'(s_1) = 0$ (and hence $\nu'(s_2) = 0$) holds if and only if the two $\mathfrak{C}$-segments corresponding to $\rho(s_1)$ and $\nu(s_2)$ makes an angle π at p.

We note that Corollaries 3.2 to 3.4 and Lemma 3.5 in the present case are verified without any modification. We have

Proposition 4.2. Let $t > 0$ be of non-exceptional mixed type. The the equations $\rho(s) = t$ and $\nu(s) = -t$ have at most a finite number of solutions s. The number of elements of $\rho^{-1}(t)$ (respectively $\nu^{-1}(-t)$) is even and constant on any interval of non-exceptional values.

Let D_+ and D_- denote the sets $\{(t,s); 0 \leqq t < \rho(s), s \in [0,L_0]\}$ and $\{(t,s); \nu(s) < t \leqq 0, s \in [0,L_0]\}$ respectively. We then define two functions L_+, L_- on $[0,\infty)$ by

$$L_+(t) := \int_0^{L_0} \chi_+(t,s)f(t,s)\, ds$$

$$L_-(t) := \int_0^{L_0} \chi_-(-t,s)f(-t,s)\, ds$$

where χ_+ and χ_- are characterisitc functions of D_+ and D_- respectively. If $t > 0$ is non-exceptional, then

$$L(t) := L_+(t) + L_-(t)$$

is nothing but the length of $S(t) = \{x \in M ; d(x,\mathfrak{C}) = t\}$.

Proposition 4.3. If $t > 0$ is non-exceptional, then there are finitely many positive numbers $\alpha_1 < \beta_1 < ,\ldots,< \alpha_k < \beta_k$ ($< \alpha_1 + L_0$) and $\lambda_1 < \mu_1 <,\ldots, < \lambda_\ell < \mu_\ell$ such that $\rho > t$ on (α_i,β_i) for $i = 1,\ldots,k$ and $\nu < -t$ on (λ_j,μ_j) for $j = 1,\ldots,\ell$ and such that $\rho < t$ on (β_i,α_{i+1}) for $i = 1,\ldots,k$ with $\alpha_{k+1} = \alpha_1 + L_0$ and $\nu > -t$ on (μ_j,λ_{j+1}) for $j = 1,\ldots,\ell$ with $\lambda_{\ell+1} = \lambda_1 + L_0$. The $S(t)$ consists of curves $z(t,s)$, $\alpha_i \leqq s \leqq \beta_i$ for $i = 1,\ldots,k$ and $z(-t,s)$, $\lambda_j \leqq s \leqq \mu_j$ for $j = 1,\ldots,\ell$, and is a set of simply closed piecewise smooth curves whose corners are at $z(t,\alpha_i)$, $z(t,\beta_i)$ and $z(-t,\lambda_j)$, $z(-t,\mu_j)$ for $i = 1,\ldots,k$ and $j = 1,\ldots,\ell$. The length of $S(t)$ is

$$L(t) = \sum_{i=1}^{k} \int_{\alpha_i}^{\beta_i} f(t,s)ds + \sum_{j=1}^{\ell} \int_{\lambda_j}^{\mu_j} f(-t,s)ds = L_+(t) + L_-(t).$$

Furthermore both L_+ and L_- are smooth on the set of non-exceptional values, and their derivatives are given as

$$L_+(t)' = \sum_{i=1}^{k} \{ \int_{\alpha_i}^{\beta_i} f_t(t,s)ds + f(t,\beta_i)\frac{d\beta_i}{dt} - f(t,\alpha_i)\frac{d\alpha_i}{dt}\},$$

$$L_-(t)' = \sum_{j=1}^{\ell} \{ -\int_{\lambda_j}^{\mu_j} f_t(-t,s)ds + f(-t,\mu_j)\frac{d\mu_j}{dt} - f(-t,\lambda_j)\frac{d\lambda_j}{dt}\},$$

$$L(t)' = 2\pi\chi(B(t)) - c(B(t)) - \sum_{i=1}^{k} [2\tan\frac{\theta_i}{2} - \theta_i] - \sum_{j=1}^{\ell} [2\tan\frac{\phi_j}{2} - \phi_j],$$

where θ_i and ϕ_j are the angles at $z(t,\alpha_i)$ and $z(-t,\lambda_j)$ between the two unit vectors tangent to the $\mathbb{C}$-segments respectively.

5. Examples. We shall provide three examples M_1, M_2 and M_3 of complete infinitely connected sufaces having total curvature $-\infty$ which are embedded in R^3. From these examples we obseve that the relations (3), (4) and (5) obtained in our Main Theorem are optimal. A point $p \in R^3$ is expressed by a canonical coordinates (x,y,z) as $p = (x(p),y(p),z(p))$. For constants $a, b, c \in R$ let $\Pi_{ax+by+cz} := \{(x,y,z) \in R^3 ; ax + by + cz = 0\}$.

Example 1. We fix numbers $\ell > 3$ and $\theta \in (0,\pi/4)$ and denote by Π_z and Π_y the planes $z = 0$ and $y = 0$ respectively. A set $W \subset \Pi_z$ which is symmetric with respect to Π_x is defined as follows. $W := \{p \in \Pi_z ; x(p) = 0, -1 \leqq y(p) \leqq 0$ or $x(p) \neq 0, y(p) = |x(p)|\cdot \cot\theta\} \cup \bigcup_{n=1}^{\infty} \{p \in \Pi_z ; y(p) > 0, \ell^n \cos\theta \leqq y(p) \leqq 2\ell^n \cos\theta, |x(p)| \leqq y(p)\cdot\tan\theta\}$. For a sufficiently small $\varepsilon > 0$ let $B_\varepsilon(W)$ be an ε-ball around W and $\partial B_\varepsilon(W)$ the boundary of this ball in R^3. This $\partial B_\varepsilon(W)$ is an infinitely connected topological surface with one end which is smooth almost everywhere. The set of non-smooth points on $\partial B_\varepsilon(W)$ forms a union of portions of ellipses and has a neighborhood

$U \subset \partial B_\varepsilon(W)$ which can be approximated to obtain a complete smooth surface $M_1 \subset R^3$ with the following properties : (i) $\partial B_\varepsilon(W) - U$ is contained entirely in M_1 and if $U_1 := M_1 - (\partial B_\varepsilon(W) - U)$ then the Gaussian curvature G does not change sign in each component of U_1, (ii) if $D := \{p \in M_1 \; ; \; y(p) \leqq -1/2\}$ then it is a disk and a surface of revolution obtained by attaching a convex cap to a flat cylinder and $\partial D = \mathfrak{C}$ is a closed geodesic, (iii) $G \leqq 0$ on $M_1 - D$ and $G < 0$ in an open set contained in $U_1 - D$, (iv) M_1 is symmetric with respect to both Π_z and Π_x.

The cut locus to $\mathfrak{C}$ is described as follows. $\mathfrak{C}$ is symmetric with respect to Π_z and Π_x and hence so is the cut locus to $\mathfrak{C}$. Let $p := (0,-1-\varepsilon,0) \in \Pi_z \cap \Pi_x \cap M_1$ and N a unit normal field along $\mathfrak{C}$ such that $N = \frac{\partial}{\partial y}$. The set of all cut points to $\mathfrak{C}$ of the ν-type consists of a single point $\{p\}$ which is degenerate. If $\gamma_0 : R \to M_1$ is the geodesic with $\gamma_0(0) = p$ and $\gamma(R)$ lies in Π_z and is symmetric with respect to Π_x, then $\gamma_0|[0,\infty)$ and $\gamma_0|(-\infty,0]$ both are rays from p and their subrays emanating from points on $\mathfrak{C}$ lying in $M_1 - D$ are $\mathfrak{C}$-rays. There is no $\mathfrak{C}$-ray other than the subrays of γ_0. The set of all cut points of the ρ-type coincides with the cut locus to p and conists of all points of $M_1 \cap (\Pi_z \cup \Pi_x) - (D \cup \gamma_0(R))$.

<u>Proposition 5.1</u>. Let $M_1 \subset R^3$ and $\mathfrak{C} \subset M_1$ be as above. Then we have

(i) M_1 admits total curvature and $s(M_1) = 0$,

(ii) $\liminf_{t \to \infty} \frac{L(t)}{t} = s(M_1) = 0$,

(iii) $\liminf_{t \to \infty} \frac{2A(t)}{t^2} > s(M_1) = 0$, and hence

$$\limsup_{t \to \infty} \frac{2A(t)}{t^2} > s(M_j).$$

Proof. (i) is clear from the construction of M_1. We may restrict to consider only cut points of the ρ-type. Recall that ρ-type cut locus to $\mathfrak{C}$ is symmetric in Π_z and Π_x and hence so is $S(t)$ for all $t > 0$. Also note that the distance function on M_1 can be approximated by the distance function on $W \subset \Pi_z$. For $n = 1, \ldots,$ let $X_{2n-1} := (\ell^n \cdot \sin\theta, \ell^n \cdot \cos\theta, 0)$ and $X_{2n} := (2\ell^n \cdot \sin\theta, 2\ell^n \cdot \cos\theta, 0)$ and $Q_{2n-1} := (0, \ell^n \cdot \cos\theta, 0)$, $Q_{2n} := (0, 2\ell^n \cdot \cos\theta, 0)$. Points on $\mathfrak{C}$ are expressed as $(\varepsilon\cos\frac{s}{\varepsilon}, -1/2, \varepsilon\sin\frac{s}{\varepsilon})$, $0 \leqq s \leqq L_0 = 2\pi\varepsilon$.

$L(t)$ and $A(t)$ are estimated as follows. If $t \in (\ell^n + 1/2, \ell^n(1 + \sin\theta) + 1/2)$, then $S(t)$ has two components and the right half of it is approximated by two copies of a portion of circle in Π_z centered at X_{2n-1} with radius $t - \ell^n - 1/2$ making an angle $\measuredangle(X_{2n}, X_{2n-1}, Q_{2n-1}) = \frac{\pi}{2} + \theta$ at X_{2n-1}. Therefore $L(t) = 4(\frac{\pi}{2} + \theta)(t - \ell^n - 1/2)$ and $A(t) > 2(\frac{\pi}{2} + \theta)(t - \ell^n - 1/2)^2 + \sum_{i=1}^{n-1} (3 \sin 2\theta) \cdot \ell^{2i}$. If $t \in (\ell^n(1 + \sin\theta) + 1/2, \ell^n(1 + \cos\theta) + 1/2)$, then $S(t)$ is connected with non-smooth points on Π_x near the segment $\overline{Q_{2n-1}Q_{2n}}$. Therefore $L(t) = 4(\frac{\pi}{2} + \theta - \phi_t)(t - \ell^n - 1/2)$ and $A(t) > 2(\frac{\pi}{2} + \theta)(\ell^n \sin\theta)^2 + \sum_{i=1}^{n-1} (3 \sin 2\theta)\ell^{2i}$, where we set $\cos\phi_t := \ell^n \sin\theta/(t - \ell^n - \frac{1}{2}) \in [\tan\theta, 1]$. If $t \in (\ell^n(1 + \cos\theta) + 1/2, 2\ell^n + 1/2)$, then $S(t)$ has three components, and setting $\cos(\psi_t/2) := \ell^n\cos\theta/(t - \ell^n - 1/2)$, we see that $L(t) = 4(\frac{\pi}{2} + \theta - \phi_t - \psi_t)(t - \ell^n - 1/2)$ and $A(t) > 2(\frac{\pi}{2} + \theta)(\ell^n \cdot \sin\theta)^2 + \sum_{i=1}^{n-1} (3 \cdot \sin 2\theta) \cdot \ell^{2i}$. If $t \in (2\ell^n + 1/2, \ell^{n+1} + 1/2)$, then $S(t)$ has two components lying in flat cylinders of radius ε, and $L(t) = 4\pi\varepsilon$ and

$A(t) > \sum_{i=1}^{n} (3 \sin 2\theta) \cdot \ell^{2i}$. (ii) is easily seen by

$$\liminf_{t \to \infty} \frac{L(t)}{t} \leqq \lim_{n\to\infty} \frac{L(2\ell^n+1)}{2\ell^n+1} = 0.$$

For every $n = 1, \ldots,$ and for $t \in (\ell^n + 1/2, \ell^{n+1} + 1/2)$ we have a function which bounds $A(t)/t^2$ from below. Thus we have

$$\liminf_{t \to \infty} \frac{A(t)}{t^2} \geqq \lim_{n\to\infty} \frac{\sum_{i=1}^{n} (3 \sin 2\theta)\ell^{2i}}{(\ell^{n+1} + 1/2)^2} = \frac{3 \sin 2\theta}{\ell^2 - 1} > 0.$$

This completes the proof of Proposition 5.1.

Example 2. We shall construct an example $M_2 \subset R^3$ which shows that the relation (5) in our Main Theorem replaced by

$\liminf_{t \to \infty} \frac{L(t)^2}{A(t)}$ instead of $\limsup_{t \to \infty} \frac{L(t)^2}{A(t)}$ does not hold. Such an M_2 is obtained by gluing $M_1 - D$ and an expanding tube V along $\mathfrak{C}$. V is constructed as follows. Let $\underline{\beta} := \liminf_{t\to\infty} \frac{A(t)}{t^2}$ for $A(t)$ in Example 1. For a positive number $c > 0$ let $\alpha \in (0,2\pi)$ be a number satisfying $2\alpha/(\alpha + \underline{\beta}) \leqq c$. V is a surface of revolution rotating around y-axis such that $\{q \in V ; y(q) \geqq -1\}$ is a flat cylinder with radius ε and such that $c(V) = -\alpha$. Then M_2 is infinitely connected with two endpoints and $s(M_2) = \alpha$. If $L_V(t)$, $L_M(t)$ and $A_V(t)$, $A_M(t)$ are the lengths of $S(t) \cap V$, $S(t) \cap (M_1 - D)$ and the areas of $B(t) \cap V$, $B(t) \cap (M - D)$ respectively, then we have $A(t) = A_V(t) + A_M(t)$, and

$$\lim_{t\to\infty} \frac{L_V(t)}{t} = \lim_{t\to\infty} \frac{2A_V(t)}{t^2} = \alpha.$$

Therefore we have

$$\liminf_{t \to \infty} \frac{L(t)^2}{A(t)} = \liminf_{t \to \infty} \frac{\{L_V(t)+L_M(t)\}^2}{A_V(t)+A_M(t)} \leqq \frac{2\alpha^2}{\alpha+\underline{\beta}} \leqq c \cdot s(M).$$

Thus we have proved the following

Proposition 5.2. For every $c > 0$ there exists a complete infinitely connected surface $M_2 \subset R^3$ having total curvature with $s(M) \in (0,2\pi)$ and a simply closed curve $\mathfrak{C}$ in M_2 such that $\liminf_{t \to \infty} \frac{L(t)^2}{A(t)} \leqq c \cdot s(M_2)$.

Example 3. The final example shows that there exists an $M_3 \subset R^3$ and a simply closed curve $\mathfrak{C} \subset M_3$ such that in the relations (3),(4) and (5) all equalities hold.

Fix an $\ell > 3$ and $\varepsilon \in (0,1/2)$ such that

$$A := \sum_{n=1}^{\infty} (\varepsilon\ell)^n < \infty.$$

Let $W \subset \Pi_z$ be defined as

$W := \{p \in \Pi_z;\ x(p) = 0,\ |y(p)| \geqq \ell$ or $y(p) = 0,\ |x(p)| \geqq \ell\} \cup \bigcup_{n=1}^{\infty} \{p \in \Pi_z;\ |x(p)| + |y(p)| = \ell^n\}$. Let $f : (\ell-1,\infty) \to (0,\varepsilon]$ be a smooth function such that $f(t) = \varepsilon^n$ for $\ell^n - 1 \leqq t \leqq \ell^n + 1$ and $f'(t) \in (-\varepsilon^n/\ell^n, 0)$ for $\ell^n + 1 < t < \ell^{n+1} - 1$. Let $B(W) \subset R^3$ be defined as $B(W) := \{q \in R^3;\ |x(q)| \geqq \ell,\ y(q)^2 + z(q)^2 \leqq f^2(x(q))$ or $|y(q)| \geqq \ell,\ x(q)^2 + z(q)^2 \leqq f^2(y(q))\} \cup \bigcup_{n=1}^{\infty} \{q \in R^3$; the distance between q and the rectangle $|x| + |y| = \ell^n$ in Π_z is not greater than $\varepsilon^n\}$. By the same manner as in Example 1, the boundary $\partial B(W)$ of $B(W)$ in R^3 can be approximated by a smooth surface M_3 with the following properties : (i) $\partial B(W) - \bigcup_{n=1}^{\infty} \{B_{4\varepsilon^n}(0, \pm\ell^n, 0) \cup B_{4\varepsilon^n}(\pm\ell^n, 0, 0)\}$ is contained entirely in M_3, (ii) the four components $B_{4\varepsilon^n}(0, \pm\ell^n, 0) \cap M_3$ and $B_{4\varepsilon^n}(\pm\ell^n, 0, 0) \cap M_3$ are congruent and have non-positive curvature for all $n = 1,\ldots,$ (iii) M_3 is symmetric with respect to Π_z, Π_y, Π_x, Π_{x+y} and Π_{x-y}.

It follows from construction that $\int_{M_3} G_+ \, dM_3 < \infty$, and hence M_3 admits total curvature and $s(M_3) = 0$, where we set $G_+ := \frac{1}{2}\{G + |G|\}$. The total area of M_3 is bounded above by $2\pi\{4 \sum_{n=1}^{\infty} [(\ell^n - \ell^{n-1})\varepsilon^n + \sqrt{2}\ell^n\varepsilon^n] = 8\pi\{(A + 1)(\ell - 1) + \sqrt{2}A\}$. Let $\mathfrak{C} \subset M_3$ be the closed geodesic of length $L_0 < 4\sqrt{2}\ell$ lying in Π_z which is symmetric with respect to Π_x, Π_y, Π_z, Π_{x+y} and Π_{x-y}. There are exactly eight $\mathfrak{C}$-rays whose images are in Π_x and Π_y. Every cut point to $\mathfrak{C}$ is of the mixed type and the cut locus to $\mathfrak{C}$ is $(\Pi_z \cup \Pi_{x-y} \cup \Pi_{x+y}) \cap (M_3 - \mathfrak{C})$. If $t \in (\ell^n, \ell^n(1 + 1/\sqrt{2}))$, then $S(t)$ has twelve components and $24\pi\varepsilon^n \leqq L(t) \leqq 24\pi\varepsilon^{n-1}$. If $t \in (\ell^n(1 + 1/\sqrt{2}), \ell^{n+1})$, then $S(t)$ has four components and $8\pi\varepsilon^n \leqq L(t) \leqq 8\pi\varepsilon^{n-1}$. $A(t)$ can be estimated from both sides. Thus we have proved the

Proposition 5.3. Let M_3 and $\mathfrak{C}$ be taken as above. Then we have

$$\liminf_{t \to \infty} \frac{L(t)}{t} = \liminf_{t \to \infty} \frac{2A(t)}{t^2} = \limsup_{t \to \infty} \frac{L(t)^2}{2A(t)} = s(M_3).$$

References.

[1] Busemann, H. The Geometry of Geodesics, Academic Press. New York, 1955.

[2] Cohn-Vossen, S. Kürzeste Wege und Totalkrümmung auf Flächen, Compositio Math., 2(1935), 69 - 133.

[3] Fiala, F. Le probléme des isopérimetres sur les surfaces à courbure positive, Comment. Math. Helv., 13(1941), 293 - 346.

[4] Hartman, P. Geodesic parallel coordinates in the large, Amer. J. Math., 86(1964), 705 - 727.

[5] Huber, A. On subharmonic functions and differential geometry in the large, Comment. Math. Helv., 32(1957), 13 - 72.

[6] Shiohama, K. Cut locus and parallel circles of a closed curve on a Riemannian plane admitting total curvature, Comment. Math. Helv., 60(1985), 125 - 138.

[7] Shiohama, K. Total curvatures and minimal areas of complete open surfaces, Proc. Amer. Math. Soc., 94(1985), 310 - 315.

[8] Shiohama, K. and Tanaka, M. A remark on the length function of geodesic parallel circles, Preprint.

Katsuhiro Shiohama
Department of Mathematics
Faculty of Science
Kyushu University
Fukuoka, 812-JAPAN

Minoru Tanaka
Department of Mathematics
Faculty of Science
Tokai University
Hiratsuka, 259 - 12 JAPAN

POSITIVELY CURVED MANIFOLDS WITH RESTRICTED DIAMETERS

Katsuhiro Shiohama and Takao Yamaguchi

1. Introduction

Let M be a connected and compact Riemannian n-manifold whose sectional curvature K_M satisfies $K_M \geq 1$. It is well known by the Bonnet classical theorem that the diameter $d(M)$ of M is less than or equal to π. The Toponogov maximal diameter theorem implies that $d(M) = \pi$ holds if and only if M is isometric to the standard unit n-sphere S^n. Grove and the first named author proved in [GS] that if $d(M) > \pi/2$, then M is homeomorphic to S^n. These results give rise to a natural question: Let $K_M \geq 1$ and let $d(M)$ be sufficiently close to π. Then is M diffeomorphic to S^n ?

The purpose of the present note is to give a partial answer to the above question. To state our result we introduce the diametrical function ρ and an invariant $\hat{d}(N)$ of a compact Riemannian manifold N. Let $\rho : N \to \mathbf{R}$ be $\rho(x) = \sup_{y \in N} d(x,y)$, where d is the distance function induced from the Riemannian metric on N. We define $\hat{d}(N) = \inf_{x \in N} \rho(x)$. If N is homogeneous, then ρ is constant and $\hat{d}(N) = d(N)$.

Our main theorem is stated as follows.

ISBN 0-12-640170-5

Main Theorem. There exists a positive number ε_n *depending only on* n *such that if* M *is a compact Riemannian n-manifold with* $K_M \geq 1$ *and* $\hat{d}(M) \geq \pi - \varepsilon_n$, *then* M *is diffeomorphic to* S^n.

Some volume pinching theorems have recently been obtained in [OSY] and [Y]. Our main theorem provides an extension of Theorem A in [OSY] from a pinching for volume to one for a distance property.

2. Proof of the Main Theorem

Throughout this note let M be a compact Riemannian n-manifold with $K_M \geq 1$. The proof of our main theorem is based on the follwing theorem. All geodesics are assumed to have unit speed.

Generalized Toponogov's Theorem ([CE]). Let γ_1, γ_2 *be geodesic segments in* M *such that* $\gamma_1(0)=\gamma_2(0)$. *Suppose that* γ_1 *is minimal and* $\text{length}(\gamma_2) \leq \pi$. *Let* $\bar{\gamma}_1$, $\bar{\gamma}_2$ *be geodesic segments in* S^n *such that* $\bar{\gamma}_1(0)=\bar{\gamma}_2(0)$, $\text{length}(\gamma_i) = \text{length}(\bar{\gamma}_i)=\ell_i$ *and* $\text{angle}(\dot{\gamma}_1(0),\dot{\gamma}_2(0)) = \text{angle}(\dot{\bar{\gamma}}_1(0),\dot{\bar{\gamma}}_2(0))$. *Then we have* $d(\gamma_1(\ell_1),\gamma_2(\ell_2)) \leq d(\bar{\gamma}_1(\ell_1),\bar{\gamma}_2(\ell_2))$.

Let δ be a positive number. Later we shall take ε sufficiently small relative to δ. Although we can obtain explicit estimates, to simplify expression we denote by $\tau(\delta|\varepsilon)$ (resp. $\tau(\varepsilon)$) a positive number depending only on δ and ε (resp. only on ε) and satisfying $\tau(\delta|\varepsilon) \to 0$ as $\varepsilon \to 0$ for each fixed δ (resp. $\tau(\varepsilon) \to 0$ as $\varepsilon \to 0$).

From now on, we assume that $\hat{d}(M) \geq \pi-\varepsilon$ for small ε, $1 \gg \varepsilon > 0$. For $m \in M$, let m' denote a unique point in M such that $\rho(m)=d(m,m')$. The uniqueness follows from the Toponogov theorem.

Lemma 1. *For a point* $x \in M$ *with* $\delta<d(m,x)<\pi-\delta$, *take minimal geodesics* γ_1 *and* γ_2 *joining* m *to* x, *and* x *to* m' *respectively. Then the angle* θ *between* γ_1 *and* γ_2 *satisfies* $|\theta-\pi| < \tau(\delta|\varepsilon)$.

This is an immediate consequence of Toponogov's theorem.

Lemma 2. *Every geodesic* γ *in* M *satisfies* $d(\gamma(0),\gamma(\pi)) > \pi-2\varepsilon$, *that is,* $\gamma|_{[0,\pi]}$ *is almost minimizing*.

Proof. Take a minimal geodesic σ joining $m=\gamma(0)$ to m', and apply Toponogov's theorem to the hinge determined by γ and σ. Then we have $d(m',\gamma(\pi)) < \varepsilon$, and hence $d(\gamma(0),\gamma(\pi)) \geq d(m,m')-d(m',\gamma(\pi)) > \pi-2\varepsilon$.

Lemma 3. *Let* x, y *be points in* M *such that* $\delta < d(x,y) < \pi-\delta$, *and* γ_1, γ_2 *geodesics joining* x *to* y. *Suppose* $\text{length}(\gamma_i) \leq \pi$ $(i=1,2)$. *Then the angles between* γ_1 *and* γ_2 *are less than* $\tau(\delta|\varepsilon)$.

Proof. Take a minimal geodesic σ joining x to y'. Since $d(y,y') > \pi-\varepsilon$, Toponogov's theorem implies

$|\text{angle}(\dot{\gamma}_i(0),\dot{\sigma}(0)) - \pi| < \tau(\delta|\varepsilon)$. This yields

$\text{angle}(\dot{\gamma}_1(0),\dot{\gamma}_2(0)) < \tau(\delta|\varepsilon)$.

To show that the Hausdorff distance $d_H(M,S^n)$ between M and S^n is close to zero, we concider a composed expoential map. For fixed points $m \in M$ and $p \in S^n$ we take a linear isometry $I : T_p(S^n) \to T_m(M)$ between the tangent spaces, and put $f = \exp_m \circ I \circ \exp_p^{-1} : B_\pi(p) \to M$, where we denote by $B_\pi(p)$ the open metric π-ball around p.

Lemma 4. The map f is a $\tau(\varepsilon)$-map, that is, it satisfies $|d(f(p_1),f(p_2)) - d(p_1,p_2)| < \tau(\varepsilon)$ for every $p_i \in B_\pi(p)$.

Proof. If $d(p,p_i) \le \delta$ or $d(p,p_i) \ge \pi-\delta$ for some i, then Lemma 2 implies that $|d(f(p_1),f(p_2)) - d(p_1,p_2)| < \tau(\delta)$. Next suppose that $\delta < d(p,p_i) < \pi-\delta$ $(i=1,2)$. Let $\bar{\gamma}_i$ be minimal geodesics joining p to p_i, and σ_1, σ minimal geodesics joining m to $f(p_1)$, m to $f(p_1)'$ respectively. Put $\gamma_i = f \circ \bar{\gamma}_i$. Since $\delta/2 < d(m,f(p_1)) < \pi - \delta/2$ by Lemma 2 it follows from Lemmas 1 and 3 that

$$|\text{angle}(\dot{\sigma}_1(0),\dot{\gamma}_2(0)) - \text{angle}(\dot{\bar{\gamma}}_1(0),\dot{\bar{\gamma}}_2(0))| < \tau(\delta|\varepsilon),$$

$$|\text{angle}(\dot{\gamma}_2(0),\dot{\sigma}(0)) - \text{angle}(\dot{\bar{\gamma}}_2(0),\dot{\bar{\gamma}}(0))| < \tau(\delta|\varepsilon),$$

where $\bar{\gamma}$ is the geodesic in S^n with $\dot{\bar{\gamma}}(0) = -\dot{\bar{\gamma}}_1(0)$. Hence Toponogov's theorem yields

$$d(f(p_1),f(p_2)) \le d(p_1,p_2) + \tau(\delta|\varepsilon)$$
$$d(f(p_2),f(p_1)') \le d(p_2,p_1^*) + \tau(\delta|\varepsilon)$$

where p_1^* is the antipodal point of p_1. Together with the fact that $d(f(p_1),f(p_1)') > \pi-\varepsilon$, the above two inequalities imply that $d(f(p_1),f(p_2)) \ge d(p_1,p_2) - \tau(\delta|\varepsilon) - \varepsilon$. Hence we have in any case $|d(f(p_1),f(p_2)) - d(p_1,p_2)| < \tau(\delta|\varepsilon)+\tau(\delta)$. Since δ is arbitrary, f must be a $\tau(\varepsilon)$-map.

The map f extends to a surjective $\tau(\varepsilon)$-map from S^n to M, which is also denoted by f. Let $\mathrm{Cut}(m)$ be the cut locus of m. Then by Lemma 4, the map $f' = \exp_p \circ I^{-1} \circ \exp_m^{-1} : M \setminus \mathrm{Cut}(m) \to S^n$ is a $\tau(\varepsilon)$-map, and extends to a $\tau(\varepsilon)$-map from M to S^n, which is also denoted by f'. Then Lemma 3 implies that $f'(M)$ is $\tau(\varepsilon)$-dense in S^n. Hence we have just proved the following:

Lemma 5. The maps f and f' give Hausdorff $\tau(\varepsilon)$-approximations between M and S^n.

Thus we have $d_H(M,S^n) < \tau(\varepsilon)$. Lemma 5 enables us to construct an embedding of M into $\mathbf{R}^{n+1}$ which is, in a sense, sufficiently C^1-close to the canonical embedding $S^n \subset \mathbf{R}^{n+1}$. This yields that the Lipschitz distance $d_L(M,S^n)$ between M and S^n is less than $\tau_n(\varepsilon)$, where $\tau_n(\varepsilon)$ depends only on n and ε and satisfies $\lim_{\varepsilon \to 0} \tau_n(\varepsilon) = 0$ (For the details see [OSY] or [Y]). In other words, there is a diffeomorphism $\Phi : M \to S^n$ such that

$$\exp(-\tau_n(\varepsilon)) < \frac{|d\Phi(\xi)|}{|\xi|} < \exp(\tau_n(\varepsilon))$$

for all tangent vectors ξ. This completes the proof of our main theorem.

From the fact $d_H(M,S^n) < \tau(\varepsilon)$, the following problem would arise quite naturally:

Problem. Is $d_L(M,S^n)$ less than $\tau(\varepsilon)$?

If this is true, the answer would provide a universal pinching constant not depending on the dimension n.

References

[CE] J. Cheeger and D. G. Ebin, Comparison Theorems in Riemannian Geometry, North-Holland Math. Library 9, 1975

[G] M. Gromov, Structures métriques pour les variétés riemanniennes, rédigé par J. Lafontaine et P. Pansu, Cedic/Fernand-Nathan Paris, 1981.

[GS] K. Grove and K. Shiohama, A generalized sphere theorem, Ann. of Math., 106(1977), 201-211.

[OSY] Y. Otsu, K. Shiohama and T. Yamaguchi, A new version of differentiable sphere theorem, to appear in Invent. Math.

[Y] T. Yamaguchi, Lipschitz convergence of manifolds of positive Ricci curvature with large volume, to appear in Math. Ann.

Katsuhiro Shiohama

Department of Mathematics
Faculty of Science
Kyushu University
Fukuoka 812
JAPAN

Takao Yamaguchi

Department of Mathematics
College of General Education
Kyushu University
Fukuoka 810
JAPAN

The ideal boundaries of complete open surfaces admitting total curvature $c(M)=-\infty$

Takashi SHIOYA

0. Introduction.

This is a continuation of the previous paper [Sy2]. Throughout this paper let M be a finitely connected, oriented, complete open Riemannian 2-manifold admitting total curvature. Here the total curvature c(M) of M is defined to be an improper integral over M of Gaussian curvature G. A well-known theorem due to Cohn-Vossen [Co1] states that if $\chi(M)$ denotes the Euler characteristic of M, then $c(M) \leq 2\pi\chi(M)$. We constructed in [Sy2] the ideal boundary $M(\infty)$, the equivalent classes of rays, of M. The purpose of this paper is to investigate the Tits metric of the ideal boundary and the visibility axiom of M having infinite total curvature. As is seen in Theorem 2.4 in [Sy2], $M(\infty)$ consists of a circle or a point if M has finite total curvature. We emphasize that if the total curvature of M is $-\infty$, then the behavior of $M(\infty)$ is quite different from the above case. We also emphasize that the visibility axiom of a surface on which the Gaussian curvature changes sign has not yet been discussed before.

We will review the results obtained in [Sy2]. In [Sy2], we defined the equivalence relation of rays and the distance d_∞ of the set of all equivalence classes of rays $M(\infty)$. Here the existence of total curvature is essential throughout the discussion (see section 1). We call the metric space $(M(\infty),d_\infty)$ the ideal boundary of M. We denote by $\gamma(\infty)$ the equivalence class containing a ray γ. If M is a Hadamard manifold (resp. an asymptotically nonnegatively curved manifold), then the metric of our ideal boundary of M coincides the Tits metric

ISBN 0-12-640170-5

due to Gromov [BGS] (resp. the metric defined by Kasue [Ks]). The equivalence relation of rays and the ideal boundary of M have the following properties.

THEOREM (5.1,[Sy2]). *If a ray* σ *of* M *is asymptotic to a ray* γ, *then* σ *and* γ *are equivalent.*

THEOREM (2.4,[Sy2]). *Assume that* M *has only one end.*

(1) If $2\pi\chi(M)-c(M)=0$, *then* $(M(\infty),d_\infty)$ *consists of a single point.*

(2) If $0<2\pi\chi(M)-c(M)<+\infty$, *then* $(M(\infty),d_\infty)$ *is isometric to a circle with the total length* $2\pi\chi(M)-c(M)$.

The metric d_∞ has natural geometric properties as follows. For a fixed simple closed smooth curve c of M and for $t\geq 0$ we set $S(t):=\{x\in M;\ d(x,c)=t\}$ and denote the inner distance of S(t) by d_t, where d is the distance function of M induced from the Riemannian metric of M. Then we have:

THEOREM (5.3,[Sy2]). *For any rays* σ *and* γ *from* c

$$\lim_{t\to\infty}\frac{d_t(\sigma(t),\gamma(t))}{t}=d_\infty(\sigma(\infty),\gamma(\infty)),$$

where a ray α *is called a ray from* c *if* $d(\alpha(t),c)=t$ *for all* $t\geq 0$.

For a ray γ the Busemann function $F_\gamma:M\to\mathbb{R}$ is defined (see, section 22, [Bu]) by

$$F_\gamma(x):=\lim_{t\to\infty}[t-d(x,\gamma(t))].$$

Then the following theorem is true.

THEOREM (5.5,[Sy2]). *For any rays* σ *and* γ *we have*

$$\lim_{t\to\infty}\frac{F_\gamma\circ\sigma(t)}{t}=\cos\min\{d_\infty(\sigma(\infty),\gamma(\infty)),\pi\}.$$

To state one of our results the following notations are needed. For a family $\{I_\lambda\}_{\lambda\in\Lambda}$ of closed intervals in $\mathbb{R}$ (possibly I_λ is a single point or an unbounded interval) we set

$$S(\{I_\lambda\}_{\lambda\in\Lambda}) := \{(z,\lambda);\ z\in I_\lambda,\ \lambda\in\Lambda\}$$

and define the distance function ρ of $S(\{I_\lambda\}_{\lambda\in\Lambda})$ by

$$\rho((z,\lambda),(w,\mu)) := \begin{cases} |z-w| & \text{if } \lambda=\mu \\ \infty & \text{if } \lambda\neq\mu \end{cases}.$$

Then we have:

THEOREM A. *Assume that* M *with one end admits total curvature* $c(M)=-\infty$. *There exists a family* $\{I_\lambda\}_{\lambda\in\Lambda}$ *of closed intervals in* $\mathbb{R}$ *such that* $(M(\infty),d_\infty)$ *is isometric to* $(S(\{I_\lambda\}_{\lambda\in\Lambda}),\rho)$.

Theorem A means that if $c(M)=-\infty$, then every connected component of $M(\infty)$ is isometric to a closed interval of $\mathbb{R}$. If $c(M)>-\infty$, it is true (see Lemma 2.3 in [Sy2]) that for any sequence $\{\gamma_j\}$ of rays converging to some ray γ, $\{\gamma_j(\infty)\}$ converges to $\gamma(\infty)$. However it does not hold in general if $c(M)=-\infty$. In fact if $c(M)=-\infty$, then there exists a sequence $\{\gamma_j\}$ converging to γ such that $\{\gamma_j(\infty)\}$ does not converge to $\gamma(\infty)$. This phenomenon makes the discussion difficult. In section 1 we extend Theorem A to the case where M has more than one end.

The notion of visibility manifolds is due to Eberlein-O'Neill. In [BGS], a Hadamard manifold X is called a visibility manifold if for any two different points $z,w\in X(\infty)$ there exists a geodesic $c:(-\infty,\infty)\to X$ with $c(-\infty)=z$ and $c(\infty)=w$, where $c(-\infty)$ denotes the class containing a ray $t\to c(-t)$. Once we have established the ideal boundary $M(\infty)$ of a complete open surface M admitting total curvature, we can define the notion of a visibility surface by the same way as developed in [BGS]. A finitely connected oriented complete open surface M admitting total curvature is called a *visibility surface* if for any two different points $z,w\in M(\infty)$ there exists a straight line $\gamma:(-\infty,\infty)\to M$ with $\gamma(-\infty)=z$ and $\gamma(\infty)=w$. Note that the total curvature of any visibility surface is $-\infty$. We have the following result.

THEOREM B. *Assume that* M *is finitely connected and admits total curvature. Then the following statements are equivalent.*

(1) M *is a visibility surface.*

(2) There exists a positive ε *such that* $d_\infty(z,w) \geq \varepsilon$ *for any different points* $z,w \in M(\infty)$.

(3) For any different points $z,w \in M(\infty)$, $d_\infty(z,w) = \infty$.

(4) For any rays σ *and* γ *with* $\sigma(\infty) \neq \gamma(\infty)$,

$$\lim_{t\to\infty} F_\gamma \circ \sigma(t) = -\infty.$$

(5) For any rays σ *and* γ *with* $\sigma(\infty) \neq \gamma(\infty)$, $F_\sigma^{-1}([a,\infty)) \cap F_\gamma^{-1}([b,\infty))$ *is bounded for all* $a,b \in \mathbb{R}$.

(6) For any rays σ *and* γ *with* $\sigma(\infty) \neq \gamma(\infty)$, $F_\sigma^{-1}([a,\infty)) \cap F_\gamma^{-1}([b,\infty)) = \emptyset$ *for some* $a,b \in \mathbb{R}$.

Note that Hadamard 2-manifolds satisfy Theorem B (see 4.14, [BGS]). By definition, a Hadamard manifold X satisfies the visibility axiom if and only if for any $p \in X$ and for any $\varepsilon > 0$ there exists a number $r(p,\varepsilon)$ with the property: if $\sigma:[a,b] \to X$ is a geodesic segment such that $d(p,\sigma) \geq r(p,\varepsilon)$, then $\measuredangle_p(\sigma(a),\sigma(b)) \leq \varepsilon$, where $\measuredangle_p(x,y)$ is an angle at p between geodesics joining p to x,y (see [EO]). Note also that a Hadamard 2-manifold satisfies the visibility axiom if and only if it is a visibility surface. However a visibility surface does not necessarily satisfy the visibility axiom. Indeed, if a visibility surface has a point p which is not a pole, then obviously the visibility surface does not satisfy the visibility axiom.

The author would like to express his thanks to Professor K. Shiohama for his assistance during the preparation of this paper.

1. Preliminaries.

We define the following convenient notation. Let D be a domain of M bounded by finitely many piecewise smooth curves $c_1,\ldots,c_n$ each of which is parametrized positively by arc length relative to D. We denote by $\kappa(D)$ the sum of curvature integrals of $c_1,\ldots,c_n$ and of the outer angles at the vertices of D. We remark the following facts.

(1.1) $\kappa(D)=-\kappa(M-D)$.

(1.2) If D *is bounded, then* $c(D)=2\pi\chi(D)-\kappa(D)$.

(1.3) Assume that there exists a compact subset K *of* M *such that* $\partial D-K$ *consists of two geodesics* σ *and* γ. *Then*

$$c(D)\leq 2\pi\chi(D)-\pi-\kappa(D).$$

(1.4) In (1.3), if there exists a constant $r\geq 0$ *such that* $d_D(\sigma(t),\gamma(t))\geq 2t-r$ *for any* $t\geq 0$, *then*

$$c(D)\leq 2\pi\chi(D)-2\pi-\kappa(D),$$

where d_D *is the inner distance of* $Cl(D)$, *the closure of* D, *induced from the Riemannian metric of* M.

(1.1) is obvious. (1.2) is due to the Gauss-Bonnet theorem. (1.3) and (1.4) are due to Cohn-Vossen's theorem [Co2].

We will define the ideal boundaries. First assume that M has only one end. Let σ and γ be arbitrary rays in M. If for some large numbers $a,b\geq 0$ $\sigma([a,+\infty))$ does not intersect $\gamma([b,+\infty))$, then we get a piecewise smooth curve $\alpha:[0,1]\to M$ joining $\sigma(a)$ to $\gamma(b)$ such that $\sigma([a,+\infty)) \cup \alpha([0,1]) \cup \gamma([b,+\infty))$ bounds a half plane D of M and an unbounded domain M-D with $\chi(M-D)=\chi(M)$, where we assume that α is parametrized positively relative to D. In this case we set

$$L(\sigma,\gamma) := \pi-\kappa(D)-c(D).$$

If for some large numbers $a,b\geq 0$, $\sigma([a,+\infty))=\gamma([b,+\infty))$ holds, then we cannot get such a curve α, and we set $L(\sigma,\gamma):=0$ and $L(\gamma,\sigma):=0$ accordingly. Obviously the following propositions (1.5),(1.6) and (1.7) hold for all rays σ and γ.

(1.5) $L(\sigma,\gamma)$ *does not depend on the choice of* α.

(1.6) $L(\sigma,\gamma)\geq 0$.

(1.7) $L(\sigma,\gamma)+L(\gamma,\sigma)=2\pi\chi(M)-c(M)$ *if there exist numbers* $a,b\geq 0$ *such that* $\sigma([a,+\infty))$ *does not intersect* $\gamma([b,+\infty))$.

For arbitrary rays σ and γ in M let c be a simple closed smooth curve bounding a closed tube U, a complete half cylinder, of M such that

(a) c intersects σ (resp. γ) at only point $\sigma(t_\sigma)$ (resp. $\gamma(t_\gamma)$),

(b) $\sphericalangle(\dot\sigma(t_\sigma),\dot c) = \sphericalangle(\dot\gamma(t_\gamma),\dot c) = \pi/2$,

(c) $\sigma([t_\sigma,+\infty))$ does not intersect $\gamma([t_\gamma,+\infty))$ otherwise $\sigma([t_\sigma,+\infty))=\gamma([t_\gamma,+\infty))$.

Here we assume that c is parametrized positively relative to U. Let $I(\sigma,\gamma)$ be the closed subarc of c from $\sigma(t_\sigma)$ to $\gamma(t_\gamma)$ and let $D(\sigma,\gamma)$ be the closed half plane in U bounded by $\sigma([t_\sigma,+\infty)) \cup I(\sigma,\gamma) \cup \gamma([t_\gamma,+\infty))$. $I(\sigma,\gamma)$ is often identified with the interval $c^{-1}(I(\sigma,\gamma))$ in $\mathbb{R}$. Then by definition we have

$$L(\sigma,\gamma) = -c(D(\sigma,\gamma)) - \int_{I(\sigma,\gamma)} \kappa ds \qquad (1.8),$$

where κ denotes the geodesic curvature of c. For any rays σ,τ and γ we get a simple closed smooth curve c bounding a tube of M such that c has the properties (a),(b) and (c) for rays σ,τ and γ. If $\sigma(t_\sigma),\tau(t_\tau)$ and $\gamma(t_\gamma)$ lie on c in this order, then (1.8) implies

$$L(\sigma,\tau)+L(\tau,\gamma)=L(\sigma,\gamma) \qquad (1.9).$$

Rays σ and γ are called equivalent if $\min\{L(\sigma,\gamma),L(\gamma,\sigma)\}=0$. From (1.9) we observe that this is an equivalence relation on the set of all rays in M.

We denote the equivalence class of a ray γ by $\gamma(\infty)$ and denote the set of all equivalence classes by $M(\infty)$. (1.9) implies that the values $\min\{L(\sigma,\gamma),L(\gamma,\sigma)\}$ does not depend on the choice of rays σ,γ in the equivalence classes $\sigma(\infty),\gamma(\infty)$. We define the function $d_\infty: M(\infty)\times M(\infty) \to \mathbb{R}\cup\{+\infty\}$ by

$$d_\infty(\sigma(\infty),\gamma(\infty)) := \min\{L(\sigma,\gamma),L(\gamma,\sigma)\}.$$

Moreover d_∞ becomes a distance function of $M(\infty)$ (see section 1 in [Sy2]). We call the metric space $(M(\infty),d_\infty)$ the ideal boundary of M.

The following proposition is obvious.

(1.10) For any straight line γ *in* M, $d_\infty(\gamma(-\infty),\gamma(\infty)) \geq \pi$.

Next we consider the case where M has more than one end. Assume that M is finitely connected with k ends. Let K be a compact domain in M such that M-Int(K) consists of disjoint closed tubes $U_1, \ldots, U_k$ and ∂K consists of k simple closed piecewise smooth curves. We set

$$s_i(M) := -c(U_i)-\kappa(U_i) \quad \text{for } i=1, \dots ,k.$$

Then by the Gauss-Bonnet theorem this value does not depend on the choice of tube U_i and we have

$$\sum_{1\le i\le k} s_i(M) = 2\pi\chi(M)-c(M).$$

Let M_i be a complete open Riemannian 2-manifold with one end such that there exists an isometric embedding $\iota_i:U_i\cup K\to M_i$ and $M_i-\iota_i(U_i\cup K)$ consists of k-1 open disk domains. Then the Gauss-Bonnet theorem implies

$$s_i(M) = 2\pi\chi(M_i)-c(M_i).$$

For any ray γ let $n(\gamma)$ be a unique number in $\{1, \dots ,k\}$ such that some subray of γ is contained in the tube $U_{n(\gamma)}$. Rays σ and γ are called equivalent if $i:=n(\sigma)=n(\gamma)$ and if two rays $\iota_i \circ \sigma_1$ and $\iota_i \circ \gamma_1$ are equivalent, where σ_1,γ_1 are subrays of σ,γ. Here we remark that there exist subrays σ_1,γ_1 of σ,γ such that $\iota_i \circ \sigma_1$ and $\iota_i \circ \gamma_1$ are rays in M_i. We denote the equivalence class of a ray γ by $\gamma(\infty)$ and denote the set of all equivalence classes by $M(\infty)$. We define the distance function $d_\infty:M(\infty)\times M(\infty)\to\mathbb{R}\cup\{+\infty\}$ by

$$d_\infty(\sigma(\infty),\gamma(\infty)) := \begin{cases} d^i_\infty(\iota_i\circ\sigma_1(\infty),\iota_i\circ\gamma_1(\infty)) & \text{if } i:=n(\sigma)=n(\gamma) \\ +\infty & \text{if } n(\sigma)\neq n(\gamma), \end{cases}$$

where σ_1,γ_1 are subrays of σ,γ and d^i_∞ is the distance of $M_i(\infty)$.

The following theorem is a direct consequence of the construction of $M(\infty)$.

THEOREM (5.2,[Sy2]). *Assume that* M *with* k *ends admits total curvature and that* M_i *for* $1\le i\le k$ *are as above. Then we have*

$$M(\infty) = M_1(\infty) \cup \dots \cup M_k(\infty) \quad \textit{(disjoint union)}.$$

2. Proof of Theorem A.

In this section we will prove Theorem A. Throughout this section we assume that M has only one end and that c is a fixed simple closed smooth curve bounding a tube U of M. In [Sy2] we have proved the following lemma which is used for the proof of Theorem A.

Lemma (2.2,[Sy2]). *For any* $x \in M(\infty)$ *there exists a ray* γ *from* c *such that* $x = \gamma(\infty)$.

Proof of Theorem A. It suffices to show that for any $x \in M(\infty)$ the connected component

$$I_x := \{ y \in M(\infty);\ d_\infty(x,y) < +\infty \}$$

of $M(\infty)$ is isometric to a closed interval in $\mathbb{R}$. Take a ray σ from c with $\sigma(\infty)=x$. Assume that $c:[0,\ell] \to M$ is parametrized positively by arc length relative to U such that $c(0)=c(\ell)=\sigma(0)$. Set

$$I_x^+ := \{ \gamma(\infty) \in M(\infty);\ \gamma \text{ is a ray from c with } 0 < L(\sigma,\gamma) < +\infty \},$$

$$I_x^- := \{ \gamma(\infty) \in M(\infty);\ \gamma \text{ is a ray from c with } 0 < L(\gamma,\sigma) < +\infty \}.$$

Then since $c(M)=-\infty$, these are disjoint subsets of I_x and satisfy $I_x = I_x^+ \cup I_x^- \cup \{x\}$. We define the function $f_\sigma : I_x^+ \cup \{x\} \to [0,+\infty)$ by

$$f_\sigma(\gamma(\infty)) := \begin{cases} L(\sigma,\gamma) & \text{if } \gamma(\infty) \neq x \\ 0 & \text{if } \gamma(\infty) = x \end{cases}$$

for each ray γ from c with $\gamma(\infty) \in I_x^+ \cup \{x\}$. Lemma 2.2 in [Sy2] and the formula (1.9) imply that the function f_σ is well-defined and is an injection. By the discussion as in the proof of Theorem 2.4 (2) in [Sy2], we conclude that the image of this function is an interval in $\mathbb{R}$.

We will prove that f_σ is an isometry. Indeed, by (1.9) for any rays τ and γ such that $\tau(\infty), \gamma(\infty) \in I_x^+ \cup \{x\}$ and $c^{-1} \circ \tau(0) < c^{-1} \circ \gamma(0)$, we have

$$d_\infty(\tau(\infty),\gamma(\infty)) = L(\tau,\gamma) = f_\sigma(\gamma(\infty)) - f_\sigma(\tau(\infty)).$$

This implies that f_σ is an isometry.

We will prove that the image of f_σ is closed. Set $b := \sup f_\sigma$. We may prove that if $b < +\infty$, then there exists a $z \in I_x^+ \cup \{x\}$ such that $f_\sigma(z) = b$. Let $\{z_i\}$ be a sequence of points in $I_x^+ \cup \{x\}$ such that $f_\sigma(z_i)$ tends to b as $i \to \infty$ and let $\{\gamma_i\}$ be a sequence of rays from c such that $\gamma_i(\infty) = z_i$ for all i. There exists a subsequence $\{\gamma_j\}$ of $\{\gamma_i\}$ such that γ_j tends to some ray γ from c. Here $0 = c^{-1} \circ \sigma(0) \leq c^{-1} \circ \gamma_j(0) \leq c^{-1} \circ \gamma(0)$. Set $z := \gamma(\infty)$. We will prove that $f_\sigma(z) = b$.

Now suppose that $L(\sigma,\gamma)=+\infty$. Then $c(D(\sigma,\gamma))=-\infty$. For an arbitrary positive ε there exists a compact set $K\subset D(\sigma,\gamma)$ such that $c(K)<-1/\varepsilon$ and

$$\int_{D(\sigma,\gamma)-K} G^+ \, dM < \varepsilon,$$

where $G^+(p):=\max\{G(p),0\}$ for $p\in M$. Since the area of $D(\gamma_j,\gamma)\cap K$ tends to zero as $j\to\infty$, $c(D(\sigma,\gamma_j)\cap K)$ tends to $c(K)$. Hence $c(D(\sigma,\gamma_j)\cap K) < -1/\varepsilon$ for all sufficiently large j. Thus

$$c(D(\sigma,\gamma_j)) = c(D(\sigma,\gamma_j)\cap K) + c(D(\sigma,\gamma_j)-K) < -1/\varepsilon + \varepsilon$$

for all sufficiently large j and $c(D(\sigma,\gamma_j))$ tends to $-\infty$. Hence by the formula (1.8), $f_\sigma(z_j)$ tends to $+\infty$. This contradicts $b<+\infty$.

Therefore $f_\sigma(z)=L(\sigma,\gamma)<+\infty$. Since $c(D(\sigma,\gamma))>-\infty$, for an arbitrary positive $\varepsilon>0$ there exists a compact set $K\subset D(\sigma,\gamma)$ such that

$$\int_{D(\sigma,\gamma)-K} |G| \, dM < \varepsilon.$$

Since the area of $D(\gamma_j,\gamma)\cap K$ tends to zero as $j\to\infty$, $c(D(\sigma,\gamma_j)\cap K)$ tends to $c(D(\sigma,\gamma)\cap K)$. On the other hand

$$| \, c(D(\sigma,\gamma)\cap K) - c(D(\sigma,\gamma)) \, | < \varepsilon$$

and

$$| \, c(D(\sigma,\gamma_j)\cap K) - c(D(\sigma,\gamma_j)) \, | < \varepsilon$$

for all j. Hence

$$| \, c(D(\sigma,\gamma_j)) - c(D(\sigma,\gamma)) \, | < 3\varepsilon$$

for all sufficiently large j. Since γ_j tends to γ,

$$\lim_{j\to\infty} \int_{I(\sigma,\gamma_j)} \kappa ds = \int_{I(\sigma,\gamma)} \kappa ds.$$

Therefore the formula (1.8) implies that

$$| \, L(\sigma,\gamma_j) - L(\sigma,\gamma) \, | < 4\varepsilon$$

for all sufficiently large j. Thus $f_\sigma(z_j)$ tends to $f_\sigma(z)$ as $j\to\infty$ and $f_\sigma(z)=b$.

Therefore $I_x^+ \cup \{x\}$ is isometric to a closed interval J^+ in $[0,+\infty)$ containing 0. In the same way we can show that $I_x^- \cup \{x\}$ is isometric to a closed interval in $(-\infty,0]$

containing 0. Thus I_x is isometric to $J^+ \cup J^+$. This completes the proof of Theorem A.

2. Proof of Theorem B.

(1) $\Rightarrow$ *(2)* is obvious by (1.10).

(2) $\Rightarrow$ *(3)* : Indeed, by Theorems 2.4 in [Sy2], 5.2 in [Sy2] and A, for any point $x \in M(\infty)$, $I_x := \{y \in M(\infty); d_\infty(x,y) < +\infty\}$ consists of a single point.

(3) $\Rightarrow$ *(1)* : For arbitrary different points $z,w \in M(\infty)$ let σ and γ be rays such that $\sigma(\infty)=z$ and $\gamma(\infty)=w$. Let $\{t_j\}$ be a monotone and divergent sequence of numbers and $\{\alpha_j:[0,\ell_j]\to M\}$ a sequence of minimizing segments such that $\alpha_j(0)=\sigma(t_j)$ and $\alpha_j(\ell_j)=\gamma(t_j)$ for all j.

If $n(\sigma)\neq n(\gamma)$, then since each α_j intersects K, some subsequence of $\{\alpha_j\}$ converges to a straight line α and two rays $\alpha|[0,+\infty)$, $t\to\alpha(-t)$ are asymptotic to σ,γ respectively, hence by Theorem 5.1 in [Sy2] we have $\alpha(-\infty)=z$ and $\alpha(\infty)=w$.

Thus we consider the case where $n(\sigma)=n(\gamma)$. Set $i:=n(\sigma)=n(\gamma)$. We may assume that c is a simple closed smooth curve bounding a tube U_i and has the properties (a),(b) and (c) in section 1 for σ and γ. We will show that some subsequence of $\{\alpha_j\}$ converges.

Suppose that each subsequence of $\{\alpha_j\}$ diverges. Since α_j for each sufficiently large j does not intersect c, without loss of generality we may assume that each α_j is contained in $D(\sigma,\gamma)$. Let D_j be a disk domain bounded by $\sigma([t_\sigma,t_j]) \cup I(\sigma,\gamma) \cup \gamma([t_\gamma,t_j]) \cup \alpha_j([0,\ell_j])$ and θ_j,φ_j be inner angles of D_j at $\sigma(t_j),\gamma(t_j)$. Since $\{D_j\}$ is a monotone increasing sequence of domains with $\cup D_j=D(\sigma,\gamma)$, $c(D_j)$ tends to $c(D(\sigma,\gamma))$ as $j\to\infty$. Hence the Gauss-Bonnet theorem implies

$$c(D(\sigma,\gamma)) = \lim_{j\to\infty} c(D_j) = \lim_{j\to\infty} [\theta_j + \varphi_j - \pi] - \int_{I(\sigma,\gamma)} \kappa ds.$$

By (1.8),

$$L(\sigma,\gamma) = \lim_{j\to\infty} [\pi - \theta_j - \varphi_j].$$

Hence $d_\infty(\sigma(\infty),\gamma(\infty))<+\infty$. This contradicts (3).

Thus there exists a subsequence $\{\alpha_k\}$ of $\{\alpha_j\}$ such that α_k tends to some straight line α. By Theorem 5.1 in [Sy2], $\alpha(-\infty)=\sigma(\infty)$ and $\alpha(\infty)=\gamma(\infty)$. This concludes (1).

(3) $\Rightarrow$ *(4)* : By Theorem 5.5 in [Sy2],

$$\lim_{t\to\infty}\frac{F_\gamma \circ \sigma(t)}{t} = -1.$$

This concludes (4).

(4) $\Rightarrow$ *(2)* : Indeed $d_\infty(\sigma(\infty),\gamma(\infty))\geq\pi/2$ by Theorem 5.5 in [Sy2].

(3) $\Rightarrow$ *(5)* : Suppose that $F_\sigma^{-1}([a,\infty)) \cap F_\gamma^{-1}([b,\infty))$ is unbounded. Then there exists an unbounded sequence $\{p_i\} \subset F_\sigma^{-1}([a,\infty)) \cap F_\gamma^{-1}([b,\infty))$ such that each subsequence of $\{p_i\}$ diverges. Note that for every sufficiently large i, p_i is not necessarily contained in some common tube U of M. Let τ_i be a ray emanating from p_i which is asymptotic to σ.

First we will show that any subsequence of $\{\tau_i\}$ diverges. Suppose that there exists a subsequence $\{\tau_j\}$ of $\{\tau_i\}$ converging to some straight line τ. For a point q on τ there exists a point q_j on τ_j such that q_j tends to q as $j\to\infty$. There are a minimizing segment $\tau_{j,t}$ joining p_j to $\sigma(t)$ and a point $q_{j,t}$ on $\tau_{j,t}$ such that $q_{j,t}$ tends to q_j as $t\to\infty$. Hence for a fixed positive ε there exist a number j_0 and a sequence $\{t_j\}$ of positive numbers such that $d(q_{j,t},q)<\varepsilon$ for all $t\geq t_j$ and for all $j\geq j_0$. Thus the triangle inequality implies that

$$\begin{aligned} d(p_j,\sigma(t)) &= d(p_j,q_{j,t}) + d(q_{j,t},\sigma(t)) \\ &\geq d(p_j,q) + d(q,\sigma(t)) - 2\varepsilon \\ &\geq t - d(q,\sigma(0)) + d(p_j,q) - 2\varepsilon \end{aligned}$$

for all $t\geq t_j$ and for all $j\geq j_0$. Therefore

$$F_\sigma(p_j) = \lim_{t\to\infty} [\, t-d(p_j,\sigma(t))\,] \leq d(q,\sigma(0)) - d(p_j,q) + 2\varepsilon$$

for all $j\geq j_0$. This contradicts $F_\sigma(p_j)\geq a$.

It follows that there exists a number i_0 such that p_i is contained in some tube $U_{n(\sigma)}$ for all $i \geq i_0$. Indeed if otherwise supposed, then each τ_i intersects $\partial U_{n(\sigma)}$ and hence a contradiction is derived by the above discussion. We may assume that c is a simple closed smooth curve bounding $U_{n(\sigma)}$ and that there is a number t_σ such that $d(\sigma(t),c)=t-t_\sigma$ for all $t \geq t_\sigma$. Let ρ_i for $i \geq i_0$ is a minimizing segment in $U_{n(\sigma)}$ joining a point on c to p_i such that $L(\rho_i) = d(p_i,c)$. Since $\{p_i\}$ diverges, there is a subsequence $\{\rho_j\}$ of $\{\rho_i\}$ converging some ray ρ from c. Without loss of generality we may assume that some subray of τ_j is contained in $D(\sigma,\rho)$ for all j. Since $\{\tau_j\}$ diverges, the domain D_j in $U_{n(\sigma)}$ bounded by σ, ρ_j, τ_j and the subarc of c from $\sigma(t_\sigma)$ to $\rho_j(0)$ tends to $D(\sigma,\rho)$ as $j \to \infty$. If θ_j is an inner angle of D_j at $\tau_j(0)$, then since $L(\sigma,\tau_j)=0$ for all j,

$$c(D(\sigma,\rho)) = \lim_{j\to\infty} c(D_j) = \lim_{j\to\infty} \theta_j - \pi - \int_{I(\sigma,\rho)} \kappa ds.$$

Hence by (1.8) we have

$$L(\sigma,\rho) = \pi - \lim_{j\to\infty} \theta_j \leq \pi \qquad \text{and hence} \qquad d_\infty(\sigma(\infty),\rho(\infty)) \leq \pi.$$

In the same way we have $d_\infty(\gamma(\infty),\rho(\infty)) \leq \pi$. Hence the triangle inequality implies $d_\infty(\sigma(\infty),\gamma(\infty)) \leq 2\pi$. This contradicts (3).

(5) ⇒ (6) follows from the fact that $\{F_\sigma^{-1}([t,\infty))\}_t$ is a monotone decreasing sequence with $\bigcap_t F_\sigma^{-1}([t,\infty)) = \emptyset$.

(6) ⇒ (2) : It follows from (6) that $F_\sigma \circ \gamma$ is bounded above. Hence by Theorem 5.5 in [Sy2], $d_\infty(\sigma(\infty),\gamma(\infty)) \geq \pi/2$.

References

[BGS] W. Ballmann, M. Gromov and V. Schroeder, *Manifolds of Nonpositive Curvature* , Progress in Math. 61, Birkhäuser, Boston-Basel Stuttgart, 1985.

[Bu] H. Busemann, *The geometry of geodesics* , Academic Press, New York, 1955.

[Co1] S. Cohn-Vossen, Kürzeste Wege und Totalkrümmung auf Flächen, Composito Math., 2(1935), 63-133.

[Co2] S. Cohn-Vossen, Totalkrümmung und geodätische Linien auf einfach zusammenhängenden offenen volständigen Flächenstücken, Recueil Math. Moscow, 43(1936), 139-163.

[EO] P. Eberlein and B. O'Neill, Visibility manifolds, Pac. J. Math. 46 (1973), 45-110.

[Fi] F. Fiala, Le problème isopérimètres sur les surface onvretes à courbure positive, Comment. Math. Helv., 13(1941), 293-346.

[Ha] P. Hartman, Geodesic parallel coordinates in the large, Amer. J. Math., 86(1964), 705-727.

[Ks] A. Kasue, A compactification of a manifold with asymptotically nonnegative curvature, Preprint.

[Md1] M. Maeda, On the existence of rays, Sci. Rep. Yokohama Nat. Univ., 26(1979), 1-4.

[Md2] M. Maeda, A geometric significance of total curvature on complete open surfaces, *Geometry of Geodesics and Related Topics* , Advanced Studies in Pure Math., 3(1984), 451-458, Kinokuniya, Tokyo, 1984.

[Md3] M. Maeda, On the total curvature of noncompact Riemannian manifolds I, Yokohama Math. J., 33(1985), 93-101.

[Og] T. Oguchi, Total curvature and measure of rays, Proc. Fac. Sci. Tokai Univ., 21(1986), 1-4.

[Ot] F. Ohtsuka, On a relation between total curvature and Tits metric, Bull. Fac. Sci. Ibaraki Univ., 20(1988).

[Sg1] K. Shiga, On a relation between the total curvature and the measure of rays, Tsukuba J. Math., 6(1982), 41-50.

[Sg2] K. Shiga, A relation between the total curvature and the measure of rays, II, Tôhoku Math. J., 36(1984), 149-157.

[Sh1] K. Shiohama, Busemann function and total curvature, Invent. Math., 53 (1979), 281-297

[Sh2] K. Shiohama, The role of total curvature on complete noncompact Riemannian 2-manifolds, Illinoies J. Math. 28(1984), 597-620.

[Sh3] K. Shiohama, Cut locus and parallel circles of a closed curve on a Riemannian plane admitting total curvature, Comment. Math. Helv., 60(1985), 125-138.

[Sh4] K. Shiohama, Total curvatures and minimal areas of complete open surfaces, Proc. Amer. Math. Soc., 94(1985), 310-316.

[Sh5] K. Shiohama, An integral formula for the measure of rays on complete open surfaces, J. Differential Geometry, 23(1986),197-205.

[SST] K. Shiohama, T. Shioya and M. Tanaka, Mass of rays on complete open surfaces, Preprint.

[Sy1] T. Shioya, On asymptotic behavior of the mass of rays, to appear in Proc. Amer. Math. Soc.

[Sy2] T. Shioya, The ideal boundaries of complete open surfaces, Preprint.

Takashi SHIOYA
Department of Mathematics
Faculty of Science
Kyushu University
Fukuoka 812
Japan

ON THE LENGTHS OF STABLE JACOBI FIELDS

Tadashi Yamaguchi

§ 0. Introduction

Rauch comparison theorem on the lengths of Jacobi fields plays basic important roles in the various places of differential geometry concerning the relations of curvatures and arc-lengths. The behavior, especially C^2-derivative, of Busemann functions of an Hadamard manifold is expressed by that of stable Jacobi fields. E. Heintze & H. C. Im Hof proved in [6] this fact and also C^2-differentiability of the Busemann function. They showed moreover the distance-estimating formula of horospheres, i.e., level hypersurfaces of Busemann function, of an Hadamard manifold whose sectional curvatures are pinched by two negative constants. The crucial matter of their arguments is the comparison theorem for stable Jacobi fields in these Hadamard manifolds, which is a refinement of [2] lecture 3 and [5] p. 117 as they remarked there.

In this paper, first of all, we modify the fundamental Rauch comparison theorem (*theorem 1*) and get a general comparison theorem for stable Jacobi fields which includes the formulas in [2], [5] and [6] as the special cases

ISBN 0-12-640170-5

(*theorem 3*). Then we apply this theorem 3 to a manifold whose sectional curvatures are pinched by a distance-function from a fixed point. A typical example of such manifolds is an asymptotically non-negatively curved manifold of [1] (*theorem 4*).

§ 1. Notations and preliminaries

Let M be a Riemannian manifold, i.e., a connected differentiable manifold without boundary of dimension $m \geq 2$ with Riemannian metric $\langle\ ,\ \rangle$ and Riemannian connection ∇. For a point $p \in M$, the sectional curvature of M with respect to a 2-dimensional subspace σ generated by two linearly independent vectors v and w of the tangent space M_p at p is denoted by K_σ or $K(v, w)$, and the set of all 2-spaces of M_p by G_p. For an n-dimensional differentiable submanifold N of M ($0 \leq n \leq m-1$) and for $p \in N$, we denote by $N_p^\perp$ the normal space of N_p in M_p, and by A_v the second fundamental tensor with respect to $v \in N_p^\perp$, that is, A_v is defined by $\langle A_v X, Y \rangle = \langle \nabla_X Y, v \rangle$ for any vector fields X, Y of M which are tangent to N on N.

In the following, geodesics are assumed to be parametrized by arc-length and a geodesic $c: [0, \infty) \longrightarrow M$ is called *a ray* if any subarcs are shortest. Given a geodesic $c: [0, l] \longrightarrow M$ ($l > 0$), a differentiable vector field X along c is called *a Jacobi field* when $X'' + R(X, \dot{c})\dot{c} = 0$ holds on c, where R is the curvature tensor of M, $\dot{c}$ the tangent vector of c and X' means the covariant derivative of X with respect to $\dot{c}$ i.e., $X' = \nabla_{\dot{c}} X$. When $p = c(0) \in N$,

$\dot{c}(0)\in N_p^{\perp}$, a Jacobi field X along $c:[0, l]\longrightarrow M$ is said to be *N-Jacobi field* if $X(0)\in N_p$, $A_{\dot{c}(0)}X(0) + X'(0) \in N_p^{\perp}$ and $\langle X, \dot{c}\rangle = 0$. The set of all N-Jacobi fields along c is denoted by $J_{c,N}$. A point $c(t_0)$ ($0<t_0\leq l$) is called *a focal point* of N along c if there exists a non-trivial N-Jacobi field X along c with $X(t_0)=0$. When $n=0$ inparticular, i.e., N = { a point p }, a focal point of { p } is called *a conjugate point* of p along c.

We give here the expressions of ($N(\lambda)$-)Jacobi fields on the manifold of constant curvature. For a fixed real number k, let $M(k)$ be an m-dimensional Riemannian manifold of constant curvature k. For a geodesic $c:[0, \infty)\longrightarrow M(k)$ and a orthonormal basis $\{ v_1, \cdots\cdots, v_{m-1}, \dot{c}(0) \}$ of $M(k)_{c(0)}$, we have parallel vector fields $X_1, \cdots\cdots, X_{m-1}, \dot{c}$ along c with $X_i(0) = v_i$. Then a differentiable vector field $Y = \sum_{i=1}^{m-1} f^i X_i + f^m \dot{c}$ along c is a Jacobi field if and only if $f^{i\prime\prime} + kf^i = 0$ ($i=1,\cdots\cdots,m-1$) and $f^{m\prime\prime} = 0$. Therefore an arbitrary Jacobi field Y along c on $M(k)$ is expressed as

$$Y(t) = \begin{cases} \sum_{i=1}^{m-1} (a^i\cos\sqrt{k}\cdot t + b^i\sin\sqrt{k}\cdot t)X_i(t) + (a^m + b^m t)\dot{c}(t) & (\text{ if } k > 0) \\ \sum_{i=1}^{m-1} (a^i + b^i t)X_i(t) + (a^m + b^m t)\dot{c}(t) & (\text{ if } k = 0) \\ \sum_{i=1}^{m-1} (a^i\cosh\sqrt{|k|}\cdot t + b^i\sinh\sqrt{|k|}\cdot t)X_i(t) + (a^m + b^m t)\cdot\dot{c}(t) & (\text{ if } k < 0) \end{cases}$$

where a^i, b^i ($i=1,\cdots\cdots,m$) are constants. Hence, when

$k>0$, $c(t)$ is a conjugate point of $c(0)$ along c if and only if $t = r\pi/\sqrt{k}$ (r is a positive integer) and when $k\leq 0$, there exist no conjugate points of $c(0)$ on c.

Moreover, for a fixed real number λ, let $N(\lambda)$ be an n-dimensional submanifold of $M(k)$ such that $1\leq n\leq m-1$, $c(0)\in N(\lambda)$, $\dot{c}(0)\in N(\lambda)^{\perp}_{c(0)}$, $\{ v_1, \cdots\cdots, v_n \}$ generates $N(\lambda)_{c(0)}$ and the eigenvalues of $A_{\dot{c}(0)}$ are λ only. Then

$$Y = \sum_{i=1}^{m-1} f^i X_i + f^m \dot{c} \in J_{c, N(\lambda)}$$

if and only if

$$f^{i\prime\prime} + kf^i = 0 \qquad (i=1, \cdots\cdots, m-1), \qquad f^m = 0,$$

$$f^i(0) = 0 \qquad (i=n+1, \cdots\cdots, m-1)$$

and

$$f^{i\prime}(0) + \lambda f^i(0) = 0 \quad (i=1, \cdots\cdots, n).$$

Therefore any $Y \in J_{c, N(\lambda)}$ is expressed as

$$Y(t) = f_{k,\lambda}(t) \sum_{i=1}^{n} a^i X_i(t) + f_k(t) \sum_{i=n+1}^{m-1} a^i X_i(t)$$

where a^i ($i=1, \cdots\cdots, m-1$) are constants and f_k, $f_{k,\lambda}$ are functions defined by

$$f_k(t) = \begin{cases} \sin \sqrt{k}\cdot t & (\text{if } k > 0) \\ t & (\text{if } k = 0) \\ \sinh \sqrt{|k|}\cdot t & (\text{if } k < 0) \end{cases}$$

and

$$f_{k,\lambda}(t) = \begin{cases} \cos \sqrt{k}\cdot t - \dfrac{\lambda}{\sqrt{k}} \sin \sqrt{k}\cdot t & (\text{if } k > 0) \\ 1 - \lambda t & (\text{if } k = 0) \end{cases}$$

$$\qquad \cosh\sqrt{|k|}\cdot t - \frac{\lambda}{\sqrt{|k|}}\sinh\sqrt{|k|}\cdot t \qquad (\text{ if } k<0).$$

Hence for any positive root t_0 of the equation;

$$\begin{cases} \cot\sqrt{k}\cdot t = \dfrac{\lambda}{\sqrt{k}} & (\text{ if } k>0) \\ \lambda t = 1 & (\text{ if } k=0) \\ \coth\sqrt{|k|}\cdot t = \dfrac{\lambda}{\sqrt{|k|}} & (\text{ if } k<0), \end{cases}$$

$c(t_0)$ is a focal point of $N(\lambda)$ along c, and $c(r\pi/\sqrt{k})$ (r is any positive integer) is also a focal point if $k>0$ and $N(\lambda)$ is not a hypersurface. There exist no other focal points. In particular when $k\leq 0$ and $\lambda\leq\sqrt{|k|}$, there exist no focal points of $N(\lambda)$ on c.

§ 2. The ratios of Jacobi fields

Let (M, N, c) be a triple of a Riemannian manifold M of $\dim M = m \geq 2$, an n-dimensional submanifold N of M ($0\leq n\leq m-1$) and a geodesic $c:[0, l]\longrightarrow M$ such that $c(0)=p\in N$, $\dot{c}(0)\in N_p^{\perp}$. We take an another similar triple $(M^{\sim}, N^{\sim}, c^{\sim})$ and denote the corresponding terms by $^{\sim}$.

We assume the following properties (K) and (A);

(K) $m\leq m^{\sim}$, $l=l^{\sim}$ and $K_{\sigma}\leq K_{\sigma^{\sim}}$ for any $t\in[0, l]$, $\dot{c}(t)\in\sigma\in G_{c(t)}$, $\dot{c}^{\sim}(t)\in\sigma^{\sim}\in G_{c^{\sim}(t)}$.

(A) If $nn^{\sim}\neq 0$, then every eigenvalue of $A_{\dot{c}(0)}$ is not greater than any one of $A_{\dot{c}^{\sim}(0)}$.

Fundamental Rauch comparison theorem may be stated in the following form;

Theorem ([9] theorem 4-3, [8] theorem 2-5)

For any triples (M, N, c), $(M^{\sim}, N^{\sim}, c^{\sim})$ satisfying (K), (A) and having no focal points of $N^{\sim}$ on $c^{\sim}([0, l])$, we take $X \in J_{c,N}$ and $X^{\sim} \in J_{c^{\sim},N^{\sim}}$ with initial conditions (1) or (2);

(1) $n=n^{\sim}=0$ and $\|X'(0)\| = \|X^{\sim\prime}(0)\|$,

(2) $n>0$, $n^{\sim}=m^{\sim}-1$ and $\|X(0)\| = \|X^{\sim}(0)\| \neq 0$,

Then $\|X(t)\| \geq \|X^{\sim}(t)\|$ holds for any $t \in [0, l]$.
Moreover, if $\|X(t_0)\| = \|X^{\sim}(t_0)\|$ for a point $t_0 \in (0, l]$, we have $\|X(t)\| = \|X^{\sim}(t)\|$, $< R(X, \dot{c})\dot{c}, X >(t) = < R(X^{\sim}, \dot{c}^{\sim})\dot{c}^{\sim}, X^{\sim} >(t)$ for any $t \in [0, t_0]$ and $< A_{\dot{c}(0)}X(0), X(0) > = < A_{\dot{c}^{\sim}(0)}X^{\sim}(0), X^{\sim}(0) >$.

The proof of this theorem essentially depends on the following simple lemma. In fact, we have only to apply the lemma for $f_1 = \|X\|^2$ and $f_2 = \|X^{\sim}\|^2$ and to prove the condition (4) by comparing their index forms.

Lemma ([9] lemma 3-4)

Let f_1 and f_2 be real-valued functions satisfying the following four conditions;

(1) f_1 and f_2 are differentiable on $[0, l]$,

(2) $f_1(t)$ and $f_2(t)$ are positive for any $t \in (0, l]$,

(3) $$\lim_{t \downarrow 0} \frac{f_1(t)}{f_2(t)} = 1,$$

(4) $$\frac{f_1'(t)}{f_1(t)} \geq \frac{f_2'(t)}{f_2(t)} \quad \text{for any } t \in (0, l].$$

Then $f_1(t) \geq f_2(t)$ holds for any $t \in [0, l]$.
Moreover, if $f_1(t_0) = f_2(t_0)$ for a point $t_0 \in [0, l]$, then

we have $f_1 = f_2$ on $[0, t_0]$.

Proof. Denoting $g_i = \frac{f'_i}{f_i}$ on $(0, l]$ for $i=1, 2$, we have

$$\frac{f_i(t)}{f_i(t_1)} = \exp \int_{t_1}^{t} g_i(s) ds \quad \text{for} \quad 0<t_1 \le t \le l,$$

hence

$$\frac{f_1(t)}{f_2(t)} = \frac{f_1(t_1)}{f_2(t_1)} \exp \int_{t_1}^{t} (g_1(s) - g_2(s)) ds$$

$$= \exp \int_{0}^{t} (g_1(s) - g_2(s)) ds$$

by (3). We apply (4) and get the conclusion.

Using this lemma we can modify slightly the above theorem and get a comparison theorem for ratios of Jacobi fields.

<u>Theorem 1</u>

Under the assumptions of the above theorem with non-trivial X and $X^\sim$, $\frac{\|X(t_1)\|}{\|X(t_2)\|} \le \frac{\|X^\sim(t_1)\|}{\|X^\sim(t_2)\|}$ holds for every $0 \le t_1 < t_2 \le l$. Moreover, if the equality holds for some $0<t_1<t_2 \le l$, we have $\|X\| = \|X^\sim\|$, $< R(X, \dot{c})\dot{c}, X > = < R(X^\sim, \dot{c}^\sim)\dot{c}^\sim, X^\sim >$ on $[0, t_2]$ and $< A_{\dot{c}(0)}X(0), X(0) > = < A_{\dot{c}^\sim(0)}X^\sim(0), X^\sim(0) >$.

Proof. In the proof of lemma we get for any $0 \le t_1 < t_2 \le l$,

$$\frac{f_1(t_1)}{f_1(t_2)} = \frac{f_2(t_1)}{f_2(t_2)} \exp \int_{t_1}^{t_2} -(g_1(s) - g_2(s)) ds \le \frac{f_2(t_1)}{f_2(t_2)}.$$

The equality condition is proved easily by this formula and the index form minimization theorem.

Corollary ([7] proposition 4-1, [6] proposition 2-3)

Let M be an m-dimensional Riemannian manifold, $c:[0, l]\longrightarrow M$ a geodesic and X a Jacobi field along c such that $X(0)=0$, $X'(0)\neq 0$ and $\langle X, \dot{c}\rangle=0$. For a constant k, we assume that $K_\sigma \leq k$ holds for any $0\leq t\leq l$, $\dot{c}(t)\in\sigma\in G_{c(t)}$, and that $l<\pi/\sqrt{k}$ when $k>0$ (respectively, $K_\sigma\geq k$ and there exist no conjugate points of $c(0)$ on c).

Then it follows that $\dfrac{\|X(t_1)\|}{\|X(t_2)\|} \leq \dfrac{f_k(t_1)}{f_k(t_2)}$ ($\geq$, resp.)

for every $0\leq t_1<t_2\leq l$, where f_k is the function given in § 1. Moreover, if the equality holds for some $0<t_1<t_2\leq l$, then we have $\|X\| = f_k$ and $K(X, \dot{c}) = k$ on $(0, t_2]$.

Proof. This is a direct consequence of theorem 1 together with the forms of Jacobi fields in the constantly-curved manifolds.

Similarly, by comparing with the submanifold in § 1, that is, with the submanifold of a constantly-curved Riemannian manifold whose second fundamental tensor at a point has only one eigen-value, we get the following;

Corollary 2

Let M be an m-dimensional Riemannian manifold and N an n-dimensional submanifold of M where $1\leq n\leq m-1$. We take a geodesic $c:[0, l]\longrightarrow M$ such that $c(0)=p\in N$ and $\dot{c}(0)\in N_p^\perp$, and a $X\in J_{c,N}$ satisfying $X(0)\neq 0$.

For two constants k and λ we assume that $K_\sigma\leq k$ for any $0\leq t\leq l$, $\dot{c}(t)\in\sigma\in G_{c(t)}$, and that $< A_{\dot{c}(0)}v, v > \leq \lambda\cdot\|v\|^2$ for any $v\in N_p$ and $l<\bar{t}_0$ (respectively, $K_\sigma\geq k$, $< A_{\dot{c}(0)}v, v > \geq \lambda\cdot\|v\|^2$, c has no focal points of N and $n=m-1$).

Then it follows that $\dfrac{\|X(t_1)\|}{\|X(t_2)\|} \leq \dfrac{f_{k,\lambda}(t_1)}{f_{k,\lambda}(t_2)}$ ($\geq$, resp.) for every $0\leq t_1<t_2\leq l$, where $f_{k,\lambda}$ is the function in § 1 and $\bar{t}_0$ is the positive minimum root of $f_{k,\lambda}(t)=0$, or $\bar{t}_0=\infty$ if there exists no such solution. Moreover, if the equality holds for some $0\leq t_1<t_2\leq l$, then we have $\|X\| = f_{k,\lambda}$, $K(X, \dot{c}) = k$ on $[0, t_2]$ and $A_{\dot{c}(0)}X(0) = \lambda\cdot X(0)$.

§ 3. Stable Jacobi Fields

In this section we denote by H an Hadamard manifold, that is, H is a complete simply-connected m-dimensional Riemannian manifold of non-positive sectional curvature. A Jacobi field X along a ray $c:[0, \infty)\longrightarrow H$ is called *stable* when $\|X(t)\|$ is bounded. The existence and uniqueness of stable Jacobi field hold;

Lemma (*[6] lemma 2-2*)

Let $c:[0, \infty)\longrightarrow H$ be a geodesic ray. Then for every $v\in H_{c(0)}$ there exists a unique stable Jacobi field X_v along c with $X_v(0)=v$.

The proof of this lemma or of the following remark (1) is contained in that of theorem 3.

Remark

(1) This lemma is valid for any geodesic $c:[0, \infty)\longrightarrow M$,

where M is an arbitrary Riemannian manifold satisfying $K_\sigma \le 0$ for all $0 \le t$, $\dot{c}(t) \in \sigma \in G_{c(t)}$.

(2) Also this lemma is valid for Riemannian manifold M withont focal points, because the divergence theorem of a Jacobi field X with $X(0)=0$, $X \ne 0$ means the uniqueness and the convergence of a family of a kind of Jacobi fields the existence. See [4].

(3) If H is of negative constant curvature $-a^2$ ($a>0$), we see in § 1 that X_v is expressed as

$$X_v(t) = e^{-at} \cdot X(t) + \langle v, \dot{c}(0) \rangle \dot{c}(t) ,$$

where X is the parallel vector field along c such that

$$X(0) = v^{\perp} = v - \langle v, \dot{c}(0) \rangle \dot{c}(0) .$$

We have a general comparison theorem for stable Jacobi fields.

Theorem 3

Let (M, c, X_v) be a triple of an m-dimensional Riemannian manifold M, a geodesic $c:[0, \infty) \longrightarrow M$ and a stable Jacobi field X_v along c such that $X_v(0) = v \perp \dot{c}(0)$. We take an another triple ($M^\sim$, $c^\sim$, $X^\sim_{v^\sim}$) and suppose that the following conditions are satisfied;

$m \le m^\sim$, $\|v\| = \|v^\sim\|$,

$K_\sigma \le K_{\sigma^\sim} \le 0$ for any $0 \le t$, $\dot{c}(t) \in \sigma \in G_{c(t)}$ and $\dot{c}^\sim(t) \in \sigma^\sim \in G_{c^\sim(t)}$.

Then $\|X_v(t)\| \le \|X^\sim_{v^\sim}(t)\|$ holds for every $0 \le t$. Moreover, if the equality holds for a $t_1 > 0$, then $\|X_v\| = \|X^\sim_{v^\sim}\|$ holds on $[0, t_1]$.

Corollary ([6] theorem 2-4)

If an Hadamard manifold H satisfies $-b^2 \leq K \leq -a^2 \leq 0$ for two constants $0 \leq a \leq b$, then it follows that

$$\|v\| e^{-bt} \leq \|X_v(t)\| \leq \|v\| e^{-at}$$

for all $t \geq 0$, $v \in TH$ and the arbitrary stable Jacobi field X_v along a geodesic $c:[0, \infty) \longrightarrow H$ such that $X_v(0) = v \perp \dot{c}(0)$.

This corollary is an immediate consequence of remark (3) together with theorem 3.

Proof of theorem 3. The idea of proof is essentially due to that of the above corollary given in [6]. We may assume $\|v\| = \|v^\sim\| \neq 0$. Since c (respectively $c^\sim$) has no conjugate points by the curvature conditions, for every positive integer $i \in \mathbb{N}$ there exists uniquely the Jacobi field X_i ($X^\sim_i$) along c ($c^\sim$) such that $X_i(0)=v$ and $X_i(i)=0$ ($X^\sim_i(0)=v^\sim$ and $X^\sim_i(i)=0$ resp.). The fundamental Rauch comparison theorem means that

$$t\|X_i'(0) - X_j'(0)\| \leq \|X_i(t) - X_j(t)\|$$

for any $i, j \in \mathbb{N}$ and $t \geq 0$. On the other hand the length $\|X_i(t)\|$ is non-increasing on $[0, i]$, because

$$\|X_i\|'' = \left(\frac{\langle X_i, X_i'\rangle}{\|X_i\|}\right)' = \frac{1}{\|X_i\|^3}\Big(\|X_i\|^2 \cdot \|X_i'\|^2 - \langle X_i, X_i'\rangle^2 - K(X_i, \dot{c}) \cdot \|X_i\|^4 \Big) \geq 0$$

for $0 \leq t < i$. Hence

$$\|X_i'(0) - X_j'(0)\| \le \frac{1}{i}\|X_i(i) - X_j(i)\| \le \frac{1}{i}\|v\|$$

holds for any $i \le j$, and this means that $(X_i'(0))$ is a Cauchy sequence and hence converges.

It is clear that $\lim_{i\to\infty} X_i(t) = X_v(t)$ for $t \ge 0$, and the same for $(X^\sim_i)$ too. Now we take $Y_i(t) := X_i(i-t)$ and $Y^\sim_i(t) := \dfrac{\|X_i'(i)\|}{\|X^\sim_i'(i)\|} X^\sim_i(i-t)$ for $t \in [0, i]$. Then Y_i (resp. $Y^\sim_i$) is a non-trivial Jacobi field along the geodesic c_i ($c^\sim_i$) such that $\langle Y_i, \dot{c}_i\rangle = 0$ and $Y_i(0)=0$ ($\langle Y^\sim_i, \dot{c}^\sim_i\rangle = 0$ and $Y^\sim_i(0)=0$). Also it satisfies $\|Y_i'(0)\| = \|-X_i'(i)\| = \|Y^\sim_i'(0)\|$ where $c_i(t) := c(i-t)$ ($c^\sim_i(t) := c^\sim(i-t)$). Therefore we can apply theorem 1 for Y_i and $Y^\sim_i$ to get $\dfrac{\|Y_i(i-t)\|}{\|Y_i(i)\|} \le \dfrac{\|Y^\sim_i(i-t)\|}{\|Y^\sim_i(i)\|}$ for all $0 \le t \le i$, namely, $\|X_i(t)\| \le \|X^\sim_i(t)\|$. Hence by $i \longrightarrow \infty$, we have $\|X_v(t)\| \le \|X^\sim_{v^\sim}(t)\|$ for any $t \ge 0$.

When there exists $t_0 > 0$ such that $\|X_v(t_0)\| < \|X^\sim_{v^\sim}(t_0)\|$, we take $d = \dfrac{\|X^\sim_{v^\sim}(t_0)\|}{\|X_v(t_0)\|} > 1$. Since $X_{du} = dX_u$ for any vector u, we have $\|X_v(t)\| < d\cdot\|X_v(t)\| = \|d\cdot X_v(t)\| = \|d\cdot X_{X_v(t_0)}(t-t_0)\| = \|X_{d\cdot X_v(t_0)}(t-t_0)\| \le \|X^\sim_{X^\sim_{v^\sim}(t_0)}(t-t_0)\| = \|X^\sim_{v^\sim}(t)\|$ for any $t \ge t_0$. Hence $\|X_v\| = \|X^\sim_{v^\sim}\|$ on $[0, t_1]$, if $\|X_v(t_1)\| = \|X^\sim_{v^\sim}(t_1)\|$ for $t_1 > 0$.

We apply theorem 3 to a manifold whose sectional curvatures are pinched by a distance-function from a fixed point and get the estimate of length of stable Jacobi field.

Theorem 4

Let $c:[0, \infty) \longrightarrow H$ be a ray of an Hadamard manifold H

and assume that there exist continuous functions k_1, $k_2 : [0, \infty) \longrightarrow (-\infty, 0]$ such that $k_1(t) \leq K_\sigma \leq k_2(t)$ for any $t \in [0, \infty)$, $\dot{c}(t) \in \sigma \in G_{c(t)}$.

Then for every stable Jacobi field X_v along c with $X_v(0) = v \perp \dot{c}(0)$ we have $\|v\| y_1(t) \leq \|X_v(t)\| \leq \|v\| y_2(t)$ for any $t \geq 0$, where for each $i = 1, 2$, $y_i : [0, \infty) \longrightarrow (0, 1]$ is the unique bounded solution of the differential equation $y_i'' + k_i y_i = 0$, $y_i(0) = 1$.

Proof. We now construct the model space $M(k)$ for any continuous function $k : [0, \infty) \longrightarrow (-\infty, 0]$ and apply theorem 3. Let $f : [0, \infty) \longrightarrow [0, \infty)$ be the unique solution of the equation $f'' + kf = 0$ with initial conditions $f(0) = 0$, $f'(0) = 1$. Using polar coordinate around the origin we give a Riemannian metric ds^2 to $\mathbb{R}^m \simeq [0, \infty) \times S^{m-1} / \{0\} \times S^{m-1}$ by

$$ds^2 = dt^2 + f(t)^2 \cdot \sum_{i, j=1}^{m-1} g_{ij}(p) dx^i dx^j$$

where $\sum_{i, j=1}^{m-1} g_{ij}(p) dx^i dx^j$ is the natural metric of the unit $m-1$ sphere S^{m-1}. We denote the C^2-differentiable Riemannian manifold $(\mathbb{R}^m, ds^2)$ by $M(k)$. (In the argument of this paper the differentiability class C^2 or C^∞ is not essential.) $M(k)$ is an Hadamard manifold, since the half line $c_0(t) = (t, p_0)$, $t \geq 0$ is evidently a ray for every fixed $p_0 \in S^{m-1}$, and by direct calculation or by [3] § 7 we have

$$K(\partial / \partial x^i, a \partial / \partial t + b \partial / \partial x^j) =$$

$$\frac{-ff''g_{ii}a^2 + (1 - f'^2)f^2(g_{ii}g_{jj} - g_{ij}{}^2)b^2}{f^2g_{ii}a^2 + f^4(g_{ii}g_{jj} - g_{ij}{}^2)b^2} \leq 0$$

for $(a, b) \neq (0, 0)$ and $i \neq j$. Especially we have $K(\partial/\partial t, \partial/\partial x^i) = -f''/f = k$ which is the curvature given previously.

We take an arbitrary vector $v \in M(k)_o$ where o is the origin of $\mathbb{R}^m$, and denote by X the parallel vector field along the above ray c_0 with $X(0) = v - \langle v, \dot{c}_0(0)\rangle \dot{c}_0(0)$. Then the stable Jacobi field X_v with $X_v(0) = v$ is expressed as

$$X_v(t) = y(t)X(t) + \langle v, \dot{c}_0(0)\rangle \dot{c}_0(t) \quad \text{on} \quad [0, \infty),$$

where the positive valued function y on $[0, \infty)$ is the unique bounded solution of $y'' + ky = 0$ with initial condition $y(0) = 1$. In fact, calculating the curvature tensor of the above warped product we have $R(X, \dot{c}_0)\dot{c}_0 = -\frac{-f''}{f}X = kX$, hence the right-hand term of the formula is a Jacobi field.

References

[1] *U. Abresch:* Lower curvature bounds, Toponogov's theorem, and bounded topology, Ann. Sci. Éc. Norm. Sup., 18 (1985) 651-670.

[2] *D. V. Anosov & Y. G. Sinai:* Some smooth ergodic systems, Russian Math. Surveys, 22 (1967) 103-167.

[3] *R. L. Bishop & B. O'Neill:* Manifolds of negative curvature, Trans. Amer. Math. Soc., 145 (1969) 1-49.

[4] *M. S. Goto:* Manifolds without focal points, J. Diff.

Geom., 13 (1978) 341-359.

[5] *L. W. Green:* The generalized geodesic flow, Duke Math. J., 41 (1974) 115-126.

[6] *E. Heintze & H. C. Im Hof:* Geometry of horospheres, J. Diff. Geom., 12 (1977) 481-491.

[7] *H. C. Im Hof & E. A. Ruh:* An equivariant pinching theorem, Comm. Math. Helv., 50 (1975) 389-401.

[8] *Y. Tsukamoto & T. Yamaguchi:* On various comparison theorems, Sūgaku, 21 (1969) 81-96. (in Japanese)

[9] *F. W. Warner:* Extension of the Rauch comparison theorem to submanifolds, Trans. Amer. Math. Soc., 122 (1966) 341-356.

Tadashi YAMAGUCHI
Department of Mathematics
College of General Education
Kyushu University
Ropponmatsu, Fukuoka 810
Japan

Chapter V
The Geometry of Laplace Operators

Riemannian manifolds p-isospectral but not (p+1)-isospectral

Akira IKEDA

0. Introduction

Let M be a compact Riemannian manifold and $Spec^p(M)$ the spectrum of Laplacian acting on the space of smooth p-forms on M. Two Riemannian manifolds M_1 and M_2 will be said to be p-isospectral if $Spec^p(M_1) = Spec^p(M_2)$, and also, in this paper, will be said to be isospectral if $Spec^p(M_1) = Spec^p(M_2)$ for all p. It is clear that there exists an isometry between M_1 and M_2, then they are isospectral. It can be considered that the collection of all $Spec^k(M)$, $k = 0, \cdots, p+1$, contains more geometric information of a manifold M than does $Spec^p(M)$, $k = 0,\cdots,p$. In 1986 Gordon [2] constructed examples of Heisenberg manifolds which are 0-isospectral but not 1-isospectral. The purpose of this paper is the followings ; for each p we give examples of Riemannian manifolds which are k-isospectral for all $k = 0,\cdots,p$ but not (p+1)-isospectral. A compact connected Riemannian manifold of constant curvature one with cyclic fundamental group is called lens space. In [5] we gave families of lens spaces which are 0-isospectral but not isometric. The manifolds in our examples are these lens

ISBN 0-12-640170-5

spaces. The Poincaré series for p-spectrum plays an important roll in constructing our examples. To get the Poincaré series we review ,in section 1, polynomial eigenforms on the sphere obtained by Ikeda-Taniguchi [3]. It becomes a rational function (Theorem 2.3). Using these forms, we compare p-spectra of lens spaces given in [5] which are 0-isospectral but not isometric. Our Main Theorem is the following (for the definition of $\mathcal{L}_p(q,n)$, see (4.2)).

Theorem 4.1. Let q be an odd prime not less than 11. Let L_1 and $L_2 \in \mathcal{L}_p(q,n)$. Assume $q_0 = (q-1)/2 = n+2$. Then
(1) L_1 is k-isospectral to L_2 for all $k = 0, 1, 2, \cdots, p$.
(2) Let $L_1 \in \mathcal{L}_p(q,n)$ and $L_2 \in \mathcal{L}_{p+1}(q,n)$. If L_1 dose not belongs to $\mathcal{L}_{p+1}(q,n)$ then L_1 is not (p+1)-isospectral to L_2.

The problem reduces to the computing the number of solutions of some congruence equations (section 4). In the last section 5, we give examples for lower prime number q in Theorem 4.1.

Throughout this paper, for a finite set A, we denote by $|A|$ the number of elements in A.

1.Polynomial forms on S^n

Let R^{n+1} be an (n+1)-dimensional real Euclidean space

and $(x^0, x^1, \ldots, x^n)$ be the standard coordinate system on R^{n+1}. Denote by $\Lambda^p(R^{n+1})$ the space of smooth p-forms on R^{n+1} and put $\Lambda^*(R^{n+1}) = \sum_{p=0}^{n} \Lambda^p(R^{n+1})$. We put $X_0 = \sum_{i=0}^{n} x^i \frac{\partial}{\partial x^i}$ and $\omega_0 = \sum_{i=0}^{n} x^i dx^i$. $r^2 = \sum_{i=0}^{n} x_i^2$. Let d_0 be the differential acting on $\Lambda^*(R^{n+1})$, δ_0 its codifferential and $\bar{\Delta} = d_0\delta_0 + \delta_0 d_0$ the Laplacian for R^{n+1}. We define a linear operator $e(\omega_0)$ of $\Lambda^*(R^{n+1})$ into itself by $e(\omega_0)\alpha = \omega_0 \wedge \alpha$ and denote by $i(X_0)$ the interior product by X_0 on $\Lambda^*(R^{n+1})$. These operators d_0, δ_0, $\bar{\Delta}$, $e(\omega_0)$ and $i(X_0)$ commute with the natural action of the orthogonal group of degree n+1, $O(n+1)$ on $\Lambda^*(R^{n+1})$. Define P_k^p $(p \geq 0, k \geq 0)$ be the set of $\alpha \in \Lambda^p(R^{n+1})$ of the forms

$$\alpha = \sum_{0 \leq i_1 < \ldots i_p \leq n} \alpha_{i_1 \ldots i_p} dx^{i_1} \wedge \ldots \wedge dx^{i_p}, \tag{1.1}$$

where $\alpha_{i_1 \ldots i_p}$ are homogeneous polynomial of degree k.

Put $P_k^p = \{0\}$ if $p < 0$ or $k < 0$. Further, we define submodules H_k^p, and $H_k^p(d)$ of P_k^p by;

$$H_k^p = \{\alpha \in P_k^p \mid \bar{\Delta}\alpha = 0,\ \delta_0\alpha = 0\}, \tag{1.2}$$

$$H_k^p(d) = \{\alpha \in H_k^p \mid d_0\alpha = 0\}, \tag{1.3}$$

$$H_k^p(\delta) = \{ \alpha \in H_k^p \mid i(X_0)\alpha = 0\}. \tag{1.4}$$

In the following, we consider these spaces P_k^p, H_k^p, $H_k^p(d)$ and $H_k^p(\delta)$ are considered as $O(n+1)$-modules.

Lemma 1.1. (See Proposition 6.2 in [3]). For $\alpha \in P_k^p$, we have

$$d_0 i(X_0)\alpha + i(X_0)d_0\alpha = (k+p)\alpha, \tag{1.5}$$

$$\delta_0 e(\omega_0)\alpha + e(\omega_0)\delta_0\alpha = -(n+1-p+k)\alpha. \tag{1.6}$$

Corollary 1.2.The following complex is exact.

$$\longrightarrow P_{k+1}^{p-1} \xrightarrow{d_0} P_k^p \xrightarrow{d_0} P_{k-1}^{p+1} \longrightarrow \qquad (p+k \neq 0). \tag{1.7}$$

Corollary 1.3. As $O(n+1)$-modules, we have

$$\sum_{i=-p}^{k} (-1)^i P_{k-i}^{p+i} = \{0\} \qquad (p+k \neq 0). \tag{1.8}$$

Lemma 1.4. (See Lemma 6.7 in [3]). For $\alpha \in H_k^p$, we have $d_0\alpha \in H_{k-1}^{p+1}$ and $i(X_0)\alpha \in H_{k+1}^{p-1}$.

Together this Lemma with Corollary 1.2, we have

Lemma 1.5. The following complex is exact.

$$\longrightarrow H_{k+1}^{p-1} \xrightarrow{d_0} H_k^p \xrightarrow{d_0} H_{k-1}^{p+1} \longrightarrow \qquad (p+k \neq 0). \tag{1.9}$$

Proposition 1.6. (See (6.1) and Proposition 6.4 in [2]). We have a direct sum decompositions;

$$H_k^p = H_k^p(\delta) \oplus H_k^p(d) \qquad (p+k \neq 0), \tag{1.10}$$

$$P_k^p = H_k^p \oplus (r^2 P_{k-2}^p + e(\omega_0) P_{k-1}^{p-1}). \tag{1.11}$$

Together (1.9) with (1.10), we have

Lemma 1.7. As $O(n+1)$-modules, we have

$$H_k^p(\delta) = \sum_{i=0}^{p} (-1)^i H_{k+i}^{p-i}. \tag{1.12}$$

Lemma 1.8. Let $\alpha \in H_{k-1}^{p-1}$ $(n+1-p+k \neq 0)$ and $\alpha \neq 0$. Then $e(\omega_0)\alpha \neq 0$.

proof. Let $\alpha \in H_{k-1}^{p-1}$. Assume $e(\omega_0)\alpha = 0$. Since $\delta_0\alpha = 0$ and $(n+1-p+k) \neq 0$, we have $\alpha = 0$ by (1.6), which is a contradiction. Q.E.D.

Proposition 1.9. We have a direct sum decompositions;

$$P_k^p = H_k^p \oplus r^2 P_{k-2}^p \oplus e(\omega_0) H_{k-1}^{p-1}. \tag{1.13}$$

Proof. By (1.11), we have

$$P_{k-1}^{p-1} = H_{k-1}^{p-1} \oplus (r^2 P_{n-3}^{p-1} + e(\omega_0) P_{k-2}^{p-2}).$$

Substituting this into (1.11), we have

$$P_k^p = H_k^p \oplus (r^2 P_{k-2}^p + e(\omega_0) H_{k-1}^{p-1}).$$

Let $h \in H_{k-1}^{p-1}$ and $h \neq 0$. Suppose $e(\omega_0)h = r^2\alpha$ for some $\alpha \in P_{k-2}^p$. Then $e(\omega_0)\alpha = 0$. By (1.6), $\alpha = e(\omega_0)\beta$ for some $\beta \in P_{k-3}^{p-1}$. Since $e(\omega_0)h = r^2\alpha$, we have $e(\omega_0)(h - r^2\beta) = 0$. By (1.6), $h - r^2\beta = e(\omega_0)\gamma$ for some $\gamma \in P_{k-2}^{p-2}$, which contradicts to (1.11).

Q.E.D.

Together (1.13) with Lemma 1.8, we have

Proposition 1.10. As O(n+1)-modules, we have

$$P_k^p - P_{k-2}^p = H_k^p + H_{k-1}^{p-1}. \tag{1.14}$$

Corollary 1.11. As O(n+1)-modules, we have

$$H_k^p = \sum_{i=0}^{k} (-1)^i (P_{k-i}^{p-i} - P_{k-i-2}^{p-i}). \tag{1.15}$$

By (1.12) and (1.15) we have,

Proposition 1.12.

$$H_k^p(\delta) = \sum_{t=0}^{p} (-1)^t (P_{k+t}^{p-t} - P_{k-t-2}^{p-t}). \tag{1.16}$$

Let S^n be the unit sphere centered at the origin of R^{n+1}. Then we have

Proposition 1.13. (See [3]) Suppose $p \leq [n/2]$ and $k > 0$. Then (1) $i^*: H_k^p(\delta) \longrightarrow i^*H_k^p(\delta) \subset \Lambda^p(S^n)$ and $i^*: H_k^p(d) \longrightarrow i^*H_k^p(d)$ are isomorphisms as $SO(n+1)$-modules. (2) $i^*H_k^p(d)$ is a d-closed eigenspace with eigenvalue $(k + p - 1)(k + n - p)$. (3) $i^*H_k^p(\delta)$ is a δ-closed eigenspace with eigenvalue $(k + p)(k + n - p - 1)$. (4) No other eigenspaces with non-zero eigenvalue of Δ_p appear in $\Lambda^p(S^n)$.

Remark 1.14. It is easy to check that the set of d-closed eigenvalues $\{(k+p-1)(k+n-p) \mid k > 0\}$ and the set of δ-closed eigenvalues $\{(k+p)(k+n-p-1) \mid k > 0\}$ are disjoint to each other.

2. Poincaré series for p-spectrum of a spherical space forms

Let R^{2n} $(n \geq 2)$ be a 2n-dimensional Euclidean space and S^{2n-1} be the unit sphere centered at the origin. A finite subgroup G of O(2n) is said to be a fixed point free if for any $g \in G$ with $g \neq 1$, 1 is not an eigenvalue of g. Then G acts on S^{2n-1} as fixed point freely and we have a spherical space form $M = S^{2n-1}/G$. Any eigenform of Δ_p on M is naturally identified with an eigenform on S^{2n-1} which is invariant for all g in G and has the same eigenvalue.

Put

$$H_k^p(d,G) = \{\alpha \in H_k^p(d) \mid g^*\alpha = \alpha \text{ for all } g \in G\}, \tag{2.1}$$

and

$$H_k^p(\delta,G) = \{\alpha \in H_k^p(\delta) \mid g^*\alpha = \alpha \text{ for all } g \in G\}. \tag{2.2}$$

We define Poincaré series $F_G^p(z)$ for p-spectrum of M by

$$F_G^p(z) = \sum_{k=1}^{\infty} (\dim H_k^p(\delta,G))z^k. \tag{2.3}$$

By Lemma 1.5 and Proposition 1.13, $F_k^p(z)$ has also the following formulae;

$$F_G^p(z) = \sum_{k=1}^{\infty} (\dim H_{k-1}^{p+1}(d,G))z^k, \tag{2.4}$$

$$F_G^p(z) = \sum_{k=0}^{\infty} (\dim H_k^{p+1}(d,G))z^{k+1}. \tag{2.5}$$

For p = 0, we have (see [5])

$$F_G^0(z) = \frac{1}{|G|}\sum_{g\in G} \frac{1-z^2}{\det(z-g)} - 1. \tag{2.6}$$

Using Remark 1.14, we have

Proposition 2.1. Let $M = S^{2n-1}/G$, $N = S^{2n-1}/H$ be spherical space spherical space forms. Then (1) M is p-isospectral to N if and only if $F_G^p(z) = F_H^p(z)$ and $F_G^{p-1}(z) = F_H^{p-1}(z)$. (2) M is k-isospectral to N for all $k \leq p$ if and only if $F_G^k(z) = F_H^k(z)$ for all $k \leq p$. (3) M is isospectral to N if and only if $F_G^p(z) = F_H^p(z)$ for all $p < n$.

Let Λ^p be the homogeneous subspace of degree p in the exterior algebra of R^{2n}. Let χ^p be the character of the natural representation of G on Λ^p and χ_k the character of the natural representation of G on P_k^0. By Corollary 1.3, we have

$$\sum_{t=0}^{p} (-t)^t \chi_t(g)\chi^{p-t}(g) = 0 \quad \text{for } g \in G \text{ and } p \neq 0. \tag{2.7}$$

By (1.16), we have

$$\dim H_k^p(\delta, G) =$$

$$\sum_{t=0}^{\infty} (-1)^t \frac{1}{|G|} \sum_{g \in G} (\chi_{k+t}(g) - \chi_{k-t-2}(g))\chi^{p-t}(g). \tag{2.8}$$

Hence

$$|G|F_G^p(z) = \sum_{t=0}^{p} (-1)^t \sum_{k=1}^{\infty} \sum_{g\in G} (\chi_{k+t}(g) - \chi_{k-t-2}(g))\chi^{p-t}(g)z^k$$

$$= \sum_{t=0}^{p} (-1)^t \sum_{g\in G} \sum_{k=1}^{\infty} (\chi_{k+t}(g)z^k - \chi_{k-t-2}(g)z^k)\chi^{p-t}(g)$$

$$= \sum_{g\in G} \sum_{t=0}^{p} (-1)^t \{ \sum_{k=1}^{\infty} (\chi_{k+t}(g)z^k - \chi_{k-t-2}(g)z^k)\chi^{p-t}(g).$$

Note that for p = 0, we have

$$\sum_{k=1}^{\infty} \chi_k(g)z^k = \frac{1}{\det(z-g)} . \tag{2.9}$$

Using (2.9) we have

$$|G|F_G^p(z) = \sum_{g\in G} \sum_{t=0}^{p} (-1)^t \{z^{-t}(\frac{1}{\det(z-g)} - \sum_{k=0}^{t} \chi_k(g)z^k)$$

$$- \frac{z^{t+2}}{\det(z-g)}\}\chi^{p-t}(g). \tag{2.10}$$

and

$$\sum_{g\in G} \sum_{t=0}^{p} (-1)^t \{z^{-t}(\frac{1}{\det(z-g)} - \sum_{k=0}^{t} \chi_k(g)z^k) - \frac{z^{t+2}}{\det(z-g)}\}\chi^{p-t}(g)$$

$$= \sum_{t=0}^{\infty} (-1)^t (z^{-t} - z^{t+2}) \sum_{g\in G} \frac{\chi^{p-t}(g)}{\det(z-g)} - R_p, \tag{2.11}$$

where $R_p = \sum_{t=0}^{p} (-1)^t z^{-t} \sum_{k=0}^{t} (\sum_{g\in G} \chi_k(g)\chi^{p-t}(g))z^k.$

Lemma 2.2. We have,

$$R_p = |G|(-1)^p z^{-p}. \tag{2.12}$$

Proof. First we compute zR_{p+1}.

$$zR_{p+1} = \sum_{t=0}^{p+1} (-1)^t z^{1-t} \sum_{k=0}^{t} \left(\sum_{g\in G} \chi_k(g)\chi^{p+1-t}(g) \right) z^k$$

$$= \sum_{t=-1}^{p} (-1)^{t+1} z^{-t} \sum_{k=0}^{t+1} \left(\sum_{g\in G} \chi_k(g)\chi^{p-t}(g) \right) z^k.$$

Together this with (2.6),

$$R_p + zR_{p+1} = \sum_{t=0}^{p} (-1)^{t+1} z^{-t} \left(\sum_{g\in G} \chi_{t+1}(g)\chi^{p-t}(g) \right) z^{t+1} +$$

$$z \sum_{g\in G} \chi^{p+1}(g)$$

$$= \sum_{g\in G} \sum_{t=0}^{p+1} (-1)^t \chi_t(g)\chi^{p+1-t}(g)$$

$$= 0.$$

Thus we have,

$$R_p = (-1)^{p-1} z^{1-p} R_1$$

$$= (-1)^{p-1} z^{1-p} (-1) z^{-1} |G|$$

$$= (-1)^p |G| z^{-p}. \qquad \text{Q.E.D.}$$

Theorem 2.3.

$$F_G^p(z) = |G|^{-1} \sum_{t=0}^{p} (-1)^t (z^{-t} - z^{t+2}) \sum_{g \in G} \frac{\chi^{p-t}(g)}{\det(z-g)} +$$

$$(-1)^{p+1} z^{-p}. \tag{2.13}$$

Put

$$F^p(G:z) = \sum_{g \in G} \frac{\chi^p(g)}{\det(z-g)} . \tag{2.14}$$

Using Theorem 2.3 and Proposition 2.1, we have

Proposition 2.4. Let $M = S^{2n-1}/G$ and $N = S^{2n-1}/H$ be spherical space forms. Let p be nonnegative integer. Then M is k-isospectral to N for all $k \leq p$ if and only if

$$F^k(G:z) = F^k(H:z) \qquad \text{for all } k \leq p. \tag{2.15}$$

Let w be an indeterminate. Note that

$$\sum_{k=0}^{2n} (-1)\ \chi^k(g) w^k = \det(w-g). \tag{2.16}$$

Using this, we have

$$\sum_{k=0}^{2n} (-1)^k F^k(G:z) w^k = \sum_{g \in G} \frac{\det(w-g)}{\det(z-g)}.$$

Hence we have,

Theorem 2.5. Let M and N be as in the Proposition 2.4. Then M is isospectral to N if and only if

$$\sum_{g \in G} \frac{\det(w-g)}{\det(z-g)} = \sum_{h \in H} \frac{\det(w-h)}{\det(z-h)}. \tag{2.16}$$

Corollary 2.6. Let M and N be as in the Proposition 2.4. Suppose there is a one to one mapping ϕ of G onto H such that for any $g \in G$, g is conjugate to $\phi(g)$. Then M is isospectral to N.

By Theorem 2.5 and Corollary 2.6, we have

Theorem 2.7. The spherical space forms with meta-cyclic fundamental group which are 0-isospectral but not isometric given in [6] are also isospectral.

Proof. The proof that these spherical space forms are isospectral is to give the one to one corresponding ϕ in Corollary 2.6. Q.E.D.

Remark. The above theorem is first pointed out by Gilkey in [1]. Here we have a direct proof using Theorem 2.5 and Corollary 2.6.

3.Lens spaces and their spectra

Let q be a positive integer and $p_1, \ldots, p_n$ integers prime to q. We define an orthogonal matrix of degree 2n by

$$g = \begin{pmatrix} R(p_1/q) & & & 0 \\ & R(p_2/q) & & \\ & & \ddots & \\ 0 & & & R(p_n/q) \end{pmatrix}. \tag{3.1}$$

where $R(\theta) = \begin{pmatrix} \cos 2\pi\theta & \sin 2\pi\theta \\ -\sin 2\pi\theta & \cos 2\pi\theta \end{pmatrix}$.

Put $\xi = \xi_p = \exp(2\pi i/q)$. Then g has 2n eigenvalues $\xi^{p_1}, \xi^{-p_1}, \dots, \xi^{p_n}, \xi^{-p_n}$ and

$$\det(z - g^t) = \prod_{i=1}^{n} (z - \xi^{tp_i})(z - \xi^{-tp_i}). \tag{3.2}$$

Let G be the cyclic group generated by g. Then G is a finite fixed point free subgroup of order q in SO(2n). We have a (2n-1)-dimensional lens space,

$$L(q:p_1, \dots, p_n) = S^{2n-1}/G. \tag{3.3}$$

Let $\tilde{\mathcal{L}}(q,n)$ be the family of all the (2n-1)-dimensional lens spaces with fundamental group of order q, and let $\tilde{\mathcal{L}}_0(q,n)$ the subfamily of $\tilde{\mathcal{L}}(q,n)$ defined by

$$\tilde{\mathcal{L}}_0(q,n) = \{L(q:p_1, \dots, p_n) \in \tilde{\mathcal{L}}(q,n) \mid p_i \not\equiv \pm p_j \pmod q)(1 \le i < j \le n)\}. \tag{3.4}$$

The set of isometry classes of $\widetilde{\mathcal{L}}_0(q,n)$ (resp. $\widetilde{\mathcal{L}}(q,n)$) is denoted by $\mathcal{L}_0(q,n)$ (resp. $\mathcal{L}(q,n)$).

For lens spaces, the following properties are known;

Theorem 3.1. (See [5]). Let $L_1 = L(q:p_1,\ldots,p_n)$ and $L_2 = L(q:s_1,\ldots,s_n)$. Then the following assertions are equivarent;

(1) L_1 is isometric to L_2.

(2) L_1 is differmorphic to L_2.

(3) L_1 is homeomorphic to L_2.

(4) There is a number ℓ and there are numbers $e_i \in \{-1,1\}$ such that $(p_1,\ldots,p_n)$ is a permutation of $(e_1\ell s_1,\ldots,e_n\ell s_n)$ (mod q).

We apply Theorem 3.1 to the case in n = 2, then we have

Proposition 3.2. (1) $L(q:p_1,p_2) \in \mathcal{L}_0(q,2)$ is isometric to $L(q:1,t)$ for some t. (2) Let $L_1 = L(q:1,s)$ and $L_2 = L(q:1,t) \in L(q,2)$. Then L_1 is isometric to L_2 if and only if $s \equiv \pm t$ or $st \equiv \pm 1 \pmod q$. (3) $|\mathcal{L}_0(q,2)| = [q_0/2]$, where $q_0 = \phi(q)/2$ with $\phi(q)$ the Euler's function.

In the followings we assume

q is prime and not less than 11. (3.4)

Then

$$q_0 = (q - 1)/2. \tag{3.5}$$

Also we assume

$$q_0 \geq n + 2. \tag{3.6}$$

Put $k = q_0 - n$. Let g be as in (3.1) and choose k-integers $\{q_1, \ldots, q_k\}$ such that (q - 1)-integers $\{p_1, -p_1, \ldots, p_n, -p_n, q_1, -q_1, \ldots, q_k, -q_k\}$ form a complete set of incongruent residues prime to q. Define an orthogonal matrix $\bar{g}$ of degree 2k by

$$\bar{g} = \begin{pmatrix} R(q_1/q) & & & 0 \\ & R(q_2/q) & & \\ & & \ddots & \\ 0 & & & R(q_k/q) \end{pmatrix}. \tag{3.7}$$

Let $\bar{G}$ be the cyclic group generated by $\bar{g}$. Then $\bar{G}$ is a finite fixed point free subgroup of order q in SO(2k) and we have (2k - 1)- dimensional lens space

$$\bar{L}(q:q_1, \ldots, q_k) = S^{2k-1}/\bar{G}. \tag{3.8}$$

Note that

$$\overline{(\overline{G})} = G. \tag{3.9}$$

It is well known that for an odd prime number q, the q-th cyclotomic polynomial $\Psi_q(z)$ is ;

$$\Psi_q(z) = (z^q - 1)/(z - 1)$$

$$= \prod_{\eta} (z - \eta) \tag{3.10}$$

where η runs through all the q-th roots of one except 1.

From (3.2) and (3.10) we have

$$\Psi_q(z) = \det(z - g^t)\det(z - \overline{g}^t) \qquad (t \not\equiv 0 \bmod q). \tag{3.11}$$

Proposition 3.3. Let $L_1 = S^{2n-1}/G_1$ and $L_2 = S^{2n-1}/G_2 \in \mathcal{L}_0(q,n)$. Then L_1 is isometric to L_2 if and only if $\overline{L}_1$ is isometric to $\overline{L}_2$.

Proof. This proposition follows easily from the fact that if (q - 1)-integers $\{\pm p_1,\dots,\pm p_n,\pm q_1,\dots,\pm q_k\}$ forms complete set of incongruent residues primes to q, then for any ℓ prime to q, $\{\pm \ell p_1,\dots,\pm \ell p_n,\pm \ell q_1,\dots,\pm \ell q_k\}$ also forms a complete set of incongruent residues prime to q. Applying this fact to Theorem 3.1, we can get Proposition 3.3. Q.E.D.

By the above Proposition we define one to one mapping ω of $\mathcal{L}_0(q,n)$ onto $\mathcal{L}_0(q,k)$ by;

$$\omega(L_1) = \overline{L}_1 \qquad L_1 \in \mathcal{L}_0(q,n). \tag{3.12}$$

Theorem 3.4. (See Theorem 3.1 in [5]). Let q be a prime not less than 11. Assume $q_0 = n + 2$. Then each lens space belonging to $\mathcal{L}_0(q,n)$ has the same 0-spectrum.

Note that if $q_0 = n + 2$, then

$$\omega(\mathcal{L}_0(q,n)) = \mathcal{L}_0(q,2), \tag{3.13}$$

lens space in $\mathcal{L}_0(q,2)$ is of 3-dimensional (3.14)

and

The number of elements in $\mathcal{L}_0(q,n)$ is $[q_0/2]$. (3.15)

Theorem 3.5. (See [4]). Let M and N be 3-dimensional lens spaces. If they have the same 0-spectrum then they are isometric to each other.

Theorem 3.6. Let $L_1 = L(q:p_1,\ldots,p_n) = S^{2n-1}/G$ and $L_2 = L(q:s_1,\ldots.s_n) = S^{2n-1}/H$ in $\mathcal{L}_0(q,n)$. Let $\overline{L}_1$ and $\overline{L}_2$ be as in (3.8). Then L_1 is isospectral to L_2 if and only if $\overline{L}_1$ is isospectral to $\overline{L}_2$.

Proof. By Theorem 2.5, L_1 is isospectral to L_2 if and only if

$$\sum_{g \in G} \frac{\det(w-g)}{\det(z-g)} = \sum_{h \in H} \frac{\det(z-h)}{\det(z-h)} . \tag{3.16}$$

Since p is prime and L_1, $L_2 \in \mathscr{L}_0(q,n)$, we have for $g \neq 1_{2n}$

$$\frac{\det(w-g)}{\det(z-g)} = \frac{\det(w-g)\det(z-\bar{g})}{\det(z-g)\det(z-\bar{g})} \tag{3.17}$$

$$= \frac{\det(w-g)\det(z-\bar{g})}{\Psi_q(z)} .$$

By interchanging w and z in the numerator of the above formula, we have

$$\sum_{g \in G} \frac{\det(w-\bar{g})\det(z-g)}{\Psi_q(z)} = \sum_{h \in H} \frac{\det(w-\bar{h})\det(w-h)}{\Psi_q(z)} . \tag{3.18}$$

Thus we have

$$\sum_{g \in G} \frac{\det(w-\bar{g})}{\det(z-\bar{g})} = \sum_{h \in H} \frac{\det(w-\bar{h})}{\det(z-\bar{h})} . \tag{3.19}$$

This implies $\bar{L}_1$ is isospectral to $\bar{L}_2$. Since $\overline{(\bar{G})} = G$, the converse of the proof is clear. Q.E.D.

Next Theorem is the first main Theorem in this paper.

Theorem 3.9. Let q be a prime not less than 11. Assume $q_0 = n+2$. Then any non-isometric lens spaces belonging to $\mathcal{L}_0(q,n)$ have the same 0-spectrum but not isospectral.

Proof. Let $L_1, L_2 \in \mathcal{L}_0(q,n)$. By Theorem 3.6, L_1 is 0-isospectral to L_2. If L_1 is not isometric to L_2, then by Proposition 3.5, $\bar{L}_1$ is not isometric to $\bar{L}_2$. Since $\bar{L}_1$ and $\bar{L}_2$ are of 3-dimensional, by Theorem 3.7, $\bar{L}_1$ is not isospectral to $\bar{L}_2$. By Theorem 3.8, L_1 is not isospectral to L_2. Q.E.D.

4. p-isospectral but not (p+1)-isospectral manifolds

In this section, we compare p-spectra of 0-isospectral lens spaces given in Theorem 3.9. And we get the main Theorem in this paper (Theorem 4.1). Therefore we assume

$$q_0 = n + 2. \tag{4.1}$$

For $r > 0$, put

$$\mathcal{L}_r(q,2) = \{L(q:t_1,t_2) \in \mathcal{L}_0(q,2) : at_1 + bt_2 \not\equiv 0 \pmod q),\ 1 \le |a| + |b| \le r+2\}. \tag{4.2}$$

Then we have a filtration of $\mathcal{L}_0(q,2)$;

$$\mathcal{L}_0(q,2) \supset \mathcal{L}_1(q,2) \supset \mathcal{L}_2(q,2) \supset \cdots\cdots . \tag{4.3}$$

Put

$$\mathcal{L}_r(q,n) = \omega^{-1}(\mathcal{L}_r(q,2)). \tag{4.4}$$

Then we also a filtration in $\mathcal{L}_0(q,n)$;

$$\mathcal{L}_0(q,n) \supset \mathcal{L}_1(q,n) \supset \mathcal{L}_2(q,n) \supset \cdots\cdot . \tag{4.5}$$

Theorem 4.1. Let q be an odd prime not less than 11. Let L_1 and $L_2 \in \mathcal{L}_p(q,n)$. Assume $q_0 = (q - 1)/2 = n + 2$. Then (1) Then L_1 is k-isospectral to L_2 for all $k = 0, 1, 2, \cdots, p$. (2) Let $L_1 \in \mathcal{L}_p(q,n)$ and $L_2 \in \mathcal{L}_{p+1}(q,n)$. If L_1 dose not belongs to $\mathcal{L}_{p+1}(q,n)$ then L_1 is not (p+1)-isospectral to L_2.

To prove this Theorem, we need some preliminaries.

Let $L_1 = L(q:r_1,\cdots,r_n) = S^{2n-1}/G \in \mathcal{L}_0(q,n)$ and $\overline{L}_1 = L(q:t_1,t_2) = S^3/\overline{G}$. Since L_1 is of dimension $(2n - 1)$, we study p-spectra only for $p < n$. Take g (resp. $\overline{g}$) a generator of G (resp. $\overline{G}$). We may assume g(resp. $\overline{g}$) has eigenvalues ξ^{r_1}, $\xi^{-r_1}, \cdots, \xi^{r_n}, \xi^{-r_n}$ (resp. $\xi^{t_1}, \xi^{-t_1}, \xi^{t_2}, \xi^{-t_2}$).

By (2.14) and (3.11), we have

$$F^{p}(G:z) = \{\sum_{t=1}^{q-1} \chi^{p}(g^{t})/\det(z-g^{t}) + \binom{2n}{p}/(z-1)^{2n}\}$$

$$= \Psi_{q}(z)^{-1} \sum_{t=1}^{q} \chi^{p}(g^{t})\det(z-\bar{g}^{t}) -$$

$$\Psi_{q}(z)^{-1}\binom{2n}{p}(z-1)^{4} + \binom{2n}{p}/(z-1)^{2n} \qquad (4.6)$$

The last two terms depends only on the dimension of the lens space but not on defining parameters of lens space.

Note that

$$\det(z - \bar{g}^{t}) = \sum_{a=0}^{4} (-1)^{a} \chi^{a}(\bar{g}^{t})z^{a}. \qquad (4.7)$$

Hence

$$\sum_{t=1}^{q} \chi^{p}(g^{t})\det(z-g^{t}) = \sum_{a=0}^{4} (-1)^{a}(\sum_{t=1}^{q} \chi^{a}(\bar{g}^{t})\chi^{p}(g^{t}))z^{a}. \qquad (4.8)$$

Since the coefficient of z^{a} is equal to z^{4-a}, we examine the coefficients of 1, z and z^{2}.

$$\sum_{t=0}^{q} \chi^{p}(g^{t}) = \sum_{d=0}^{[p/2]} \{\binom{n}{d} \sum{}' (\sum_{t=1}^{q} \xi^{t(\pm r_{j_1} \pm \cdots \pm r_{j_{p-2d}})})\},$$

$$(4.9.0)$$

$$\sum_{t=0}^{q} \chi^{p}(g^{t})\chi^{1}(\bar{g}^{t}) =$$

$$\sum_{d=0}^{[p/2]} \binom{n}{d} \{ \sum_{i=1}^{2} \sum{}' (\sum_{t=1}^{q} \xi^{t(\pm r_{j_1} \pm \cdots \pm r_{j_{p-2d}} \pm t_i)})\}, \qquad (4.9.1)$$

$$\sum_{t=0}^{q} \chi^{p}(g^{t})\chi^{2}(\bar{g}^{t}) =$$

$$\sum_{d=0}^{[p/2]} \binom{n}{d} \{ \sum{}' (\sum_{t=1}^{q} \xi^{t(\pm r_{j_1} \pm \cdots \pm r_{j_{p-2d}} \pm t_1 \pm t_2)})\} + 2 \sum_{t=1}^{q} \chi^{p}(g^{t}), \qquad (4.9.2)$$

where the summation in $\sum'$ runs over all integers $r_{j_1}, \cdots, r_{j_{p-2d}} \in \{r_1, \cdots, r_n\}$ with $1 < j_1 < \cdots < j_{p-2d} \leq n$.

To compare p-spectra for lens spaces in $\mathscr{L}_0(q,2)$, the following elementary fact is used;

$$\begin{aligned} \sum_{t=1}^{q} \xi^{t} &= 0 \qquad \text{if } t \not\equiv 0 \pmod q, \\ &= q \qquad \text{if } t \equiv 0 \pmod q. \end{aligned} \qquad (4.10)$$

We introduce some notations. For $r \geq 1$, we put

$$K_r = \{(u_1,\cdots,u_r)\in \mathbb{Z}^r : 1 \le u_1<\cdots< u_r<q,$$
$$u_i+ u_j\not\equiv 0 \pmod q \text{ for all } i,j \text{ with } i \ne j\}. \tag{4.11}$$

$$K_r(t_1,t_2) = \{(u_1,\cdots,u_r) \in K_r: u_i \not\equiv \pm t_j \pmod q$$
$$\text{for } 1 \le i \le r,\ j = 1,2.\}, \tag{4.12}$$

$$A_r(s) = \{(u_1,\cdots,u_r) \in K_r: u_1+\cdots+u_r\equiv s \pmod q)\} \tag{4.13}$$

and

$$A_r(t_1,t_2) = K_r(t_1,t_2) \cap A_r(0)$$
$$= \{(u_1,\cdots,u_r)\in K_r:u_i \not\equiv \pm t_j \pmod q,$$
$$u_1+ \cdots + u_r \equiv 0 \pmod q)\}. \tag{4.14}$$

For $s \not\equiv 0$, we also put

$$A_r(t_1,t_2) = K_r(t_1,t_2) \cap (A_r(s)\cup A_r(-s))$$
$$= \{(u_1,\cdots,u_r)\in K_r:u_i \not\equiv \pm t_j \pmod q,$$
$$u_1+ \cdots + u_r \equiv \pm s \pmod q)\}. \tag{4.15}$$

Take another lens space $L_2 = L(q:s_1,\cdots,s_n) \in \mathcal{L}_0(q,n)$ with $\overline{L}_2 = L(q:v_1,v_2)$. Then we have

Proposition 4.2. Let L_1 and L_2 be as in the above. Then L_1 is k-isospectral to L_2 for all $k \le p$ if and only if it satisfies

(1) $|A_r(t_1,t_2)| = |A_r(v_1,v_2)|$ for all $r = 1,\cdots,p$,

(2) $|A_r(t_1,t_2;t_1)| + |A_r(t_1,t_2;t_2)| =$

$|A_r(v_1,v_2;v_1)| + |A_r(v_1,v_2;v_2)|$ for all $r = 1,\cdots,p$

and

(3) $|A_r(t_1,t_2;t_1+t_2)| + |A_r(t_1,t_2;t_1-t_2)| =$

$|A_r(v_1,v_2;v_1+v_2)| + |A_r(v_1,v_2;v_1-v_2)|$ for all $r = 1,\cdots,p$.

Proof. This follows from Proposition 2.4, (4.9.0), (4.9.1) and (4.9.2). Q.E.D.

Lemma 4.3. (1) $|A_r(s)| = |A_r(1)|$ $s \not\equiv 0 \pmod q$.

(2) $|A_r(0)| \neq |A_r(1)|$.

Proof. (1) For an integer u, we choose $\bar{u}$ with $0 \leq \bar{u} \leq q$ and $u \equiv \bar{u} \pmod q$. For any $(u_1,\cdots,u_r) \in A_r(1)$, the r-tuple $(\overline{su_1},\cdots,\overline{su_r})$ belongs to $A_1(s)$ up to a permutation. This correspondence is clearly one to one of $A_r(1)$ onto $A_r(s)$, which proves (1). (2) We have $\sum_{s=1}^{q} |A_r(s)| = |K_r| = \binom{q-1}{r}$.

Using (1), $|A_r(0)|+(q-1)|A_r(1)|=\binom{q-1}{r}$. If $|A_r(0)|=|A_r(1)|$, then $q|A_r(0)|= \binom{q-1}{r}$. Since q is prime , this can not happen. Q.E.D.

Lemma 4.4. For $r \geq 3$, we have

$$|A_r(t_1,t_2)| = |A_r(0)| - |A_{r-1}(t_1,t_2;t_1)| - |A_{r-1}(t_1,t_2;t_2)|$$
$$- |A_{r-2}(t_1,t_2;t_1+t_2)| - |A_{r-2}(t_1,t_2;t_1-t_2)|.$$

Proof. We decompose the set $A_r(0)$ into five disjoint subsets as follows; the subset of all elements of $A_r(0)$ which satisfies that one of the u_i's equal to $\pm t_1$ (mod q)(resp. $\pm t_2$ (mod q)) and non of the u_i's equals to $\pm t_2$ (mod q) (resp. $\pm t_1$ (mod q)) is identified with $A_{r-1}(t_1,t_2;t_1)$ (resp. $A_{r-1}(t_1,t_2;t_2)$). As the same way, the subset of all elements $A_r(0)$ which satisfies that one of the u_i's equals to $\pm t_1$ and other one equals to $\pm t_2$ (resp. $\not\equiv t_2$) is identified with $A_{r-2}(t_1,t_2;t_1+t_2)$ (resp. $A_{r-2}(t_1,t_2;t_1-t_2)$). The subset of remainder elements is $A_r(t_1,t_2)$. Q.E.D.

As the same way as above, we have the following lemmas 4.5, 4.6 and 4.7.

Lemma 4.5. Let $L(q:t_1,t_2) \in \mathcal{L}_r(q,2)$. Then for $r \geq 3$, we have

(1) $|A_r(t_1,t_2;t_1)| = |A_r(1)| - |A_{r-1}(t_1,t_2;2t_1)|$

$- 2|A_{r-1}(t_1,t_2)| - |A_{r-1}(t_1,t_2;t_1+t_2)| - |A_{r-1}(t_1,t_2;t_1-t_2)|$

$$- |A_{r-2}(t_1,t_2;2t_1+t_2)| - |A_{r-2}(t_1,t_2;2t_1-t_2)|$$

$$- 2|A_{r-2}(t_1,t_2;t_2)|.$$

(2) $|A_r(t_1,t_2;t_2)| = |A_r(1)| - |A_{r-1}(t_1,t_2;2t_2)|$

$$- 2|A_{r-1}(t_1,t_2)| - |A_{r-1}(t_1,t_2;t_1+t_2)| - |A_{r-1}(t_1,t_2;t_1-t_2)|$$

$$- |A_{r-2}(t_1,t_2;t_1+2t_2)| - |A_{r-2}(t_1,t_2;t_1-2t_2)|$$

$$- 2|A_{r-2}(t_1,t_2;t_1)|.$$

(3) $|A_r(t_1,t_2;t_1+t_2)| = |A_r(1)| - |A_{r-1}(t_1,t_2;2t_1+t_2)|$

$$- |A_{r-1}(t_1,t_2;t_1+2t_2)| - |A_{r-1}(t_1,t_2;t_1)| - |A_{r-1}(t_1,t_2;t_2)|$$

$$- |A_{r-2}(t_1,t_2;2t_1+2t_2)| - |A_{r-2}(t_1,t_2;2t_1)|$$

$$- |A_{r-2}(t_1,t_2;2t_2)|. - 2|A_{r-2}(t_1,t_2)|,$$

(4) $|A_r(t_1,t_2;t_1-t_2)| = |A_r(1)| - |A_{r-1}(t_1,t_2;2t_1-t_2)|$

$$- |A_{r-1}(t_1,t_2;t_1-2t_2)| - |A_{r-1}(t_1,t_2;t_1)| - |A_{r-1}(t_1,t_2;t_2)|$$

$$- |A_{r-2}(t_1,t_2;2t_1-2t_2)| - |A_{r-2}(t_1,t_2;2t_1)|$$

$$- |A_{r-2}(t_1,t_2;2t_2)| - 2|A_{r-2}(t_1,t_2)|,$$

(5) $|A_r(t_1,t_2;s)| = |A_r(1)| - |A_{r-1}(t_1,t_2;s+t_1)|$

$- |A_{r-1}(t_1,t_2;s-t_1)| - |A_{r-1}(t_1,t_2;s+t_2)| - |A_{r-1}(t_1,t_2;s-t_2)|$

$- |A_{r-2}(t_1,t_2;s+t_1+t_2)| - |A_{r-2}(t_1,t_2;s+t_1+t_2)|$

$- |A_{r-2}(t_1,t_2;s-t_1+t_2)| - |A_{r-2}(t_1,t_2;s-t_1-t_2)|,$

if $s \not\equiv 0, \pm t_1, \pm t_2, \pm(t_1 \pm t_2) \pmod q$.

The following two lemmas are easy to check.

Lemma 4.6. Let $L(q:t_1,t_2) \in \mathcal{L}_0(q,2)$. Then we have

(1) $|A_1(t_1,t_2)| = 0,$

(2) $|A_1(t_1,t_2:t_j)| = 0 \quad (j = 1,2),$

(3) $|A_1(t_1,t_2:s)| = 2 \quad$ if $s \not\equiv 0, \pm t_1, \pm t_2 \pmod q$.

Lemma 4.7. Let $L(q:t_1,t_2) \in \mathcal{L}_1(q,2)$. Then we have

(1) $|A_2(t_1,t_2)| = 0,$

(2) $|A_2(t_1,t_2:t_j)| = |A_2(1)| - 6 \quad (j = 1,2),$

(3) $|A_2(t_1,t_2;t_1+t_2)| = |A_2(1)| - |A_1(t_1,t_2;2t_1+t_2)|$

$- |A_1(t_1,t_2;t_1+2t_2)| - 2,$

(4) $|A_2(t_1,t_2;t_1-t_2)| = |A_2(1)| - |A_1(t_1,t_2;2t_1-t_2)|$

$- |A_1(t_1,t_2;t_1-2t_2)| - 2,$

(5) $|A_2(t_1,t_2;s)| = |A_2(1)| - |A_1(t_1,t_2;s+t_1)|$

$- |A_1(t_1,t_2;s-t_1)| - |A_1(t_1,t_2;s+t_2)| - |A_1(t_1,t_2;s-t_2)|$

if $s \not\equiv 0, \pm t_1, \pm t_2, \pm(t_1 \pm t_2) \pmod q$.

The following Proposition 4.8 is the special case of Theorem 4.1.

Proposition 4.8. (1) Let $L_1 \in \mathcal{L}_0(q,2)$ and $L_2 \in \mathcal{L}_1(q,2)$. Suppose $L_1 \notin \mathcal{L}_1(q,2)$. Then L_1 is not 1-isospectral to L_2. (2) Let $L_1 \in \mathcal{L}_1(q,2)$ and $L_2 \in \mathcal{L}_2(q,2)$. Suppose $L_1 \notin \mathcal{L}_2(q,2)$. Then L_1 is 1-isospectral to L_2 but not 2-isospectral to L_2.

Proof. If $L(q:t_1,t_2) \in \mathcal{L}_1(q,2)$, then $t_1 \pm t_2 \not\equiv \pm t_i \pmod q$ $(i=1,2)$. Hence by lemma 4.6, $|A_1(t_1,t_2;t_1+t_2)| = |A_1(t_1,t_2;t_1-t_2)| = 2$. On the contrary, $L(q;t_1,t_2) \in \mathcal{L}_0(q,2)$ but $\notin \mathcal{L}_1(q,2)$, then $|A_1(t_1,t_2;t_1+t_2)|$ or $|A_1(t_1,t_2;t_1-t_2)|$ vanishes by lemma 4.6. This proves (1). Suppose that $L(q:t_1,t_2) \in \mathcal{L}_2(q,2)$. Then $3t_1 \pm t_2 \not\equiv 0$, $t_1 \pm 3t_2 \not\equiv 0$, $2t_1 \pm 2t_2 \not\equiv 0$, $2t_1 \pm t_2 \not\equiv 0$ and $t_1 \pm 2t_2 \not\equiv 0 \pmod q$. Hence we have, $2t_1 \pm t_2 \not\equiv \pm t_i$, $t_1 \pm 2t_2 \not\equiv \pm t_i$ and $t_1 \pm t_2 \not\equiv \pm t_i \pmod q$ $(i=1,2)$. Thus $|A_1(t_1,t_2;t_1+t_2)| = |A_1(t_1,t_2;t_1-t_2)| = |A_2(1)| - 2$. On the other hand, if $L(q:t_1,t_2) \in \mathcal{L}_2(q,2)$, then $2t_1 \pm t_2 \not\equiv -t_1$ or $\pm t_1 \pm t_2 \not\equiv t_2 \pmod q$. Hence by lemma 4.6 and 4.7, we have $|A_1(t_1,t_2;t_1+t_2)| + |A_1(t_1,t_2;t_1-t_2)| < |A_2(1)| - 4$. By Proposition 4.2, this shows (2). Q.E.D.

Using Lemma 4.5, 4.6 and 4.7, we can get the following Proposition.

Proposition 4.9. For $r \geq 3$, let $L(q:t_1,t_2) \in \mathcal{L}_r(q,2)$. Then we have (1) $|A_r(t_1,t_2)| = C_{r,0} +$

$$\sum_{\substack{1\leq a+b\leq r-1\\ a,b>0}} C_{r,0,a,b}\{|A_1(t_1,t_2;at_1+bt_2)|+|A_1(t_1,t_2;at_1-bt_2)|\},$$

(2) $|A_r(t_1,t_2;t_1)| + |A_r(t_1,t_2;t_2)| = C_{r,0} +$

$$\sum_{\substack{1\leq a+b\leq r\\ a,b>0}} C_{r,1,a,b}\{|A_1(t_1,t_2;at_1+bt_2)|+|A_1(t_1,t_2;at_1-bt_2)|\}$$

and

(3) $|A_r(t_1,t_2;t_1+t_2)| + |A_r(t_1,t_2;t_1-t_2)| = C_{r,2} +$

$$\sum_{\substack{1\leq a+b\leq r+1\\ a,b>0}} C_{r,2,a,b}\{|A_1(t_1,t_2;at_1+bt_2)|+|A_1(t_1,t_2;at_1-bt_2)|\}$$

where the constants $C_{r,i}$ $(1=0,1,2)$ and $C_{r,i,a,b}$ $(i=0,1,2$ and $a,b > 0$ $1\leq a+b\leq r+i)$ are independent of $L(q:t_1,t_2) \in \mathcal{L}_r(q,2)$. Moreover for any fixed i $(i=0,1,2)$, the constants $C_{r,i,a,r+i-a}$ $(0 < a < r+i)$ do not vanish and have the same sign.

Proof of Theorem 4.1.

(1) Let $L_1 \in \mathscr{L}_p(q,n)$ with $\overline{L}_1 = L(q:t_1,t_2)$. Then $at_1 \pm bt_2 \not\equiv 0 \pmod q$ for all integers a, b with $1 \leq |a|+|b| \leq p+2$. Hence if $1 \leq |a|+|b| \leq p+1$, then $at_1 + bt_2 \not\equiv 0, \pm t_1, \pm t_2 \pmod q$. Using Proposition 4.2 and Proposition 4.9, we can see easily the first part of Theorem 4.1.

(2) To prove (2) we use Proposition 4.9 for the case r = p+1. Let $L_1 \in \mathscr{L}_p(q,n)$ but $\notin \mathscr{L}_{p+1}(q,n)$ with $\overline{L}_1 = L(q:t_1,t_2)$. Then there is a unique pair of integers a_0, b_0 with $a_0 > 0$, $a_0t_1 + b_0t_2 \equiv 0 \pmod q$ and $a_0 + |b_0| = p+3$. Then we have $(a_0-1)t_1 + b_0t_2 \equiv -t_1 \pmod q$ and $a_0t_1 + (b_0-1)t_2 \equiv -t_2 \pmod q$. For other pair of (a,b) with $a + |b| = p+2$ and $a > 0$, we have $at_1 + bt_2 \not\equiv 0, \pm t_1, \pm t_2 \pmod q$. Together these facts with Proposition 4.9, we can complete the proof of (2). Q.E.D.

Theorem 4.10. For each $p \geq 0$, we have lens spaces which are k-isospectral to each other for all $k=0,1,\cdots,p$ but not (p+1)-isospectral.

Proof. For a given p, we have a prime number q with $(p+2)(p+3)+1 < q$. Let $L_1 = L(q:1,p+2)$ and $L_2 = L(q:1,p+3)$. It is easy to see $L_1 \notin \mathscr{L}_{p+1}(q,2)$, by the definition. For $a,b \geq 0$ with $1 \leq a + b \leq p + 2$, we have $|a \pm b(p+2)| \leq (p+1)(p+2)+1 < q$. Hence if $a \pm b(p+2) \equiv 0 \pmod q$, then $a = b(p+2)$. Since $1 \leq a,b < p+2$, this can not happen. Thus $L_1 \in \mathscr{L}_p(q,2)$. As the same way, we have $L_2 \in \mathscr{L}_{p+1}(q,2)$. Applying Theorem 4.1 for lens spaces $\bar{L}_1$ and $\bar{L}_2$, we get the Theorem. Q.E.D.

5.examples

We retain the notaions as in section **4**. To give examples of lens spaces which are k-isospectral for all $k \le p$ but not (p+1)-isospectral , it suffices to determine the lens spaces belonging to $\mathscr{L}_p(q,n)$. By Proposition 3.5, it suffices to determine lens spaces belonging to $\mathscr{L}_p(q,2)$. Using Proposition 3.2 and (4.2), we can determine easily $\mathscr{L}_p(q,2)$ for lower prime numbers q. We give the table of $\mathscr{L}_p(q,2)$ for prime numbers $q < 100$. The table should be read as follows; for prime number q, t(p) in the row means that the lens space L(q:1,t) which belongs to $\mathscr{L}_p(q,2)$ but does not belong to $L_{p+1}(q,2)$.

q	t(p)
11	2(0) 3(1)
13	2(0) 3(1) 5(2)
17	2(0) 3(1) 4(2) 5(2)
19	2(0) 3(1) 4(2) 7(2)
23	2(0) 3(1) 4(2) 5(3) 7(2)
29	2(0) 3(1) 4(2) 5(3) 8(4) 9(2) 12(4)
31	2(0) 3(1) 4(2) 5(3) 7(4) 11(2) 12(4)
37	2(0) 3(1) 4(2) 5(3) 6(4) 7(4) 8(5) 10(4) 13(2)

q	t(q)
41	2(0) 3(1) 4(2) 5(3) 6(4) 9(6) 11(4) 12(5) 13(2) 16(4)
43	2(0) 3(1) 4(2) 5(3) 6(4) 8(5) 9(4) 10(4) 12(6) 15(2)
47	2(0) 3(1) 4(2) 5(3) 6(4) 7(5) 9(4) 10(5) 11(4) 13(6) 15(2)
53	2(0) 3(1) 4(2) 5(3) 6(4) 7(5) 8(6) 10(5) 11(4) 12(6) 14(4) 17(2) 23(6)
59	2(0) 3(1) 4(2) 5(3) 6(4) 7(5) 8(6) 9(7) 11(6) 14(4) 18(5) 19(2) 24(4) 25(6)
61	2(0) 3(1) 4(2) 5(3) 6(4) 7(5) 8(6) 9(6) 11(8) 13(6) 16(4) 17(7) 21(2) 22(5) 24(4)
67	2(0) 3(1) 4(2) 5(3) 6(4) 7(5) 8(6) 9(7) 10(7) 12(8) 13(4) 14(5) 16(4) 18(6) 23(2) 29(6)
71	2(0) 3(1) 4(2) 5(3) 6(4) 7(5) 8(6) 11(8) 15(6) 16(8) 17(4) 20(6) 21(8) 22(5) 23(2) 26(7) 28(4)
73	2(0) 3(1) 4(2) 5(3) 6(4) 7(5) 8(6) 10(7) 11(8) 13(8) 14(5) 15(4) 16(8) 17(6) 19(4) 25(2) 27(8) 31(6)
79	2(0) 3(1) 4(2) 5(3) 6(4) 7(5) 8(6) 9(7) 11(6) 12(9) 14(8) 15(6) 18(8) 19(4) 23(7) 27(2) 28(5) 29(8) 32(4)
83	2(0) 3(1) 4(2) 5(3) 6(4) 7(5) 8(6) 9(7) 10(8) 11(9) 13(8) 16(5) 17(4) 18(9) 19(8) 20(4) 22(6) 24(6) 27(2) 30(7)
89	2(0) 3(1) 4(2) 5(3) 6(4) 7(5) 8(6) 9(7) 12(9) 13(6) 14(8) 16(10) 17(6) 20(8) 23(4) 24(8) 25(7) 27(8) 28(5) 29(2) 34(10) 36(4)
97	2(0) 3(1) 4(2) 5(3) 6(4) 7(5) 8(6) 9(7) 10(8) 11(8) 13(10) 17(8) 18(9) 19(4) 20(5) 21(10) 22(10) 23(6) 25(4) 26(8) 28(6) 30(7) 33(2) 35(8)

We describes our examples more explicitly for lower prime numbers q using the table.

(1) $q = 11$, $n = 3$.

Put $L_1 = L(11:1,2)$ and $L_2 = L(11:1,3)$. Then $L_1 \in \mathcal{L}_0(11,2) \setminus \mathcal{L}_1(11,2)$ and $L_2 \in \mathcal{L}_1(11,2) \setminus \mathcal{L}_2(11,2)$. Then $\bar{L}_1 = L(11:3,4,5) = L(11:1,2,5)$ and $\bar{L}_2 = L(11:2,4,5) = L(11:1,2,3)$. By Theorem 4.1, Then $\bar{L}_1$ is 0-isospectral to $\bar{L}_2$ but not 1-isospectral.

As the same way as the above,

(2) $q = 13$, $n = 4$.

Put $L_1 = L(13:1,2)$, $L_2 = L(13:1,3)$ and $L_3 = L(13:1,5)$. Then $\bar{L}_1 = L(13:3,4,5,6)$, $\bar{L}_2 = L(13:2,4,5,6)$ and $\bar{L}_3 = L(13:2,3,4,6)$. In this case, $\bar{L}_1 \in \mathcal{L}_0(13,4) \setminus \mathcal{L}_1(13,4)$, $\bar{L}_2 \in \mathcal{L}_1(13,4) \setminus \mathcal{L}_2(13,4)$ and $\bar{L}_3 \in \mathcal{L}_2(13,4) \setminus \mathcal{L}_3(13,4)$. Thus $\bar{L}_1$, $\bar{L}_2$ and $\bar{L}_3$ are 0-isospectral to each other. $\bar{L}_1$ is not 1-isospectral to $\bar{L}_2$ and $\bar{L}_3$. $\bar{L}_2$ and $\bar{L}_3$ are 1-isospectral but not 2-isospectral.

References

[1]P.B.Gilkey, *On spherical space forms with metacyclic*

fundamental groups which are isospectral but not equivariant cobordant, Compositio Math. **56**(1985), 171-200.

[2] C.S.Gordon, *Riemannian manifolds isospectral on functions but not 1-forms*, J.Diff.Geom. **24**(1986), 79-96.

[3] A.Ikeda and Y.Taniguchi, *Spectra and Eigenforms of the Laplacian on S^N and $P^N(\mathbb{C})$*, Osaka J. Math. **15**(1978), 515-546.

[4] A.Ikeda and Y.Yamamoto, *On the spectra of 3-dimensional lens spaces*, Osaka J. Math. **16**(1978), 447-469.

[5] A.Ikeda, *On lens spaces which are isospectral but not isometric*, Ann.Scient.Éc.Nor.Sup (1980), 303-315.

[6] A.Ikeda, *On spherical space forms which are isospectral but not isometric*, J.Math. Soc. Japan **17**(1983), 437-444.

Department of Mathematics

Faculty of School Education

Hiroshima university

Shinonome Hiroshima 734,

Japan

On the Holomorphicity of Pluriharmonic Maps

Dedicated to Professor Shingo Murakami on his 60th birthday

Hisashi NAITO

Abstract. *In this paper, we prove a condition for the holomorphicity of pluriharmonic maps from a compact Kähler manifold with non-positive Ricci curvature to a compact Riemann surface. This condition is an extension of Eells-Wood's condition for the holomorphicity of harmonic maps between Riemann surfaces.*

1. Introduction

Let (M, g) be a compact Kähler manifold with non-positive Ricci curvature and (N, h) be a compact Riemann surface. A smooth map $f : M \longrightarrow N$ is called a harmonic map if the second fundamental form ∇df of the map f satisfies $\operatorname{trace} \nabla df = 0$. Furthermore f is called a pluriharmonic map if ∇df satisfies $\nabla df(Z, \overline{Z}) = 0$ for any (1,0)-vector Z on M. Clearly, a pluriharmonic map is harmonic and a $\pm$-holomorphic map is pluriharmonic. By "the map f is $\pm$-holomorphic" we means that f is either holomorphic or antiholomorphic. Conversely, when is a harmonic map or a pluriharmonic map $\pm$-holomorphic? One of the answers to this problem is the result of J. Eells and J. C. Wood [**4**].

Theorem [**4**] *Let X and Y be compact Riemann surfaces. If a harmonic map $f : X \longrightarrow Y$ satisfies*

$$\chi(X) + |\deg(f) \cdot \chi(Y)| > 0,$$

then f is $\pm$-holomorphic. Here $\chi(X)$ and $\chi(Y)$ denote the Euler number of X and Y, respectively.

Recently, K. Ono [**13**] obtained a condition for holomorphicity of a harmonic map from a compact Kähler manifold with negative Ricci curvature to a hyperbolic Riemann surface. In the case of a pluriharmonic map, Y. Ohnita [**12**] shows that any stable harmonic maps from the n-dimensional complex projective space with Fubini-Study metric to a Riemannian manifold is pluriharmonic.

The purpose of this paper is to prove that a pluriharmonic map from M to N is $\pm$-holomorphic under a certain condition.

1980 *Mathematics Subject Classification* (1985 *Revision*). 58E20.

Key words and phrases. Harmonic maps.

ISBN 0-12-640170-5

Theorem 1.1 *Let $f : M \longrightarrow N$ be a pluriharmonic map. If f satisfies*

$$c_1(M) \cup \Omega(M)[M] + |f^* c_1(N) \cup \Omega(M)[M]| > 0,$$

where $\Omega(M) := [\omega_M^{m-1}] \in H^{2(m-1)}(M, \mathbf{R})$ and $m = \dim_{\mathbf{C}} M$, then f is $\pm$-holomorphic. Here ω_M denotes the Kähler form of M, and $c_1(M)$ and $c_1(N)$ denote the first Chern class of M and N, respectively.

As a corollary of Theorem 1.1, we state a condition of holomorphicity of a harmonic map from a compact Kähler manifold with non-positive Ricci curvature to a hyperbolic Riemann surface.

Corollary 1.2 *Let M and N be as above. Suppose N is hyperbolic, namely N equippes the constant curvature -1. If a harmonic map $f : M \longrightarrow N$ satisfies the condition in Theorem 1.1, then f is $\pm$-holomorphic.*

We remark that if $\dim_{\mathbf{C}} M = 1$ then the above result is just J. Eells and J. C. Wood's one.

2. Behaviour of the energy density

For a smooth map $f : M \longrightarrow N$, we have a complexified differential $d^{\mathbf{C}} f : T^{\mathbf{C}} M \longrightarrow T^{\mathbf{C}} N$. The complexified tangent spaces $T^{\mathbf{C}} M$ and $T^{\mathbf{C}} N$ split as $T^{\mathbf{C}} M = T^{'} M \oplus T^{''} M$ and $T^{\mathbf{C}} N = T^{'} N \oplus T^{''} N$, respectively. Thus we have

$$\partial f : T^{'} M \longrightarrow T^{'} N,$$

and

$$\overline{\partial} f : T^{''} M \longrightarrow T^{'} N.$$

A map f is called holomorphic if and only if $\overline{\partial} f \equiv 0$ and antiholomorphic if and only if $\partial f \equiv 0$. The Hermitian inner product on $T^{\mathbf{C}} M$ is denoted by $\langle\langle \cdot, \cdot \rangle\rangle$.

Using the complex structures of M and N, we define the partial energy densities $e^{'}(f)$ and $e^{''}(f)$ as follows:

$$e^{'}(f) = |\partial f|^2 = g^{i\overline{j}} h(f) \partial_i f \partial_{\overline{j}} \overline{f},$$

and

$$e^{''}(f) = |\overline{\partial} f|^2 = g^{i\overline{j}} h(f) \partial_i \overline{f} \partial_{\overline{j}} f.$$

The energy density $e(f) = \frac{1}{2}|df|^2$ satisfies $e(f) = e^{'}(f) + e^{''}(f)$. The total energy of f is defined by $E(f) = \int_M e(f)$.

A smooth map f is called a harmonic map whenever f is a stationary point of the functional E. Clearly, f is harmonic if and only if

$$\operatorname{trace} \nabla df = g^{i\bar{j}} \nabla_i \partial_{\bar{j}} f = 0.$$

Moreover, a map f is called pluriharmonic whenever

$$\nabla df = \nabla_i \partial_{\bar{j}} f = 0, \quad \text{for all} \quad i, j.$$

Throughout this paper, we use the following notations. The Ricci forms on M and N are denoted by Ψ_M and Ψ_N, respectively. For the 2-forms Ω_1 and Ω_2 on M, the pairing of Ω_1 and Ω_2 will be defined by $\langle \Omega_1, \Omega_2 \rangle = \Omega_1 \wedge *\Omega_2 = (\Omega_1)^{i\bar{j}} (\Omega_2)_{i\bar{j}}$. The square norm of $\nabla_{1,0} \partial f$ is denoted by $|\nabla_{1,0} \partial f|^2 = g^{i\bar{j}} g^{k\bar{l}} h(f) \nabla_i \partial_k f \nabla_{\bar{j}} \partial_{\bar{l}} \bar{f}$.

With the above notations, we observe the behaviours of $e'(f)$ and $e''(f)$.

Lemma 2.1 *If $f : M \longrightarrow N$ is a harmonic map, we have*

$$\Delta e'(f) \geq |\nabla_{1,0} \partial f|^2 + |\nabla_{1,0} \bar{\partial} f|^2 + e'(f) \langle \omega_M, \Psi_M \rangle + e'(f) \langle \omega_M, f^* \Psi_N \rangle, \tag{2.1}$$

and

$$\Delta e''(f) \geq |\nabla_{0,1} \bar{\partial} f|^2 + |\nabla_{0,1} \partial f|^2 + e''(f) \langle \omega_M, \Psi_M \rangle - e''(f) \langle \omega_M, f^* \Psi_N \rangle. \tag{2.2}$$

In particular, if f is pluriharmonic, we obtain

$$\Delta e'(f) \geq |\nabla_{1,0} \partial f|^2 + e'(f) \langle \omega_M, \Psi_M \rangle + e'(f) \langle \omega_M, f^* \Psi_N \rangle, \tag{2.3}$$

and

$$\Delta e''(f) \geq |\nabla_{0,1} \bar{\partial} f|^2 + e''(f) \langle \omega_M, \Psi_M \rangle - e''(f) \langle \omega_M, f^* \Psi_N \rangle. \tag{2.4}$$

(*Proof*) We prove only (2.2).

An elementary calculation (for example using holomorphic normal coordinates) yields

$$\begin{aligned} \Delta e''(f) = & g^{k\bar{l}} g^{i\bar{j}} h(f) \left(\nabla_k \partial_{\bar{j}} f \nabla_{\bar{l}} \partial_i \bar{f} + \nabla_{\bar{l}} \partial_{\bar{j}} f \nabla_k \partial_i \bar{f} \right) \\ & + g^{k\bar{l}} g^{i\bar{j}} h(f) \left(\nabla_{\bar{l}} \nabla_k \partial_{\bar{j}} f \partial_i \bar{f} + \partial_{\bar{j}} f \nabla_{\bar{l}} \nabla_k \partial_i \bar{f} \right). \end{aligned} \tag{2.5}$$

The first term of the right hand side is equal to $|\nabla_{0,1} \bar{\partial} f|^2 + |\nabla_{0,1} \partial f|^2$.

The Ricci formula and the harmonicity of f imply that

$$
\begin{aligned}
&g^{k\bar{l}}g^{i\bar{j}}h(f)\left(\nabla_{\bar{l}}\nabla_k\partial_{\bar{j}}f\partial_i\bar{f}+\partial_{\bar{j}}f\nabla_{\bar{l}}\nabla_k\partial_i\bar{f}\right)\\
=&g^{i\bar{j}}h(f)^M R_i^m \partial_m\bar{f}\partial_{\bar{j}}f\\
&-g^{k\bar{l}}\left(g^{i\bar{j}}h(f)\partial_i\bar{f}\partial_{\bar{j}}f\right){}^N R\left(\partial_k f\partial_{\bar{l}}\bar{f}-\partial_k\bar{f}\partial_{\bar{l}}f\right)\\
=&\langle\!\langle \mathrm{Ric}_M(\overline{\partial} f),\overline{\partial} f\rangle\!\rangle - e''(f)\langle\omega_M, f^*\Psi_N\rangle\\
=&e''(f)\,\mathrm{Ric}_M\left(\frac{\overline{\partial} f}{|\overline{\partial} f|},\frac{\overline{\partial} f}{|\overline{\partial} f|}\right)-e''(f)\langle\omega_M, f^*\Psi_N\rangle.
\end{aligned}
\tag{2.6}
$$

Here ${}^M R_m^i$ denotes the Ricci tensor of M and ${}^N R$ denotes the curvature tensor of N. Since $\dim_{\mathbf{C}} N = 1$, there is a unique component of the curvature tensor of N different from zero.

If M has non-positive Ricci curvature, then we get

$$
\mathrm{Ric}_M\left(\frac{\overline{\partial} f}{|\overline{\partial} f|},\frac{\overline{\partial} f}{|\overline{\partial} f|}\right)\geq \text{Scalar curvature of } M = g^{i\bar{j}\,M}R_{i\bar{j}} = \langle\omega_M,\Psi_M\rangle. \tag{2.7}
$$

Therefore, (2.5)–(2.7) yield (2.2). The similar argument shows other inequalities. ∎

If $\dim_{\mathbf{C}} M = 1$, or $\mathrm{Ric}_M \equiv 0$ then the equality holds in (2.1)–(2.4). In the case of $\dim_{\mathbf{C}} M = 1$, we can find these equalities in R. Schoen and S. -T. Yau [**14**] and J. Jost [**7**].

3. Proof of Theorem

In this section, we will prove the main theorem. To do so, first we prepare a lemma.

Lemma 3.1 *If $f : M \longrightarrow N$ is a pluriharmonic map, then at any point where ∂f or $\overline{\partial} f$ is non zero, we have*

$$
\Delta \log |\partial f|^2 \geq \langle\omega_M,\Psi_M\rangle + \langle\omega_M, f^*\Psi_N\rangle, \tag{3.1}
$$

or

$$
\Delta \log |\overline{\partial} f|^2 \geq \langle\omega_M,\Psi_M\rangle - \langle\omega_M, f^*\Psi_N\rangle. \tag{3.2}
$$

(*Proof*) For any positive smooth function φ on M, we can show

$$
\Delta \log\varphi = \frac{\Delta\varphi}{\varphi} - \frac{1}{\varphi^2}g^{i\bar{j}}\partial_i\varphi\partial_{\bar{j}}\varphi. \tag{3.3}
$$

For the sake of simplicity, we abbreviate $g^{i\bar{j}}\partial_i\varphi\partial_{\bar{j}}\varphi$ to $|d\varphi|^2$.

First we observe

$$|\nabla_{1,0}\partial f|^2|\partial f|^2 - \left|d|\partial f|^2\right|^2 \geq 0, \tag{3.4}$$

and

$$|\nabla_{0,1}\overline{\partial} f|^2|\overline{\partial} f|^2 - \left|d|\overline{\partial} f|^2\right|^2 \geq 0, \tag{3.5}$$

provided f is pluriharmonic.

To obtain (3.5), we remark

$$\begin{aligned}
0 \leq & g^{i\bar{j}}g^{k\bar{l}}g^{s\bar{t}}h(f)^2\left(\nabla_i\partial_k\overline{f}\partial_s\overline{f} - \nabla_i\partial_s\overline{f}\partial_k\overline{f}\right)\left(\nabla_{\bar{j}}\partial_{\bar{l}}f\partial_{\bar{t}}f - \nabla_{\bar{j}}\partial_{\bar{t}}f\partial_{\bar{l}}f\right) \\
= & g^{i\bar{j}}g^{k\bar{l}}g^{s\bar{t}}h(f)^2\left(\nabla_i\partial_k\overline{f}\partial_s\overline{f}\nabla_{\bar{j}}\partial_{\bar{l}}f\partial_{\bar{t}}f + \nabla_i\partial_s\overline{f}\partial_k\overline{f}\nabla_{\bar{j}}\partial_{\bar{t}}f\partial_{\bar{l}}f\right. \\
& \left. - \nabla_i\partial_s\overline{f}\partial_k\overline{f}\nabla_{\bar{j}}\partial_{\bar{l}}f\partial_{\bar{t}}f - \nabla_i\partial_k\overline{f}\partial_s\overline{f}\nabla_{\bar{j}}\partial_{\bar{t}}f\partial_{\bar{l}}f\right) \\
= & 2\left(|\partial_{0,1}\overline{\partial} f|^2|\overline{\partial} f|^2 - \left|d|\overline{\partial} f|^2\right|^2\right),
\end{aligned} \tag{3.6}$$

provided f is pluriharmonic.

From (2.4), at point where $\overline{\partial} f$ is non zero,

$$\frac{\Delta|\overline{\partial} f|^2}{|\overline{\partial} f|^2} \geq \frac{|\nabla_{0,1}\overline{\partial} f|^2}{|\overline{\partial} f|^2} + \langle\omega_M, \Psi_M\rangle - \langle\omega_M, f^*\Psi_N\rangle. \tag{3.7}$$

Combining (3.3), (3.5) and (3,7), we obtain

$$\Delta \log|\overline{\partial} f|^2 \geq \langle\omega_M, \Psi_M\rangle - \langle\omega_M, f^*\Psi_N\rangle. \tag{3.8}$$

A similar argument implies (3.1). ∎

Now we prove the following proposition thanks to the above Lamma.

Proposition 3.2 *Let M be a compact Kähler manifold with non-positive Ricci curvature and N be a compact Riemann surface. If $f : M \longrightarrow N$ is a pluriharmonic map which satisfies the condition:*

$$\int_M \langle\omega_M, \Psi_M\rangle + \left|\int_M \langle\omega_M, f^*\Psi_N\rangle\right| > 0,$$

then f is $\pm$-holomorphic.

(*Proof*) The map f is pluriharmonic implies f is a harmonic map. Therefore, by similarity principle, the zero set of $|\partial f|^2$ or $|\overline{\partial} f|^2$ consists of a finite number of analytic subvarieties of M. (c.f. E. Heinz [**5**], J. Jost and S. -T. Yau [**8**]). For the sake of clearness, we mention that pluriharmonicity of f implies ∂f (respectively $\overline{\partial} f$) is a holomorphic section of a suitable bundle.

The standard residue argument yields from (3.1) or (3.2),

$$0 \geq \int_M \langle \omega_M, \Psi_M \rangle + \int_M \langle \omega_M, f^* \Psi_N \rangle, \quad \text{unless} \quad |\partial f|^2 \equiv 0, \tag{3.9}$$

or

$$0 \geq \int_M \langle \omega_M, \Psi_M \rangle - \int_M \langle \omega_M, f^* \Psi_N \rangle, \quad \text{unless} \quad |\overline{\partial} f|^2 \equiv 0. \tag{3.10}$$

(c.f. [**9**]).

Therefore, if a pluriharmonic map f satisfies

$$\int_M \langle \omega_M, \Psi_M \rangle + \left| \int_M \langle \omega_M, f^* \Psi_N \rangle \right| > 0,$$

then f is $\pm$-holomorphic. ∎

To prove Theorem 1.1 we remark that $\int_M \langle \omega_M, f^* \Psi_N \rangle$ is a smooth homotopy invariant (c.f. J. Eells and L. Lemaire [**1**]). That is to say, $\frac{d}{dt} \int_M \langle \omega_M, f_t^* \Psi_N \rangle = 0$ for a smooth family of maps $f_t : M \longrightarrow N$.

It is easily shown that

$$\begin{aligned} \langle \omega_M, \Psi_M \rangle &= \omega_M \bigwedge * \Psi_M = \Psi_M \bigwedge \frac{\omega_M^{m-1}}{(m-1)!}, \\ \langle \omega_M, f^* \Psi_N \rangle &= \omega_M \bigwedge * f^* \Psi_N = f^* \Psi_N \bigwedge \frac{\omega_M^{m-1}}{(m-1)!}. \end{aligned} \tag{3.11}$$

Therefore we get

$$\begin{aligned} \int_M \langle \omega_M, \Psi_M \rangle &= \frac{1}{(m-1)!} \int_M \mathrm{c}_1(M) \bigwedge \Omega(M), \\ \int_M \langle \omega_M, f^* \Psi_N \rangle &= \frac{1}{(m-1)!} \int_M f^* \mathrm{c}_1(N) \bigwedge \Omega(M). \end{aligned} \tag{3.12}$$

Under these hypotheses, Proposition 3.2 implies that if a pluriharmonic map $f : M \longrightarrow N$ satisfies

$$\mathrm{c}_1(M) \cup \Omega(M)[M] + |f^* \mathrm{c}_1(N) \cup \Omega(M)[M]| > 0,$$

then f is $\pm$-holomorphic.

Finally, we memtion the proof of Corollary 1.2. Using $\partial\overline{\partial}$-Bochner-Kodaira formula of Y. -T. Siu, we can see that a harmonic map from compact Kähler manifold to a hyperbolic Riemann surface is pluriharmonic, (Y. -T. Siu [**15**], N. Mok [**11**]).

Acknowledgements

The author would like thank to A. Katsuda, T. Kohno, Y. Isaka and S. Mukai for their helpful suggestion and continuous encouragement. He also thank to K. Ono who reads his the first manuscript and suggests some valuable advice to him.

REFERENCES

[1] J. Eells and L. Lemaire, "Selected Topics in Harmonic Maps," C. B. M. S. Regional Conference Serise in Math. 50, 1983.
[2] J. Eells and L. Lemaire, *A report on harmonic maps*, Bull. London Math. Soc. **10** (1978), 1–68.
[3] J. Eells and L. Lemaire, *Another report on harmonic maps*, Bull. London Math. Soc. **20** (1988), 385–524.
[4] J. Eells and J. C. Wood, *Restrictions on harmonic maps of surfaces*, Topology **15** (1976), 263–266.
[5] E. Heinz, *On certain nonlinear elliptic differential equations and univalent mappings*, J. Anal. Math. **5** (1956/57), 179–272.
[6] J. Jost, "Harmonic Maps between Surfaces," L. N. M. 1062, Springer–Verlag, Berlin–Heiderberg –New York, 1984.
[7] J. Jost, "Harmonic Maps between Riemannian Manifolds," Proc. Centre for Math. Anal., Australian National University, 1983.
[8] J. Jost and S. -T. Yau, *Harmonic mappings and Kähler manifolds*, Math. Ann. **262** (1983), 145–166.
[9] S. Kobayashi and K. Nomizu, "Foundations of Differential Geometry I, II," Wiley (Intersceince), New York, 1963, 1969.
[10] A. Lichnerowicz, *Applications harmoniques et variétés kählerinnes*, Symp. Math. III (Bologna 1970), 341–402.
[11] N. Mok, *Foliation techniques and vanishing theorems*, Comtemporary Math. **49** (1986), 79–118.
[12] Y. Ohnita, *On pluriharmonicity of stable harmonic maps*, J. London Math. Soc. (2) **35** (1987), 563–568.
[13] K. Ono, *On the holomorphicity of harmonic maps from compact Kähler manifolds to hyperbolic Riemann surfaces*, Proc. Amer. Math. Soc. **102** (1988), 1071–1076.
[14] R. Schoen and S. -T. Yau, *On univalent harmonic maps*, Invent. Math. **44** (1978), 265–278.
[15] Y. -T. Siu, *The complex-analyticity of harmonic maps and strong rigidity of compact Kähler manifolds*, Ann. of Math. **112** (1980), 73–111.

DEPARTMENT OF MATHEMATICS
NAGOYA UNIVERSITY
NAGOYA 464, JAPAN

ON THE INSTABILITY OF MINIMAL SUBMANIFOLDS AND HARMONIC MAPS

Takashi Okayasu

§0. Introduction

Let $f : M \to \overline{M}$ be an isometric minimal immersion. Let $D \subset M$ be a domain with a compact closure $\overline{D}$ and a smooth boundary ∂D. We call D *stable* if the second variation of its volume is nonnegative for every deformation of D that leaves ∂D fixed. We call D *unstable* if D is not stable. In [**10**] we asked the following question.

QUESTION. *Let $\overline{M}^n$ be a compact Riemannian manifold with the sectional curvature $K_{\overline{M}}$ satisfying $0 < \delta \leq K_{\overline{M}} \leq 1$ (or more generally, $Ric_{\overline{M}} \geq (n-1)\delta > 0$). Let $f : M^p \to \overline{M}^n$ be an isometric minimal immersion of a p-dimensional Riemannian manifold M^p. Let $D \subset M$ be a relatively compact domain with $\lambda_1(D) < p\delta$, where $\lambda_1(D)$ is the first eigenvalue of the Laplacian of D with the Dirichlet boundary condition. Is D unstable?*

When $p = 1$ the question is none other than the classical theorem of Bonnet and Myers (cf. Theorem 1.26 in [**3**]). When $p = n-1$ and $M^p, \overline{M}^n$ are orientable, the question can be solved affirmatively ([**10**]). However, without any further assumption, we have a counter example as follows. Let $M = CP^1$ be the totally geodesic submanifold of CP^m which has the Fubini-Study metric with its sectional curvature in $[1/4, 1]$. Let B_r be the geodesic ball of radius r in CP^1. Since B_r is a complex submanifold of CP^m, B_r is stable. However, $\lim_{r \to \pi} \lambda_1(B_r) = 0$ ([**2**, p.50]).

So in this paper we consider the above question under the condition $\delta > 1/4$. In [**10**], we gave partial positive answers to the question (Theorems B, 3.2 and 3.4 in [**10**]). In particular, we proved the existence of a constant C depending on n, p, δ and $\sup_D |B|^2$ (where B is the second fundamental form of M in $\overline{M}$) such that if $\lambda_1(D) < C$, then D is unstable. In this paper we use the method of Howard [**4**] to get a constant depending only on n, p and δ with the above property (Theorem 1.1).

In section 2, we consider harmonic maps. We use the method of [**9**], [**12**] to prove the Liouville type theorem for minimizing harmonic maps and stable harmonic maps from the Euclidean spaces to compact minimal submanifolds of δ-pinched spheres. As a corollary we get some informaion on low dimensional homotopy groups of compact minimal submanifolds.

§1. Instability of minimal submanifolds

To state our main theorem, we recall some notation of [**4**] and define a constant $C(n, p, \delta)$.

- $c_\lambda(t) = \cos(\sqrt{\lambda}t)$, $s_\lambda(t) = \sin(\sqrt{\lambda}t)/\sqrt{\lambda}$ $\qquad (\lambda > 0)$.
- $\tilde{g}_1(t, \delta) =$ middle value of $\left\{ c_1(t), 0, \dfrac{s_1(t)c_\delta(t)}{s_\delta(t)} \right\}$.

ISBN 0-12-640170-5

- $g_1(t,\delta) = (\tilde{g}_1(t,\delta))^2, \qquad 0 \le t \le \pi.$
- $g_2(t,\delta) = \max\left\{c_1(t)^2, \left(\frac{s_1(t)c_\delta(t)}{s_\delta(t)}\right)^2\right\}, \qquad 0 \le t \le \pi.$
- $g_1(t,\delta) = g_2(t,\delta) = 0, \qquad \pi < t.$
- $$F_{n,p}(\delta) = \int_0^\pi \{2pg_2(t,\delta)s_\delta(t)^{n-1} - (p+1)g_1(t,\delta)s_1(t)^{n-1} - (n-1)\delta c_1(t)^2 s_1(t)^{n-1}\}dt.$$

  $$\left(F_{n,p}(1) = -(n-p)\int_0^\pi \cos^2(t)\sin^{n-1}(t)dt < 0\right)$$
- $\delta(n,p) = \inf\{\delta; \delta > 1/4 \text{ and } F_{n,p}(\delta) \le 0\}.$
- $$C(n,p,\delta) = p^2 F_{n,p}(\delta)\left\{(p+1)\int_0^\pi \sqrt{g_2(t,\delta)}s_\delta(t)^{n-1}dt + \left[\left((p+1)\int_0^\pi \sqrt{g_2(t,\delta)}s_\delta(t)^{n-1}dt\right)^2 - pF_{n,p}(\delta)\int_0^\pi s_\delta(t)^{n-1}dt\right]^{1/2}\right\}^{-2}.$$

  $$\left(\lim_{n\to\infty} C(n,p,1) = p\right)$$

A Riemannian manifold $\overline{M}$ is calld δ-pinched ($0 < \delta \le 1$) if the sectional curvature $K_{\overline{M}}$ satisfies $\delta \le K_{\overline{M}} \le 1$.

THEOREM 1.1. *Let $\overline{M}^n$ be an n-dimensional simply-connected δ-pinched Riemannian manifold with $\delta > \delta(n,p)$. Let $f : M^p \to \overline{M}^n$ be an isometric minimal immersion of a p-dimensional Riemannian manifold M^p. Let $D \subset M$ be a relatively compact domain with $\lambda_1(D) < C(n,p,\delta)$. Then D is unstable.*

To prove Theorem 1.1, we need some Lemmas. Let $\overline{\nabla}$, $\overline{R}$ be the Riemannian connection and the curvature tensor of $\overline{M}$, respectively. Let Ω_M, $\Omega_{\overline{M}}$ be the Riemannian measures of M and $\overline{M}$, respectively. For any vector field V on $\overline{M}$ define $\mathcal{A}^V \in Hom(T\overline{M}, T\overline{M})$ by

$$\mathcal{A}^V(X) = \overline{\nabla}_X V.$$

Extend $\mathcal{A}^V$ to $Hom(\wedge^p T\overline{M}, \wedge^p T\overline{M})$ by defining

$$\mathcal{A}^V(X_1 \wedge \cdots \wedge X_p) = \sum_{i=1}^p X_1 \wedge \cdots \wedge \quad {}^V X_i \wedge \cdots \wedge X_p.$$

Note that if $V = \nabla f$ for some function f, then $\mathcal{A}^V$ is self-adjoint. We define $\overline{R}_V(X) = \overline{R}(X,V)V$. Extend $\overline{R}_V$ also to $\wedge^p T\overline{M}$ as a derivation.

Let ξ be the field of unit p-vectors on M (defined up to sign) such that at any $x \in M$, $\xi_x = e_1 \wedge \cdots \wedge e_p$ for some orthonormal basis $e_1, \cdots, e_p$ of T_xM. Suppose that V is a vector field of $\overline{M}$ such that $V|_{\partial D} \equiv 0$. Let φ_t be the flow generated by V. Set $\mathcal{V}(t)$ = volume of $\varphi_t(D)$. Then we have the following second variation formula (cf. p.21, p.22 in [6]).

Lemma 1.2.

$$\left.\frac{d^2\mathcal{V}}{dt^2}\right|_{t=0}(:= I(V,V)) = \int_D \{-\langle \mathcal{A}^V\xi,\xi\rangle^2 + \langle \mathcal{A}^V\mathcal{A}^V\xi,\xi\rangle + \langle \mathcal{A}^V\xi,\, \mathcal{A}^V\xi\rangle - \langle(\mathcal{A}^V)^2\xi,\xi\rangle - \langle \overline{R}_V\xi,\xi\rangle\}\Omega_M.$$

Let φ be a smooth function with $\varphi|_{\partial D} = 0$. We calculate the integrand of the second variational formula for the variational vector field φV. We denote by V^T the orthogonal projection of V to TM, and set $V^N = V - V^T$.

Lemma 1.3.

(a)
$$\langle \mathcal{A}^{\varphi V}\xi,\xi\rangle^2 = \langle V,\nabla\varphi\rangle^2 + 2\varphi\langle V,\nabla\varphi\rangle\langle \mathcal{A}^V\xi,\xi\rangle + \varphi^2\langle \mathcal{A}^V\xi,\xi\rangle^2.$$

(b)
$$\langle \mathcal{A}^{\varphi V}\xi,\, \mathcal{A}^{\varphi V}\xi\rangle = \varphi^2\langle \mathcal{A}^V\xi,\, \mathcal{A}^V\xi\rangle + 2\varphi\langle V,\nabla\varphi\rangle\langle \mathcal{A}^V\xi,\xi\rangle + \langle V,\nabla\varphi\rangle^2 + |\nabla\varphi|^2|V^N|^2 + 2\varphi\langle \mathcal{A}^V(\nabla\varphi), V^N\rangle$$

(c)
$$\langle(\mathcal{A}^{\varphi V}\mathcal{A}^{\varphi V} - (\mathcal{A}^{\varphi V})^2)\xi,\xi\rangle = 2\varphi\langle V,\nabla\varphi\rangle\langle \mathcal{A}^V\xi,\xi\rangle - 2\varphi\langle \mathcal{A}^V(V^T),\nabla\varphi\rangle + \varphi^2\langle(\mathcal{A}^V\mathcal{A}^V - (\mathcal{A}^V)^2)\xi,\xi\rangle.$$

(d)
$$\langle \overline{R}_{\varphi V}\xi,\xi\rangle = \varphi^2\langle \overline{R}_V\xi,\xi\rangle.$$

Proof. In proving (c) we use

$$\begin{aligned}
&\langle(\mathcal{A}^E\mathcal{A}^E - (\mathcal{A}^E)^2)\xi,\xi\rangle \\
&= \sum_{i\neq j}\langle e_1\wedge\cdots\wedge\overline{\nabla}_{e_i}E\wedge\overline{\nabla}_{e_j}E\wedge\cdots\wedge e_p, e_1\wedge\cdots\wedge e_p\rangle \\
&= \sum_{i\neq j}\{\langle\overline{\nabla}_{e_i}E,e_i\rangle\langle\overline{\nabla}_{e_j}E,e_j\rangle - \langle\overline{\nabla}_{e_i}E,e_j\rangle\langle\overline{\nabla}_{e_j}E,e_i\rangle\}.
\end{aligned}$$

The proofs of (a), (b), (c) and (d) are straightforward calculation, so we omit them.

Proof of Theorem 1.1. We choose φ to be the first eigenfunction of Laplacian of D with the Dirichlet boundary condition. As in [4] define a function $f : \mathbf{R} \to \mathbf{R}$ by

$$f(t) = \begin{cases} -\cos(t) = -c_1(t) & |t| \leq \pi, \\ 1 & |t| \geq \pi. \end{cases}$$

For any $x \in \overline{M}$ let $\rho_x(y) =$ geodesic distance of y from x and let $V_x(f)$ be the vector field defined by

$$V_x(f) = \nabla(f\circ\rho_x) = f'(\rho_x)\nabla\rho_x.$$

Then from Lemma 1.2, we get

$$
\begin{aligned}
&I(\varphi V_x(f),\varphi V_x(f)) \\
&\quad= \int_D \Big\{\varphi^2\Big[\langle \mathcal{A}^{V_x(f)}\xi, \mathcal{A}^{V_x(f)}\xi\rangle - \langle \mathcal{A}^{V_x(f)}\xi,\xi\rangle^2 + \langle \mathcal{A}^{V_x(f)}\mathcal{A}^{V_x(f)}\xi,\xi\rangle \\
&\qquad\qquad - \langle (\mathcal{A}^{V_x(f)})^2\xi,\xi\rangle - \langle \overline{R}_{V_x(f)}\xi,\xi\rangle\Big] \\
&\qquad\quad + |\nabla\varphi|^2|V_x(f)^N|^2 + 2\varphi\langle V_x(f),\nabla\varphi\rangle\langle \mathcal{A}^{V_x(f)}\xi,\xi\rangle \\
&\qquad\quad + 2\varphi\langle \mathcal{A}^{V_x(f)}(\nabla\varphi), V_x(f)^N - V_x(f)^T\rangle\Big\}\Omega_M.
\end{aligned} \tag{1.1}
$$

Fix a point $x_0 \in M$. Our object is to give a upper bound on the integrand of (1.1) and express the integral in terms of the distance of x from x_0. Let $U\overline{M}_{x_0}$ be the unit sphere of $T\overline{M}_{x_0}$ and $\rho = \rho_{x_0}$. Then denote by (ρ, u) polar coordinates on $\overline{M}$ centered at x_0. Since $|V_x(f)| \le 1$, $|\langle \mathcal{A}^{V_x(f)}\xi,\xi\rangle| \le p\sqrt{g_2(\rho(x),\delta)}$ and $|\langle \mathcal{A}^{V_x(f)}(\nabla\varphi), V_x(f)^T\rangle| \le |\nabla\varphi|\sqrt{g_2(\rho(x),\delta)}$ (cf. p.328 in [4]), we get

$$
|\nabla\varphi|^2|V_x(f)^N|^2 \le |\nabla\varphi|^2, \tag{1.2}
$$

$$
\begin{aligned}
&2\varphi\langle V_x(f),\nabla\varphi\rangle\langle \mathcal{A}^{V_x(f)}\xi,\xi\rangle + 2\varphi\langle \mathcal{A}^{V_x(f)}(\nabla\varphi), V_x(f)^N - V_x(f)^T\rangle \\
&\quad \le 2|\varphi||\nabla\varphi|(p+1)\sqrt{g_2(\rho(x),\delta)} \le (p+1)\Big\{\frac{\varphi^2}{k} + k|\nabla\varphi|^2\Big\}\sqrt{g_2(\rho(x),\delta)},
\end{aligned} \tag{1.3}
$$

where k is a positive constant fixed later. Integrating (1.1) on $\overline{M}$ and using Fubini's Theorem, we obtain

$$
\begin{aligned}
&\int_{\overline{M}} I(\varphi V_x(f),\varphi V_x(f))\Omega_{\overline{M}}(x) \\
&= \int_D\int_{\overline{M}} \Big\{\varphi^2\Big[\langle \mathcal{A}^{V_x(f)}\xi, \mathcal{A}^{V_x(f)}\xi\rangle - \langle \mathcal{A}^{V_x(f)}\xi,\xi\rangle^2 + \langle \mathcal{A}^{V_x(f)}\mathcal{A}^{V_x(f)}\xi,\xi\rangle \\
&\qquad\qquad - \langle (\mathcal{A}^{V_x(f)})^2\xi,\xi\rangle - \langle \overline{R}_{V_x(f)}\xi,\xi\rangle\Big] \\
&\qquad + |\nabla\varphi|^2|V_x(f)^N|^2 + 2\varphi\langle V_x(f),\nabla\varphi\rangle\langle \mathcal{A}^{V_x(f)}\xi,\xi\rangle \\
&\qquad + 2\varphi\langle \mathcal{A}^{V_x(f)}(\nabla\varphi), V_x(f)^N - V_x(f)^T\rangle\Big\}\Omega_{\overline{M}}(x)\Omega_M(x_0) \\
&\le p\mathrm{vol}(S^{n-1})F_{n,p}(\delta)\int_D \varphi^2\Omega_M + \mathrm{vol}(S^{n-1})\int_D |\nabla\varphi|^2\Omega_M\int_0^\pi s_\delta(t)^{n-1}dt \\
&\qquad + \mathrm{vol}(S^{n-1})(p+1)\int_D\Big(\frac{\varphi^2}{k} + k|\nabla\varphi|^2\Big)\Omega_M\int_0^\pi \sqrt{g_2(t,\delta)}s_\delta(t)^{n-1}dt \\
&= \mathrm{vol}(S^{n-1})\Big(\int_D \varphi^2\Omega_M\Big)\Big\{pF_{n,p}(\delta) + \lambda_1(D)\int_0^\pi s_\delta(t)^{n-1}dt \\
&\qquad + (p+1)\Big(\frac{1}{k} + k\lambda_1(D)\Big)\int_0^\pi \sqrt{g_2(t,\delta)}s_\delta(t)^{n-1}dt\Big\},
\end{aligned} \tag{1.4}
$$

where at the last inequality we used (1.2), (1.3) and (3-4), (4-15) in [4]. From (1.4) we see that if

$$
\begin{aligned}
\lambda_1(D) <&\Big\{-pF_{n,p}(\delta) - \frac{p+1}{k}\int_0^\pi \sqrt{g_2(t,\delta)}s_\delta(t)^{n-1}dt\Big\} \\
&\times\Big\{\int_0^\pi s_\delta(t)^{n-1}dt + (p+1)k\int_0^\pi \sqrt{g_2(t,\delta)}s_\delta(t)^{n-1}dt\Big\}^{-1},
\end{aligned} \tag{1.5}
$$

then

$$\int_{\overline{M}} I(\varphi V_x(f), \varphi V_x(f))\Omega_{\overline{M}}(x) < 0.$$

That is, for some $x \in \overline{M}$, $I(\varphi V_x(f), \varphi V_x(f)) < 0$, and consequently D is unstable. To finish the proof, we choose $k > 0$ to maximize the right hand side of (1.5) and obtain

$$\begin{aligned} \lambda_1(D) < & p^2 F_{n,p}(\delta)\Big\{ (p+1)\int_0^{\pi} \sqrt{g_2(t,\delta)}s_\delta(t)^{n-1}dt \\ & + \Big[\Big((p+1)\int_0^{\pi}\sqrt{g_2(t,\delta)}s_\delta(t)^{n-1}dt\Big)^2 - pF_{n,p}(\delta)\int_0^{\pi} s_\delta(t)^{n-1}dt\Big]^{1/2}\Big\}^{-2}. \end{aligned}$$

§2. Instability of harmonic maps

Every smooth harmonic map $u : S^{p-1} \to N$ defines a homogeneous *tangent map* $\overline{u} : \mathbf{R}^p \to N$ by putting $\overline{u}(x) = u(x/|x|)$. If $u \neq$ const, then $0 \in \mathbf{R}^p$ will be a singular point of $\overline{u}$. We consider the following condition (C_l): any tangent map $\overline{u}$ that minimize energy, in $L_1^2(\mathbf{R}^p, N)$, on compact subsets of $\mathbf{R}^p$ are constant for $3 \leq p \leq l$ $(l \geq 3)$. In [11], Schoen and Uhlenbeck proved the following theorem.

Theorem 2.1. *Suppose that a Riemannian manifold N satisfies the condition C_l for some $l \geq 3$. If $u \in L_1^2(M^n, N)$ is energy minimizing with $u(x) \in N_0$ a.e. for a compact subset $N_0 \subset N$, then the Hausdorff dimension of the singular points of u is at most $n - l - 1$. In particular, if $n \leq l$ such maps are always smooth.*

It is easy to see that if the hypothesis of Theorem 2.1 is satisfied, then $\pi_k(M) = 0$ for $2 \leq k \leq l - 1$ ([9]).

In this section we study minimizing tangent maps into minimal submanifolds of δ-pinched spheres.

Theorem 2.2. *Let $(\overline{M}^{N-1}, h)$ be a complete simply-connected δ-pinched $(N-1)$-dimensional Riemannian manifold with $k_3(\delta)^2 \leq 4(2n+\delta-1)/N(1+\delta)$ (see [10] for the definition of $k_3(\delta)$). Suppose that $f : (M^n, g) \to (\overline{M}^{N-1}, h)$ is an isometric minimal immersion of an n-dimensional compact Riemannian manifold (M^n, g) with $Ric_M \geq (n-1)\rho$. If either*

(a) $3 \leq p \leq 5$ *and*

$$\begin{aligned} \frac{2(p-2)n}{1+2(p-1)n} \leq \frac{1}{n}\Big\{ & \frac{4\rho}{1+\delta} - \frac{2n+\delta-1}{1+\delta} \\ & - \frac{N}{4}k_3(\delta)^2 - \Big(\frac{N}{1+\delta}(2n-2\rho+\delta-1)\Big)^{1/2} k_3(\delta)\Big\}, \end{aligned}$$

or

(b) $p = 6$ *and*

$$\begin{aligned} \frac{4}{5} < \frac{1}{n}\Big\{ & \frac{4\rho}{1+\delta} - \frac{2n+\delta-1}{1+\delta} \\ & - \frac{N}{4}k_3(\delta)^2 - \Big(\frac{N}{1+\delta}(2n-2\rho+\delta-1)\Big)^{1/2} k_3(\delta)\Big\}, \end{aligned}$$

then there is no non-constant minimizing tangent map $\overline{u} : \mathbf{R}^p \to M$.

COROLLARY 2.3. *Under the assumption of the theorem, there is no non-constant harmonic map $u : \mathbf{R}^p \to M$ which minimizes energy on each compact subset of $\mathbf{R}^p$.*

COROLLARY 2.4. *Under the assumption of the theorem, we have $\pi_k(M) = 0$ for $2 \leq k \leq p-1$.*

THEOREM 2.5. *Let $(\overline{M}^{N-1}, h)$ be a complete simply-connected δ-pinched $(N-1)$-dimensional Riemannian manifold with $k_3(\delta)^2 \leq 4(2n+\delta-1)/N(1+\delta)$. Suppose that $f : (M^n, g) \to (\overline{M}^{N-1}, h)$ is an isometric minimal immersion of an n-dimensional compact Riemannian manifold (M^n, g) with $Ric_M \geq (n-1)\rho$. If either*

(a) $p = 2$ *and*

$$\frac{4\rho}{1+\delta} - \frac{2n+\delta-1}{1+\delta} - \frac{N}{4}k_3(\delta)^2 - \left(\frac{N}{1+\delta}(2n-2\rho+\delta-1)\right)^{1/2} k_3(\delta) > 0,$$

or

(b) $3 \leq p \leq 4$ *and*

$$\frac{p-1}{p} \leq \frac{1}{n}\left\{\frac{4\rho}{1+\delta} - \frac{2n+\delta-1}{1+\delta} - \frac{N}{4}k_3(\delta)^2 - \left(\frac{N}{1+\delta}(2n-2\rho+\delta-1)\right)^{1/2} k_3(\delta)\right\},$$

then there is no non-constant stable harmonic map $u : \mathbf{R}^p \to M$.

PROOF OF THEOREM 2.2. We use the notation of [10]. Let V be a parallel section of E with respect to ∇'. We calculate the second variation of $\overline{u}$ for the variational vector field φV^T, where φ is a function with compact support in $\mathbf{R}^p \backslash \{0\}$. Then

$$\begin{aligned} I(\varphi V^T, \varphi V^T) = \int_{\mathbf{R}^p} \Big\{ & |\nabla\varphi|^2 |V^T|^2 + \frac{1}{2}\sum_a \nabla_{e_a}(\varphi^2)\nabla_{e_a}\langle V^T, V^T\rangle \\ & + \varphi^2 \sum_a \Big(|\nabla_{\overline{u}_* e_a} V^T|^2 - \langle R(V^T, \overline{u}_* e_a)\overline{u}_* e_a, V^T\rangle \Big) \Big\}. \end{aligned} \tag{2.1}$$

We set $\mathcal{W} = \{V \in \Gamma(E); \nabla' V = 0\}$. Let $V_1, \cdots, V_N$ be an orthonormal basis of $\mathcal{W}$. We compute $\sum_{i=1}^N I(\varphi V^T, \varphi V^T)$. From (2.1) and the inequality at the bottom of page 219 in [10], we obtain

$$\begin{aligned} \sum_{i=1}^N I(\varphi V_i^T, \varphi V_i^T) \leq \int_{\mathbf{R}^p} \Big\{ & n|\nabla\varphi|^2 + \varphi^2 \Big[\frac{N}{4}k_3(\delta)^2 + \frac{2n+\delta-1}{1+\delta} \\ & - \frac{4\rho}{1+\delta} + \left(\frac{N}{1+\delta}(2n-2\rho+\delta-1)\right)^{1/2} k_3(\delta) \Big] |d\overline{u}|^2 \Big\} \end{aligned} \tag{2.2}$$

Using (2.2) instead of (3.1) in [9], we can follow the proof of Theorem (2.1) in [9].

Since the proofs of Corollary 2.3, Corollary 2.4 and Theorem 2.5 are same as those of [9], [12], we omit them.

When M is not a minimal submanifold of $\overline{M}$, we have the following result (compare this with Corollary 2.4).

Theorem 2.6. *Let $f : M^n \to \overline{M}^m$ be an isometric immersion of an n-dimensional simply-connected compact Riemannian manifold M^n $(n \geq 4)$ into an m-dimensional compact δ-pinched Riemannian manifold $\overline{M}^m$. Suppose that $|B|^2 < (16/3)(\delta - 1/4)$, where B is the second fundamental form of M in $\overline{M}$. Then M is homeomorphic to S^n.*

Proof. We use the notation of [8]. Let $R, \overline{R}$ be the curvature tensors of $M, \overline{M}$, respectively. Let $\mathcal{R}$ be the curvature operator of M defined by

$$\langle \mathcal{R}(x \wedge y), u \wedge v\rangle = \langle R(x, y)v, u\rangle, \quad \text{for } x, y, u, v \in TM.$$

We extend the curvature operator $\mathcal{R}$ to a complex linear map from $\wedge^2 TM \otimes \mathbf{C}$ to $\wedge^2 TM \otimes \mathbf{C}$. We also extend the Riemannian metric of M to a Hermitian inner product $\langle\!\langle\ ,\ \rangle\!\rangle$ on $TM \otimes \mathbf{C}$. Let $e_1, \cdots, e_4$ be orthonormal tangent vectors of TM and set $z = e_1 + ie_2$, $w = e_3 + ie_4$. Then using the Gauss equation and (1.2) in [8], we obtain

$$\begin{aligned}
\langle\!\langle \mathcal{R}(z \wedge w), z \wedge w\rangle\!\rangle &= \langle R(e_1, e_3)e_3, e_1\rangle + \langle R(e_2, e_4)e_4, e_2\rangle \\
&\quad + \langle R(e_1, e_4)e_4, e_1\rangle + \langle R(e_2, e_3)e_3, e_2\rangle \\
&\quad + 2\langle R(e_1, e_2)e_3, e_4\rangle \\
&\geq 4\delta + \langle B(e_1, e_1), B(e_3, e_3)\rangle - |B(e_1, e_3)|^2 \\
&\quad + \langle B(e_2, e_2), B(e_4, e_4)\rangle - |B(e_2, e_4)|^2 \\
&\quad + \langle B(e_1, e_1), B(e_4, e_4)\rangle - |B(e_1, e_4)|^2 \\
&\quad + \langle B(e_2, e_2), B(e_3, e_3)\rangle - |B(e_2, e_3)|^2 \\
&\quad + 2\langle \overline{R}(e_1, e_2)e_3, e_4\rangle + 2\langle B(e_1, e_4), B(e_2, e_3)\rangle \\
&\quad - 2\langle B(e_1, e_3), B(e_2, e_4)\rangle \\
&\geq 4\delta + 2\langle \overline{R}(e_1, e_2)e_3, e_4\rangle - |B|^2 \\
&\geq 4\delta - \frac{4}{3}(1 - \delta) - |B|^2 > 0,
\end{aligned}$$

where at the third inequality we used an inequality of Berger [1]

$$\left|\langle \overline{R}(e_1, e_2)e_3, e_4\rangle\right| \leq \frac{2}{3}(1 - \delta).$$

Thus M has positive curvature on totally isotropic two-planes. By the Main Theorem of [8], M^n is homeomorphic to S^n.

REMARKS. 1. Under the weaker condition on Ricci curvature than in Corollary 2.4, we can conclude $\pi_2(M) = 0$ ([**10**]).
2. When $\overline{M}^m = S^m(1)$ in Theorem 2.6, there are sharp results ([**7**], [**5**]). For example, if $|B|^2 < \min(n-1, 2\sqrt{n-1})$, then M^n is homeomorphic to S^n.

REFERENCES

[1] M. Berger, Sur quelques variétés riemanniennes suffisament pincées, Bull. Soc. Math. France **88**(1960), 57-71.

[2] I. Chavel, *Eigenvalues in Riemannian Geometry*, Academic Press, 1984.

[3] J. Cheeger and D. Ebin, *Comparison theorems in Riemannian Geometry*, North -Holland, 1975.

[4] R. Howard, The nonexistence of stable submanifolds, varifolds, and harmonic maps in sufficiently pinched simply connected Riemannian manifolds, Michigan Math. J. **32**(1985), 321-334.

[5] R. Howard and S. W. Wei, On the existence and non-existence of stable submanifolds and currents in positively curved manifolds and the topology of submanifolds in Euclidean spaces, Preprint.

[6] H. B. Lawson, Jr., *Minimal varieties in real and complex geometry*, University of Montreal Press, 1974.

[7] H. B. Lawson, Jr. and J. Simons, On stable currents and their application to global problems in real and complex geometry, Ann. of Math. **98**(1973), 427-450.

[8] M. Micallef and J.-D.Moore, Minimal two-spheres and the topology of manifolds with positive curvature on totally isotropic two-planes, Ann. of Math. **127**(1988), 199-227.

[9] H. Nakajima, Regularity of minimizing harmonic maps into certain Riemannian manifolds, Preprint.

[10] T. Okayasu, On the instability of minimal submanifolds in Riemannian manifolds of positive curvature, To appear in Math. Z..

[11] R. Schoen and K. Uhlenbeck, A regularity theory for harmonic maps, J. Diff. Geom. **17**(1982), 307-335.

[12] R. Schoen and K. Uhlenbeck, Regularity of minimizing harmonic maps into the sphere, Invent. Math. **78**(1984), 84-100.

Takashi OKAYASU

Department of Mathematics
Faculty of Science
Hirosaki University
Hirosaki 036
Japan

Spectra of Riemannian Manifolds without Focal Points

Hajime Urakawa

Introduction and Statement of Results. In this paper, we study relationship of the following geometric/analytic quantities (1) ~ (6) of a complete non-compact Riemannian manifold (M,g) without focal points :

(1) (*bottom of spectrum*)

$$\lambda_0(M) := \inf\{ \frac{\int_M f\,\Delta f\, dv_g}{\int_M f^2\, dv_g} \; ; f \in C_c^\infty(M) \},$$

where Δ is the (non-negative) Laplacian of (M,g). For any fixed point $p \in M$, let $B(r) := \{ x \in M ; d(x,p) < r \}$ be the geodesic ball around p with radius r. Then we define :

(2) (*bottom of mean curvature*)

$$m(M) := \inf\{ m(r) ; 0<r<\infty \},\ m(r) := \inf\{\text{mean curvature of } \partial B(r)\},$$

(3) (*Cheeger's constant*) $\quad h(M) := \inf\{\mathrm{Vol}_{n-1}(\partial D)/\mathrm{Vol}_n(D) ; D \subset M\},$

(4) (*exponential growth*) $\quad \mu(M) := \limsup_{r\to\infty} (\log V(r))/r,$

(5) (*isoperimetric growth*) $\quad \bar{h}(M) := \limsup_{r\to\infty} S(r)/V(r),$

(6) (*derivative isoperimetric growth*) $\quad \bar{\bar{h}}(M) := \limsup_{r\to\infty} S'(r)/V'(r),$

where $V(r) := \mathrm{Vol}_n(B(r))$, $S(r) := \mathrm{Vol}_{n-1}(\partial B(r))$, and $S'(r)$, $V'(r)$ are the derivatives with respect to r. Then we obtain :

Theorem 1. *For a simply connected complete Riemannian manifold (M,g) without focal points,*

$$0 \leqq m(M) \leqq h(M) \leqq \sqrt{4\lambda_0(M)} \leqq \mu(M) \leqq \bar{h}(M) \leqq \bar{\bar{h}}(M).$$

ISBN 0-12-640170-5

Remark. It is known (cf.[C], [B]) that the relations $h(M) \leqq \sqrt{4\lambda_0(M)} \leqq \mu(M)$ hold for any complete non-compact Riemannian manifold (M,g).

Theorem 2. *For a simply connected complete Riemannian manifold* (M,g) *without focal points, we assume the Ricci curvature* $\mathrm{Ric}(\partial/\partial r)$ *along* $\partial/\partial r$ *satisfies* $\mathrm{Ric}(\partial/\partial r) \leqq -(n-1)k$, *for a positive constant* k. *Then*

$$\sqrt{(n-1)k} \leqq m(M).$$

Theorem 3. *Let* (M,g) *be a simply connected Riemannian symmetric space* $M = G/K$ *of non-compact type whose metric* g *is the* G*-invariant one induced from the Killing form of the Lie algebra of* G. *Then*

$$\sqrt{4\lambda_0(M)} = \mu(M) = \bar{h}(M) = \bar{\bar{h}}(M) = 2\,\|\rho\|\ ;$$

$$m(M) = \inf\{<2\rho,H>; H\in\mathfrak{a}^+,\ \|H\| = 1\}.$$

In particular, if M *is of rank one,*

$$m(M) = \sqrt{4\lambda_0(M)} = \mu(M) = \bar{h}(M) = \bar{\bar{h}}(M) = 2\,\|\rho\|\ ,$$

where $\mathfrak{a}^+$ *is potitive restricted Weyl chamber, and* ρ *is half sum of positive restricted roots of* M.

Acknowledgement:. I would like to thank for helpful discussions with my colleagues Makoto Kaneko and Yoshihiko Suyama.

§1. Preliminaries to the Inequality $\mu(M) \leqq \bar{h}(M)$.

1.1. L'Hospital Theorem. We first show the following which maybe known but we are not able to find in the literature :

Lemma 4. *For real valued differentiable functions* f, h *on the open interval* $(0,\infty)$, *satisfying* $\lim_{x\to\infty} f(x) = \lim_{x\to\infty} h(x) = \infty$. *Then*

(*i*) $\limsup_{x\to\infty} f(x)/h(x) \leqq \limsup_{x\to\infty} f'(x)/h'(x)$,

(*ii*) $\liminf_{x\to\infty} f(x)/h(x) \geqq \liminf_{x\to\infty} f'(x)/h'(x)$.

Remark. The assertions (i), (ii) of Lemma 4 yield the so-called *L'Hospital theorem*. One can show the similar type inequalities for the other cases.

Proof (with M. Kaneko). We only show (i) here. One can show (ii) by the same way. Given $\varepsilon>0$ and $n\in\mathbb{N}$, there exists x_n such that for all $x\geqq x_n$,

$$(1-\frac{f(n)}{f(x)})/(1-\frac{h(n)}{h(x)})>1-\varepsilon.$$

Since x_n can be arbitrarily large, we may assume that x_n is monotone increasing and $x_n\geqq n$. Then for $x>x_n$, there exists $n<\xi<x$ such that

$$(1-\varepsilon)\frac{f(x)}{h(x)}<\frac{f(x)}{h(x)}\frac{1-\frac{f(n)}{f(x)}}{1-\frac{h(n)}{h(x)}}=\frac{f(x)-f(n)}{h(x)-h(n)}=\frac{f'(\xi)}{h'(\xi)}\leqq\sup\{\frac{f'(\xi)}{h'(\xi)};t\geqq n\}.$$

Here we may assume $f(x)/h(x)>0$ since $\lim_{x\to\infty}f(x)=\lim_{x\to\infty}h(x)=\infty$. Then

$$\frac{f(x)}{h(x)}<\frac{1}{1-\varepsilon}\sup\{\frac{f'(\xi)}{h'(\xi)};t\geqq n\},$$

for all $x>x_n$. Thus we obtain

$$\sup\{\frac{f(x)}{h(x)};x\geqq x_n\}<\frac{1}{1-\varepsilon}\sup\{\frac{f'(\xi)}{h'(\xi)};t\geqq n\}.$$

Letting $t\to\infty$, we get $\limsup_{x\to\infty}f(x)/h(x)\leqq(1-\varepsilon)^{-1}\limsup_{x\to\infty}f'(x)/h'(x)$. Since $\varepsilon>0$ is arbitrary, we obtain (i). □

1.2. Differentiability of V(t). In this subsection, we study which complete Riemannian manifold (M,g) has differentiable $V(t)=\mathrm{Vol}_n(B(t))$ in t. For a fixed point $p\in M$, identifying the unit sphere of the tangent space of M at p with $S^{n-1}\subset\mathbb{R}^n$, define $\Theta;S^{n-1}\times\mathbb{R}\to M$ by $\Theta(\theta,t)=\exp_p(t\theta)$. Then

$$\Theta^*dv_g=\sqrt{g}(t,\theta)\,d\theta\wedge dt,$$

where dv_g is the canonical measure of (M,g) and $d\theta$ is the canonical measure of S^{n-1}. For $\theta\in S^{n-1}$, let $t_0=C(\theta)$ be the *cut value* of θ , i.e., being γ_θ the geodesic of (M,g) with $\gamma_\theta(0)=p$ and $\dot{\gamma}_\theta(0)=\theta$,

$$d(p,\gamma_\theta(t))=t\text{ for all }t\leqq t_0\text{ and }d(p,\gamma_\theta(t))<t\text{ for all }t>t_0.$$

Then C is continuous on S^{n-1} and

$$V(t)=\mathrm{Vol}_n(B(t))=\int_{S^{n-1}}F(t,\theta)\,d\theta,\text{ where }F(t,\theta):=\int_0^{\min\{t,C(\theta)\}}\sqrt{g}(s,\theta)ds.$$

Then we obtain :

Proposition 5. *Let (M,g) be a complete Riemannian manifold. Assume that (cut locus \ conjugate locus)$\cap\partial B(t)$ has measure zero in $\partial B(t)$ for each $t>0$. Then* $V(t) = \mathrm{Vol}_n(B(t))$ *is differentiable everywhere and*

$$\frac{dV}{dt} = \int_{S^{n-1}} \frac{\partial F}{\partial t}(t,\theta)d\theta = \int_{C_{\_}(t)} \sqrt{g}(t,\theta)d\theta,$$

where $C\_(t) := \{\theta \in S^{n-1} ;\ t< C(\theta)\}$.

Remark. (i) If (M,g) does not satisfy the condition of Proposition 5, there is a counter example, in fact, for $(M,g) = (\mathbf{P}^2(\mathbb{R}),\mathrm{can})$, $V(t)$ is not differentiable at $t = \pi/2$ and (*cut locus \ conjugate locus*)$\cap\partial B(\pi/2) = \partial B(\pi/2)$. (ii) On the other hand, a lot of Riemannian manifolds, for example, Blaschke manifolds, negative curved mani-folds, Riemannian manifolds with a pole, satisfy the condition.

Proof. Let $0<t_0<\infty$. For $|\varepsilon|<t_0$, consider

$$\frac{V(t_0+\varepsilon) - V(t_0)}{\varepsilon} = \int_{S^{n-1}} \frac{F(t_0+\varepsilon,\theta) - F(t_0)}{\varepsilon} d\theta .$$

Denoting by $F_\varepsilon(t_0,\theta)$, the integrand, we get the following :

(i) there exists a positive constant K such that $|F_\varepsilon(t_0,\theta)|\leqq K$ for all $\theta \in S^{n-1}$ and $|\varepsilon|<t_0$,

(ii) $\lim_{\varepsilon\to 0}F_\varepsilon(t_0,\theta) = \partial F/\partial t(t_0,\theta)$, outside the measure zero set $C^{-1}(\{t\})\cap(S^{n-1}\setminus D_t)$, where $D_t = \{\theta\in S^{n-1}; t = C(\theta), \sqrt{g}(t,\theta) = 0\}$.

(iii) $\partial F/\partial t(t_0,\theta)$ is integrable in θ on S^{n-1} because it is 0 if $t_0>C(\theta)$, and it is $\sqrt{g}(t_0,\theta)$ if $t_0<C(\theta)$.

Indeed, for (i) , note that

$$|F_\varepsilon(t_0,\theta)| = \frac{1}{|\varepsilon|}\left|\int_{\min\{t_0,C(\theta)\}}^{\min\{t_0+\varepsilon,C(\theta)\}} \sqrt{g}(s,\theta)ds\right| \leqq \frac{|\min\{t_0+\varepsilon,C(\theta)\}-\min\{t_0,C(\theta)\}|}{|\varepsilon|} K,$$

where $K:= \max\{\sqrt{g}(s,\theta) ;\ 0\leqq s\leqq\max\{2t_0,C\}\}$, $C := \max\{C(\theta) ;\ \theta\in S^{n-1}\}$. The inequality $|\min\{t_0+\varepsilon,C(\theta)\}-\min\{t_0,C(\theta)\}| \leqq |\varepsilon|$ implies (i). For (ii), note that $C^{-1}(\{t\})\cap(S^{n-1}\setminus D_t)$ has measure zero by the assumption, and (iii) follows by definition.

Thus Lebesgue's bounded convergence theorem implies the desired results.

□

Corollary 6. *Let (M,g) be a complete non-compact Riemannian manifold.*

Assume that (cut locus \ conjugate locus) $\cap \partial B(t)$ *has measure zero in* $\partial B(t)$ *for each* $t>0$. *Then*

$$\mu(M) \leqq \limsup_{t\to\infty} V'(t)/V(t).$$

§2. Riemannian Manifolds without Focal Points.

2.1. Focal Points. Following [O], [E], we explain focal points. Let $Q \subset M$ be a submanifold, and $\tau ; [0,1] \to M$ be a geodesic segment in M perpendicular to Q at $p = \tau(0)$. A vector field Y along τ is a *Q-Jacobi field* if it is the variation vector field of a variation of τ through geodesics which are initially perpendicular to Q. A point $p = \tau(1)$ is a *focal point* of Q along τ if there is a non-trivial Q-Jacobi field along τ vanishing at p. A Riemannian manifold (M,g) is said to have *no focal points* if any imbedded geodesic segment in M, regarded as a submanifold Q, has no focal points in M. Then :

Lemma 7. (1) (O'Sullivan [O, p.298]) *A complete Riemannian manifold* (M,g) *has :*

non-poistive curvature $\Rightarrow$ *no focal points* $\Rightarrow$ *no conjugate points.*

(2) (Eschenburg [E, p.80]) *Let* (M,g) *be a simply connected complete Riemannian manifold without focal points. Then the distance function from any point* p *in* M, $r(x) := d(p,x)$, *is convex,* i.e., $\mathrm{Hess}(r)$ *is positive semi-definite.*

2.2. Distance Function. A Riemannian manifold (M,g) has a *pole* p if there is no conjugate point in M, i.e., $\exp_p ; T_pM \to M$ is regular everywhere. It is wellknown that, if p is a pole of a simply connected complete Riemannian manifold (M,g), then $\exp_p$ is a diffeomorphism of T_pM onto M. In what follows, we assume (M,g) is a simply connected Riemannian manifold with a pole p. A global geodesic polar coordinate (r,θ) on $M \setminus \{p\}$ is given by $M \setminus \{p\} \ni x = \exp_p(r\theta)$, $r = d(p,x)$, $\theta \in S^{n-1} \cong \{\theta \in T_pM;\ \|\theta\| = 1\}$. Taking a local coordinate $(\theta_2, \dots, \theta_n)$ in S^{n-1}, define a local coordinate $(x_1, \dots, x_n) = (r, \theta_2, \dots, \theta_n)$ in M. Then

$$g = dr^2 + \sum_{i,j=2}^{n} g_{ij}(r,\theta)\, dx_i dx_j ,$$

and then

$$\Delta r = -\frac{1}{\sqrt{g}} \frac{\partial}{\partial r} \sqrt{g} = -(\log \sqrt{g})' . \tag{2.1}$$

2.3. Mean Curvature of $\partial B(r)$. The second fundamental form α of $\partial B(r)$ with respect to the inward normal vector $-\partial/\partial r$ is given by $\alpha = \mathrm{Hess}(r)$, and then the mean curvature $m(r,\theta)$ of $\partial B(r)$ at (r,θ) is given by

$$(2.2) \quad m(r,\theta) = \mathrm{trace}\, \alpha = \mathrm{trace}(\mathrm{Hess}(r)) = -\Delta r.$$

2.4. Ricci Curvature. The curvature tensor $R(X,Y)Z = \nabla_X\nabla_Y Z - \nabla_Y\nabla_X Z - \nabla_{[X,Y]}Z$, satisfies

$$R^{1}_{\bullet j 1 \ell} = -\frac{1}{2}\frac{\partial^2}{\partial r^2} g_{j\ell} + \frac{1}{4} g^{ab} \frac{\partial g_{jb}}{\partial r}\frac{\partial g_{\ell b}}{\partial r}, \; 2 \leqq j,\ell \leqq n,$$

in particular, the Ricci curvature $\mathrm{Ric}(\partial/\partial r)$ of $\partial/\partial r$ is given by

$$Ric(\frac{\partial}{\partial r}) = R^{1 \bullet \bullet j}_{\bullet j 1 \bullet} = -\frac{\partial}{\partial r}\mathrm{trace}(\mathrm{Hess}(r)) - \| \mathrm{Hess}(r) \|^2.$$

Then we get:

Lemma 8 (cf. Avez([**A**]). *Let (M,g) be a complete simply connected Riemannian manifold wthout focal points. Then*

$$\| \mathrm{Hess}(r) \|^2 \leqq (\mathrm{trace}(\mathrm{Hess}(r)))^2 \leqq n \| \mathrm{Hess}(r) \|^2.$$

§3. Proof of Theorems.

Proof of Theorem 1. (1) By Lemma 7, $m(r,\theta) \geqq 0$, which implies $m(M) \geqq 0$.

(2) : $m(M) \leqq h(M)$. For a relatively compact domain $D \subset M$,

$$m(M)\int_D dv_g \leqq \int_D m(r,\theta)\, dv_g = \int_D \mathrm{div}\,\mathrm{grad}\,(r)\, dv_g = \int_{\partial D} \langle \mathrm{grad}(r), \mathbf{n} \rangle$$

$$\leqq \int_{\partial D} \| \mathrm{grad}(r) \| \, \| \mathbf{n} \| = \mathrm{Vol}_{n-1}(\partial D),$$

where $\mathbf{n}$ is the unit normal vector field of ∂D. Then we obtain

$$m(M) \leqq \inf_{D \subset M} \frac{\mathrm{Vol}_{n-1}(\partial D)}{\mathrm{Vol}_n(D)} = h(M).$$

(3) : $h(M) \leqq \sqrt{4\lambda_0(M)}$. By Cheeger [C], for a relatively compact domain $D \subset M$,

$$\lambda_0(D) \geqq \tfrac{1}{4} h(D)^2,$$

where $h(D) := \inf_{D' \subset D} \mathrm{Vol}_{n-1}(\partial D')/\mathrm{Vol}_n(D') \geqq h(M)$. Then we get

$$\lambda_0(M) = \inf_{D\subset M} \lambda_0(D) \geqq \tfrac{1}{4}h(M)^2.$$

(4) : $\sqrt{4\lambda_0(M)} \leqq \mu(M)$, see Brooks [B].

(5) : $\mu(M) \leqq \bar{h}(M)$. Since (M,g) has no conjugate points, $V(r)$ has a continuous derivative $S(r)$. Then by Lemma 4, we get this inequality.

(6) : $\bar{h}(M) \leqq \bar{\bar{h}}(M)$. By Lemma 7, $\sqrt{g}\,' \geqq 0$, $S(r) = \int_{S^{n-1}} \sqrt{g}(r,\theta)\, d\theta$ is smooth and monotone increasing in r. If $\lim_{r\to\infty} S(r) = \infty$, $\lim_{r\to\infty} V(r) = \infty$. Then we get $\bar{h}(M) \leqq \bar{\bar{h}}(M)$. If $\lim_{r\to\infty} S(r) < \infty$, $\bar{h}(M) = 0$. Because, since $\sqrt{g}\,' \geqq 0$ implies $\sqrt{g} \geqq C$ (a positive constant) and M is diffeomorphic to $\mathbb{R}^n$,

$$\lim_{r\to\infty} V(r) = \lim_{r\to\infty} \int_{S^{n-1}} d\theta \int_0^r \sqrt{g}(s,\theta) ds = \infty .$$

Proof of Theorem 2. Under the assumptions, we get by the facts in §2,

$$(3.1) \qquad (n-1)k \leqq \frac{\partial}{\partial r} m(r,\theta) + m(r,\theta)^2 .$$

On the other hand, $m(r,\theta) > \sqrt{(n-1)k}$, if r is near 0 , since $m(r,\theta) = (\log\sqrt{g})' \sim (n-1)/r$ as $r \to \infty$. By (3.1),

$$\sqrt{(n-1)k} \geqq m(r,\theta) \quad \Rightarrow \quad 0 \leqq \frac{\partial}{\partial r} m(r,\theta).$$

Thus $m(r,\theta) \geqq \sqrt{(n-1)k}$. Otherwise, there exists a positive number r_0 such that

$$\frac{\partial}{\partial r} m(r_0,\theta) < 0 \quad , \text{ and } \quad m(r_0,\theta) < \sqrt{(n-1)k} ,$$

which is a contradiction. □

Proof of Theorem 3. Let $M = G/K$ be a simply connected Riemannian symmetric space of non-compact type , $\mathfrak{g}$, $\mathfrak{k}$, the Lie algebras of G, K, and $\mathfrak{g} = \mathfrak{k} + \mathfrak{p}$, the Cartan decomposition. Let $< , >$ be the inner product of $\mathfrak{p}$ induced from the Killing form of $\mathfrak{g}$. Taking $\{X_i; i=1, \ldots ,n\}$, an orthonormal basis of $(\mathfrak{p}, < , >)$, the Riemannian metric g on M can be expressed(cf.[H]) in terms of the global coordinate, $M \ni \exp(\Sigma_i y_i X_i) \cdot o \mapsto (y_1, \ldots ,y_n)$, $o = \{K\} \in M$, as

$$g(\frac{\partial}{\partial y_i}, \frac{\partial}{\partial y_j})(x\cdot o) = <(\sum_{m=0}^{\infty} \frac{(-\mathrm{ad}X)^m}{(m+1)!}(X_i))_{\mathfrak{p}}, (\sum_{m=0}^{\infty} \frac{(-\mathrm{ad}X)^m}{(m+1)!}(X_j))_{\mathfrak{p}}> , \ x = \exp X, X \in \mathfrak{p},$$

where $Z_{\mathfrak{p}}$, $Z \in \mathfrak{g}$, is the $\mathfrak{p}$-component with respect to the Cartan decomposition. In particular, for $X = \mathrm{Ad}(k)H$, $k \in K$, $H \in \mathfrak{a}^+$,

$$\sqrt{\det(g(\frac{\partial}{\partial y_i}, \frac{\partial}{\partial y_j}))(x\cdot o)} = \prod_{a\in\Sigma_+} (\frac{\sinh a(H)}{a(H)})^{m_a},$$

where $\mathfrak{a}^+$ is a positive Weyl chamber of a maximal abelian subspace of $\mathfrak{p}$, Σ_+ is the set of positive restricted roots and m_a , $a \in \Sigma_+$, is its multiplicity. Taking the geodesic polar coordinate $(x_1, \ldots, x_n)$ with $x_1 = r$, around the origin o, we get

$$\sqrt{\det(g(\frac{\partial}{\partial x_i}, \frac{\partial}{\partial x_j}))} = r^{n-1}\sqrt{\det(g(\frac{\partial}{\partial y_i}, \frac{\partial}{\partial y_j}))}.$$

Thus we obtain

$$(\log\sqrt{g})' = \frac{\ell-1}{r} + \sum_{a\in\Sigma_+} m_a\, a(H_\theta)\coth r a(H_\theta),$$

where ℓ is rank of M, and $H_\theta = H / \| H \|$. Therefore $m(M) = \inf\{ <2\rho,H> ;\ H\in\mathfrak{a}^+,\ \| H \| = 1\}$.

On the other hand,

$$(3.2) \qquad h(M) \leqq \sup_{\theta\in S^{n-1}} \limsup_{r\to\infty} \frac{\sqrt{g}'}{\sqrt{g}}(r,\theta).$$

Indeed, denoting by $\hat{h}(M)$, the right hand side, given $\varepsilon>0$ and $\theta\in S^{n-1}$, there exists $r_0(\theta)$ such that for all $r \geqq r_0(\theta)$,

$$\hat{h}(M)+\varepsilon \geqq (\sqrt{g}'/\sqrt{g})(r,\theta).$$

Putting $r_0 := \max_{\theta\in S^{n-1}} r_0(\theta)$, for all $\theta\in S^{n-1}$ and $r \geqq r_0$,

$$(\hat{h}(M)+\varepsilon)\sqrt{g}(r,\theta) \geqq \sqrt{g}'(r,\theta).$$

Integrating over S^{n-1} and letting r to infinity, we obtain

$$(3.3) \quad \limsup_{r\to\infty} \frac{S'(r)}{V'(r)} \leqq \hat{h}(M).$$

Then to prove (3.2), we only show

$$(3.4) \quad h(M) \leqq \limsup_{r\to\infty} \frac{S'(r)}{V'(r)}.$$

It follows immediately from Lemma 4, since $\lim_{r\to\infty} V(r) = \lim_{r\to\infty} S(r) = \infty$.

Now we get $\hat{h}(M) = 2\| \rho \|$ by the above calculation of $(\log\sqrt{g})'$. On the other hand, since we already have got $\sqrt{4\lambda_0(M)} \geqq 2\| \rho \|$ (cf. [U, p.150]), we obtain the the desired equality. □

References

[A] A. Avez, Variétés riemanniennes sans points focaux, C.R. Acad. Sci. Paris, **270**(1970), 188-191.

[B] R. Brooks, A relation between growth and the spectrum of the Laplacian, Math. Zeit., **178**(1981), 501-508.

[C] J. Cheeger, A lower bound for the smallest eigenvalue of the Laplacian, Problems in Analysis, A Symposium in Honor of S. Bochner, Princeton,(1969), 195-199.

[D] H. Donnelly, On the essential spectrum of a complete Riemannian manifold, Topology, **20**(1981), 1-14.

[E] J.H. Eschenburg, Stabilitätsverhalten des Geodätisahen Flusses Riemannscher Mannigfaltigkeiten, Bonner Math. Schrift., **87**(1976).

[H] S. Helgason, *Differential Geometry and Symmetric Spaces,* Academic Press, New York,1965.

[M] H.P. McKean, An upper bound to the spectrum of Δ on a manifold of negative curvature, J. Diff. Geometry, **4**(1970), 359-366.

[On] R. Osserman, Bonnesen-style isoperimetric inequalities, Amer. Math. Monthly, **86**(1979), 1-29.

[O] J.J. O'Sullivan, Manifolds without conjugate points, Math. Ann., **210**(1974), 311-295.

[P] M.A. Pinsky, The spectrum of the Laplacian on a manifold of negative curvature I, J. Diff. Geometry, **13**(1978), 87-91.

[U] H. Urakawa, On the least positive eigenvalue of the Laplacian for the compact quotient of a certain Riemannian symmetric space, Nagoya Math. J., **78**(1980), 137-152.

[Y] S.T. Yau, Isoperimetric constants and the first eigenvalue of a compact Riemannian manifold, Ann. scient. Éc. Norm. Sup., **8**(1975), 487-507.

Hajime URAKAWA
Department of Mathematics
College of General Education
Tohoku University
Kawauchi, Sendai, 980,
Japan

Chapter VI
Yang-Mills Connections

COMPACTIFYING THE MODULI SPACE OF 1-INSTANTONS ON HP^n

By Hideo DOI and Takayuki OKAI

Introduction

The moduli space of Yang-Mills connections carries a natural Riemannian metric γ. This metric plays an interesting role in the differential geometric approach of the study of moduli spaces. For example, D. Groisser and T.H. Parker [3] have shown that, for the moduli space of 1-instantons on a 1-connected compact Riemannian 4-manifold with positive definite intersection form, the completion by the distance function induced from γ coincides with the compactification in the sense of S. K. Donaldson [2].

In this paper, we give a correspondence between γ and the compactifications of the moduli spaces $\mathcal{M}_n$ and $\mathcal{N}_c$, where $\mathcal{M}_n$ denotes the moduli space of 1-instantons on HP^n, and $\mathcal{N}_c$ denotes that of null correlation bundles on CP^{2n+1}.

In §1, we give an integral representation of γ. Using this, we investigate the behavior of γ near the boundary of $\mathcal{M}_n$. We shall show that the completion $\bar{\mathcal{M}}_n$ by γ coincides with the compactification $\hat{\mathcal{M}}_n$ in the sense of Donaldson. We also show that $\hat{\mathcal{M}}_n$ becomes a closed ball $B^{n(2n+3)}$. This is a generalization of the famous fact that $\hat{\mathcal{M}}_1 = B^5$. In particular, $\hat{\mathcal{M}}_n$ has finite volume and finite diameter with respect to γ.

In §2, we treat with several compactifications of $\mathcal{N}_c$. Using the information about $\mathcal{M}_n$, we shall see that both the compactification of $\mathcal{N}_c$ (in the sense of Donaldson) and the completion of $\mathcal{N}_c$ by γ are homeomorphic to $CP^{n(2n+3)}$. We also show that the compactification of $\mathcal{N}_c$ due to M. Maruyama [5] is isomorphic to $CP^{n(2n+3)}$, too.

1. Metric completion and compactification of $\mathcal{M}_n$

Let $E = \{(x,v) \in HP^n \times H^{n+1};\ {}^t x v = 0\}$, where † denotes the Hermitian conjugation. Let $\mathcal{M}_n$ be the moduli space of $Sp(n)$-connections on E attaining the minimum of the Yang-Mills functional. Let $p : CP^{2n+1} \to HP^n$ be the canonical projection. For any $D \in \mathcal{M}_n$, p^*D attains the minimum of the Yang-Mills functional on p^*E since the $Sp(n+1)$-invariant connection ∇ on E belongs to $\mathcal{M}_n$. Hence p^*D is an Einstein-Hermitian connection on p^*E [4]. Then it is easy to see that $\mathcal{M}_n = R^\times Sp(n+1)\backslash GL(n+1,H)$ (cf. [1, Theorem 1.1]). Let $\mathcal{D}_n = \{(D,S);\ D = \lim D_j$ in C^∞_{loc} on $E|_{HP^n \setminus S}$ for some $D_j \in \mathcal{M}_n$, and S is a minimal closed exceptional set$\}$, where "minimal" means that there do not exist a subsequence $\{D_i\}$ of $\{D_j\}$,

ISBN 0-12-640170-5

gauge transformations g_i or a proper closed subset $\Sigma \subset S$ such that $\lim g_i^* D_i$ exists on $E|_{HP^n \setminus \Sigma}$. For (D,S), $(D',S') \in \mathcal{D}_n$, we write $(D,S) \sim (D',S')$ if $S = S'$ and D is gauge equivalent to D', and we set $\hat{\mathcal{M}}_n = \mathcal{D}_n/\sim$. Then we have $\hat{\mathcal{M}}_n = \{ Y \in gl(n+1,H);\ {}^t\bar{Y} = Y,\ Y \geq 0,\ Y \neq 0\}/R_+^\times$ [1, Theorem 1.2].

Now we give a brief review about the Riemannian structure on the moduli space of connections. Let G be a compact Lie group and let $P \to M$ be a principal G-bundle over a compact Riemannian manifold. $\mathcal{C}$ and $\mathcal{G}$ denote the affine space consisting of all G-connections and the gauge transformation group on P, respectively. It is needless to say that we should work on suitable Sobolev spaces. If $D \in \mathcal{C}$ is irreducible, we may identify $T_D(\mathcal{C}/\mathcal{G})$ with $\{v \in \Gamma(M, T^*M \otimes (P \times_{Ad} g))\,;\ D^*v = 0\}$. Now we set for $v, w \in T_D(\mathcal{C}/\mathcal{G})$,

$$\gamma(v,w) = \int_M (v,w) * 1,$$

where (v,w) is the usual metric on $T^*M \otimes (P \times_{Ad} g)$ and $*$ is the Hodge operator.

An element $X \in gl(n+1,H)$ defines vector fields $\partial_t L_{\exp tX}|_{t=0}$ on HP^n, H^{n+1} and $\partial_t R_{\exp tX}|_{t=0}$ on $R^\times Sp(n+1)\backslash GL(n+1,H) = \mathcal{M}_n$. We denote all of them by X. Also for a connection D, we denote its curvature by $F(D)$.

PROPOSITION 1.1 *If $X \in gl(n+1,H)$ is Hermitian, then we have*

$$\gamma(X,X) = \int_{HP^n} |\iota_X F(D)|^2 * 1 \qquad at\ D \in \mathcal{M}_n.$$

PROOF. First we have $\partial_t R_{\exp tX} D|_{t=0} = \iota_X F(D) + D\xi$ for some $\xi \in \Gamma(HP^n, sp(E))$. Let $(\ ,\ )$ be the Riemannian metric on HP^n and set $X^\vee = (X,\ ) \in \Gamma(HP^n, T^*HP^n)$. Then $*\iota_X F(D) = -X^\vee \wedge *F(D)$. The canonical projection $q : H^{n+1}\backslash\{0\} \to HP^n$ is a Riemannian submersion if we consider $H^{n+1} \setminus \{0\}$ as a Riemannian manifold with $|dz|^2/|z|^2$. Also we decompose ${}^t\bar{z}\,dz/|z|^2 = \omega_1 + i\omega_2 + j\omega_3 + k\omega_4$ with $\omega_i \in \Gamma(H^{n+1}\backslash\{0\}, T^*H^{n+1})$. Then $\omega_1 = d|z|^2/2|z|^2$ and $\omega = \omega_1\wedge\omega_2\wedge\omega_3\wedge\omega_4$ is the volume element along a fiber of q. Because $q^*X^\vee = \{\mathrm{Re}({}^t\bar{z}X dz) - {}^t\bar{z}Xz\, d|z|^2/2|z|^2\}/|z|^2$, $\omega_1 \wedge * q^*F(D) = 0$ and $\mathrm{Re}(d\,{}^t\bar{z} \wedge X dz) = 0$, we see that $\omega \wedge q^*D(X^\vee \wedge *F(D)) = dq^*X^\vee \wedge * q^*F(D) = 0$. Thus $\iota_X F(D) \in \mathrm{Ker} D^*$. □

Let $\Lambda^+ = \{\lambda = \mathrm{diag}(\lambda_0, \lambda_1, \cdots, \lambda_n);\ 1 = \lambda_0 > \lambda_1 > \cdots > \lambda_n > 0\}$. For $\mu \in Cl\Lambda^+$, let $K(\mu)$ denote the centraliser of μ in $Sp(n+1)$ and let K stand for $K(\lambda)$ with $\lambda \in \Lambda^+$. Setting $\nu(\lambda, Kg) = R^\times Sp(n+1)\lambda g$ for $\lambda \in \Lambda^+$ and $g \in Sp(n+1)$, we have an open embedding $\nu : \Lambda^+ \times K\backslash Sp(n+1) \to R^\times Sp(n+1)\backslash GL(n+1,H) = \mathcal{M}_n$. Then the Riemannian metric γ on $\mathcal{M}_n$ splits as $\nu^*\gamma = \alpha(\lambda) + \beta(\lambda)$, where $\alpha(\lambda)$ is a metric on $\Lambda^+ \subset R^n$ and $\beta(\lambda)$ is an $Sp(n+1)$-invariant metric on $K\backslash Sp(n+1)$.

Here we note that $T_1(K\backslash Sp(n+1)) = T_1(K\backslash K(\mu)) \oplus T_1(K(\mu)\backslash Sp(n+1))$ in $sp(n+1)$.

LEMMA 1.2. *Let $\lambda \in \Lambda^+$ and $\mu \in Cl\Lambda^+$.*

(1) $\limsup_{\lambda\to\mu} \nu^*\gamma < \infty$.

(2) $\liminf_{\lambda\to\mu} \beta(\lambda)|T_1(K(\mu)\backslash Sp(n+1)) > 0$.

(3) $\lim_{\lambda\to\mu} \beta(\lambda)|T_1(K\backslash K(\mu)) = 0$.

(4) $\liminf_{\lambda\to\mu} \rho^*\alpha > 0$, *where* $\rho(\lambda) = \lambda^{1/2}$ *for* $\lambda \in \Lambda^+$.

PROOF. Let $\{e_0, \cdots, e_n\}$ be the standard basis of H^{n+1}, and for $z \in H^{n+1}$ let z_i denote the i-th component i.e. $z = \sum_{0\le i\le n} e_i z_i$. For $X \in gl(n+1, H)$, $\mu \in Cl\Lambda^+$ and $z \in H^{n+1}$, we set $\psi(\mu, X)(z) = |\mu z|^{-4}|P\mu \operatorname{Im}(Xz\, d\,{}^t z)\mu P|^2$, where $P = 1 - u\,{}^t u$ with $u = |\mu z|^{-1}\mu z$. If $X = {}^t X$ and $\lambda \in \Lambda^+$, then we have $4\psi(\lambda, X) = |\iota_X q^*\lambda^* F(\nabla)|^2$ on $S^{4n+3} = \{z \in H^{n+1};\ |z| = 1\}$.

First we assume $\operatorname{rank}\mu > 1$. Then $|\mu z|^{-4} \in L^1(S^{4n+3})$. Therefore (1), (2) and (3) follow immediately from Lebesgue's dominated convergence theorem. Also $|\mu z|^{-6} \in L^1(S^{4n+3})$ and $\lambda_i^{-4}\psi(\lambda, e_i\,{}^t e_i) = |\lambda z|^{-4}|(\lambda_i^{-1}e_i - u\bar{z}_i|\lambda z|^{-1})\operatorname{Im}(z_i d\,{}^t z)\lambda P|^2$. Hence we have (4).

Second we assume $\operatorname{rank}\mu = 1$. In the following, c_1, c_2 and c_3 will stand for some constants. For $i > 0$, $\psi(\lambda, e_i\,{}^t e_j) \le c_1|\lambda z|^{-4}\lambda_i^2$ on S^{4n+3}. Let $s = e_0 + x$ for $x \in H^n = \sum_{1\le i\le n} e_i H$ and let d^{4n} be the standard volume element on $R^{4n} = H^n$. Then

$$\begin{aligned}\int_{S^{4n+3}} \lambda_i^2|\lambda z|^{-4} * 1 &= 2\pi^2 \int_{H^n} \lambda_i^2({}^t s\lambda^2 s)^{-2}(1+|x|^2)^{-2n} d^{4n} \\ &\le c_2 \int_0^\infty \lambda_i^2(1+\lambda_i^2 r^2)^{-2}(1+r^2)^{-2} r^3 dr.\end{aligned}$$

Hence we see that $\lim_{\lambda\to\mu} \int_{S^{4n+3}} \psi(\lambda, e_i\,{}^t e_j) * 1 = 0$, and

$$\begin{aligned}&\limsup_{\lambda\to\mu} \int_{S^{4n+3}} \psi(\lambda, e_0\,{}^t e_i) * 1 \\ &= \limsup_{\lambda\to\mu} \int_{S^{4n+3}} |\lambda z|^{-4}|(e_0 - u\bar{u}_0)\operatorname{Im}(z_i d\bar{z}_0)(e_0 - u_0\,{}^t u)|^2 * 1 \\ &= \limsup_{\lambda\to\mu} 3\int_{S^{4n+3}} |\lambda z|^{-8}(|\lambda_1 z_1|^2 + \cdots + |\lambda_n z_n|^2)^2 |z_i|^2 * 1 \\ &< \infty.\end{aligned}$$

Thus we have $\limsup_{\lambda\to\mu}\beta(\lambda)<\infty$ and (3). Also we see that

$$\liminf_{\lambda\to\mu}\int_{S^{4n+3}}\psi(\lambda,e_0\,{}^te_i)*1\geq\liminf_{\lambda\to\mu}\int_{S^{4n+3}}|\lambda z|^{-8}|\lambda_1 z_1|^4|z_i|^2*1>0.$$

This implies (2). Since $\lambda_i^{-2}\psi(\lambda,e_i\,{}^te_i)\leq c_3|\lambda z|^{-4}(|u_0u_iz_i|^2+\lambda_1^2)$, we have $\limsup_{\lambda\to\mu}\alpha(\lambda)<\infty$. Note that $\psi(\lambda,e_i\,{}^te_i)=4|\lambda z|^{-4}\,|\lambda_iz_i|^2\,\{\lambda_i^2+|\lambda uu_i|^2-2|\lambda_iu_i|^2+2(\mathrm{tr}\lambda^2-|\lambda u|^2)\,(1-|u_i|^2)\}$. Hence we have

$$\liminf_{\lambda\to\mu}\lambda_i^{-4}\int_{S^{4n+3}}\psi(\lambda,e_i\,{}^te_i)*1=\infty.$$

This implies (4). □

THEOREM 1.3. *Let $\bar{\mathcal{M}}_n$ be the completion with respect to the distance function induced from γ. Then we have*

$\bar{\mathcal{M}}_n=\hat{\mathcal{M}}_n=B^{n(2n+3)}$, *a closed ball of dimension $n(2n+3)$.*

PROOF. Let $f:\Lambda^+\to\mathcal{M}_n$ denote the map $f(\lambda)=\lambda^*\nabla$. Lemma 1.2 (1) and (4) imply that f is uniquely extended to a homeomorphism $Cl\Lambda^+\to Clf(\Lambda^+)\subset\bar{\mathcal{M}}_n$. We denote this also by f. By Lemma 1.2 (3), we can define a map $f:\hat{\mathcal{M}}_n\to\bar{\mathcal{M}}_n$ by $f(\mu\cdot g)=f(\mu)\cdot g$ with $\mu\in Cl\Lambda^+$ and $g\in Sp(n+1)$. Clearly f is surjective. Also Lemma 1.2 (2) implies that f is injective. Thus we have a homeomorphism $f:\hat{\mathcal{M}}_n\to\bar{\mathcal{M}}_n$.

We notice that $\hat{\mathcal{M}}_n=\{Y\in gl(n+1,H);\ {}^tY=Y,\ Y\geq0$ and $\mathrm{tr}\,Y=n+1\}=1_{n+1}+\{y\in gl(n+1,H);\ {}^ty=y,\ m(y)=$ [the smallest eigenvalue of y] ≥-1 and $\mathrm{tr}\,y=0\}$. Then $1+y\mapsto|m(y)|y/|y|$ gives a homeomorphism $\hat{\mathcal{M}}_n\to B^{n(2n+3)}=\{y\in gl(n+1,H);\ {}^ty=y,\ \mathrm{tr}\,y=0$ and $|y|\leq1\}$. □

2. Compactifying the moduli space of null correlation bundles

A holomorphic vector bundle N over CP^{2n+1} is called a null correlation bundle if it has a resolution $0\to\mathcal{O}(-1)\to\Omega(1)\to N\to0$. Let $\mathcal{N}_c$ be the moduli space of null correlation bundles. Then $\mathcal{N}_c=\{\varphi\in GL(2n+2,C);\ {}^t\varphi=-\varphi\}/C^\times=C^\times Sp(n+1,C)\backslash GL(2n+2,C)$ [6]. Let $\eta(m)=2n\binom{m+2n+1}{2n+1}-\sum_{0\leq i\leq 2n-1}\binom{m+i}{i}$. We denote by $\mathcal{S}$ the moduli space of stable sheaves on CP^{2n+1} with the Hilbert poynomial η. Then $\mathcal{S}$ becomes a projective variety by [5]. Since $\mathcal{N}_c\subset\mathcal{S}$, we can consider the closure $Cl\mathcal{N}_c$ as a compactification of $\mathcal{N}_c$.

Note that $GL(n+1, H)$ is naturally embedded in $GL(2n+2, C)$. In particular we regard Λ^+ as a subset of $GL(2n+2, C)$. Then we have the Cartan decompositions $GL(n+1, H) = R^\times Sp(n+1)\Lambda^+ Sp(n+1)$ and $GL(2n+2, C) = C^\times Sp(n+1, C)\Lambda^+ U(2n+2)$. If a null correlation bundle N is corresponding to λg ($\lambda \in \Lambda^+$, $g \in U(2n+2)$), then $g^* p^* \lambda^* \nabla$ is an Einstein-Hermitian connection on N. Hence $\mathcal{N}_c$ is identified with a subset of the moduli space of Einstein-Hermitian connections on p^*E. Define $\hat{\mathcal{N}_c}$ just the same as $\hat{\mathcal{M}}_n$ (see §1). Also $\bar{\mathcal{N}_c}$ denotes the completion with respect to the distance function induced from the Riemannian metric γ.

THEOREM 2.1. $\hat{\mathcal{N}_c} = \bar{\mathcal{N}_c} = Cl\mathcal{N}_c = CP^{n(2n+3)}$.

PROOF. We notice that $CP^{n(2n+3)} = \{\varphi \in gl(2n+2, C)\,;\, {}^t\varphi = -\varphi,\ \varphi \neq 0\}/C^\times$. By Theorem 1.3 and the Cartan decompositions, we have $\hat{\mathcal{N}_c} = CP^{n(2n+3)}$. From the proof of Proposition 1.1, it follows that the inclusion map $\mathcal{M}_n \to \mathcal{N}_c$ is preserving the Riemannian metric up to a constant factor. Because $T_m\mathcal{M}_n \otimes C = T_m\mathcal{N}_c$ at $m \in \mathcal{M}_n \subset \mathcal{N}_c$, we have $\bar{\mathcal{N}_c} = CP^{n(2n+3)}$.

Now we will show $Cl\mathcal{N}_c = CP^{n(2n+3)}$. Let $V = C^{2n+2}$ and $V^\vee = \mathrm{Hom}(V, C)$. We identify V with $\Gamma(CP^{2n+1}, \mathcal{O}(1))$. Then we have two exact sequences

$$0 \longrightarrow \Omega(1) \longrightarrow V \otimes \mathcal{O} \longrightarrow \mathcal{O}(1) \longrightarrow 0 \quad \text{(the twisted dual Euler sequence)},$$

$$0 \to \mathcal{O}(-1) \to V^\vee \otimes \mathcal{O}.$$

For $t = (t_i) \in C^{n+1}\backslash\{0\}$, we set $j_t = \sum_{0\leq i\leq n}(e_{n+1+i}\otimes e_i - e_i \otimes e_{n+1+i})t_i \in \mathrm{Hom}(V^\vee, V)$, where $\{e_0, \cdots, e_{2n+1}\}$ is the standard basis. Then $j_t \otimes 1 : V^\vee \otimes \mathcal{O} \to V \otimes \mathcal{O}$ defines a homomorphism $J_t : \mathcal{O}(-1) \to \Omega(1)$. Let $N_t = \mathrm{Coker}\, J_t$. In view of the Cartan decomposition for $GL(2n+2, C)$, it is enough to show the following

LEMMA 2.2. *$\{N_t;\ t \in C^{n+1}\backslash\{0\}\}$ is a flat family of stable sheaves.*

PROOF. Let $CP^{2n+1} = \mathrm{Proj}\, C[x_0, \cdots, x_{2n+1}]$, $z_i = x_i/x_0$ and $\zeta_i = e_i - z_i e_0$. We will work on an affine open subscheme $Z = \mathrm{Spec}\, C[z_1, \cdots, z_{2n+1}]$. Then $\{\zeta_1, \cdots, \zeta_{2n+1}\}$ is a basis of $\Omega(1)$ as $\mathcal{O}_Z$-module. Since $J_t(1/x_0) = \sum_{0\leq i\leq n}(\zeta_{n+1+i}z_i - \zeta_i z_{n+1+i})t_i$, we see that $\Omega(1)/\mathcal{O}_Z J_t(1/x_0)$ is torsion free.

Let $W = \mathrm{Spec}\, C[w_0, \cdots, w_n] \setminus \{\text{the origin } 0\}$. We set $J = \sum_{0\leq i\leq n}(e_{n+1+i} \otimes e_i - e_i \otimes e_{n+1+i}) \otimes w_i$. Clealy for any $\mathcal{O}_W$-module $\mathcal{B}$, $J : \mathcal{O}(-1)\otimes\mathcal{B} \to \Omega(1)\otimes\mathcal{B}$ is injective. This implies that $\{N_t;\ t \in C^{n+1} \setminus \{0\}\}$ is a flat family.

By Theorem 1.3, N_t with $t \in R^{n+1} \setminus \{0\}$ has an Einstein-Hermitian connection outside the singularity set. Hence N_t with $t \in C^{n+1} \setminus \{0\}$ is a stable sheaf. □

References

[1] H.Doi & T.Okai : Moduli space of 1-instantons on a quaternionic projective space HP^n, Hiroshima Math. J. to appear.

[2] S.K.Donaldson : Connections, cohomology and the intersection forms of 4-manifolds, J. Differential Geom. **24** (1986) 275 - 341.

[3] D.Groisser & T.H.Parker : The geometry of the Yang-Mills moduli space for definite manifolds, J. Differential Geom. to appear.

[4] N.Koiso : Yang-Mills connections and moduli space, Osaka Math. J. **24** (1987) 147 -171.

[5] M.Maruyama : Moduli of stable sheaves, II, J. Math. Kyoto Univ. **18** (1978) 557 - 614.

[6] H.Spindler : Holomorphe Vectorbündel auf P_n mit $c_1 = 0$ und $c_2 = 1$, Manuscripta Math. **42** (1983) 171 - 198.

Hideo Doi

Department of Mathematics
Faculty of Science
Hiroshima University
Hiroshima, 730
Japan

Takayuki Okai

Department of Mathematics
Faculty of Science
Hiroshima University
Hiroshima, 730
Japan

Based Anti-Instantons and Gravitational Instantons

MITSUHIRO ITOH

Dedicated to Professor Akio Hattori on his sixtieth birthday

Introduction

In this article three geometrical investigations on the moduli space of anti-instantons are proceeded.

One is on the moduli space over the standard 4-sphere S^4. Namely we show that the moduli space becomes in a canonical sense a hyperkähler manifold.

The second investigation concerns geometrical structure defined on asymptotically locally Euclidean (ALE) 4-manifolds which are gravitational instantons. Actually we verify the simply-connectivity of ALE hyperkähler 4-manifolds which are typical examples of gravitational instantons.

The third is on the moduli space of anti-instantons of finite curvature action integral over ALE hyperkähler 4-manifolds containing especially the Euclidean 4-space. In fact in §3 we present the following result. The moduli space is endowed naturally with a hyperkähler structure.By using Donaldson's argument, we investigate the compactification of the moduli space when the instanton number is small.

Denote by P an $SU(2)$-bundle over S^4 of $c_2 = k$ and by $\mathcal{M}_k$ the quotient space of all anti-instantons on P divided by the gauge transformation group $\mathcal{G}$.

ISBN 0-12-640170-5

By supplementing frames one defines the framed moduli space $\mathcal{M}_k^\vee$ over $\mathcal{M}_k$. By using the twistor fibration $\pi : P^3(\mathbb{C}) \to S^4$ and the imbedding $P^2(\mathbb{C}) \to P^3(\mathbb{C})$, Donaldson showed that $\mathcal{M}_k^\vee$ is isomorphic to the moduli of holomorphic vector bundles over $P^2(\mathbb{C})$ being trivial along the fibre $P^1(\mathbb{C}) = \pi^{-1}(\infty)$, $\infty \in S^4$([**5**]). It carries then a complex structure.

Any anti-instanton is transformed by a suitable gauge transformation into an anti-instanton vanishing at the north pole ∞, that is, a based anti-instanton. Those based anti-instantons are divided also by the subgroup of $\mathcal{G}$ to define the moduli space of based anti-instantons $\mathcal{M}_k^\infty$ which is isomorphic with $\mathcal{M}_k^\vee$.

On the other hand, from the conformality of the anti-selfduality a based anti-instanton can be regarded as an anti-instanton over $\mathbb{R}^4 \approx S^4 \backslash \{\infty\}$. The Euclidean 4-space $\mathbb{R}^4$ admits, as the simplest geometrical space, a hyperkähler structure, namely, covariantly constant almost complex structures I, J, K satisfying the quaternion relation $I\,J = -J\,I = K$. This hyperkähler structure is inherited canonically over the moduli space of anti-instantons over $\mathbb{R}^4$. The way to define the hyperkähler structure over the moduli space is very natural. Refer for it to [**16**]. See also [17] and [18] for the moment map method.

The hyperkähler structure introduced above is well transferred over $\mathcal{M}_k^\infty$ so that it turns out to be an $8\,k$ dimensional hyperkähler manifold.

In §2 we define an ALE gravitational instanton in a Riemannian analogy with the notion of Yang-Mills (anti-)instanton. Every simply connected ALE gravitational instanton is geometrically a hyperkähler manifold and has an end $S^3/\Gamma \times \mathbb{R}$ where Γ is a discrete subgroup of $SU(2)$.

One of our results in § 2 is the simply-connectivity of ALE hyperkähler 4-manifolds. Every ALE hyperkähler 4-manifold is proved case by case to be simply connected by using the classification of discrete subgroups Γ and the theorem due to Kronheimer(Theorem 2. 5). However, our result is verified directly under a general situation.

As the Yang-Mills moduli space over a compact 4-manifold has a fruitful implication, it is of interest to consider the moduli space over complete open 4-manifolds. In fact, in order to define his graded homology for homology 3-sphere Σ^3 Floer introduces the (anti-)instanton field equations over the open 4-manifold $\Sigma^3 \times \mathbb{R}$([9]).

We restrict ourself in §3 to ALE hyperkähler 4-manifolds (X, h). Like just the $\mathbb{R}^4$ case the moduli space over (X, h) relative to asymptotical gauge admits a natural hyperkähler structure. On the other hand, an essentially different point which the moduli space possesses compared with the compact case is that the moduli space carries rational instanton number. This corresponds to the fact that the conformal compactification $\hat{X}$ of the (X, h) is an orbifold with one point singularity and an anti-instanton over (X, h) of finite curvature action integral extends to an anti-instanton over a pseudo bundle(see [8] and [11]).

The author expresses his thanks to H. Nakajima for useful discussion.

1. Based instantons

Let P be a G-bundle over S^4 of $c_2 = k > 0$. For simplicity we assume G is SU(2).

A connection A on P is called an anti-instanton if the curvature $F(A)$ is anti-selfdual with respect to the standard metric of S^4. Then from Chern-Weil theorem a connection is an anti-instanton if and only if $\int |F(A)|^2 dv = 8\pi^2 k$.

The group of gauge transformations $\mathcal{G}$ acts on the space $\mathcal{A}_-$ of anti-instantons invariantly to define the moduli space $\mathcal{M}_k$ which parametrizes gauge-orbits in $\mathcal{A}_-$.

The moduli space $\mathcal{M}_k$ is a real analytic smooth manifold of dimension $8k - 3$([1]).

The framed moduli space $\mathcal{M}_k^\vee$ is defined as $\mathcal{M}_k^\vee = \mathcal{A}_- \times P_\infty / \mathcal{G}$, where P_∞ is the fibre of P over $\infty \in S^4$, and then is a fibre space over $\mathcal{M}_k$ with fibre G/Z (Z is the center of G).

Consider the closed normal subgroup $\mathcal{G}^\infty = \{g \in \mathcal{G}; g|_{P\infty} = id\}$ and take the quotient $\mathcal{M}_k^\infty = \mathcal{A}_-/\mathcal{G}^\infty$. Then $\mathcal{M}_k^\infty$ is a fibre bundle over $\mathcal{M}_k$ with fibre G/Z.

There exists an identification between $\mathcal{M}_k^\vee$ and $\mathcal{M}_k^\infty$:

$$\begin{array}{ccc} \mathcal{M}_k^\vee & \xrightarrow{f} & \mathcal{M}_k^\infty \\ \downarrow & & \downarrow \\ \mathcal{M}_k & \xrightarrow{identity} & \mathcal{M}_k. \end{array}$$

The identification f is given by the assignment of a $\mathcal{G}^\infty$-equivalence class $[g_u(A)]^\infty$ to gauge equivalence class $[(A, u)], A \in \mathcal{A}_-, u \in P_\infty$, where $g_u \in \mathcal{G}$ is a gauge transformation sending u to the identity e in P_∞

THEOREM 1.1. *The moduli space $\mathcal{M}_k^\infty$ defined by the based gauge transformations admits naturally a Riemannian structure and a covariantly constant quaternion structure.*

From the conformal invariance $\mathcal{M}_k^\infty$ presents gauge equivalence classes of anti-instantons over $\mathbb{R}^4$. So, it becomes a hyperkähler manifold relative to the natural quaternion structure of $\mathbb{R}^4 \cong \mathbb{H}$.

For the proof of this theorem we need some preparation.

For the bundle $P \longrightarrow S^4$ we choose a trivialization U containing the infinity ∞ : $P|_U \simeq U \times G$.

We denote by g_U and A_U local expressions on U of gauge transformation g and connection A, respectively. Then $\mathcal{G}^\infty$ is $\{g \in \mathcal{G}; g_U(\infty) = e\}$ and has a normal subgroup $\mathcal{G}^{\infty,1} = \{g \in \mathcal{G}^\infty; d(g_U) = 0 at \infty\}$, which fixes invariantly the space of *based* anti-instantons $\mathcal{A}_-^\infty = \{A \in \mathcal{A}_-; A_U = 0 \ at \infty \ \}$.

PROPOSITION 1.2. *The moduli space $\mathcal{M}_k^\infty$ has the smooth fibration over the ordinary moduli space $\mathcal{M}_k$ with fibre G/Z. Moreover $\mathcal{M}_k^\infty$ is identified with the moduli space of based anti-instantons $\mathcal{A}_-^\infty/\mathcal{G}^{\infty,1}$.*

PROOF: The assignment of the $\mathcal{G}$-equivalence class $[A]$ to each $\mathcal{G}^\infty$-equivalence class $[A]^\infty$ gives the projection $\pi; \mathcal{M}_k^\infty \to \mathcal{M}_k$. The local triviality is shown by the slice lemma as follows.

Consider the diagram:

$$\begin{array}{ccc} \mathcal{A}_- & \xrightarrow{p^\infty} & \mathcal{M}_k^\infty \\ p\downarrow & & \downarrow \pi \\ \mathcal{M}_k & \xrightarrow{identity} & \mathcal{M}_k \end{array}$$

For each $[A] \in \mathcal{M}_k$ there exists a slice S_A at A in $\mathcal{A}_-$ such that $\mathcal{U} = p(S_A)$ is a neighborhood of $[A]$ in $\mathcal{M}_k$ and the map : $S_A \times (\mathcal{G}/Z) \to p^{-1}(\mathcal{U}), (A+\alpha, g) \mapsto g(A+\alpha)$ is a homeomorphism. Further any gauge transformation which maps an element of S_A into S_A reduces to the identity up to the center factor([**10**, Theorem 3.2]).

The slice S_A has the coordinate in the first cohomology group H_A^1.

By this slice lemma we see on the above diagram

$$\pi^{-1}(\mathcal{U}) = p^\infty(p^{-1}(\mathcal{U})) \simeq S_A \times (\mathcal{G}/Z)/\mathcal{G}^\infty.$$

Since $Z \cap \mathcal{G}^\infty = \{id\}$, $(\mathcal{G}/Z)/\mathcal{G}^\infty = G/Z$. So, $\mathcal{M}_k^\infty$ turns out to be a smooth fibration over $\mathcal{M}_k$.

It is easily seen that $\mathcal{M}_k^\infty$ is in one-to-one correspondence with the space $\mathcal{A}_-^\infty/\mathcal{G}^{\infty,1}$.

Remark. We have a *slice* for $\mathcal{A}_-^\infty$ just like the slice for $\mathcal{A}_-$. There is indeed a smooth map : $\mathcal{S}_A \to \mathcal{G}^\infty; A+\alpha \mapsto g = g(\alpha)$ such that $g(A+\alpha) = g(\alpha)(A+\alpha)$ lies inside $\mathcal{A}_-^\infty$.

The map : $\mathcal{S}_A \subset \mathcal{A}_- \to \mathcal{A}_-^\infty; A+\alpha \mapsto g(\alpha)(A+\alpha)$ is then a smooth embedding by the above slice lemma and its image $\mathcal{S}_A^\infty$ gives a slice at A inside $\mathcal{A}_-^\infty$ for $\mathcal{A}_-$. We take for all $a \in G$ $g_a \in \mathcal{G}$ which satisfies $(g_a)_U \equiv a$ over a ball $B \subset U, \infty \in B$. Let $\mathcal{S}_A^\infty(a) = \{g_a(A+\beta); A+\beta \in \mathcal{S}_A^\infty\}$. Then $\mathcal{S}_A^\infty(a) \subset \mathcal{A}_-^\infty$ and $\mathcal{S}_A^\infty(a) \cap \mathcal{S}_A^\infty(a') = \emptyset$ for $a \neq a'$. Set $\mathcal{S}_A^\infty(G) = \coprod\{\mathcal{S}_A^\infty(a); a \in G\}$.

Obviously the map : $\mathcal{S}_A^\infty(G) \times \mathcal{G}^{\infty,1} \to \mathcal{A}_-^\infty$ has the slice property. Since g_a depends smoothly on each local value a, the slice $\mathcal{S}_A^\infty(G)$ for$\mathcal{A}_-^\infty$ is smoothly parametrized in terms of the slice $\mathcal{S}_A$ for $\mathcal{A}_-$ and the group G.

From Proposition 1.2 we will identify $\mathcal{M}_k^\infty$ and $\mathcal{A}_-^\infty/\mathcal{G}^{\infty,1}$.

We set for $S^4 = \mathbb{R}^4 \cup \{\infty\}$ $U = S^4 \backslash \{\infty\}$ and $V = S^4 \backslash \{0\}$. We denote by $y; U \to \mathbb{R}^4 \cong \mathbb{H}$, $x; V \to \mathbb{R}^4 \cong \mathbb{H}$ the stereographic projections at $\infty, 0 \in S^4$, respectively.

A bundle P over S^4 has the transition function $\varphi; U \cap V \cong \mathbb{H}\backslash\{0\} \to G$.

For any smooth connection $A = (A_U, A_V)$ on P one has

$$A_V(x) = \varphi^{-1}(x)\ d\varphi(x) + \varphi^{-1}(x) A_U^*(y(x))\varphi(x), x \in \mathbb{H}\backslash\{0\}, \qquad (1.1)$$

where the one-form $A_U^*(y(x))$ is the pull-back of $A_U(y)$ by the inversion $y = x^{-1}$.

We make the following asymptotic assumption; the m-th derivative of $\varphi(x)$, $\varphi^{-1}(x)$ and $\varphi^{-1}(x)\ \partial\varphi/\partial x^{\mu}$, $\mu = 1, ..., 4$ are $O(\frac{1}{|x|^{m+1}})$ as $|x| \to \infty$,for all $m > 0$.

The map $\varphi_k(x) = (x/|x|)^k : \mathbb{H}\backslash\{0\} \to Sp(1) \cong SU(2)$ satisfies this assumption. So, one gets an SU(2) bundle P of $c_2 = k$.

We obtain for this bundle P the following

PROPOSITION 1.3. *(i) For* $A = (A_U, A_V) \in \mathcal{A}_-^{\infty}$ *the m-th derivatives of components of* $A_V(x)$ *are* $O(\frac{1}{|x|^{m+1}})$ $(|x| \to \infty)$. *(ii) For any difference of connections* $\alpha = (\alpha_U, \alpha_V) = A' - A$, $A, A' \in \mathcal{A}_-^{\infty}$ *the m-th derivatives of* $\alpha_V(x)$ *are* $O(\frac{1}{|x|^{m+3}})$ *and (iii) for* $g = (g_U, g_V) \in \mathcal{G}^{\infty,1}$ *the m-th derivatives of* $g_V(x) - id$ *are also* $O(\frac{1}{|x|^{m+4}})$.

Conversely from α_V and g_V satisfying the asymptotic conditions one gets $\alpha = (\alpha_U, \alpha_V)$ and $g = (g_U, g_V)$ defined globally on S^4 with $\alpha_U(0) = 0$, $g_U(0) = e$ and $d(g_U)(0) = 0$, respectively.

From this proposition the space W_l consisting of $\alpha_V \in \Omega^1(\mathbb{R}^4; su(2))$ which is the local expression on V of some difference $A - A' \in \mathcal{A}_-^{\infty}$ is within the Sobolev space $L_l^2(\Omega^1(\mathbb{R}^4; su(2)))$ and the completion of W_l coincides with $L_l^2(\Omega^1(\mathbb{R}^4; su(2)))$ for all $l \geq 0$.

For A,α, g we denote their local expression on V by the same symbols.

Fix an anti-instanton A in $\mathcal{A}_-^{\infty}$. We set $\mathcal{A}_{-,l}^{\infty} = \{A + \alpha \in \mathcal{A}_-; \alpha \in L_l^2(\Omega^1(\mathbb{R}^4; su(2)))\}$ and $\mathcal{G}_{l+1}^{\infty,1} = \{g : \mathbb{R}^4 \to G; g(x) - id \in L_{l+1}^2\}$. Here l is a large integer. From the Sobolev lemma the exponential map induces then a smooth exponential map : $L_{l+1}^2(\Omega^0(\mathbb{R}^4; su(2))) \to \mathcal{G}_{l+1}^{\infty,1}$.

So, the tangent space at A to the orbit of asymptotical gauge transformations is the image of $d_A : L_{l+1}^2(\Omega^0(\mathbb{R}^4; su(2))) \to L_l^2(\Omega^1(\mathbb{R}^4; su(2)))$.

Therefore, by linearizing the anti-instanton equations the quotient $\mathrm{Ker}\, d_A^+/\mathrm{Im}\, d_A$ gives the tangent space at $[A]$ to the moduli space of L_l^2-

Sobolev anti-instantons over $\mathbb{R}^4$, $\mathcal{A}^{\infty}_{-,l}/\mathcal{G}^{\infty,1}_{l+1}$, where d_A^+ is the selfdual part of d_A. By using the Euclidean metric formal adjoint d_A^*, the tangent space is identified with the kernel of the operator $\mathcal{D}_A = (d_A^*, d_A^+)$: $L_l^2(\Omega^1(\mathbb{R}^4; su(2))) \rightarrow L_{l+1}^2(\Omega^0(\mathbb{R}^4; su(2))) \oplus L_{l-1}^2(\Omega_+^2(\mathbb{R}^4; su(2))) \subset L_{l-1}^2((\Omega^0 \oplus \Omega_+^2)(\mathbb{R}^4; su(2)))$.

PROPOSITION 1.4. *The operator $\mathcal{D}_A$ is Fredholm for all $A \in \mathcal{A}_-^{\infty}$.*

To prove this proposition we apply the criterion on the Fredholm property in [3].

Consider a differential operator of order m on $C^{\infty}(\mathbb{R}^n; W)$,

$$D = \sum_{|\alpha| \leq m} a_{\alpha}(x)(\partial/\partial x)^{\alpha},$$

where $a_{\alpha}(x)$ are smooth functions on $\mathbb{R}^n$ which are endomorphisms of a vector space W, $\dim W < \infty$.

THEOREM 1.5 ([3]). *Assume*

$$(\partial/\partial x)^{\beta}\ a_{\alpha}(x) = O(\frac{1}{|x|^{|\beta|}}),\ (|x| \rightarrow \infty)$$

and the total symbol $\sigma(x,\xi) = \sum_{|\alpha| \leq m} a_{\alpha}(x)\xi^{\alpha}$ is nonsingular on large enough sphere $\sum_{\mu=1}^{n}(x_{\mu}^2 + \xi_{\mu}^2) = K$. Then $D : L_l^2(\mathbb{R}^n; W) \rightarrow L_{l-m}^2(\mathbb{R}^n; W)$ is a Fredholm operator.

PROOF OF PROPOSITION 1.4: Since $\mathcal{D}_A$ is first order, the total symbol σ is written as $\sigma = \sigma_1 + \sigma_0 : \Lambda_x^1 \otimes su(2) \rightarrow (\Lambda_x^0 \oplus \Lambda_{+,x}^2) \otimes su(2)$ where

$$\sigma_1(x,\xi)\ \alpha = (-\sum_{\mu} \xi_{\mu}\alpha_{\mu}, \omega^1(\xi,\alpha), \omega^2(\xi,\alpha), \omega^3(\xi,\alpha)) \tag{1.2}$$

and

$$\sigma_0(x,\xi)\ \alpha = (-\sum_\mu [A_\mu, \alpha_\mu], \psi^1(A,\alpha), \psi^2(A,\alpha), \psi^3(A,\alpha)) \tag{1.3}$$

for $A = \sum A_\mu dx^\mu$ and $\alpha = \sum \alpha_\mu dx^\mu$.

Here $\omega^i(\xi,\alpha)$ and $\psi^i(A,\alpha)$ are bilinear forms, i=1,2,3. We have then

$$|\sigma_1(x,\xi)\alpha|^2 = |\xi|^2|\alpha|^2, \tag{1.4}$$

and

$$|\sigma_0(x,\xi)\alpha|^2 = \sum_{\mu,\nu} |[A_\mu(x), \alpha_\nu]|^2, \xi \in T_x\mathbb{R}^4, \alpha \in \Lambda^1_x \otimes su(2). \tag{1.5}$$

On the domain where $\xi \neq 0$ in the large sphere $\Sigma_{(x,\xi)}(K)$ we have

$$|\sigma(x,\xi)\alpha|^2 = |\sigma_1\alpha + \sigma_0\alpha|^2 > 0.$$

To apply Theorem 1.5 we need to estimate $|\sigma_0(x,\xi)\alpha|^2$ on the closed set $\Sigma_{(x,0)}(K)$.

Since in (1.1) the pure gauge term $\varphi^{-1}(x)\ d\varphi(x)$ is a main term at infinity, it suffices to show $\sum_{\mu,\nu} |[\varphi^{-1}(x)\partial\varphi/\partial x^\mu, \alpha_\nu]|^2 > 0$ for nonzero α of $su(2)$, in other words, $[\varphi^{-1}(x)\ \partial\varphi/\partial x^\mu, \alpha_\nu] = 0$ implies $\alpha_\nu = 0$. But this is derived by the simple form of φ.

On the other hand we have

PROPOSITION 1.6. *For all* $A \in \mathcal{A}^\infty_{-,l}$,

$$\operatorname{Coker} \mathcal{D}_A \cong \operatorname{Ker} \mathcal{D}^*_A \cap L^2_{l-1}((\Omega^0 \oplus \Omega^2_+)(\mathbb{R}^4; su(2))) = \{0\}.$$

PROOF: Apply the Weitzenböck formula to the L_2-adjoint $\mathcal{D}_A^*$. Then we see

$$\mathcal{D}_A \mathcal{D}_A^*(\psi, \Psi) = (\nabla_A^* \nabla_A \psi, \nabla_A^* \nabla_A \Psi),$$

because $\mathbb{R}^4$ is flat([15]). So, $(\psi, \Psi) \in \operatorname{Ker} \mathcal{D}_A^*$ must be covariantly constant and hence vanishes.

From Propositions 1.4, 1.6 the $\mathcal{G}_{l+1}^{\infty,1}$-slice lemma is valid also for $\mathcal{A}_{-,l}^{\infty}$,and an ϵ-neighborhood in Ker $\mathcal{D}_A$ gives by the Kuranishi deformation theory at each $[A]$ a neighborhood of $\mathcal{M}_{-,l}^{\infty}$. As a consequence, $\mathcal{M}_{-,l}^{\infty}$ carries a smooth manifold structure.

Through the inverse map of the Kuranishi map the ϵ-neighborhood of Ker $\mathcal{D}_A$ is mapped real analytically onto $\mathcal{S}_{A,l}^{\infty} = \{A + \alpha \in \mathcal{A}_{-,l}^{\infty}; d_A^* \alpha = 0, |\alpha|_l < \epsilon_1\}$. From the regularity theorem on solutions of elliptic equations each element of $\mathcal{S}_{A,l}^{\infty}$ becomes smooth and satisfies the decay conditions at infinity, and hence naturally extends to an anti-instanton on the bundle P over S^4.

Therefore, decomposing $\mathcal{S}_{A,l}^{\infty} \subset \mathcal{A}_{-,l}^{\infty}$ into slices $\mathcal{S}_A^{\infty}(G) \times \mathcal{G}^{\infty,1}$ for based anti-instantons $\mathcal{A}_-^{\infty}$ over S^4 and dividing them by $\mathcal{G}^{\infty,1}$ induces a smooth map : $\mathcal{S}_{A,l}^{\infty} \longrightarrow \mathcal{S}_A^{\infty}(G)$.

Conversely, in the slice $\mathcal{S}_A^{\infty}(G)$ in $\mathcal{A}_-^{\infty}$ the anti-instanton A has two types of effective infinitesimal deformations. One is coming from the ordinary slice $\mathcal{S}_A$ and another is caused by proper gauge transformations tending to constant $(\neq e)$ at infinity. These types of infinitesimal deformations are exactly inside of Ker $\mathcal{D}_A$ in $L_l^2(\Omega^1(\mathbb{R}^4; su(2)))$, the tangent space to $\mathcal{S}_{A,l}^{\infty}$ at A.So, one sees from these arguments that $\mathcal{M}_k^{\infty}$ is diffeomorphic with$\mathcal{M}_{-,l}^{\infty}$.

We define next a hyperkähler structure on $\mathcal{M}_{-,l}^{\infty}$. The covariantly constant quaternion structure $\{I, J, K\}$ of $\mathbb{R}^4$ induces in a natural way a quaternion structure on each tangent space Ker $\mathcal{D}_A$, which is proved covariantly constant(see **[16]** for the definition and the proof).

The Riemannian metric on the moduli space is defined by the L^2-inner product with respect to the Euclidean metric.

Thus, Theorem 1.1 is completed.

2. Gravitational instantons.

Let (X, h) be an oriented Riemannian 4-manifold. Denote by (e_a) a local orthonormal frame of (X, h) and by (θ^a) its dual frame.

Let $R = (R^a_b)$ be the Riemannian curvature tensor; $R^a_b = \frac{1}{2}\sum_{c,d} R^a_{bcd}\theta^c \wedge \theta^d$.

Regarding R as a local 2-form taking values in 4×4 skew symmetric matrices, we define another 2-form $*R$ with values in 4×4 skew symmetric matrices by

$$*R = \frac{1}{2}\sum R^a_{bcd} * (\theta^c \wedge \theta^d), \tag{2.1}$$

for the Hodge operator $*$. In the analogy with the Yang-Mills selfduality equations we call the following as gravitational self-dual equations:

$$R^a_b = \pm(*R)^a_b, \tag{2.2}$$

DEFINITION 2.1. *A connected oriented complete Riemannian 4-manifold (X, h) is said to be ALE(asymptotically locally Euclidean) if there exist a compact $K \subset X$ and a diffeomorphism $\phi : X\backslash K \to \mathbb{R} \times S^3/\Gamma$ such that (i) Γ is a discrete subgroup of $SO(4)$, (ii) the induced metric $(\phi^{-1})^*h = (h_{\mu\nu})$ is over $\mathbb{R} \times S^3/\Gamma$ asymptotic to the standard flat metric up to order $O(\frac{1}{r^2})$ $(r \to \infty)$, ($r = r(x)$ denotes the first factor of $\mathbb{R} \times S^3/\Gamma$).*

When $\Gamma = \{e\}$, such a (X, h) is called an AE manifold.

In [**12**], [**2**] for analytical arguments of ALE manifolds further asymptotical conditions are posed on derivatives of $(h_{\mu\nu})$.

DEFINITION 2.2. *A connected oriented Riemannian 4-manifold is called a selfdual (or anti-selfdual) gravitational instanton if the metric satisfies (2.2).*

Eguchi and Hanson constructed a gravitational instanton which is ALE([**7**]). Their example is actually a 4-manifold diffeomorphic to the cotangent bundle of $P^1(\mathbb{C})$ and has a boundary at infinity $P^3(\mathbb{R})$. One easily observes that any gravitational instanton is Ricci flat and half-conformally flat.

Remarks. Each non-flat ALE gravitational instanton has only one end at infinity. Only the Euclidean 4-space is an AE gravitational instanton. More generally any AE Ricci flat 4-manifold is proved to be the standard Euclidean 4-space ([**27**]).

By the aid of holonomy argument we have in [**19**]

PROPOSITION 2.3. *Let (X, h) be a simply connected complete Riemannian 4-manifold. Then the following conditions are equivalent. (i) (X, h) is a gravitational instanton, (ii) (X, h) is Ricci flat Kähler, (iii) (X, h) is hyperkähler.*

Typical examples of ALE gravitational instanton, for example, Eguchi-Hanson metric, multi-center Taub-Nut metrics are all simply-connected and thus hyperkähler.

Conversely we have

THEOREM 2.4. *Let (X, h) be a non-flat ALE gravitational instanton. If*

(X, h) is hyperkähler, then X is simply connected.

PROOF: We make use of the idea of Y.Tsukamoto appeared in **[25]**.

Assume X is not simply connected. Then X has a closed geodesic $\gamma = \gamma(s)$, not homotopic to a constant curve, of shortest length in the homotopy class $[\gamma]$. In fact, since (X, h) is ALE, there is a large $r_0 > 0$ such that any closed curve c in X is homotopic to some closed curve c' in $X_{r_0} = K \cup \{x; r(x) \leq r_0\}$ with $L(c') \leq L(c)$.

Then from the usual argument on the shortest curve the second variation of the length is nonnegative at the shortest curve γ. Let $\{I, J, K\}$ be the quaternion structure on (X, h). Let $V = d\gamma/ds$ be the velocity vector. Then, the second variation along variation vectorsIV, JV and KV are expressed as $-\int_0^1 R(V \wedge IV)ds$, $-\int_0^1 R(V \wedge JV)ds$ and $-\int_0^1 R(V \wedge KV)ds$, respectively, where $R(V \wedge IV)$ is the sectional curvature $h(R(V, IV)IV, V)$.

Since (X, h) is Ricci flat, the following curvature identity holds for all vectors U, W:

$$h(R(U, W)W, U) + h(R(U, IW)IW, U) + \tag{2.3}$$

$$h(R(U, JW)JW, U) + h(R(U, KW)KW, U) = 0.$$

So, from this the bilinear form $\mathcal{J}$ associated to the second variation satisfies

$$\mathcal{J}(IV, W) = \mathcal{J}(JV, W) = \mathcal{J}(KV, W) = 0, \tag{2.4}$$

for all variation vectors W along γ. Thus the parallel vector fields IV, JV, KV are Jacobi fields and hence satisfy

$$R(V, IV)V = R(V, JV)V = R(V, KV)V = 0. \tag{2.5}$$

Vector fields $Y = aIV + bJV + cKV$ with $a^2 + b^2 + c^2 = 1$, normal to γ, are parallel Jacobi fields so that the metric h is flat in γ-direction.

For such a vector Y we can define a geodesic variation of γ, $\alpha(s,t)$, $|t| < \epsilon$, $0 \leq s \leq 1$, $\alpha(s,0) = \gamma(s)$, $\partial/\partial t\ \alpha(s,0) = Y(s)$. See Lemma 14.4 in [22]. While an argument is given in a local form there, we can apply it to our case because of the uniqueness of solutions.

Put $\alpha_t(.) = \alpha(.,t)$. Differentiate by t the energy integral $E(\alpha_t) = \int |\partial\alpha_t/\partial s|^2 ds$. We have then $d/dtE(\alpha_t) = 0$ and then $L(\alpha_t)^2 = E(\alpha_t) = E(\gamma) = L(\gamma)^2$, namely, α_t is shortest for $|t| < \epsilon$. So h is flat in a tubulor neighborhood of γ. Since (X,h) is analytic, it is entirely flat. Thus we complete the proof.

Let (X,h) be an ALE gravitational instanton. At infinity (X,h) is $\mathbb{R} \times S^3/\Gamma$ and $\Gamma \subset SO(4)$ is a finite subgroup.

The group $SO(4)$ has the double covering $\varphi : Sp(1) \times Sp(1) \to SO(4); q \mapsto q_1.q.q_2^{-1}$, $q \in \mathbb{R}^4 = \mathbb{H}$, where q_1, q_2 are unit quaternions. When the left (or right) factor of $\varphi^{-1}(\Gamma)$ is $\{\pm e\}$, $\Gamma \subset SO(4)$ is said to act on the right (or on the left). Each $\Gamma \subset SO(4)$ acting on the right is identified through φ with a finite subgroup in $Sp(1) \cong SU(2)$. It is known that finite subgroups in $Sp(1)$ are exhausted by cyclic groups Z_n, binary dihedral groups D_n^*, binary tetrahedral group T^*, binary octahedral group O^* and binary icosahedral group I^* ([13]).

The following was conjectured first by Hitchin([14]) and shown by Kronheimer in terms of the moment map([19]).

THEOREM 2.5([19]). *Let Γ be a finite subgroup of $SU(2)$. Then there exists an ALE hyperkähler 4-manifold with end S^3/Γ at infinity. Here Γ acts by right as isometries on $\mathbb{R}^4$. Conversely, every ALE hyperkähler 4-manifold can be represented in this way.*

Relating to this theorem one can raise the problem. Is every Ricci flat ALE 4-manifold with end S^3/Γ, $\Gamma \subset SU(2)$, a gravitational instanton ?

For this problem the following is obtained by H.Nakajima([**23**]).

THEOREM 2.6. *Let (X, h) be a Ricci flat ALE 4-manifold. If it is spin and has an end S^3/Γ where $\Gamma \subset SU(2)$ acts on the right, then (X, h) is hyperkähler.*

Just similarly to the compactification of $\mathbb{R}^4$ we can define for an ALE 4-manifold (X, h) a conformal compactification $\hat{X}$ at infinity. Identify the end and $\mathbb{R} \times S^3/\Gamma$, and regard h as a metric being defined over $\mathbb{R} \times S^3/\Gamma$. Consider the induced metric $\overline{h}$ given by the natural covering $\mathbb{R} \times S^3 \to \mathbb{R} \times S^3/\Gamma$. This is then almost Euclidean and Γ-invariant.

Obviously $\Gamma \subset SO(4)$ acts isometrically on the conformal compactification $(S^4, \hat{h})$ of $(\mathbb{R} \times S^3, \overline{h})$ and fixes the north pole ∞. We choose a geodesic ball D^4 centered at ∞ such that D/Γ has the boundary $\partial(D/\Gamma) = S^3/\Gamma$. We attach it to the boundary ∂X_r of X_r to get the compact space $X_r \cup D/\Gamma$ which we call the conformal compactification $\hat{X}$. The space $\hat{X}$ is a Riemannian orbifold ([11]);

$$\hat{X} = \{(X_r, \rho h, \{e\}), (D, \hat{h}, \Gamma)\}$$

for a positive smooth function ρ.

The integration over $\hat{X}$ $\int_{\hat{X}} \alpha \wedge \beta$ for two-forms α, β induces the intersection form of $\hat{X}$.

PROPOSITION 2.7. *Let (X, h) be an anti-selfdual ALE gravitational instanton. Then over the conformal compactification $\hat{X}$ all selfdual harmonic forms vanish. So, $H^2_+(\hat{X}) = 0$, that is, the intersection form is negative definite.*

Let α be a selfdual harmonic form on $\hat{X}$. From the conformal invariance α induces a two-form α_X over (X,h) which is of finite L^2-norm and is selfdual. Apply the Weitzenböck formula to this form. Then α_X turns out to be parallel so that α_X vanishes and Proposition 2.7 is proved.

3. Yang-Mills fields on gravitational instanton.

Let (X,h) be an ALE hyperkähler 4-manifold with an end S^3/Γ for a finite subgroup Γ of $SU(2)$ acting on the right.

Let A be an $SU(2)$-anti-instanton on the trivial bundle over X. Assume the curvature action integral $\int_X |F(A)|^2 dv$ is finite.

Before investigating anti-instantons, we have to state a Γ-equivariant version of Uhlenbeck's removability theorem([**26**]) as follows. Its proof is immediate from the original proof.

THEOREM 3.1. *Let $B \subset \mathbb{R}^4$ be an open ball, $o \in B$, with a metric not necessarily a Euclidean metric. Let Γ be isometries of B with $|\Gamma| < \infty$ fixing o and acting freely on $B\backslash\{o\}$. Suppose A is a Yang-Mills connection on a G-bundle over $B\backslash\{o\}$ with finite $\int_{B\backslash\{o\}} |F(A)|^2 dv$. If the connection form A is Γ-invariant (i.e., $\gamma^*(A) = A$ for all $\gamma \in \Gamma$), then there exists a Γ-invariant gauge transformation $g : B \to G$ such that $g^*(A)$ extends smoothly over B. Here $g^*(A)$ is a Γ-invariant Yang-Mills connection on B.*

From this theorem any $SU(2)$ anti-instanton A of finite curvature action integral extends to an anti-instanton on $\hat{X}$. So, A is seen to be asymptotically flat over X.

Now make the following assumption; there is a gauge transformation g outside of a large ball in X such that $g(A)(x) = g^{-1}(x)dg(x) + g^{-1}(x)A(x)g(x)$ attenuates sufficiently fast as $r(x) \to \infty$.

Then, the action integral $\int_X -Tr(F \wedge *F) = \int_X Tr F \wedge F$ reduces to the boundary integral at infinity $-\frac{1}{3}\lim_{r\to\infty}\int_{\partial X_r}(g^{-1})^*(Tr\theta^3)$, θ being the Maurer-Cartan form of $SU(2)$. On the other hand $Tr\theta^3 = 12\, dv$ and $(g^{-1})^* dv = -deg(g)\, dv_\Gamma$ (dv, dv_Γ are the standard volume elements of $S^3, S^3/\Gamma$, respectively). Thus the action integral of A is $8\pi^2 deg(g)/|\Gamma|$.

For each integer $k > 0$ we consider the set of anti-instantons with curvature action integral $8\pi^2 k/|\Gamma|$ and divide it by asymptotical gauges to get the moduli space $\mathcal{M}(X, k/|\Gamma|)$ of anti-instantons of instanton number $k/|\Gamma| \in \mathbb{Q}$.

The following theorem is a goal of this section.

THEOREM 3.2. *The moduli space $\mathcal{M}(X, k/|\Gamma|)$ is a hyperkähler manifold.*

Similarly as in the argument of §1, one can assert the following for any anti-instanton A

LEMMA 3.3. *The operator $\mathcal{D}_A$ given by*

$$\mathcal{D}_A : L_l^2(\Omega^1(X; su(2))) \to L_{l-1}^2((\Omega^0 \oplus \Omega_+^2)(X; su(2))),$$

$$\mathcal{D}_A(\alpha) = (d_A^* \alpha, d_A^+ \alpha).$$

is Fredholm so that $\operatorname{Ker} \mathcal{D}_A$ *and* $\operatorname{Ker} \mathcal{D}_A^*$ *are both of finite dimensional.*

Remark. To this ALE case we can apply also the Fredholm criterion stated in Theorem 1.5, because X is asymptotically flat (see **[21]**). On the other hand, instead of Theorem 1.5, another type of criterion on the Fredholm property over open manifolds is obtained in **[20]** in terms of the principal symbol. For this, Lockhart and McOwen utilize weighted Sobolev spaces. We can of course apply it.

LEMMA 3.4. $\operatorname{Ker} \mathcal{D}_A^* = 0$ *for all* A.

PROOF: The Weitzenböck formula is applied([15]);

$$\mathcal{D}_A \mathcal{D}_A^*(\phi, \Phi) = (\nabla_A^* \nabla_A \phi, \nabla_A^* \nabla_A \Phi + scal_h \Phi) \qquad (3.$$

Then for $(\phi, \Phi) \in \operatorname{Ker} \mathcal{D}_A^* \subset L_{l-1}^2$

$$\int_X |\mathcal{D}_A^*(\phi, \Phi)|^2 dv = \int_X < \mathcal{D}_A \mathcal{D}_A^*(\phi, \Phi), (\phi, \Phi) > dv,$$

being zero, is written as $\int_X (|\nabla_A \phi|^2 + |\nabla_A \Phi|^2) dv$ so that the norms $|\phi|$ and $|\Phi|$ are constant and hence (ϕ, Φ) vanishes.

From these lemmas the operator $\mathcal{D}_A$ together with the Kuranishi map induces for each $[A] \in \mathcal{M}(X, k/|\Gamma|)$ a neighborhood of o in $\operatorname{Ker} \mathcal{D}_A$ as a chart around $[A]$. Therefore, the moduli space is a smooth manifold.

Since the quaternion structure (I, J, K) of (X, h) operates on $\Omega^1(X; su(2))$ canonically, we have Theorem 3.2 just as the Euclidean moduli space.

From this theorem the dimension of the moduli space is proved to be a multiple of four. On the other hand, like the moduli space of based anti-instantons over S^4 it is regarded as a fibre bundle with fibre over the moduli space $\mathcal{M}(\hat{X}; k/|\Gamma|)$ of anti-instantons on a pseudo bundle $\hat{P}$ of $c_2 = k/|\Gamma|$ over the conformal compactification $\hat{X}$. The fibre is $Z(\rho)/Z$ for the centralizer $Z(\rho)$ of the isotropy representation ρ of $\hat{P}$ at ∞, $\Gamma \to SU(2)$. This is because gauge transformation extends to $\hat{P}$ over

$\hat{X}$ only when it tends at infinity to an element in $Z(\rho)$.

$$\begin{array}{c} \mathcal{M}(X,c_2) \simeq \mathcal{M}^{\infty}(\hat{X},c_2) \\ \downarrow \\ \mathcal{M}(\hat{X},c_2) \end{array}$$

We will now consider the compactification of $\mathcal{M}(X,q), q \in \mathbb{Q}$. To see this one can use the compactification of $\mathcal{M}(\hat{X},q)$.

An idealized anti-instaton is a tuple $(A; x_1, \ldots, x_n)$ with an anti-instanton A of instanton number $q' \in \mathbb{Q}$ and with unordered n points of $\hat{X}$, not necessarily distinct([6]).

The curvature action density of $(A; x_1, \ldots, x_n)$ is given by the distribution with delta functions attaching at each x_j

$$|F(A)|^2 + 8\pi^2 \sum_{x_j \neq \infty} \delta(x_j) + 8\pi^2/|\Gamma| \times \sharp\{j; x_j = \infty\}\ \delta(\infty).$$

Then the idealized anti-instanton has instanton number $q = q' + l + (n - l)/\Gamma$ where $l = \sharp\{j; x_j \in X\}$.

The gauge action is defined on the set of idealized anti-instantons to yield the moduli space $\mathcal{M}^{id}(\hat{X},q)$ of idealized anti-instantons.

Then by definition $\mathcal{M}(\hat{X},q), \mathcal{M}(\hat{X},q-1) \times X$ and for example $\mathcal{M}(\hat{X},q-1/|\Gamma|) \times \{\infty\}$ are subsets of $\mathcal{M}^{id}(\hat{X},q)$.

The following theorem asserts that the closure of $\mathcal{M}(\hat{X},q)$ in $\mathcal{M}^{id}(\hat{X},q)$ is compact in a suitable topology.

THEOREM 3.5([11]). *Let $\{[A_i]\}$ be a sequence in $\mathcal{M}(\hat{X},q)$. Then, (i) there exist a subsequence $\{i'\}$ of $\{i\}$ and finite points $x_1, \ldots, x_n \in \hat{X}$,* *possibly empty, and a sequence of gauge transformations* $\{g_{i'}\}$ *such that*

$$g_{i'}(A_{i'}) \to A_\infty \ (i' \to \infty)$$

on $\hat{X}\backslash\{x_1, ..., x_n\}$ *in the* C^∞*-sense for an anti-instanton* A_∞ *of instanton number* q_∞, *(ii) for each* $x_j \in X$, $j = 1,...,n$ *there is an* $SU(2)$*-bundle* P_j *over* S^4 *with an anti-instanton and for* $x_j = \infty$ *a* Γ*-invariant* $SU(2)$*-bundle* P_∞ *over* S^4 *with a* Γ*-invariant anti-instanton, and (iii) the instanton number satisfies* $q = q_\infty + l + l'/|\Gamma|$ *where* $l = \sum_j c_2(P_j)$, $l' = c_2(P_\infty)$.

In summary, every sequence $\{[A_i]\}$, not converging inside $\mathcal{M}(\hat{X}, q)$, goes to a limit point $([A_\infty]; x_1, ..., x_n)$ in $\mathcal{M}^{id}(\hat{X}, q)$.

As a direct consequence we obtain

COROLLARY 3.6. *Let* (X, h) *be an ALE hyperkähler 4-manifold with an end* $\mathbb{R} \times S^3/\Gamma$ *and* Γ *act on the right. Assume* $0 < k < |\Gamma|$. *Then* $\mathcal{M}(\hat{X}, k/|\Gamma|)$ *has in its compactification stratified components of the following type* $\mathcal{M}(\hat{X}, (k-1)/|\Gamma|) \times \{\infty\}$, $\mathcal{M}(\hat{X}, (k-2)/|\Gamma|) \times \{\infty^2\}$,...,$\mathcal{M}(\hat{X}, 0) \times \{\infty^k\}$, *where* ∞^j *means a* Γ*-invariant anti-instanton of instanton number* j *concentrating at* ∞ *and* $\mathcal{M}(\hat{X}, 0) = \{[\theta]\}$, θ *being the trivial connection.*

From this the compactification of $\mathcal{M}(\hat{X}, 1/|\Gamma|)$ of the least instanton number is in general a one point compactification by gluing the basic anti-instanton of curvature concentrating at ∞ and then the moduli space $\mathcal{M}(X, 1/|\Gamma|)$ of anti-instantons over X is compactified as a $Z(\rho)/Z$-fibre space over this compactification.

Recently Nakajima obtained the following ALE completeness theorem([**24**])

THEOREM 3.7. *If* $\dim \mathcal{M}(X, k/|\Gamma|) = 4$, *then* $\mathcal{M}(X, k/|\Gamma|)$ *is a complete hyperkähler Riemannian manifold and each of its noncompact components is ALE hyperkähler.*

REFERENCES

[1] M.Atiyah, N.Hitchin, I.Singer, *Self-duality in four-dimensional Riemannian geometry,* Proc. Roy. Soc. London **A,362** (1978), 425-461.

[2] S. Bando, A. Kasue, H.Nakajima, *On a construction of coordinates at infinity on manifolds with fast curvature decay and maximal volume growth,* preprint.

[3] R. Bott, R. Seeley, *Some remarks on the paper of Callias,* Commun. Math. Phys. **62** (1978), 235-245.

[4] S. Donaldson, *An application of gauge theory to four dimensional topology,* Jour. Diff. Geom. **18** (1983), 279-315.

[5] ———, *Instantons and geometric invariant theory,* Commun. Math. Phys. **93** (1984), 453-460.

[6] ———, *Connections, cohomology and the intersection forms of 4-manifolds,* Jour. Diff. Geom. **24** (1986), 275-341.

[7] T. Eguchi, A. Hanson, *Self-dual solutions to Euclidean gravity,* Ann. Phys. **120** (1979), 82-106.

[8] R. Fintushel, R. Stern, *Pseudofree orbifolds,* Ann. Math. **122** (1985), 335-364.

[9] A. Floer, *An instanton-invariant for 3-manifolds,* Commun. Math. Phys. **118** (1988), 215-240.

[10] D. Freed, K. Uhlenbeck, "Instantons and 4-manifolds," Math. Sci. Res. Inst. Publication, Springer, New York, 1984.

[11] M. Furuta, *On self-dual pseudo-connections on some orbifolds*, preprint.

[12] G. Gibbons, C. Pope, *The positive action conjecture and asymptotically Euclidean metrics in quantum gravity*, Commun. Math. Phys. **66** (1979), 267-290.

[13] A. Hattori, *On 3-dimensional elliptic space forms(in Japanese)*, Sugaku **12** (1961), 164-167.

[14] N. Hitchin, *Polygons and gravitons*, Math. Proc. Camb. Phil. Soc. **83** (1979), 465-476.

[15] M. Itoh, *On the moduli space of anti-self-dual Yang-Mills connections on Kähler surfaces*, Publ. R. I. M. S. **19** (1983), 15-32.

[16] ———, *Quaternion structure on the moduli space of Yang-Mills connections*, Math. Ann. **276** (1987), 581-593.

[17] M.Itoh, H.Nakajima, *Yang-Mills connections and Einstein -Hermitian metrics*, preprint.

[18] S. Kobayashi, "Differential geometry of complex vector bundles," Iwanami, Tokyo, 1987.

[19] P. Kronheimer, *The construction of gravitational instantons as hyperkähler quotients*, preprint.

[20] R. Lockhart, R. McOwen, *Elliptic differential operators on noncompact manifolds*, Ann. Scuola Norm Sup. Pisa **3** (1985), 409-447.

[21] R. McOwen, *Fredholm theory of partial differential equations on complete Riemannian manifolds*, Pacific J. Math. **87** (1980), 169-185.

[22] J. Milnor, "Morse theory," Princeton Univ., Princeton, 1963.

[23] H.Nakajima, *Self-duality of ALE Ricci-flat 4-manifolds and positive mass theorem*, preprint.

[24] ———, *Moduli spaces of anti-self-dual connections on ALE*

gravitational instantons, preprint.

[25] Y. Tsukamoto, *On Kaehlerian manifolds with positive holomorphic sectional curvature*, Proc. Japan Acad. **33** (1957), 333-335.

[26] K. Uhlenbeck, *Removable singularities in Yang-Mills fields*, Commun. Math. Phys. **83** (1982), 11-29.

[27] E. Witten, *A new proof of the positive energy theorem*, Commun. Math. Phys. **80** (1981), 381-402.

Institute of Mathematics,
University of Tsukuba, 305
Japan

THE 2 DIMENSIONAL BETTI NUMBER OF MODULI SPACE OF INSTANTONS.

YASUHIKO KAMIYAMA

§1 Introduction and statement of results.

We shall denote by M_k and $\tilde{M}_k$ the moduli space of SU(2) instantons with instanton number $C_2 = k$ and the corresponding framed moduli space respectively.
The following results are known concerning M_k and $\tilde{M}_k$.

(1) [7] M_k and $\tilde{M}_k$ are connected.

(2) [6] $\pi_1(\tilde{M}_k) = \mathbb{Z}/_2$.

$$\pi_1(M_k) = \begin{cases} \mathbb{Z}/_2 & k:\text{even} \\ 0 & k:\text{odd.} \end{cases}$$

(3) [1] M_1 is diffeomorphic to $\mathbb{R}^5$.

(4) [2] M_2 has the same homotopy type as $G_2(\mathbb{R}^5)$ where $G_2(\mathbb{R}^5)$ is the Grasmann manifold of 2 planes in $\mathbb{R}^5$.

(5) [5] The Euler characteristic of M_k equals to the number of divisors of k.

In this note we shall indicate a proof of the following theorem.

THEOREM For $k \geqq 3$ $H_2(\tilde{M}_k; \mathbb{Q}) = 0$.

COROLLARY For $k \geqq 3$ $H_2(M_k; \mathbb{Q}) = 0$.

In fact it is well known that there is a principal bundle

ISBN 0-12-640170-5

$$SO(3) \to \tilde{M}_k \to M_k.$$

Hence Corollary follows immediately from Theorem.

Remark From (3) and (4) it follows that $H_2(M_k;\mathbb{Q}) = 0$ for $k=1,2$.

§2 Proof of Theorem.

The proof of theorem depends on Donaldson's description of $\tilde{M}_k$.

Let $M(i,j;\mathbb{C})$ be the set of matrices of size $i \times j$ and if $i=j$ we write $M_i(\mathbb{C})$ for $M(i,i;\mathbb{C})$.

We define F to be the set of elements $(A,B,a,b) \in M_k(\mathbb{C}) \times M_k(\mathbb{C}) \times M(2,k;\mathbb{C}) \times M(k,2;\mathbb{C})$ satisfying the following conditions.

(i) $[A,B] + ba = 0$.

(ii) For all $\lambda, \mu \in \mathbb{C}$

$\begin{pmatrix} A + \lambda \\ B + \mu \\ a \end{pmatrix}$ is injective and $(A + \lambda \quad B + \mu \quad b)$ is surjective.

Proposition ([4])

$\tilde{M}_k$ is a quotient of F by the action of $GL(k;\mathbb{C})$

$$\begin{aligned} A &\mapsto PAP^{-1} \\ B &\mapsto PBP^{-1} \\ a &\mapsto aP^{-1} \\ b &\mapsto Pb \end{aligned} \qquad P \in GL(k;\mathbb{C}).$$

Main steps for computing $H_2(\tilde{M}_k;\ \mathbb{Q})$ are the following.

1∘ Let $\tilde{F}$ be the set of matrices (A,B,a,b) satisfying the following conditions.

(i) $[A,B] + ba = 0$.

(ii) A and B has no common eigenvectors.

(iii) A^t and B^t has no common eigenvectors.

(iv) rank $a = 2$, rank $b = 2$.

Then we show that $\tilde{F}$ is a Zariski open set of F and the complex codimension of $F - \tilde{F}$ in F is $k-1$.

2∘ Let S be the set of pairs of matrices (A,B) satisfying the following conditions.

(i) rank $[A,B] = 2$.

(ii) A and B has no common eigenvectors.

(iii) A^t and B^t has no common eigenvectors.

Define $\pi : \tilde{F} \to S$ by $\pi\,(A,B,a,b) = (A,B)$.
Then we show that $\pi : \tilde{F} \to S$ is a principal bundle with structure group $GL(2;\mathbb{C})$.

3∘ Let T be the elements (A,B) of S such that the eigenpolynomial of A has no common root.
Then we show that T is an irreducible complex hypersurface in S and the following principal bundle exists.

$$GL(k;\mathbb{C})/_{\mathbb{C}^*} \to T \to T/_{GL(k;\mathbb{C})}.$$

4∘ Let N be the set of pairs of matrices (A,B) satisfying the following conditions.

(i) A is a diagonal matrix which has no common diagonal elements.

(ii) rank $[A,B] = 2$.

Let Σ_k be the symmetric group of k elements.

Then we show the following covering space exists.

$$\Sigma_k \to N/(\mathbb{C}^*)^k \to N/(\mathbb{C}^*)^k/\ \Sigma_k.$$

and $N/(\mathbb{C}^*)^k/\Sigma_k$ is homeomorphic to $T/GL(k;\mathbb{C})$.

5∘ Let Ω be the set of diagonal matrices A which have no common diagonal elements and let G be the set of matrices C satisfying the following conditions.

(i) All of the diagonal elements of C are zero.

(ii) For each row and column there exists a nontrivial elements.

(iii) rank $G = 2$.

Then we show that $H^*(\Omega \times G/(\mathbb{C}^*)^k;\mathbb{Q}) = H^*(N/(\mathbb{C}^*)^k;\mathbb{Q})$ as representation spaces of Σ_k.

Now, following the above steps we can compute $H^2(\tilde{M}_k;\mathbb{Q})$. In 5∘ it can be proved that $H^q(G/(\mathbb{C}^*)^k;\mathbb{Q}) = 0$ for $q = 1,2$ and $H^*(\Omega;\mathbb{Q})$ is computed in [3] that $H^1(\Omega;\mathbb{Q}) = V$ and $H^2(\Omega;\mathbb{Q}) = 0$ where V is a representation space of Σ_k such that $V^{\Sigma_k} = \mathbb{Q}$. These gives us $H^1(N/(\mathbb{C}^*)^k;\mathbb{Q}) = \mathbb{Q}$ and $H^2(N/(\mathbb{C}^*)^k;\mathbb{Q}) = 0$. Then spectral sequence argument shows that $H^1(N/(\mathbb{C}^*)^k/\Sigma_k;\mathbb{Q}) = \mathbb{Q}$ and $H^2(N/(\mathbb{C}^*)^k/\Sigma_k;\mathbb{Q}) = 0$.

By 4∘ it follows that $H^1(T/GL(k,\mathbb{C});\mathbb{Q}) = \mathbb{Q}$ and $H^2(T/GL(k,\mathbb{C});\mathbb{Q}) = 0$.

By the spectral sequence argument to the fibration in 3∘ we can see that $H^1(T;\mathbb{Q}) = \mathbb{Q}$ and $H^2(T;\mathbb{Q}) = 0$.

Let $\tilde{F}|T$ denote $\pi^{-1}(T)$ then we can see in 2∘ that $H^1(\tilde{F}|T;\mathbb{Q}) = \mathbb{Q} + \mathbb{Q}$ and $H^2(\tilde{F}|T;\mathbb{Q}) = 0$.
If we define the action of $GL(k,\mathbb{C})$ on $\tilde{F}|T$ as a subset of F we have the following principal bundle

$$GL(k,\mathbb{C}) \to \tilde{F}|T \to \tilde{F}|T/\ GL(k,\mathbb{C}).$$

Note that the action of $GL(k,\mathbb{C})$ on F is free.
By using the spectral sequence argument it can be shown that
$H^1(\tilde{F}|T/GL(k,\mathbb{C});\mathbb{Q}) = \mathbb{Q}$ or $\mathbb{Q}+\mathbb{Q}$ and if the first occurs then $H^2(\tilde{F}|T/GL(k;\mathbb{C});\mathbb{Q}) = 0$.

Now the general position argument shows that
$H^q(F/GL(k,\mathbb{C});\mathbb{Q}) = H^q(\tilde{F}/GL(k,\mathbb{C});\mathbb{Q})$ for $q= 1,2$ and $k \geqq 3$.
As T is a complement of an irreducible complex hypersurface in S so $\tilde{F}|T/GL(k,\mathbb{C})$ is in $\tilde{F}/GL(k;\mathbb{C})$.
Therefore $H^2(\tilde{F}/GL(k,\mathbb{C}),\tilde{F}|T/GL(k,\mathbb{C});\mathbb{Q}) = \mathbb{Q}$.
Using this fact and $H^1(\tilde{M}_k;\mathbb{Q}) = 0$ [6] the cohomology long exact sequence of $(\tilde{F}/GL(k,\mathbb{C}),\ \tilde{F}|T/GL(k,\mathbb{C}))$ shows that
$H^1(\tilde{F}|T/GL(k,\mathbb{C});\mathbb{Q}) = \mathbb{Q}$
and so $H^2(\tilde{F}/GL(k,\mathbb{C});\mathbb{Q}) = 0$.
This completes the proof of Theorem.

References

[1] M.F.Atiyah,V.G.Drinfeld,N.J.Hitchen and Yu.I.Manin Constructions of instantons. Phys.Lett.65 A (1978).

[2] H.Aupetit and A.Douady Fibrés stables de rang 2 sur $P^3\mathbb{C}$ avev $C_1=0$, $C_2=2$. Ecole Normale Supérieure (1977-1978).

[3] R.Cohen,J.Lada and P.May The homology of iterated loop spaces. Springer Lecture Note in Math.533 (1976).

[4] S.K.Donaldson Instantons and geometric invariant theory. Commun.Math.Phys.93 (1984).

[5] M.Furuta Euler number of moduli space of instantons. Proceedings of the Japan academy 63A (1987).

[6] J.Hurtubise Instantons and jumping lines. Commun.Math.Phys.105 (1986).

[7] H.Taubes Path-connected Yang-Mills moduli spaces. J.Diff.Geo.19 (1984).

Yasuhiko KAMIYAMA

Department of Mathematics

Faculty of Science

University of Tokyo

Hongo, Tokyo 113

Japan

ON EQUIVARIANT YANG-MILLS CONNECTIONS

Hiromichi MATSUNAGA

Dedicated to Professor Shôrô Araki on his sixtieth birthday

In this note we study an equivariant analogue of Yang-Mills connections. For a compact Lie group Γ, a Γ-vector bundle over a smooth semi-free Γ-manifold yields a simple structure. Generalizing this, we define a Γ-vector bundle of diagonal type and an equivariant connection, and further give some of their elementary properties in Section 1.

In Section 2, for a linear S^1-action on the 4-sphere, we prove the equivariance of ± 1-instantons, and in Section 3 prove the equivariance of k-instantons. We use this in Section 5.

In Section 4 we discuss the space of infinitesimal deformations of equivariant (anti-) self-dual connections, equivariant elliptic complexes and obtain an equivariant Lefschetz formula.

In the last section we determine the dimension of the invariant moduli space of self-dual connections. As an application we give an example of an S^1-action on S^4 which does not lift to an action on a vector bundle over S^4. We also give an example of an equivariant anti-self-dual connection.

Finally we note in a remark that Example 3 in Section 5 gives a monopole, and so in this case our equivariant connection is essentially the same as the invariant connection due to Atiyah [9].

I am grateful to my colleages, especially to Professor M.Itoh for

ISBN 0-12-640170-5

useful discussions during the symposium on Geometry 1988 in Japan.

1. Vector bundles of diagonal type and equivariant connections

Let Γ be a compact Lie group, M a Γ-space, and $E \longrightarrow M$ be a real or complex Γ-vector bundle with structure group G.

<u>Definition</u>. A Γ-vector bundle $p\colon E \longrightarrow M$ is of diagonal type if and only if there exist a covering $M = \bigcup U_i$ consisting of Γ-invariant open sets, homomorphisms $\alpha_i\colon \Gamma \longrightarrow G$ and equivariant local trivialities $\phi_i\colon U_i \times V \longrightarrow p^{-1}(U_i)$ for a G-vector space V such that the identity

$$\phi_i(\gamma x, \alpha_i(\gamma)(v)) = \gamma\phi_i(x,v) \quad \text{for } \gamma \in \Gamma, \quad (x,v) \in U_i \times V$$

holds.

For a Γ-vector bundle, the transition functions $\{f_{ji}\}$ satisfies

$$\alpha_j(\gamma) f_{ji}(x) = f_{ji}(\gamma x)\alpha_i(\gamma) \tag{1}$$

then by the commutative diagram

$$\begin{array}{ccc}
U_i \times G \ni (x,g) & \cong & (x, f_{ji}(x)g) \in U_j \times G \\
\downarrow & & \downarrow \\
\gamma(x,g) & & \\
= (\gamma x, \alpha_i(\gamma)g) & \cong & (\gamma x, \alpha_j(\gamma) f_{ji}(x) g) \\
 & & = (\gamma x, f_{ji}(\gamma x)\alpha_i(\gamma) g),
\end{array}$$

we obtain the associated Γ-G principal bundle.

From now on, we treat only smooth vector bundles. Let $E \longrightarrow M$ be an S^1-vector bundle of diagonal type, and $\{A_i\}$ be a connection.

<u>Proposition 1</u>. There exists an S^1-action on the space of connections which is given by

$$(\gamma A_i)(x) = \mathrm{Ad}(\alpha_i(\gamma))A_i(\gamma^{-1}x) \quad \text{for } \gamma \in S^1, \ x \in U_i \tag{2}$$

Proof. The proposition is obtained by equalities :

$$A_j(x) = f_{ij}(x)^{-1}df_{ij}(x) \quad + f_{ij}(x)^{-1}A_i(x)f_{ij}(x),$$

$$\mathrm{Ad}(\alpha_j(\gamma))(f_{ij}(\gamma^{-1}x)^{-1}df_{ij}(\gamma^{-1}x) + f_{ij}(\gamma^{-1}x)^{-1}A_i(\gamma^{-1}x)f_{ij}(\gamma^{-1}x))$$

$$= f_{ij}(x)^{-1}df_{ij}(x) + f_{ij}(x)^{-1}\mathrm{Ad}(\alpha_i(\gamma))A_i(\gamma^{-1}x)f_{ij}(x)$$

$$= f_{ij}(x)^{-1}df_{ij}(x) + f_{ij}(x)^{-1}(\gamma A_i)(x)f_{ij}(x)$$

$$= (\gamma A_j)(x).$$

The group $\mathcal{G}$ of gauge transformations consists of cross-sections $\{f_i\}$ of the bundle $P\times_{\mathrm{Ad}(G)}G$, where P denotes the associated principal bundle.

Proposition 2. Let $\{f_i(x)\}$ be a gauge transformation, then $\{\mathrm{Ad}(\alpha_i(\gamma))f_i(\gamma^{-1}x)\}$ is so.

Proof.

$$\alpha_j(\gamma)f_j(\gamma^{-1}x)\alpha_j(\gamma)^{-1} = \alpha_j(\gamma)f_{ji}(\gamma^{-1}x)f_i(\gamma^{-1}x)f_{ji}(\gamma^{-1}x)^{-1}\alpha_j(\gamma)^{-1}$$

$$= f_{ji}(x)(\mathrm{Ad}(\alpha_i(\gamma))f_i(\gamma^{-1}x))f_{ji}(x)^{-1}.$$

Set $\mathrm{Ad}(\alpha_i(\gamma))f_i(\gamma^{-1}x) = (\gamma f_i)(x)$.

Corollary. The moduli space $\mathcal{A}/\mathcal{G}$ admits an induced S^1-action.

Proof. $A_i(x) \sim f_i(x)^{-1}df_i(x) + f_i(x)^{-1}A_i(x)f_i(x)$, then

$$(\gamma A_i)(x) \sim (\gamma f_i)(x)^{-1}d((\gamma f_j)(x)) + (\gamma f_i)(x)^{-1}(\gamma A_i)(x)(\gamma f_i(x)).$$

Definition. A connection $A_i(x)$ is S^1-equivariant if and only if $\gamma A_i = A_i$ for each i and $\gamma \in S^1$.

Now let M be a compact connected S^1-4 manifold. We choose an S^1-invariant Riemannian metric on M. Then we have

Proposition 3. The Yang-Mills functional $\frac{1}{2}\int_M |F(A)|^2 dv$ is S^1-invariant.

Proof. We have

$$d(\gamma A_i)(x) + (\gamma A_i)(x)\wedge(\gamma A_i)(x)$$
$$= Ad(\alpha_i(\gamma))(dA_i(\gamma^{-1}x) + A_i(\gamma^{-1}x)\wedge A_i(\gamma^{-1}x)),$$

and the Killing form is $Ad(\alpha_i(\gamma))$ invariant, so we obtain the proposition.

Now we remark that the Hodge star operator $* : \Lambda^2 \longrightarrow \Lambda^2$ is conformally invariant, then each element γ of the group S^1 transforms (anti-) self-dual connections to the similar ones, and acts on the space $\mathcal{A}_{\pm}/\mathcal{G}$. Next we have

Proposition 4. Let M be a compact semi-free S^1-manifold, then any S^1-vector bundle over M is of diagonal type.

Proof. If $S^1(x)$ is a principal orbit, then it admits an invariant neighborhood N, and the orbit is an S^1-deformation retract of N and the restriction $E|N$ of the bundle is S^1-isomorphic to the product $N\times V$ for a trivial S^1-vector space V. If x is a fixed point, then there exists an invariant neighborhood U(x) and the restriction $E|U(x)$ is S^1-isomorphic to the product $U(x)\times V$ for an S^1-vector space V. Thus we have the proposition.

2. Equivariance of ± 1-instantons

Let $x = z_1 + z_2 j$ be an element of the symplectic group Sp(1). We have the isomorphism

$$Sp(1)\ni x \longrightarrow \begin{bmatrix} \bar{z}_1 & \bar{z}_2 \\ -z_2 & z_1 \end{bmatrix} \in SU(2), \text{ the special unitary group.}$$

By this a characteristic map $\chi^- : S^3 \longrightarrow SU(2)$ is given, which determines a principal SU(2)-bundle over the 4-sphere S^4, and its instanton number is -1. Similarly

$$Sp(1)\ni x \longrightarrow \begin{bmatrix} \bar{z}_1 & -\bar{z}_2 \\ z_2 & \bar{z}_1 \end{bmatrix} \in SU(2), \text{ say } \chi^+,$$

gives the +1-instanton bundle. Here the 4-sphere S^4 is obtained from the disjoint union $H \cup H'$ of two quaternions by the identification

$$H - 0 \ni x \longrightarrow x^{-1} \in H' - 0,$$

then on the equator, the identification is given by

$$z_1 + z_2 j \longrightarrow \bar{z}_1 - z_2 j.$$

Define an S^1-action on the space H by $z_1 + z_2 j \longrightarrow \gamma^a z_1 + \gamma^b z_2 j$ for $\gamma \in S^1$, then on the lower half sphere the action is of type (-a,b).

Now we have

<u>Theorem</u> 1. If $a = \pm b$, then ± 1-instanton is S^1-equivariant.

<u>Proof</u>. We have

$$\chi^-(\gamma x) = \begin{bmatrix} \gamma^{-a} & 0 \\ 0 & \gamma^b \end{bmatrix} \chi^-(x) \begin{bmatrix} 1 & 0 \\ 0 & \gamma^{-b-a} \end{bmatrix} = \begin{bmatrix} 1 & 0 \\ 0 & \gamma^{a+b} \end{bmatrix} \chi^-(x) \begin{bmatrix} \gamma^{-a} & 0 \\ 0 & \gamma^{-b} \end{bmatrix}$$

$$\chi^+(\gamma x) = \begin{bmatrix} \gamma^a & 0 \\ 0 & \gamma^b \end{bmatrix} \chi^+(x) \begin{bmatrix} 1 & 0 \\ 0 & \gamma^{-a+b} \end{bmatrix} = \begin{bmatrix} 1 & 0 \\ 0 & \gamma^{b-a} \end{bmatrix} \chi^+(x) \begin{bmatrix} \chi^a & 0 \\ 0 & \gamma^{-b} \end{bmatrix}.$$

For +1-instanton bundle, on the upper hemi-sphere, 1-forms

$$\phi_1^+ = -x_1 dx_2 + x_2 dx_1 - x_3 dx_4 + x_4 dx_3$$

$$\phi_2^+ = -x_1 dx_3 + x_3 dx_1 - x_4 dx_2 + x_2 dx_4$$

$$\phi_3^+ = -x_1 dx_4 + x_4 dx_1 - x_2 dx_3 + x_3 dx_2$$

are transformed to

$$\begin{bmatrix} 1 & 0 & 0 \\ 0 & \cos(a+b)\theta & -\sin(a+b)\theta \\ 0 & \sin(a+b)\theta & \cos(a+b)\theta \end{bmatrix} \begin{bmatrix} \phi_1^+ \\ \phi_2^+ \\ \phi_3^+ \end{bmatrix},$$

and on the lower hemisphere, to

$$\begin{bmatrix} 1 & 0 & 0 \\ 0 & \cos(-a+b)\theta & -\sin(-a+b)\theta \\ 0 & \sin(-a+b)\theta & \cos(-a+b)\theta \end{bmatrix} \begin{bmatrix} \phi_1^+ \\ \phi_2^+ \\ \phi_3^+ \end{bmatrix}.$$

For the basis of the Lie algebra,

$$e_1 = \frac{\sqrt{-1}}{2}\begin{bmatrix}1 & 0\\0 & -1\end{bmatrix}, \quad e_2 = \frac{\sqrt{-1}}{2}\begin{bmatrix}0 & -i\\i & 0\end{bmatrix}, \quad e_3 = \frac{\sqrt{-1}}{2}\begin{bmatrix}0 & 1\\1 & 0\end{bmatrix},$$

the transformation can be described by

$$\mathrm{Ad}\begin{bmatrix}1 & 0\\0 & \bar{\gamma}^{a+b}\end{bmatrix} : (e_1,e_2,e_3) \longrightarrow (e_1,e_2,e_3)\begin{bmatrix}1 & 0 & 0\\0 & \cos(a+b)\theta & \sin(a+b)\theta\\0 & -\sin(a+b)\theta & \cos(a+b)\theta\end{bmatrix}$$

$$\mathrm{Ad}\begin{bmatrix}\bar{\gamma}^{a} & 0\\0 & \bar{\gamma}^{b}\end{bmatrix} : (e_1,e_2,e_3) \longrightarrow (e_1,e_2,e_3)\begin{bmatrix}1 & 0 & 0\\0 & \cos(a-b)\theta & \sin(a-b)\theta\\0 & -\sin(a-b)\theta & \cos(a-b)\theta\end{bmatrix},$$

respectively on S^4_+ and S^4_-. Therefore the +1-instanton $A^+ = e_1\phi_1^+ + e_2\phi_2^+ + e_3\phi_3^+$ satisfies the relation $\mathrm{Ad}(\alpha(\bar{\gamma}))A^+(\gamma x) = A^+(x)$ if $a = b$. In the case $a = -b$ we use the another representation of $\chi^+(\gamma x)$ to get the equivariance. Similarly we can do for -1-instanton.

3. Equivariance of k-instantons

Let $b_1,b_2,\ldots,b_k$ be the k-th roots of 1 multiplied by a sufficiently large real number. By the theorem in Chap. II [1], we have a self-dual k-instanton given by

$$A_k(x) = (1 + \Sigma_{n=1}^{k} |b_n - x|^{-2})^{-1} \operatorname{Im} \Sigma_{n=1}^{k} \overline{(b_n - x)}^{-1} d(b_n - x)^{-1},$$

where we have choosed $\lambda = (1,1,\ldots,1)$. Let γ be one of the k-th roots of unity, then up to permutations, $(b_1\gamma,\ldots,b_k\gamma) = (\gamma b_1,\ldots,\gamma b_k) = (b_1,\ldots,b_k)$. Then we have

$$A_k(x\gamma) = A_k(x), \quad A_k(\gamma x) = \gamma A_k(x)\,\bar{\gamma} = \mathrm{Ad}\begin{bmatrix}\gamma & 0\\0 & \bar{\gamma}\end{bmatrix} A_k(x),$$

for the su(2)-basis. Thus we have proved

<u>Proposition</u> 5. If the k-instanton bundle admits a lifting action such that one of the chracteristic maps $\chi_k^+(x)$ satisfies

$$\chi_k^+(x\gamma) = \begin{bmatrix}\gamma & 0\\0 & \bar{\gamma}\end{bmatrix} \chi_k^+(x) \begin{bmatrix}1 & 0\\0 & 1\end{bmatrix}$$

then the k-instanton is Z_k-equivariant.

In Section 4 we prove that the bundle can not admits such a lifting.

Next we consider the S^1-action which is given by

$$(*) \quad S^3 \ni x \longrightarrow \gamma x \bar{\gamma} \in S^3 \quad \text{for } \gamma \in S^1, \text{ then}$$

$$\chi^+(\gamma x \bar{\gamma}) = \begin{bmatrix} \bar{\gamma} & 0 \\ 0 & \gamma \end{bmatrix} \chi^+(x) \begin{bmatrix} \gamma & 0 \\ 0 & \bar{\gamma} \end{bmatrix}, \quad (\chi^+(\gamma x \bar{\gamma}))^k = \begin{bmatrix} \bar{\gamma} & 0 \\ 0 & \gamma \end{bmatrix} (\chi^+(x))^k \begin{bmatrix} \gamma & 0 \\ 0 & \bar{\gamma} \end{bmatrix}.$$

Now we have

$$\overline{[\gamma(b_n-x)\bar{\gamma}]^{-1}} = \gamma\overline{(b_n-x)^{-1}}\bar{\gamma}, \quad d[\gamma[b_n-x)^-]^{-1} = \gamma d(b_n-x)^{-1}\bar{\gamma},$$

Thus we have proved

Theorem 2. Concernig the action (*), the k-instanton bundle admits an S^1-equivariant self-dual connection.

4. The space of infinitesimal deformations of S.D.(A.S.D.)-connections

Let M be a compact S^1-4 manifold with positive scalar curvature, further, be self-dual or a compact Kahler surface. Consider a self-dual or anti-self-dual S^1-equivariant connection $\{A_U\}$ on an S^1-G principal bundle P of diagonal type, where G is a compact semi-simple Lie group. Following [4],[5] we use the notations,

$\mathfrak{g}_P = P \times_{AdG} \mathfrak{g}$, $\quad A^k(\mathfrak{g}_P) = \Gamma(M, \Lambda^k \otimes \mathfrak{g}_P)$,

$A^k_{S^1}(\mathfrak{g}_P)$: S^1-equivariant $\mathfrak{g}_P$-valued k-forms, $\quad d^{(0)} = \nabla$: the connection,

$d^\nabla : A^1(\mathfrak{g}_P) \longrightarrow A^2(\mathfrak{g}_P)$: covariant exterior derivative,

$d^\nabla_{S^1}$: the restriction of d^∇ onto $A^1_{S^1}(\mathfrak{g}_P)$,

$p_{\pm} : A^2(\mathfrak{g}_P) \longrightarrow A^2_{\pm}(\mathfrak{g}_P)$, the projection onto S.D.(A.S.D.) part.

By the section 6 in [4] and (2.8) in [5], we have the following elliptic complex,

$$E_{\mp} : 0 \longrightarrow A^0(\mathfrak{g}_P) \longrightarrow A^1(\mathfrak{g}_P) \xrightarrow{p_{\mp} d^\nabla} A^2_{\mp}(\mathfrak{g}_P) \longrightarrow 0.$$

Then we have

<u>Lemma</u> 1. The operators ∇ and $p_{\mp}d^{\nabla}$ are S^1-equivariant.

<u>Proof</u>. Let U be an invariant neighborhood as in Section 1. For $\psi \in A^0(\mathfrak{g}_P)_U$, the restriction of $A^0(\mathfrak{g}_P)$ on U, and $\gamma \in S^1$,

$$\begin{aligned}\nabla(\gamma\psi(x)) &= \nabla(Ad(\alpha_U(\gamma))\psi(\gamma^{-1}x)) = d(Ad(\alpha_U(\gamma))\psi(\gamma^{-1}x)) \\ &\qquad + [A_U(x), Ad(\alpha_U(\gamma))\psi(\gamma^{-1}x)] \\ &= Ad(\alpha_U(\gamma))(d\psi(\gamma^{-1}x) + Ad(\alpha_U(\gamma))[A_U(\gamma^{-1}x), \psi(\gamma^{-1}x)] \\ &= Ad(\alpha_U(\gamma))(d\psi(\gamma^{-1}x) + [A_U(\gamma^{-1}x), \psi(\gamma^{-1}x)]) = (\gamma\cdot\nabla\psi)(x).\end{aligned}$$

For $\Psi \in A^1(\mathfrak{g}_P)_U$,

$$d^{\nabla}(\gamma\cdot\Psi(x)) = d(Ad(\alpha_U(\gamma))\Psi(\gamma^{-1}x) + [A(x)_\wedge Ad(\alpha_U(\gamma))\Psi(\gamma^{-1}x)] = (\gamma\cdot d^{\nabla}\Psi)(x),$$

and operator * commute with the action, then we obtain the lemma.

By the lemma, we have

<u>Corollary</u>. $(A^0_{S^1}(\mathfrak{g}_P)) \subset A^1_{S^1}(\mathfrak{g}_P)$, $p_{\pm}d^{\nabla}(A^1_{S^1}(\mathfrak{g}_P)) \subset A^2_{\pm\, S^1}(\mathfrak{g}_P)$.

Now we have

<u>Theorem</u> 3. The invariant space of infinitesimal deformations of an irreducible S^1-equivariant (anti-) self-dual connection is given by $H^1_{S^1}(E) = Ker(p_{\pm}d^{\nabla})_{S^1} / \operatorname{Im} \nabla_{S^1}$ and it is a linear subspace of the space $H^1(E)$.

<u>Proof</u>. By the corollary above and an analogy of the proof of the theorem 6.1 in [4], we have the half of the assertion. Let $i_* : H^1_{S^1}(E) \longrightarrow H^1(E)$ be the homomorphism induced from the inclusion map $Ker(p_{\pm}d^{\nabla})_{S^1} \subset Ker(p_{\pm}d^{\nabla})$. Let $\psi \in Ker(p_{\pm}d^{\nabla})$ be an element which satisfies $\gamma\psi = \gamma$ for each $\gamma \in S^1$ and $i_*([\psi]) = 0$. Then there exists $\psi_0 \in A^0(\mathfrak{g}_P)$ such that $\psi = \nabla\psi_0$. By Lemma 1, $\nabla(\gamma\psi_0) = \nabla\psi_0$ for $\gamma \in S^1$. Set $\hat{\psi}_0 = \int_\Gamma(\gamma\psi_0)d\gamma$, then $\nabla\hat{\psi}_0 = \nabla\psi_0$, therefore $\psi \in \nabla(A^0_{S^1}(\mathfrak{g}_P))$ and $[\psi] = 0$. Thus the

homomorphism i_* is monomorphic. Hence the proof is completed.

Next let γ be a generator of the topologically cyclic group S^1. In the Lefschetz number $L(\gamma,E) = tr(\gamma|H^0(E))-tr(\gamma|H^1(E))+tr(\gamma|H^2(E_{\mp}))$, if we set $\gamma = 0,1$, then we have

<u>Lemma</u> 2. $H^1_{S^1}(E) = -L(0,E)$, $H^1(E) = -L(1,E)$.

Let $V_{\pm}$ be the complex spinor bundles of $\pm\frac{1}{2}$ spinors on M. By the section 1 in [4], $V_+ \otimes V_- = \Lambda^1_C$, $V_{\pm}\otimes V_{\pm} = \Lambda^2_{\pm C} \oplus \Lambda^0_C$. Then by the theorem 3.1 in [2], we obtain

<u>Theorem</u> 4. If the fixed point sets are isolated, then

$$- L(\gamma,E_{\mp}) = \Sigma_i \frac{ch_\gamma((V_{\pm} - V_{\mp})V_{\mp})}{ch_\gamma \Lambda_{-1}(T_{p_i} M\otimes C)} ch_\gamma(\mathcal{G}^C_P)_{p_i},$$

where $\{p_i\}$ is the fixed point sets.

Here we give an application of this theorem to a lifting problem of an S^1-action. In Section 2 we have seen that +1 instanton bundle admits a lifting of an S^1-action with type $(a,b) = (1,-1)$, and +1 instanton is equivariant. Now we have

<u>Corollary</u>. For $k \geq 2$, the k-instanton bundle can not admit such a lifting that $\overset{+}{\chi}_k(x\gamma) = \begin{bmatrix}\gamma & 0\\ 0 & \bar{\gamma}\end{bmatrix}\overset{+}{\chi}_k(x)\begin{bmatrix}1 & 0\\ 0 & 1\end{bmatrix}$.

<u>Proof</u>. Near the north pole <u>n</u> the action is of type (1,-1) and near the south pole <u>s</u>, (-1,-1). Assume that the k-instanton bundle admits such a lifting, then there exists a lifting of the cyclic subgroup Z_k-action. By Proposition 5 we can apply Theorem 4 and for a generator γ of Z_k,

$$H^1(E_-) = -L(\gamma,E_-)$$

$$= \frac{ch_\gamma((V_+-V_-)V_-)}{ch_\gamma(\Lambda_{-1}(T_n \otimes C)} ch_\gamma(E \otimes E^*-C) + \frac{ch_\gamma((V_+-V_-)V_-)}{ch_\gamma(\Lambda_{-1}(T_s \otimes C))} ch_\gamma(E \otimes E^*-C)$$

$$= [(1-\gamma)^2(1-\bar{\gamma})^2]^{-1}((2-\gamma-\bar{\gamma})(\gamma+\bar{\gamma})3 + (\gamma+\bar{\gamma}-2)2((\gamma+\bar{\gamma})^2-1))$$

$$= 2(\gamma+\bar{\gamma}) + 1,$$

hence by the dimensional reason k must be 1.

5. Examples and further discussions

Example 1. Let a=b=1 in Section 2, then

$$\chi^+(\gamma x) = \begin{bmatrix} 1 & 0 \\ 0 & 1 \end{bmatrix} \chi^+(x) \begin{bmatrix} \gamma & 0 \\ 0 & \bar{\gamma} \end{bmatrix}, \text{ by Theorem 4,}$$

$$H^1(E_-) = [(1-\gamma)^2(1-\bar{\gamma})^2]^{-1}[(\gamma+\bar{\gamma}-2)2((\gamma+\bar{\gamma})^2-1) + (2-\gamma-\bar{\gamma})(\gamma+\bar{\gamma})3]$$

$$= 2(\gamma+\bar{\gamma}) + 1, \text{ and } \dim H^1_{S^1} = 1.$$

Example 2. Here we consider a reducible connection. On the complex projective plane $P_2(C)$, we give an S^1-action by

$$\gamma([z_0,z_1,z_2]) = [z_0,\gamma^2 z_1,\gamma z_2] \text{ for } \gamma \in S^1.$$

The action is liftable to the principal U(1)-bundle $S^5 \longrightarrow P_2(C)$. Denote by L the associated complex line bundle $S^5 \times_{S^1} C \longrightarrow P_2(C)$. Then the associated principal bundle of the plane bundle $L \oplus L^{-1}$ is reducible to an $S(U(1) \times U(1))$-bundle, and the associated su(2)-bundle is isomorphic to the Whitney sum $L^2 \oplus 1$. Since the curvature of the hermitian connection is locally given by $\sqrt{-1}\Omega$, where Ω is the fundamental form of the Fubini-Study metric, the connection is self-dual. The connection is locally given by

$$A = [1+|\xi_1|^2+|\xi_2|^2]^{-1}2(\bar{\xi}_1 d\xi_1+\bar{\xi}_2 d\xi_2),$$

then invariant under the S^1-action, and the Lie algebra $\sqrt{-1}R$ is invariant under the adjoint action. Thus the connection is S^1-equivariant. The fixed points in $P_2(C)$ are [1,0,0],[0,1,0],[0,0,1]. Now we are ready to calculate the Lefschetz number, $-L(\gamma,E_-)$,

$$= [(1-\gamma)(1-\bar{\gamma})(1-\gamma^2)(1-\bar{\gamma}^2)]^{-1}(\gamma^{3/2}+\bar{\gamma}^{3/2}-\gamma^{1/2}-\bar{\gamma}^{1/2})(3+\gamma^4+\bar{\gamma}^4+1)$$
$$+ [((1-\gamma)(1-\bar{\gamma}))^2]^{-1}((\gamma+\bar{\gamma}-2)(\gamma+\bar{\gamma})(\gamma^2+\bar{\gamma}^2+1))$$
$$= \gamma^3+\bar{\gamma}^3+\gamma^2+\bar{\gamma}^2+\gamma+\bar{\gamma}-1.$$

Since the connection is reducible, $H^0(E_-) = C$, therefore

$$H^1(E_-) = \gamma^3+\bar{\gamma}^3+\gamma^2+\bar{\gamma}^2+\gamma+\bar{\gamma}.$$

Thus the connection is isolated in the moduli space.

Example 3. We consider an S^1-action on the 4-sphere with non isolated fixed point set. As in Section 3, we give an S^1-action on the quaternions H by $x = z_1+z_2j \longrightarrow \gamma x\bar{\gamma} = z_1+\gamma^2 z_2 j$. Then we have to treat a non faithful action given by $S^1 \ni \gamma \longrightarrow \gamma^2 \in S^1$. Let $P \longrightarrow M$ be an S^1-G principal bundle with a non faithful S^1-action on M. Let $S^1(x)$ be a principal orbit, then it is an equivariant deformation retract of an invariant neighborhood N, and $P|N = N \times_{S^1(x)} (S^1 \times_{Z_2} G)$ for some Z_2-action on G. If the Z_2-action is trivial, then the bundle is of diagonal type. The k-instanton bundle $S^7 \times_{Sp(1)} H \longrightarrow S^4$ admits an S^1-action which is given by $\gamma(q_1,q_2) = (\gamma q_1,\gamma q_2)$ for $\gamma \in S^1$, $(q_1,q_2) \in S^7$. For a positive <u>even</u> integer k, the assumption for the Z_2-action is satisfied, and the bundle is of diagonal type. Now by Theorem 2 we have an equivariant connection. Since the 4-sphere S^4 is a Spin manifold, we can apply the formulas in [3]. Let $H \longrightarrow S^2$ be the complex Hopf line bundle over the fixed sphere S^2. By the restriction

$$(\chi^+(\gamma x\bar{\gamma}))^k = \begin{bmatrix}\bar{\gamma} & 0\\ 0 & \gamma\end{bmatrix}\begin{bmatrix}z_1 & -\bar{z}_2\\ z_2 & \bar{z}_1\end{bmatrix}\begin{bmatrix}\gamma & 0\\ 0 & \bar{\gamma}\end{bmatrix} \longrightarrow (\chi^+(\gamma z_1\bar{\gamma}))^k$$
$$= \begin{bmatrix}\bar{\gamma} & 0\\ 0 & \gamma\end{bmatrix}\begin{bmatrix}z_1^k & 0\\ 0 & \bar{z}_1^k\end{bmatrix}\begin{bmatrix}\gamma & 0\\ 0 & \bar{\gamma}\end{bmatrix},$$

we obtain the bundle $H^k(\bar{\gamma}) \oplus \bar{H}^k(\gamma)$ on the fixed sphere S^2, where $\bar{H}$ denotes the conjugate bundle of the bundle H. Let $P_k \longrightarrow S^4$ be the principal bundle with characteristic map $(\chi^+)^k$, and $P_{\underline{k}|} \longrightarrow S^2$ be the restriction. Then

$$P_{\underline{k}|} \times_{Ad(SU(2))} (su(2) \otimes C) = (H^k(\bar{\gamma}) \oplus \bar{H}^k(\gamma))^2 - 1.$$

Since $C_1(H)[S^2] = 1$, we have

$$H^1(E_-) = (\bar{\gamma}+\gamma)(\bar{\gamma}-\gamma)^{-1} 2k(-\gamma^2+\bar{\gamma}^2) = 4k + 2k(\gamma^2+\bar{\gamma}^2).$$

Therefore $H^1_{S^1} = 4k$.

Example 4. Here we present an equivariant anti-self-dual connection of index -1 on the space $\bar{P}_2(C)$ obtained from the space $P_2(C)$ by the reversing the orientation. First we seek an equivariant self-dual connection of index 1 on the space. As in Example 2, we give an S^1-action by

$$P_2(C) \ni [z_1, z_2, z_3] \longrightarrow [z_1, \gamma^2 z_2, \gamma z_3] \in P_2(C) \quad \text{for } \gamma \in S^1.$$

Let A_0 be the flat connection on the product complex line bundle on the space $P_2(C)$. Then A_0 is equivariant. We choose an S^1-invariant Riemann-ian metric on the space $P_2(C)$ and follow equivariantly the construction in [6]. Using the notations in [6], the Gaussian coordinate chart $\phi : V_1 \longrightarrow R^4$ around $m = [0,0,1]$ is given by $[z_1, z_2, z_3] \longrightarrow (z_1/z_3, z_2/z_3)$, then the action is given by $(\bar{\gamma}\xi_1, \gamma\xi_2)$ in R^4. It is seen that the connection A^λ is equivariant. For the stereographic projection form the north pole $\bar{s} : R^4 \longrightarrow S^4 - (p)$, the map $\bar{s}$ can be extended to the map $f : P_2(C) \ni [z_1, z_2, z_3] \longrightarrow (2z_1\bar{z}_3, 2z_2\bar{z}_3, 2|z_3|^2 - 1)$. Then the bundle $P^\lambda \longrightarrow P_2(C)$ is the bundle induced from the bundle with characteristic map χ^+ by the map f. The L_k-norm on $\Lambda^1(\mathcal{G}_P)$ given by

$$\|\Phi\|_{L_k} = \left(\int_{P_2(C)} (\Phi, \Phi)^{k/2} \sqrt{|g|}\, dx\right)^{1/k}$$

is S^1-invariant, where g denotes the Riemannian metric. Therefore we can do an equivariant analogy to the construction in [6]. There exist three fixed points, $p_1 = [1,0,0]$, $p_2 = [0,1,0]$, $p_3 = [0,0,1] = m$. Reversing the orientation in the space $P_2(C)$, at these points we have S^1-module structures of tangent spaces $\bar{T}$, the half spinors $\bar{V}_+, \bar{V}_-$ and fibres such that

$$p_1 : \bar{T} = \gamma^2+\bar{\gamma},\ \bar{V}_+ = \gamma^{1/2}+\bar{\gamma}^{1/2},\ \bar{V}_- = \gamma^{3/2}+\bar{\gamma}^{3/2},\ E_{p_1} = \gamma+\bar{\gamma},$$

$p_2 : \bar{T} = \bar{\gamma}^2+\gamma$, and the other data are the same to the ones at p_1,

$$p_3 : \bar{T} = \gamma+\bar{\gamma},\ \bar{V}_+ = \gamma+\bar{\gamma},\ \bar{V}_- = 2,\ E_{p_3} = 2.$$

Hence by a calculation $H^1(E_+) = 2(\gamma+\bar{\gamma}) + 1$ and $\dim H^1_{S^1} = 1$.

Remark. Here we show that by Theorem 2 and Example 3 we can obtain monopoles as in [9]. Let R^2 be the fixed axis in $H = R^4$ under the action (*) in Section 3. Any $x \in H-R^2$ can be represented in the manner $x = z+re^{i\theta}j$ with the complex variable z and the polar coordinate (r,θ). Then for the complex numbers $b_n, n=1,\ldots,k$, we have

$$\begin{aligned}(b_n-x)^{-1} &= ((\bar{b}_n-\bar{z}) - re^{i\theta}j)(|b_n-z|^2 + r^2)^{-1} \\ &= z_n(r,z,b_n) + r_n(r,z,b_n)e^{i\theta}j,\end{aligned}$$

where z_n and r_n denotes the obvious functions. Therefore

$$\mathrm{Im}\,\overline{(b_n-x)}^{-1}d(b_n-x)^{-1} = \mathrm{Im}\ \bar{z}_n dz_n - ir_n^2 d\theta + (\bar{z}_n dr_n + i\bar{z}_n r_n d\theta - r_n d\bar{z}_n)e^{i\theta}j.$$

The k-instanton A can be described as $A(\theta,r,z) = A_0 d\theta + A_1 dr + A_2 dx_1 + A_3 dx_2$ for $z = x_1+ix_2$. We consider the instanton A near the subspace $\theta = 0$, then $g = e^{-i\theta}$ can be considered as an elment of the gauge group, and $(g^*A)(2\theta,r,z) = -id\theta + A(0,r,z)$. It is easy to see that $\partial_\theta (g^*A)_m = 0$ for $m = 0,1,2,3$. Then by the next lemma we can see that

$$((g^*A)_1 dr + (g^*A)_2 dx_1 + (g^*A)_3 dx_2,\ (g^*A)_0)$$

gives a monopole.

Lemma ([10]). If an instanton $A = A_0 d\theta + A_1 dr + A_2 dx + A_3 dy$ satisfies the condition $\partial_\theta A_m = 0, m = 0,1,2,3$, then $(A_1 dr + A_2 dx + A_3 dy, A_0)$ is a monopole.

Proof. Set $\theta = x_0$, $r = x_1$ $x = x_2$, $y = x_3$, then the curvature is given by $F_{i0} = \partial_i A_0 + [A_i, A_0] = (\nabla_A A_0)_i$, $i = 0,1,2,3$. Since the operator * is conformally invariant, $F_{01} = F_{23}$, $F_{02} = f_{31}$, $F_{03} = F_{12}$, therefore we have the Bogomolny equation $\nabla_A(A_0) = *F$.

References

[1] M.F.Atiyah, The geometry of Yang-Mills fields, Fermi Lectures, Scuola Normale Superiore, Pisa (1979).

[2] M.F.Atiyah and G.B.Segal, The index of elliptic operators, II, Ann. of Math. 87(1968), 531-545.

[3] M.F.Atiyah and F.Hirzebruch, Spin-manifolds and group actions, Essays on topology and related topics, Springer, 1969.

[4] M.F.Atiyah, N.J.Hitchin and I.M.Singer, Self-duality in four-dimen -sional Riemannian geometry, Proc. R. Soc. Lond. A 362(1978),425-461.

[5] M.Itoh, On the moduli space of anti-self-dual Yang-Mills connections , Publ. RIMS., Kyoto Univ., 19(1983), 15-32.

[6] M.Itoh, Self-dual Yang-Mills equations and Taubes' theorem, Tsukuba J. Math. 8(1)(1984), 1-24.

[7] M.Itoh and I.Mogi, Differential geometry and gauge fields, Kyoritsu, 1986, in Japanese.

[8] H.Matsunaga and H.Minami, Forgetful homomorphisms in equivariant K-theory, Publ., RIMS., Kyoto Univ., 22(1)(1986), 143-150.

[9] M.F.Atiyah, Instantons in two and four dimensions, Commun. Math. Phys., 93(1984), 437-451.

[10] M.Itoh, Yang-Mills equations, Instantons and monopoles, Sugaku, 37(4)(1985), 322-337, in Japanese.

Department of Mathematics
Faculty of Science
Shimane University
Matsue Japan

Asymptotical Stability of Yang-Mills' Gradient Flow

TAKEYUKI NAGASAWA

1 Introduction.

We study the existence and asymptotical stability of Yang-Mills' gradient flow.

Let $J(\cdot)$ be a functional on some functional space X. The gradient flow $u(t)$ of $J(\cdot)$ with the initial value a is, if exists, a C^1-flow satisfying

$$\begin{cases} \dfrac{du(t)}{dt} = -\operatorname{grad} J(u(t)) & t \in (0, \infty), \\ u(0) = a, \end{cases} \tag{1.1}$$

where $-\operatorname{grad} J(\cdot)$ is the Euler-Lagrange operator of $J(\cdot)$.

If $u_0 \in X$ is a critical point of $J(\cdot)$, then it is a stationary solution of (1.1), and *vice versa*. Therefore the investigation of gradient flow is intimately related to the variational problem: Find u satisfying

$$\operatorname{grad} J(u) = 0. \tag{1.2}$$

We find one of examples in which the gradient flow takes effect in the paper by Eells and Sampson [2], where they showed the existence of harmonic maps, *i.e.*, the critical maps of the energy integral defined on maps $f : M \to N$ between two Riemannian manifolds

$$J(f) = \frac{1}{2} \int_M \|df\|^2.$$

ISBN 0-12-640170-5

We say that the critical point u_0 is asymptotically stable if there exists a neighborhood $U(u_0)$ of u_0 in X such that for any $a \in U(u_0)$ the gradient flow of $J(\cdot)$ with the initial value a exists and converges to u_0 as $t \to \infty$ in some topology.

We shall make researches on the gradient flow of the Yang-Mills functional which is given by the square integral of the curvature R^∇ associated to a metric connection ∇ on a Riemannian vector bundle E over a Riemannian manifold M:

$$J(\nabla) = \mathcal{YM}(\nabla) = \frac{1}{2}\int_M \langle R^\nabla, R^\nabla \rangle_x. \tag{1.3}$$

Let ∇_0 be a fixed base connection. For every connection ∇, the difference

$$A = \nabla - \nabla_0 \tag{1.4}$$

is a global cross section of $\Omega^1(\mathfrak{G}_E)$ (for the definition of $\Omega^1(\mathfrak{G}_E)$ etc., see §2.1). Using fundamental calculation (Propositions in §2, below), we find that (1.2) for (1.3) is written in a system of second-order partial differential equations (the Yang-Mills equation) of A with the principal term $\delta^{\nabla_0} d^{\nabla_0} A$, where d^{∇_0} is the covariant derivation operator of ∇_0 and δ^{∇_0} is its formal adjoint operator. The operator $-\delta^{\nabla_0} d^{\nabla_0}$, however, is not elliptic type. Hence the Yang-Mills equation itself is not in the framework of elliptic partial differential equations.

A similar situation occurs in (1.1) for (1.3), *i.e.*, (1.1) is not parabolic system for (1.3). To avoid this difficulty, we use the gauge invariancy of the Yang-Mills functional (Proposition 2.2.1 (2)). We take

$$A = g^{-1} \circ \nabla \circ g - \nabla_0, \quad g \in \mathcal{G} \; : \; \text{the gauge group} \tag{1.5}$$

instead of (1.4). We can choose a "good" g so that it recovers the ellipticity of the principal term of (1.1) for (1.3). The "goodness" of g is written in certain differential equation. Under this condition, (1.1) is reduced to a system of semi-linear parabolic equations. By virtue of the standard technique for the parabolic system, we shall show the asymptotical stability of some critical points of the Yang-Mills functional.

2 Formulation of Problem.

2.1 Notation and Definition.

First we introduce terminology used here (basically we follow the notation in [8]). Let (M, g) be a smooth n-dimensional Riemannian manifold, where $n \geq 2$. Suppose that $(E, \langle,\rangle)$ is a Riemannian vector bundle over (M, g) of rank m. We denote the space of all smooth metric connections on E by $\mathcal{C}_E$. For $\nabla \in \mathcal{C}_E$ we can define a naturally induced connection on $\mathrm{Hom}(E, E) \simeq E^* \otimes E$ in a canonical way. Namely, for $\nabla \in \mathcal{C}_E$ and a section $L \in \mathrm{Hom}(E, E)$, we define $\nabla(L)$ by

$$\nabla(L)(\phi) = \nabla(L\phi) - L(\nabla\phi) \quad \text{for any} \quad \phi \in \Gamma(E).$$

The $\mathrm{Hom}(E, E)$-valued 2-form R^∇ defined as follows is called the *curvature* of a connection ∇:

$$R^\nabla_{V,W} = \nabla_V \nabla_W - \nabla_W \nabla_V - \nabla_{[V,W]}$$

for any smooth vector fields V, W on M.

Definition 2.1.1. The *Yang-Mills functional* $\mathcal{YM} : \mathcal{C}_E \to [0, \infty]$ is given by

$$\mathcal{YM}(\nabla) = \frac{1}{2} \int_M \langle R^\nabla, R^\nabla \rangle_x.$$

Since we do not assume compactnesss of M, the range of the functional contains ∞.

To calculate the Euler-Lagrange operator $-\operatorname{grad} \mathcal{YM}(\nabla)$, we define the gauge group.

Definition 2.1.2. G_E and $\mathfrak{G}_E$ denote the bundles defined by

$$G_E = \{L \in \mathrm{Hom}(E, E) \; ; \; \langle L\phi, L\psi \rangle = \langle \phi, \psi \rangle \quad \text{for all} \quad \phi, \psi \in E\},$$

$$\mathfrak{G}_E = \{L \in \mathrm{Hom}(E, E) \; ; \; \langle L\phi, \psi \rangle = -\langle \phi, L\psi \rangle \quad \text{for all} \quad \phi, \psi \in E\}.$$

$\mathcal{G}$ and $\mathcal{Y}$ are spaces of all smooth sections of G_E and $\mathfrak{G}_E$ respectively. $g \in \mathcal{G}$ acts on $\nabla \in \mathcal{C}_E$ in the following way:

$$g(\nabla) = g \circ \nabla \circ g^{-1}. \tag{2.1.1}$$

$\mathcal{G}$ is called the *gauge group.*

Before defining the Yang-Mills connection, we state some elementary facts without proofs (for proofs, see [8, 1]).

Proposition 2.1.1. *For every two connection ∇, $\nabla' \in \mathcal{C}_E$, the difference $A = \nabla' - \nabla$ is a global cross section of $\Omega^1(\mathfrak{G}_E)$. Conversely $\nabla + A \in \mathcal{C}_E$ for any $\nabla \in \mathcal{C}_E$ and any $A \in \Omega^1(\mathfrak{G}_E)$.*

Proposition 2.1.2. *The curvature $R^{\nabla'}$ of $\nabla' = \nabla + A$ ($\nabla \in \mathcal{C}_E$, $A \in \Omega^1(\mathfrak{G}_E)$) is expressed in the form*

$$R^{\nabla'} = R^{\nabla} + d^{\nabla}A + [A, A].$$

Let $\Omega_0^1(\mathfrak{G}_E)$ be the subset of $\Omega^1(\mathfrak{G}_E)$ consisting of all elements with compact support. By Propositions 2.1.1 and 2.1.2 we find that if $\mathcal{YM}(\nabla) < \infty$, then for $\nabla^{\varepsilon} = \nabla + \varepsilon A$ ($A \in \Omega_0^1(\mathfrak{G}_E)$)

$$\frac{d}{d\varepsilon}\mathcal{YM}(\nabla^{\varepsilon})\Big|_{\varepsilon=0} = \int_M \langle R^{\nabla}, d^{\nabla}A\rangle_x = \int_M \langle \delta^{\nabla}R^{\nabla}, A\rangle_x.$$

Keeping this in mind, we define grad $\mathcal{YM}(\nabla)$ by

$$\operatorname{grad} \mathcal{YM}(\nabla) = \delta^{\nabla}R^{\nabla}$$

even for ∇ with $\mathcal{YM}(\nabla) = \infty$.

Definition 2.1.3. A connection $\nabla \in \mathcal{C}_E$ is called the *Yang-Mills connection*, if

$$\delta^{\nabla}R^{\nabla} = 0$$

is satisfied.

2.2 The Equations of Yang-Mills' Gradient Flow.

We consider the equations of Yang-Mills' gradient flow

$$\frac{d\nabla(t)}{dt} = -\,\delta^{\nabla(t)} R^{\nabla(t)}, \tag{2.2.1}$$

which follows from Definition 2.1.3, around a fixed base connection ∇_0.

The action (2.1.1) of $\mathcal{G}$ on $\mathcal{C}_E$ yields the following facts, proofs of which are by direct calculations.

Proposition 2.2.1. (1) *The curvature $R^{g(\nabla)}$ of $g(\nabla) = g \circ \nabla \circ g^{-1}$ ($\nabla \in \mathcal{C}_E$, $g \in \mathcal{G}$) is expressed in the form*

$$R^{g(\nabla)} = g \circ R^{\nabla} \circ g^{-1}.$$

(2) *$\langle R^{g(\nabla)}, R^{g(\nabla)} \rangle_x = \langle R^{\nabla}, R^{\nabla} \rangle_x$ holds, and therefore the Yang-Mills functional is invariant under the gauge action:*

$$\mathcal{YM}(g(\nabla)) = \mathcal{YM}(\nabla).$$

Taking (2) of the proposition into consideration, we set

$$\nabla(t) = g(t)(\nabla_0 + A(t)) = g(t) \circ (\nabla_0 + A(t)) \circ g^{-1}(t),$$

where $A(t) \in \Omega^1(\mathfrak{G}_E)$, $g(t) \in \mathcal{G}$ (cf. (1.5)). We want to write down (2.2.1) by use of $A(t)$ and $g(t)$. To this end, we need the following proposition.

Proposition 2.2.2.

$$\delta^{\nabla + A} S = \delta^{\nabla} S - [A, S] \quad \textit{for} \quad S \in \Omega^2(\mathfrak{G}_E)$$

and

$$\delta^{g(\nabla)} R^{g(\nabla)} = g \circ \delta^{\nabla} R^{\nabla} \circ g^{-1}$$

are valid.

The proof can be found in [8, 1].

After some calculations using Propositions 2.1.2, 2.2.1-2, (2.2.1) is rewritten into

$$
\begin{aligned}
\frac{dA(t)}{dt} = & -\delta^{\nabla_0} d^{\nabla_0} A(t) - \delta^{\nabla_0} R^{\nabla_0} - \delta^{\nabla_0}[A(t), A(t)] \\
& + [\nabla_0 + A(t), Y(t)] + [A(t), R^{\nabla_0}] \\
& + [A(t), d^{\nabla_0} A(t)] + [A(t), [A(t), A(t)]],
\end{aligned} \tag{2.2.2}
$$

where $Y(t) = g^{-1}(t)\dfrac{dg(t)}{dt}$. As stated in Introduction, the operator $-\delta^{\nabla_0} d^{\nabla_0}$ in the principal term is not elliptic. To recover ellipticity, we impose certain gauge condition. Since both $A(t)$ and $g(t)$ are unknown in (2.2.2), in order to solve it we need some additional condition.

To put it plainly, we add the term $-d^{\nabla_0}\delta^{\nabla_0} A(t)$ to the right-hand side. The justification of the addition is in two methods. If $A(t)$ satisfies the *Coulomb gauge condition*

$$\delta^{\nabla_0} A(t) = 0,$$

then such an addition has no problem. Kozono, Maeda and Naito investigated the stability of the Yang-Mills connections under this condition in [6, 11]

Yokotani utilized another method. Noting $Y(t) \in \Omega^0(\mathfrak{G}_E)$ because of $g(t) \in \mathcal{G}$, he imposed the condition

$$g^{-1}(t)\frac{dg(t)}{dt} = Y(t) = -\delta^{\nabla_0} A(t) \tag{2.2.3}$$

on $g(t)$. This makes $-d^{\nabla_0}\delta^{\nabla_0} A(t)$ of the term $[\nabla_0, Y(t)]$. He constructed under this method a local solutions of (2.2.2) and (2.2.3) in [12].

We employed Yokotani's method and studied the stability problem around the flat connection ∇_0 in [5] and the regularity of the flow in [10]. The following sections are summary of them.

In what follows, we restrict ourselves to the case where M is the Euclidian space $\mathbf{R}^n$ or a bounded domain $\Omega \subset \mathbf{R}^n$ with smooth boundary, where $n \geq 2$. Suppose that E is the trivial Riemannian vector bundle over (M, g_0) of rank m, where g_0 is the standard metric on $\mathbf{R}^n$. We denote by ∇_0 a cannonical flat connection determined by the trivialization of the bundle. Trivially ∇_0 is a Yang-Mills connection.

3 Analysis.

In this section, we shall investigate the abstract results on the partial and ordinary differential equations, mainly the unique existence and stability of their solutions, which are applicable to (2.2.2) and (2.2.3).

Notation of various functional spaces and their norms in this and next sections follow one in [7]. The rigorous proofs of results in this section are found in [5][1] and [10].

3.1 The Semi-Linear Heat Equations.

First we consider the following system of semi-linear heat equations:

$$\begin{cases} u_t &= \Delta u + F_1(u, \partial u) + F_2(u) \quad \text{on} \quad M \times (0, \infty), \\ u(0) &= a, \\ u|_{\partial M} &= 0 \ \text{ if } \ \partial M \neq \emptyset, \\ u &= (u^1, \cdots, u^N), \quad N \in \mathbf{N}, \end{cases} \tag{3.1.1}$$

where

$$\begin{cases} F_1(u, \partial u) &= \sum\limits_{i,j,k} a_{ijk} u^i \partial_j u^k, \\ F_2(u) &= \sum\limits_{i,j,k} b_{ijk} u^i u^j u^k, \end{cases} \tag{3.1.2}$$

[1]In this paper, proofs are written under the situation that a_{ijk} and b_{ijk} in (3.1.2) are constants. However it is developed in our case with minor changes.

and

$$\partial_j = \frac{\partial}{\partial x_j}, \quad \Delta = \sum_{j=1}^{n} \partial_j^2.$$

Taking (2.2.2) into account, we suppose that a_{ijk} and b_{ijk} are bounded together with their derivatives ∂a_{ijk}, ∂b_{ijk}.

First we construct a solution u to (3.1.1).

Theorem 3.1.1. *Let $a \in L_n(M)$. Then there exists a positive constant λ such that if $\|a\|_{n,M} < \lambda$ then there exists a unique solution $u(t) \in \overset{\circ}{W}{}_n^1(M) \cap W_n^2(M)$ for $t > 0$ to (3.1.1) satisfying properties*

$$\begin{cases} t^{(1-\frac{n}{p})/2} u(t) \in BC([0,\infty); L_p(M)) & \text{for} \quad n \le p < \infty, \\ t^{(1-\frac{n}{2q})} \partial u(t) \in BC([0,\infty); L_q(M)) & \text{for} \quad n \le q < \infty \end{cases} \tag{3.1.3}$$

with values

$$\begin{cases} t^{(1-\frac{n}{p})/2} u(t)\Big|_{t=0} = \begin{cases} a & p = n, \\ 0 & n < p < \infty, \end{cases} \\ t^{(1-\frac{n}{2q})} \partial u(t)\Big|_{t=0} = 0 & n \le q < \infty. \end{cases} \tag{3.1.4}$$

Moreover $u(t)$ belongs to $C^0([0,\infty); L_n(M)) \cap C^1((0,\infty); L_n(M))$.

Proof. To begin with, we construct the *mild solution* satisfying

$$u(t) = e^{t\Delta} a + \int_0^t e^{(t-\tau)\Delta} \{F_1(u(\tau), \partial u(\tau)) + F_2(u(\tau))\} d\tau, \tag{3.1.5}$$

where $\{e^{t\Delta}\}_{t \ge 0}$ is a strongly continuous semigroup generated by Δ with the domain $\mathcal{D}(\Delta) = \overset{\circ}{W}{}_p^1(M) \cap W_p^2(M)$. $e^{t\Delta}$ satisfies the L_p-L_q-estimates:

$$\begin{cases} \|e^{t\Delta} a\|_{p,M} \le C(p,q,n) t^{-(\frac{n}{q}-\frac{n}{p})/2} \|a\|_{q,M} \\ \qquad\qquad (1 < q \le p < \infty), \\ \|\partial e^{t\Delta} a\|_{p,M} \le C(p,q,n) t^{-(1+\frac{n}{q}-\frac{n}{p})/2} \|a\|_{q,M} \\ \qquad\qquad (1 < q \le p < \infty). \end{cases} \tag{3.1.6}$$

The solution to (3.1.5) is constructed by a successive approximation

$$\begin{cases} u_0(t) &= e^{t\Delta}a, \\ u_{m+1}(t) &= u_0(t) + \int_0^t e^{(t-\tau)\Delta}\{F_1(u_m(\tau), \partial u_m(\tau)) + F_2(u_m(\tau))\}d\tau, \\ & \qquad m = 0,\ 1,\ 2,\ \cdots. \end{cases} \tag{3.1.7}$$

Let us define the norm $|||\cdot|||$ by

$$|||u||| = \max\left\{\sup_{t\geq 0} t^{\left(1-\frac{n}{r}\right)/2}\|u(t)\|_{r,M},\ \sup_{t\geq 0} t^{\frac{1}{2}}\|\partial u(t)\|_{n,M}\right\}.$$

By virtue of (3.1.6), for $r \in (\frac{3}{2}n, 3n)$ the sequence $\{u_m\}_{m=0,1,\cdots}$ is well-defined, *i.e.*, $|||u_m||| < \infty$ and if $\|a\|_{n,M}$ is sufficiently small, then $|||u_m|||$ is uniformly bounded:

$$\{u_m\} \subset X := \{u\ ;\ |||u||| \leq K < 1\}.$$

In a similar manner we get the facts that the map

$$\Phi\ :\ u \mapsto \Phi(u)(t) = e^{t\Delta}a + \int_0^t e^{(t-\tau)\Delta}\{F_1(u(\tau), \partial u(\tau)) + F_2(u(\tau))\}d\tau$$

is defined on X into itself, and that Φ is contractive on X.

Therefore u_m converges to some u in X, and it is a unique solution to (3.1.5) on X. u satisfies properties

$$\|u\|_{r,M} \leq Ct^{-\left(1-\frac{n}{r}\right)/2},\quad \|\partial u\|_{n,M} \leq Ct^{-\frac{1}{2}},\quad C \to 0 \text{ as } \|a\|_{n,M} \to 0.$$

Making use of these decay properties and (3.1.6) appropriately, we get (3.1.3) and (3.1.4) for the mild solution $u(t)$.

In a similar argument to [3, Proposition 2.4], we have $u \in \mathcal{D}(\Delta)$ for $t > 0$ and $u \in C^0([0,\infty); L_n(M)) \cap C^1((0,\infty); L_n(M))$. □

We apply the theorem to (2.2.2) to construct a flow $A(t)$. The properties (3.1.3) for $A(t)$, however, are not sufficient for solving (2.2.3). Hence we impose more integrability on initial data.

Theorem 3.1.2. *We assume the hypothesis in Theorem 3.1.1 and* $a \in \overset{\circ}{W}{}^1_n(M)$. *Then the solution* $u(t)$ *constructed by Theorem 3.1.1 satisfies*

$$\begin{cases} t^{\frac{1}{2}}\partial u(t) \in BC([0,\infty); L_\infty(M)) \quad \text{for } M = \mathbf{R}^n, \\ t^{\frac{1}{4}+\frac{\beta}{2}}\partial u(t) \in BC([0,\infty); L_\infty(M)) \quad \text{for } M = \Omega \\ \qquad\qquad\qquad \left(\frac{1}{2} < \beta < 1\right). \end{cases} \tag{3.1.8}$$

Moreover $u(t)$ *belongs to* $C^0([0,\infty); \overset{\circ}{W}{}^1_n(M)) \cap C^1((0,\infty); \overset{\circ}{W}{}^1_n(M))$.

Proof. First we show Theorem for $M = \mathbf{R}^n$. In this case we have no boundary to worry about, and therefore the operator ∂ can commute with $e^{t\Delta}$. Hence u_m's satisfy

$$\partial u_{m+1}(t) = e^{t\Delta}\partial a + \int_0^t e^{(t-\tau)\Delta}\partial\{F_1(u_m(\tau), \partial u_m(\tau)) + F_2(u_m(\tau))\}d\tau,$$

$$m = 0,\ 1,\ 2,\ \cdots.$$

By help of (3.1.6), we have

$$\|\partial u_m(t)\|_{2n,M} \le Ct^{-\frac{1}{4}}, \quad \|\partial^2 u_m(t)\|_{2n,M} \le Ct^{-\frac{3}{4}},$$

where C is independent of m, if $\|a\|_{n,M}$ is sufficiently small. Since u_m converges to u, u also has the same estimates. By virtue of the Gagliado-Nirenberg inequality

$$\|\partial u\|_{\infty,M} \le C\|\partial^2 u\|_{2n,M}^{\frac{1}{2}}\|\partial u\|_{2n,M}^{\frac{1}{2}},$$

we obtain (3.1.8) for $M = \mathbf{R}^n$.

When $M = \Omega$ (*i.e.*, $\partial M \ne \emptyset$), the operator ∂ does not commute with $e^{t\Delta}$, but the fractional power of $(-\Delta)^\alpha$ of $(-\Delta)$ of arder α $(0 < \alpha < 1)$ does. Hence we use $(-\Delta)^{\frac{1}{2}}$ instead of ∂. The $\mathcal{D}((-\Delta)^\alpha)$-norm is defined by

$$\|a\|_{\mathcal{D}((-\Delta)^\alpha)} = \|(-\Delta)^\alpha a\|_{n,M},$$

which is equivalent to the graph norm $\|a\|_{n,M} + \|(-\Delta)^\alpha a\|_{n,M}$ for $\alpha > 0$ when Ω is bounded. For the fractional power operator, the L_p-L_q-

estimate is also valid:

$$\begin{cases} \|(-\Delta)^\alpha e^{t\Delta} a\|_{p,M} \le C(p,q,n,\alpha) t^{-\left(2\alpha+\frac{n}{q}-\frac{n}{p}\right)/2} \|a\|_{q,M} \\ \qquad\qquad (1 < q \le p < \infty). \end{cases} \tag{3.1.9}$$

We denote the norm of the Bessel potential space $\mathcal{L}^{k,p}(M)$ by $\|\cdot\|^{(k)}_{p,M}$. It is well-known that

$$\|a\|^{(k)}_{p,M} \le C\|(-\Delta)^{\frac{k}{2}} a\|_{p,M}.$$

and that $\overset{\circ}{W}{}^1_n(M)$ and $\mathcal{D}((-\Delta)^{\frac{1}{2}})$ coincide as vector spaces and carry equivalent norms. Therefore our hypotheses imply $(-\Delta)^{\frac{1}{2}} a \in L_n(M)$. Operating $(-\Delta)^{\frac{1}{2}+\alpha}$ $\left(0 \le \alpha < \frac{1}{2}\right)$ to both sides of the second equation of (3.1.7), we have

$$(-\Delta)^{\frac{1}{2}+\alpha} u_{m+1}(t) = (-\Delta)^\alpha e^{t\Delta} (-\Delta)^{\frac{1}{2}} a$$
$$+ \int_0^t (-\Delta)^{\frac{1}{2}+\alpha} e^{(t-\tau)\Delta} \{F_1(u_m(\tau), \partial u_m(\tau)) + F_2(u_m(\tau))\} d\tau,$$
$$m = 0,\ 1,\ 2,\ \cdots.$$

We consider the case $\alpha = 0$. Utilizing (3.1.9), we can obtain

$$\|(-\Delta)^{\frac{1}{2}} u_m(t)\|_{2n,M} \le C t^{-\frac{1}{4}}$$

with a constant C independent of m. Combining this estimate with (3.1.3) and (3.1.9), we can show

$$\|(-\Delta)^{\frac{1}{2}+\alpha} u_m(t)\|_{2n,M} \le C t^{-\frac{1}{4}-\alpha}$$

for $0 \le \alpha < \frac{1}{2}$. Here C is independent of m, and therefore the same estimate for u holds also.

Sobolev's imbedding theorem gives

$$\|\partial u(t)\|_{\infty,M} \le C(\beta) \|\partial u(t)\|^{(\beta)}_{2n,M} \le C(\beta) \|u(t)\|^{(1+\beta)}_{2n,M} \text{ if } \beta > \frac{1}{2}.$$

Consequently we get

$$\|\partial u(t)\|_{\infty,M} \le C(\beta) t^{-\frac{1}{4}-\frac{\beta}{2}} \text{ for } \frac{1}{2} < \beta < 1.$$

The assertion $u(t) \in C^0([0,\infty); \overset{\circ}{W}{}_n^1 (M)) \cap C^1((0,\infty); \overset{\circ}{W}{}_n^1 (M))$ is shown in a similar manner to Theorem 3.1.1. □

Next we discuss the regularity of u, which yields from the integrability on $M \times (0,\infty)$ or $M \times (0,T)$.

We denote $M \times (0,\infty)$ and $M \times (0,T)$ by Q_∞ and Q_T respectively. For indices p_1, p_2, q, r and s which fulfill

$$\frac{1}{q} = \left(\frac{1}{s} - \frac{1}{p_1}\right)\frac{n}{2}, \quad \frac{1}{r} = \left(\frac{1}{n} + \frac{1}{s} - \frac{1}{p_2}\right)\frac{n}{2},$$

$$p_1,\ q,\ r > s > 1, \quad p_2 > \left(\frac{1}{n} + \frac{1}{s}\right)^{-1},$$

the following $L_{p,q}$-estimates for $\{e^{t\Delta}\}_{t\geq 0}$ hold:

$$\begin{cases} \|e^{t\Delta}a\|_{p_1,q,Q_\infty} \leq C(p_1,q,s,n,M)\|a\|_{s,M}, \\ \\ \|\partial e^{t\Delta}a\|_{p_2,r,Q_\infty} \leq C(p_2,r,s,n,M)\|a\|_{s,M}. \end{cases} \tag{3.1.10}$$

These follows from (3.1.6) with help of the Marcinkiewicz interpolation theorem.

Using (3.1.10) and the Hardy-Littlewood-Sobolev inequality, under the hypothesis in Theorem 3.1.1, we can show

$$\|u\|_{n+2,Q_\infty} + \|\partial u\|_{\frac{n+1}{2},n+1,Q_\infty} \leq C,$$
$$C \to 0 \ \text{ as } \ \|a\|_{n,M} \to 0.$$

The hypothesis $a \in \overset{\circ}{W}{}_n^1 (M)$ in Theprem 3.1.2 gives us $a \in \bigcap_{s\geq n} L_s(M)$ by Sobolev's imbedding theorem. Hence in a similar manner we have

$$\|u\|_{\frac{s(n+2)}{n},Q_\infty} + \|\partial u\|_{p,r,Q_\infty} \leq C,$$

for

$$s \geq n,$$
$$p > \max\left\{\frac{n+2}{n}, \left(\frac{1}{n} + \frac{1}{s}\right)^{-1}\right\},$$
$$r > \max\left\{\frac{n+2}{n+1}, s\right\},$$

provided

$$\frac{1}{p}-\frac{1}{n}<\frac{1}{n+2}\left(2+\frac{n}{s}\right)<\frac{1}{p}+\frac{1}{n}.$$

Combining these estimates, we have $F_1(u,\partial u)+F_2(u)\in L_{p_1}(Q_T)$ for any $T\in(0,\infty)$ and some $p_1\in\left(\frac{n+2}{3},\frac{n+2}{2}\right)$. A priori estimates of $W_p^{2,1}(Q_T)$-type [7, IV, Theorem 9.1 or VII, Theorem 10.4] give $u\in W_{p_1}^{2,1}(Q_T)$ provided $a\in W_{p_1}^{2-\frac{2}{p_1}}(M)$. Using [7, II, Lemma 3.3], we have $F_1(u,\partial u)+F_2(u)\in L_{p_2}(Q_T)$ for some $p_2>\frac{n+2}{2}$, therefore $u\in W_{p_2}^{2,1}(Q_T)$ provided $a\in W_{p_2}^{2-\frac{2}{p_2}}(M)$. By the same procedure we obtain $u\in W_{p_3}^{2,1}(Q_T)$ provided $a\in W_{p_3}^{2-\frac{2}{p_3}}(M)$ for some $p_3>n+2$. By virtue of [7, II, Lemma 3.3] again, the Hölder continuity of non-linear terms follows if coefficients a_{ijk} and b_{ijk} have the Hölder continuity. Finally the Schauder estimate [7, IV, Theorem 5.1/5.2 or VII, Theorem 10.1/10.2] gives the fact $u\in H^{\alpha+2,\frac{\alpha}{2}+1}(\overline{Q_T})$ for some $\alpha\in(0,1)$ provided that a has the Hölder continuity up to its second derivatives. Hence we have

Theorem 3.1.3. *We assume that a_{ijk} and b_{ijk} are Hölder continuous in $\overline{Q_\infty}$. If a belongs to $\bigcap_{s\geq n} W_s^{2-\frac{2}{s}}(M)$, $\|a\|_{n,M}$ is small, and it is Hölder continuous up to its second derivatives with $a|_{\partial M}=0$ if $\partial M\neq\emptyset$, then there exists a unique global classical solution to* (3.1.1).

Using a standard bootstrap argument, we get

Theorem 3.1.4. *Assume the hypotheses of the previous theorem and C^∞-smoothness of a_{ijk} , b_{ijk} and a. If the compatibility conditions of any order between initial and boundary data hold, then the solution is also C^∞.*

Remark. We have an interior regularity result in a similar manner when a_{ijk} and b_{ijk} and a have only interior smoothness.

3.2 Linear Ordinary Differential Equation.

We rewrite (2.2.3) into

$$\frac{dg(t)}{dt} = -\, g(t)\delta^{\nabla_0}A(t), \tag{3.2.1}$$

which is dealt with as a $\mathcal{G}$-valued linear system of ordinary differential equations. In this subsection we investigate some class of linear ordinary differential equations, coefficient of which has singularity at $t = 0$, including (3.2.1). The solution can be constructed by succesive approximation as usual if the singularity is not so strong.

We consider the following system:

$$\dot{X}(t) = -X(t)B(t) - B(t), \quad X(0) = O, \tag{3.2.2}$$

where $X(t)$ is an unknown $k \times k$-matrix-valued function on M, and $B(t)$ is a known $k \times k$-matrix-valued function on M satisfying

$$\|XB(t)\|_{\infty,M} \leq Ct^{-\gamma}\|X\|_{\infty,M} \quad \text{for} \quad k \times k\text{-matrix } X \tag{3.2.3}$$

for some $\gamma \in (0,1)$.

We set $\{X_m\}_{m=0,1,\cdots}$ by

$$\begin{cases} X_0(t) &= -\displaystyle\int_0^t B(\tau)d\tau, \\ X_{m+1}(t) &= -\displaystyle\int_0^t X_m(\tau)B(\tau)d\tau + X_0(t), \qquad m = 0,\ 1,\ 2,\ \cdots. \end{cases}$$

It follows form (3.2.3) that

$$\|X_0(t)\|_{\infty,M} \leq \frac{Ct^{1-\gamma}}{1-\gamma},$$

where C is a positive constant in (3.2.3), and that by induction on m

$$\|X_{m+1}(t) - X_m(t)\|_{\infty,M} \leq \left(\frac{Ct^{1-\gamma}}{1-\gamma}\right)^{m+2} \Big/ (m+2)!$$

with the same constant C. Consequently X_m converges to some X, which satisfies (3.2.2) and

$$\|X(t)\| \leq \exp\left\{\frac{Ct^{1-\gamma}}{1-\gamma}\right\} - 1.$$

This implies $X \in C^0([0,\infty); L_\infty(M)) \cap C^1((0,\infty); L_\infty(M))$.

Suppose that both $X(t)$ and $\tilde{X}(t)$ satisfy (3.2.2). Then

$$X(t) - \tilde{X}(t) = -\int_0^t (X(\tau) - \tilde{X}(\tau))B(\tau)d\tau$$

holds, and therefore

$$\|X(t) - \tilde{X}(t)\|_{\infty,M} \leq C \sup_{0\leq\tau\leq t} \|X(\tau) - \tilde{X}(\tau)\|_{\infty,M} \int_0^t \tau^{-\gamma} d\tau$$

is valid by (3.2.3). From this, we have $X(t) \equiv \tilde{X}(t)$ on small interval $[0,t]$. Let $[0,T^*)$ be the maximal interval on which $X(t) \equiv \tilde{X}(t)$ holds, and suppose $T^* < \infty$. Using a similar argument

$$\|X(t) - \tilde{X}(t)\|_{\infty,M} \leq C \sup_{T^*\leq\tau\leq T^*+\varepsilon} \|X(\tau) - \tilde{X}(\tau)\|_{\infty,M} \int_{T^*}^{T^*+\varepsilon} \tau^{-\gamma} d\tau$$

for $t \in [T^*, T^*+\varepsilon]$. If ε is sufficiently small, then the contradiction to the maximality of T^* occurs.

Thus we have

Theorem 3.2.1. *There exists a unique solution $X(t)$ to (3.2.2) which belongs to $C^0([0,\infty); L_\infty(M)) \cap C^1((0,\infty); L_\infty(M))$.*

4 Main Theorem.

When ∇_0 is a flat connection, under the condition (2.2.3), (2.2.2) is reduced to the following system of heat equations:

$$\begin{aligned} \frac{dA(t)}{dt} &= \Delta A(t) - \delta^{\nabla_0}[A(t), A(t)] + [A(t), -\delta^{\nabla_0} A(t)] \\ &\quad + [A(t), d^{\nabla_0} A(t)] + [A(t), [A(t), A(t)]]. \end{aligned} \tag{4.1}$$

We consider this system for $A(t)$, vanishing on ∂M if $\partial M \neq \emptyset$, with given initial data $A(0)$. It is obvious to see that the arguments in §3.1 is applicable to (4.1).

Theorems 3.1.1 and 3.1.2 yield the fact if $A(0) \in \overset{\circ}{W}{}^1_n(M)$ and $||A(0)||_{n,M}$ is sufficiently small, then there exists a unique solution $A(t) \in W^2_n(M)$ for $t > 0$ to (4.1) with initial value $A(0)$ in $C^0([0,\infty); \overset{\circ}{W}{}^1_n(M)) \cap C^1((0,\infty); \overset{\circ}{W}{}^1_n(M))$. And this solution has properties

$$\begin{cases} ||A(t)||_{p,M} \leq Ct^{-(1-\frac{n}{p})/2} & \text{for} \quad n \leq p < \infty, \\ ||\partial A(t)||_{q,M} \leq Ct^{-(1-\frac{n}{2q})} & \text{for} \quad n \leq q < \infty. \end{cases}$$

By virtue of the Gagliado-Nirenberg inequality $||u||_{\infty,M} \leq C||u||^{\frac{1}{2}}_{2n,M} \times ||\partial u||^{\frac{1}{2}}_{2n,M}$, we find that the above estimate for $||A(t)||_{p,M}$ also holds for $p = \infty$.

Moreover $||\delta^{\nabla_0} A(t)||_{\infty,M}$ is dominated by $Ct^{-\frac{1}{2}}$ for $M = \mathbf{R}^n$, $C(\beta)t^{-\frac{1}{4}-\frac{\beta}{2}}$ for $M = \Omega$ $\left(\frac{1}{2} < \beta < 1\right)$ by Theorem 3.1.2.

Next we consider (3.2.1), *i.e.*,

$$\frac{dg(t)}{dt} = -\,g(t)\delta^{\nabla_0} A(t), \tag{4.2}$$

with initial condition

$$g(0) = \text{identity}. \tag{4.3}$$

If we put $h(t) = g(t) - \text{identity}$ and if $g(t)$ is a solution to (4.2) - (4.3), then $h(t)$ must satisfy

$$\frac{dh(t)}{dt} = -\,h(t)\delta^{\nabla_0} A(t) - \delta^{\nabla_0}, \quad h(0) = O.$$

Conversely $h(t)$ is a solution of the above problem, then $g(t) = h(t) +$ identity satisfies (4.2) - (4.3).

The estimate for $||\delta^{\nabla_0} A(t)||_{\infty,M}$ is a type of (3.2.3). Therefore we have a unique solution $h(t)$, and therefore $g(t)$, by Theorem 3.2.1 in $C^0([0,\infty); L_\infty(M)) \cap C^1((0,\infty); L_\infty(M))$.

The regularity of $A(t)$ follows from Theorem 3.1.4 or Remark after it provided $A(0)$ satisfies the conditions there. If $A(t)$ is smooth, then the smoothness of $g(t)$ follows from the theorems of regularity and continuous dependence on parameters of ordinary differential equations.

Therefore we have the following stability result on the flat connection ∇_0.

Theorem 4.1. *Let a Riemannian manifold (M, g_0) and a Riemannian vector bandle $(E, \langle , \rangle)$ be as stated in the last paragraph of §2. For $\varepsilon > 0$ we denote a neiborhood*

$$\left\{\nabla \in \mathcal{C}_E\,;\, \nabla - \nabla_0 \in \overset{\circ}{W}{}^1_n(M) \cap \bigcap_{s \geq n} W_s^{2-\frac{2}{s}}(M), \|\nabla - \nabla_0\|_{n,M} < \varepsilon\right\}$$

of $\nabla_0 \in \mathcal{C}_E$ by U_ε. Then there exists a positive constant $\varepsilon > 0$ such that for any $\nabla \in U_\varepsilon(\nabla_0)$ there exist a $\mathcal{C}_E$-valued smooth function $\nabla(t)$ and a $\mathcal{G}$-valued smooth function $g(t)$ satifying

$$\begin{cases} \dfrac{d\nabla(t)}{dt} &= -\operatorname{grad} \mathcal{YM}(\nabla(t)), \quad t \in (0, \infty), \\ \nabla(0) &= \nabla, \\ g(0) &= \text{identity}, \end{cases}$$

and

$$\lim_{t \to \infty} g^{-1}(t) \circ \nabla(t) \circ g(t) = \nabla_0$$

in $L_p(M)$ for $n < p \leq \infty$ with decay rate $t^{-\left(1-\frac{n}{p}\right)/2}$.

Proof. What we have not shown yet is the fact $A(t) \in \Omega^1(\mathfrak{G}_E)$ and $g(t) \in \mathcal{G}$ for $t > 0$.

We take transpose of both sides of (4.1) and put $-{}^tA(t) = B(t)$. Then it is easy to see that $B(t)$ satisfies (4.1) with the replacement $A(t)$ by $B(t)$. Since $B(0) = A(0)$ and since the solution of (4.1) is unique under the smallness condition $\|A(0)\|_{n,M} = \|B(0)\|_{n,M} < \varepsilon$, $A(t)$ is skew-symmetric.

The solution $g(t)$ of (4.2) constructed by Theorem 3.2.1 is explicitly expressed by the Peano-Baker series:

$$g(t) = \sum_{m=0}^{\infty} \Phi_m(t),$$

where

$$\begin{cases} \Phi_0(t) &= \text{identity}, \\ \Phi_{m+1}(t) &= -\int_0^t \Phi_m(\tau)\delta^{\nabla_0}A(\tau)d\tau, \qquad m = 0,\ 1,\ 2,\ \cdots. \end{cases}$$

Let $\{\Psi_m(t)\}_{m=0,1,2,\cdots}$ be

$$\begin{cases} \Psi_0(t) &= \text{identity}, \\ \Psi_{m+1}(t) &= \int_0^t \delta^{\nabla_0}A(\tau)\Psi_m(\tau)d\tau, \qquad m = 0,\ 1,\ 2,\ \cdots. \end{cases}$$

We can show that the series $\tilde{g}(t) = \sum_{m=0}^{\infty} \Psi_m(t)$ converges and that $\tilde{g}(t)$ satisfies

$$\frac{d\tilde{g}(t)}{dt} = \delta^{\nabla_0}A(t)\tilde{g}(t), \qquad \tilde{g}(t) = \text{identity}.$$

Since $\frac{d}{dt}(g(t)\tilde{g}(t)) = 0$ and since $g(0)\tilde{g}(0) = \text{identity}$, $\tilde{g}(t)$ is the inverse of $g(t)$. By virtue of $A(t) \in \Omega^1(\mathfrak{G}_E)$, ${}^t\tilde{g}(t)$ is a solution of (4.2) - (4.3). Because the solution is unique, we can conclude $g(t) \in \mathcal{G}$. □

References

[1] Bourguignon, J.-P. and H. B. Lawson, Jr., *Stability and isolation phenomena for Yamg-Mills fields*, Comm. Math. Phys. **79** (1981), 189-230.

[2] Eells, Jr., J. and J. H. Sampson, *Harmonic mappings of Riemannian manifolds*, Amer. J. Math. **86** (1964), 109-160.

[3] Giga, Y. and M. Miyakawa, *Solutions in L_r of the Navier-Stokes initial value problem*, Arch. Rational Mech. Anal. **89** (1985), 267-281.

[4] Kono, K., "Weak Asymptotical Stability of Yang-Mills Fields," Master Thesis, Keio Univ., 1988 (Japanese).

[5] Kono, K. and T. Nagasawa, *Weak asymptotical stability of Yang-Mills' gradient flow*, Tokyo J. Math. **11** (1988), 339-357.

[6] Kozono, H. and Y. Maeda, *On asymptotic stability for the Yang-Mills gradient flow*, preprint.

[7] Ladyženskaja, O. A., V. A. Solonnikov and N. N. Ural'ceva, "Linear and Quasi-Linear Equations of Parabolic Type," Transl. Math. Monographs **23**, Amer. Math. Soc., Providence, R. I., 1968.

[8] Lawson, Jr., H. B., "The Theory of Gauge Fields in Four Dimensions," Regional Conf. Ser. in Math. **58**, Amer. Math. Soc., Providence, R. I., 1985.

[9] Mogi, I. and M. Itoh, "Differential Geometry and Gauge Theory," Kyōritsu Shuppan, Tokyo, 1986 (Japanese).

[10] Nagasawa, T., *A note on $L_{p,q}$-estimates for some semilinear parabolic equations*, preprint.

[11] Naito, H., H. Kozono and Y. Maeda, *A stable manifold theorem for the Yang-Mills gradient flow*, preprint.

[12] Yokotani, M., *Local existence of the Yang-Mills gradient flow*, preprint.

Mathematical Institute, Tôhoku University
Sendai 980, Japan

Perspectives in Mathematics

Vol. 1 J.S. Milne, *Arithmetic Duality Theory*
Vol. 2 A. Borel et al., *Algebraic D-Modules*
Vol. 3 Ehud de Shalit, *Iwasawa Theory of Elliptic Curves with Complex Multiplication*
Vol. 4 M. Rapoport, N. Schappacher, P. Schneider, editors, *Beilinson's Conjectures on Special Values of L-Functions*
Vol. 5 Paul Günther, *Huygens' Principle and Hyperbolic Equations*
Vol. 6 Stephen Gelbart and Freydoon Shahidi, *Analytic Properties of Automorphic L-Functions*
Vol. 7 Mark Ronan, *Lectures on Buildings*
Vol. 8 K. Shiohama, editor, *Geometry of Manifolds*